Jurgen Schulz-Schaeffer

Cytogenetics

Plants, Animals, Humans

With 219 Figures

Springer-Verlag
New York   Heidelberg   Berlin

JURGEN SCHULZ-SCHAEFFER
Professor of Agronomy and Genetics
Department of Plant and Soil Science
and Institute of Genetics,
Montana State University
Bozeman, Montana 59717, USA

Library of Congress Cataloging in Publication Data
Schulz-Schaeffer, Jurgen.
    Cytogenetics, plants, animals, humans.
    Bibliography: p.
    Includes index.
    1. Cytogenetics.   I. Title.
QH430.S38   1980   574.87'3223   79-25520

© 1980 by Springer-Verlag New York Inc.

Softcover reprint of the hardcover 1st edition 1980

ISBN-13: 978-1-4612-6062-2   e-ISBN-13: 978-1-4612-6060-8

DOI: 10.1007/978-1-4612-6060-8

Preface

Since 1961 the author has taught a course in Cytogenetics at Montana State University. Undergraduate and graduate students of Biology, Chemistry, Microbiology, Animal and Range Science, Plant and Soil Science, Plant Pathology and Veterinary Science are enrolled. Therefore, the subject matter has been presented in an integrated way to correlate it with these diverse disciplines. This book has been prepared as a text for this course. The most recent Cytogenetics text was published in 1972, and rapidly developing research in this field makes a new one urgently needed.

This book includes many aspects of Cytogenetics and related fields and is written for the college student as well as for the researcher. It is recommended that the student should have taken preparatory courses in Principles of Genetics and Cytology. The content is more than is usually taught during one quarter of an academic year, thus allowing an instructor to choose what he or she would like to present to a class. This approach also allows the researcher to obtain a broad exposure to this field of biology. References are generously supplied to stimulate original reading on the subject and to give access to valuable sources. The detailed index is intended to be of special assistance to researchers.

Individual chapters were carefully reviewed and constructively criticized by Drs. PENELOPE ALLDERDICE (Memorial University, New Foundland, Canada), CHARLES BURNHAM (University of Minnesota), STEPHAN CHAPMAN (Clemson University, South Carolina), DOUGLAS DEWEY (USDA, SEA, Utah State University), FRIEDRICH EHRENDORFER (University of Vienna, Austria), FRED ELLIOTT (Michigan State University), STEVE FRANSEN (South Dakota State University), WERNER GOTTSCHALK (University of Bonn, Fed. Rep. of Germany), WAYNE HANNA (Georgia Costal Plain Experiment Station), M. L. H. KAUL

(University of Bonn, Fed. Rep. of Germany), ALEXANDER MICKE (International Atomic Energy Agency, Vienna, Austria), ROSALIND MORRIS (University of Nebraska), ROBERT NILAN (Washington State University), PHILLIP PALLISTER (Shodair Hospital, Helena, Montana), RALPH RILEY (Plant Breeding Institute, Cambridge, Great Britain), OWEN ROGERS (University of New Hampshire), VAL SAPRA (Alabama A and M University), ERNEST SEARS (USDA, SEA, University of Missouri), ROBERT SHERWOOD (USDA, SEA, Pennsylvania State University), ROBERT SOOST (University of California, Riverside), G. LEDYARD STEBBINS, Jr. (University of California, Davis), J. SYBENGA (Agricultural University, Wageningen, Netherlands), TAKUMI TSUCHIYA (Colorado State University), BRUCE YOUNG (USDA, SEA, Grassland Forage Research Center, Temple, Texas), CHENG YU (Lousiana State University Medical Center), DAVID CAMERON, HELEN CAMERON, THOMAS CARROLL, SCOTT COOPER, EUGENE HOCKETT, ERHARD HEHN, HOMER METCALF, PIERCE MULLEN, JOHN RUMELY, DAVID SKAAR, ERNEST VYSE, and GUYLYN WARREN (Montana State University). I appreciate the efforts of all these scientists, many of whose suggestions have been incorporated into the text. However, I am aware that continued improvement is always possible and I will therefore, be grateful for further corrections and suggestions.

I also thank all others who helped with the typing, proofreading, reviewing and indexing. Finally, I thank my loving family, RUTH, IRIS, ROSE, HEIDI, and CHRISTINA, who, through their patience and help, have made this book possible.

August, 1980   JURGEN SCHULZ-SCHAEFFER

Contents

PART I   INTRODUCTION

Chapter 1: History of Cytogenetics   2

PART II   STRUCTURE OF CHROMOSOMES

Chapter 2: Gross Morphology of Chromosomes   32

2.1   Mitotic Metaphase Chromosomes   33
    2.1.1   Total Length of Chromosomes   33
    2.1.2   The Centromere   34
    2.1.3   The Nucleolus Organizer Region   36
2.2   Meiotic Pachytene Chromosomes   38
    2.2.1   Heterochromatin vs. Euchromatin   38
    2.2.2   The Chromomeres   39
    2.2.3   The Centromere   40
    2.2.4   The Telomeres   41
    2.2.5   The Nucleolus Organizer Region   41
    2.2.6   The Knobs   42
2.3   Banding Patterns in Mitotic Metaphase Chro-
    mosomes   43
    2.3.1   C-Bands   44
    2.3.2   G-Bands   47
    2.3.3   Q-Bands   48
    2.3.4   R-Bands   48
    2.3.5   Miscellaneous Bands   50

Chapter 3: Fine Structure of Chromosomes   52

3.1   The Structure of DNA   52
3.2   The Structure of RNA   53
3.3   Nucleoproteins   54

3.4   Models of Chromosome Ultrastructure   55
      3.4.1   The Folded Fiber Model   56
      3.4.2   The Molecular Chromosome Model......   57
      3.4.3   The Multistranded Chromosome Model ..   57
      3.4.4   General Chromosome Model   59
3.5   Ultrastructure of the Centromere   60

PART III   FUNCTION OF CHROMOSOMES

Chapter 4: Function of Autosomes   62

4.1   Linkage   62
4.2   The Mechanism of Crossing Over   63
      4.2.1   The Partial Chiasmatype Theory   63
      4.2.2   The Belling Hypothesis   63
      4.2.3   The Copy-Choice Hypothesis   64
      4.2.4   The Polaron Hybrid DNA Model of Cross-
              ing Over   64
4.3   The Cytological Basis of Crossing Over   65
4.4   Locating Genes on Chromosomes and Genetic
      Mapping   68

Chapter 5: Function of Sex-Chromosomes   78

5.1   The X-Y System   78
      5.1.1   Bridges' Balance Theory   79
      5.1.2   Goldschmidt's Theory   80
      5.1.3   Pipkin's Theory   81
5.2   The Function of the Y Chromosome   82
5.3   Dosage Compensation   83
      5.3.1   The Single Active X Hypothesis.........   83
      5.3.2   Sex Chromatin and Drumsticks   84
5.4   Sex Linkage   88

PART IV   MOVEMENT OF CHROMOSOMES

Chapter 6: Chromosomes During Mitosis   90

6.1   Interphase....................................   90
6.2   Preparation for Mitosis   92
      6.2.1   The Centrosome   93
6.3   Prophase   95
6.4   Metakinesis...................................   97
      6.4.1   Chromosome Congression   97

6.4.2   Centromere Orientation 97
6.4.3   Chromosome Distribution 98
6.5   Metaphase 98
6.6   Anaphase 99
6.7   Telophase 101

Chapter 7: Chromosomes During Meiosis 102

7.1   Premeiotic Interphase 105
7.2   Prophase I 106
7.2.1   Leptotene 108
7.2.2   Zygotene 108
7.2.3   Pachytene 113
7.2.4   Diplotene 114
7.2.5   Diakinesis 118
7.3   Metaphase I 120
7.4   Anaphase I 122
7.5   Telophase I and Interkinesis 123
7.6   Prophase II 125
7.7   Metaphase, Anaphase, and Telophase II 127

Chapter 8: Chromosomes During Sexual Reproduction   128

8.1   Sexual Reproduction in Plants 131
8.1.1   Microsporogenesis and Spermatogenesis ..   131
8.1.2   Megasporogenesis and Syngamy 132
8.2   Sexual Reproduction in Animals 134
8.2.1   Spermotogenesis 134
8.2.2   Spermiogenesis 136
8.2.3   Oogenesis and Syngamy 139

PART V   VARIATION IN CHROMOSOME TYPES

Chapter 9: Polyteny and Lampbrush Chromosomes ...   148

9.1   Polyteny vs. Endopolyploidy 148
9.2   Morphological Characteristics of Polytene Chromosomes 150
9.3   Puffing 152
9.4   Super Chromosomes 154
9.5   Somatic Synapsis 156
9.6   Lampbrush Chromosomes 156

Chapter 10: Ring-Chromosomes, Telocentric Chromosomes, Isochromosomes, and B Chromosomes   159

10.1 Ring-Chromosomes 159
10.2 Telocentric Chromosomes 165

10.3 Isochromosomes   167
10.4 B Chromosomes   170
　　　10.4.1 B Chromosome Structure and Genetic
　　　　　　　Constitution   171
　　　10.4.2 Numerical Distribution, Variability, and
　　　　　　　Effects of B Chromosomes   172
　　　10.4.3 Meiotic and Postmeiotic Behavior of B
　　　　　　　Chromosomes   173

PART VI   VARIATION IN CHROMOSOME STRUC-
TURE

Chapter 11: Chromosome Deletions   176

11.1 Breakage-Reunion and Exchange Hypotheses   176
11.2 Spontaneous and Induced Chromosome and Chro-
　　　matid Aberrations   179
11.3 Terminal Deficiencies   183
11.4 Interstitial Deletions   185
11.5 Breakage-Fusion-Bridge Cycle   186
11.6 Genetic and Cytological Tests of Deletions   188
11.7 Human Deletion Syndromes   190

Chapter 12: Chromosome Duplications   194

12.1 Types of Chromosome Duplications   194
12.2 Origin of Chromosome Duplications   195
12.3 Position Effect   197
　　　12.3.1 The Ac-Ds System   199
12.4 Other Phenotypic Effects   200
12.5 Human Chromosome Duplication Syndromes   201

Chapter 13: Chromosome Inversions   203

13.1 Pericentric Inversions   203
13.2 Paracentric Inversions   208
13.3 Complex Inversions   213
13.4 Inversions as Crossover Suppressors   216
13.5 The Schultz-Redfield Effect   218

Chapter 14: Chromosome Translocations   219

14.1 Types of Translocations   219
14.2 Origin of Translocations   221
14.3 Reciprocal Translocations   223
14.4 Translocations in Humans   228
14.5 Complex Heterozygosity   233

14.6 *Oenothera* Cytogenetics   234
14.7 Other Systems with Complex Heterozygosity   237
14.8 Chromosome Mapping via Translocations   239

PART VII   VARIATION IN CHROMOSOME NUMBER

Chapter 15: Haploidy, Diploidy, and Polyploidy   244

15.1 Haploidy   244
    15.1.1 Origin of Haploids   244
    15.1.2 Meiotic Behavior of Monohaploids   249
    15.1.3 Meiotic Behavior of Polyhaploids   249
    15.1.4 Possible Use of Haploids   251
15.2 Diploidy   251
    15.2.1 Diploidization   252
15.3 Polyploidy   252
    15.3.1 Classification of Polyploidy   252
    15.3.2 Autopolyploidy   254
    15.3.3 Segmental Allopolyploidy   266
    15.3.4 Genome Allopolyploidy   267
    15.3.5 Complications with Polyploidy in Man and
           Animals   270

Chapter 16: Aneuploidy   272

16.1 Euploidy   272
16.2 Aneuploidy   273
    16.2.1 Nullisomy   273
    16.2.2 Monosomy   274
    16.2.3 Telosomy   279
    16.2.4 Trisomy   280
    16.2.5 Tetrasomy   291

PART VIII   VARIATION IN CHROMOSOME FUNC-
TION AND MOVEMENT

Chapter 17: Variation in Function of Autosomes   294

17.1 Somatic Segregation   294
    17.1.1 Somatic Crossing Over   294
    17.1.2 Chromosomal Chimeras   296
    17.1.3 Chromosomal Mosaics   299
    17.1.4 Polysomaty   299
    17.1.5 Somatic Reduction   300
17.2 Variations in Mitosis   300

17.3 Variations in Meiosis   301
    17.3.1 Asynapsis and Desynapsis   302
    17.3.2 Variation in Crossing Over   303
    17.3.3 Variation in Chromosome Size   304
    17.3.4 Variation in Spindle Formation   304
    17.3.5 Other Variations in Meiosis   305
17.4 Male Sterility   306
17.5 Preferential Segregation of Chromosomes   307

Chapter 18: Variation in Function of Sex Chomosomes   311

18.1 Variation in Sex Ratio   311
18.2 Different Sex Chromosome Systems   312
18.3 Cytogenetics of *Sciara*   314

Chapter 19: Apomixis and Parthenogenesis   318

19.1 Apomixis in Plants   318
    19.1.1 Vegetative Reproduction   320
    19.1.2 Agamospermy   320
19.2 Parthenogenesis in Animals   323
    19.2.1 Experimental Induction of Parthenogenesis
       in Animals   326

PART IX   EXTRACHROMOSOMAL INHERITANCE

Chapter 20: Plastids, Mitochondria, Intercellular Symbionts, and Plasmids   330

20.1 Plastids   330
20.2 Mitochondria   331
20.3 Intracellular Symbionts   332
    20.3.1 The P-Particles of *Paramecium*   332
    20.3.2 The Sigma Virus in *Drosophila*   333
    20.3.3 The Maternal Sex-Ratio Condition in
       Drosophila   334
    20.3.4 The Milk Factor in the Mouse (MTV) . . .   335
    20.3.5 Cytoplasmic Male Sterility (CMS)   335
20.4 Plasmids, Episomes and Transposable Elements . .   336
    20.4.1 The F-Episomes   337
    20.4.2 Colicinogenic Factors   339
    20.4.3 Resistance Factors   339
    20.4.4 Bacteriophages   340
    20.4.5 IS-Elements and Transposons   342

REFERENCES

General References 348
Specific References 349

Index... 419

Part I   Introduction

Chapter 1
History of Cytogenetics*

Many reviews of the history of biology have been written. Different approaches have been used in writing these reviews. This is not just another list of dates to be added to the multitude that has already been published. Instead, the author presents some of the people who have contributed to the advancement of a science, the motives that led them to choose their subject of investigation, the reasons for choosing those subjects, and the prior advances that made their contributions possible.

Cytogenetics was developed from two originally separate sciences—cytology and genetics. Cytogenetics deals with the study of heredity through the methods of cytology and genetics. The science is concerned with the structure, number, function, and movement of chromosomes and the numerous variations of these properties as they relate to the transmission, recombination, and expression of the genes. It also deals with nonchromosomal hereditary factors.

To fully understand the history of cytogenetics, one has to look at its roots. Consequently, this history of cytogenetics includes the histories of cytology and genetics as well as cytogenetics. The men who were chosen to be featured in this historical sketch made significant contributions to these sciences and, in this sense, represent milestones. Many other important contributions were made by other men, all of whom could not be mentioned here.

Johannes Sachariassen and **Zacharias** (1588–1631) **Janssen,** two Dutch eyeglass makers, father and son, between the years 1591 and 1608 produced the first operational compound microscope. They combined two double convex lenses in a tube. The magnification was not more than ten times, but it nevertheless caused great excitement.

William Harvey (1578–1657), in 1651 put forward the concept that all living things, including man, originate from eggs and that the semen has a vitalizing role in the reproduction process. Harvey was a court physician to King Charles I of England; later he became a professor at Oxford. While he was the king's physician, though, he once dissected a doe in the King's forests, his privilege as the King's physician. He found a fetus in the uterus of the doe. This initiated his interest in the conception of life. He eventually dissected more than 80 different species of animals.

*Altered and reprinted with permission of the publisher from: Schulz-Schaeffer, J., 1976. A Short History of Cytogenetics. Biol. Zentralbl. 95:193–221.

Since he did not possess a microscope, he was, naturally, unable to observe the eggs of mammals, but he presupposed their existence on theoretical grounds, a conclusion that was confirmed long afterward. Harvey also developed the *theory of epigenesis,* which states that in the course of embryonic development new structures and organisms develop from an originally undifferentiated mass of living material.

Marcello Malpighi (1628–1694), a professor at Bologna, Italy, and later private physician to the Pope, discovered the microscopic anatomy of both animals and plants. In 1661, he discovered the capillaries in the lungs of animals and, thus, completed the story of the circulation of the blood as told by Harvey. He was interested in the development of plant seeds, the structure of plant stems and roots, and the function of leaves as well as in the anatomy of the silkworm, the development of the chick embryo, and the microscopic structure of glands and tissues.

Robert Hooke (1635–1703), an architect as well as a microscopist and the first curator of the Royal Society of London, in 1665 described cork and other cells and introduced the term **cell.** His was the first drawing ever made of cells.
Microscopes at that time magnified 100 to 200 times with a distortion of shape and color that increased with magnification. Nevertheless, these microscopes revealed many new things. Still, it was necessary to wait for better lenses to see anything more. Scientists waited for 160 years, and during this period they, naturally, argued about what they had seen.

Régnier de Graaf (1641–1673), a young surgeon from Delft, Netherlands, in 1672 discovered follicles in the human ovary and identified them, incorrectly, as eggs. They were named after him—**Graafian follicles.** He discovered these follicles after he observed that the progeny of mammals presents characteristics of both the mother and the father; therefore, he reasoned, both sexes must transmit agents of heredity. In search of some physical evidence for this observation, he studied sections of ovaries prepared for examination. He found that before conception there were small watery lumps on the surface of the ovaries. He observed them first in rabbits, then in ewes, and finally in human beings. Although Graaf did not see the eggs, it now seemed certain that female mammals and women produced eggs like the ones laid by birds or fishes. It also seemed certain that the egg contained within itself the role and universal principle of heredity for all life. Those who supported this view became known as **ovists.**

Nehemiah Grew (1628–1711), an English physician and plant anatomist, worked with Hooke in London and described **bladders** and **pores** in wood and pith. Grew was an ardent student of plant structure. He published two illustrated volumes on the microscopic anatomy of plants (1672, 1682). In these volumes, he advanced a theory that the pistil in plants corresponds to the female, and the stamen, with its pollen, to the male. These were the first consistent studies leading to an understanding of the reproductive parts of plants. In all, he published well over one hundred engravings made from drawings of what he viewed through his microscope.

Like Hooke, Grew was a curator of the Royal Society. Malpighi and Grew contributed so much to plant anatomy that little addition was made to their work for more than a century. It was nearly two centuries later that the fundamental value of their achievements was fully appreciated. Great credit is due to these scientists who guessed that here at hand was information of the highest importance for the future of science.

Anthony van Leeuwenhoek (1632–1723), a fellow townsman of de Graff from Delft, Netherlands, improved the microscope lens system through grinding. He was an enterprising man and was also competent at selling linen, gauging wind, and surveying buildings. The lenses he ground for his microscopes magnified specimens several hundred times. One of his lenses, which is still in existence, is said to magnify as much as 270 times.

Starting in 1674, he reported on his studies of spermatozoa, bacteria, and protozoa. He observed, but did not identify, nuclei in blood cells. Just how Leeuwenhoek managed to see spermatozoa, bacteria, and the nuclei of blood corpuscles is still not fully understood. The answer in part is that he chose a simple type of microscope, one that needed infinite patience to use but negated many of the optical aberrations of the contemporary compound form.

Leeuwenhoek also observed the association of sperm with eggs in frogs and fishes and considered the sperm to furnish the essential life-giving properties, while the egg merely provided the proper environment for nutrition and development of the embryo. Those who supported this view became known as **animalculists.** Ovists, motivated by de Graaf, and animalculists carried on a controversy for decades. In 1680 Leeuwenhoek was elected a Fellow of the Royal Society of London.

Jan Swammerdam (1637–1680), from Amsterdam, also using the microscope, in 1679 reported his studies of the development of insects. He visualized development as a simple enlargement from a minute but preformed animal to the adult. Later, after this idea had been applied widely and supplemented by imagination, it became a general explanation of development known as the **preformation theory.** According to this theory, the egg, sperm, or zygote contains a folded, preformed adult in miniature which unfolds during development. This theory was opposed to the **theory of epigenesis** put forth by Harvey which stated that new structures arise in the course of development (Swammerdam, 1752).

Rudolf Jacob Camerarius (1665–1721), a German professor of medicine, in 1694 reported on early pollination experiments and the existence of sexual reproduction in flowering plants. He showed that in the maize plant, seeds are not produced unless pollen is applied to the pistils. He therefore concluded that the pollen is the "male" element and the pistils are the "female" element, and he discussed their connection with a number of theories on sexuality and fertilization in general.

Although he did not contribute anything of special value from a theoretical point of view, his work made possible the experimental approach to plant hybridization. He is also credited with the first artificially produced plant hybrid on record; he produced a plant from a cross between hemp and hop.

Pierre Louis Moreau de Maupertuis (1698–1759), President of King Frederick the Great's Academy of Sciences in Berlin and a member of the Royal Society of London, described an autosomal dominant pattern of inheritance in the polydactyly of men and discussed a concept of segregation. His study of human pedigrees was a novel enterprise at that time. He described this enterprise no less than three different times (1745, 1751, and 1752). His studies were also the basis upon which he founded his theory of the formation of the foetus and the nature of heredity, a theory that brilliantly anticipated the discoveries of Mendel and de Vries. He applied the mathematical theory of probability to genetics a century before Mendel and undertook experiments in animal breeding to throw light on his theories.

Joseph Gottlieb Kölreuter (1733–1806), German botanist, during 1761–1766 published information about hybrids between plant varieties that might resemble one parent or the other or present a combination of their features. Camerarius was the first to experiment in this field. For a number of years Kölreuter crossed different types of tobacco with one another. Later, he crossed other plant genera such as pinks, *Aquilegia, Verbascum,* and others. One of his most valuable observations on reciprocal crosses showed the equality of contributions from the two parents. Thus, he provided clear evidence that in reciprocal crosses, the hereditary contribution of the two parents to their offspring was equal.

Jean Baptiste Pierre Antoine de Monet, better known to history as the Chevalier de Lamarck (1744–1829) was a discharged lieutenant who at the age of fifty was made professor of zoology in Paris, France, and attained lasting fame, despite his lack of formal scientific training. In 1809, he theorized that species can change gradually into new ones through a constant strengthening and perfecting of adaptive characteristics and that these **acquired characteristics** are transmitted to the offspring. This theory is often called Lamarckism. He also stated the importance of the cell in the living organism.

Karl Ernst von Baer (1792–1876), a professor at Königsberg, Germany, discovered the mammalian egg and published two famous embryological works (1827, 1828). He found that there are microscopic specks of jelly inside the Graafian follicles on the surface of the ovary of the rabbit. He had also found that there are similar specks in the oviduct entering the womb. He had rightly concluded that these specks, one eightieth of a centimeter across, which happen to be the largest cells in the body, were the female germ cells. They were the eggs long ago imagined by Graaf. But it was not before 1854 that a sperm was seen making its way into the egg of a frog, and it was not until the following year that the same process was seen in various weeds and algae.

Robert Brown (1773–1858), a Scottish botanist, in 1828 discovered the cell nucleus in the flowering plant *Tradescantia.* Although he practiced medicine as a surgeon for five years, he later abandoned this and turned his efforts toward botanical sciences. He was librarian to the Linnaean Society and curator at the British

Museum. His remarkable account (1828) of the properties and behavior of the nucleus stand unmodified and without correction. He was a very skillful and careful observer. He also observed the random thermal motion of small particles, still known as **Brownian movement.**

Hugo von Mohl (1805–1872), a German medical doctor and professor of physiology and later of botany, is generally mentioned as the creator of modern plant-cytology. In 1835, in his doctoral dissertation, he described cell division and emphasized the importance of protoplasm. He upheld clearly and convincingly that the cells in algae and even higher plants arise through partition-walls being formed between previously existing cells. These partition-walls he investigated and described with great accuracy.

Matthias Jacob Schleiden (1804–1881) is credited, together with Schwann, with proposing the **cell theory,** which states that the cell is a unit of biological organization. Schleiden studied law and took up practice as an advocate. Later he became a professor of botany at Jena, Germany. He became famous with the publication of his *Contributions to Phytogenesis* in 1838. He recognized the importance of Brown's discovery of the cell nucleus, which Brown himself had failed to do, and sought to reconstruct the course of development of the cell, for which he wisely chose the embryonic cell as the starting point of his study. He also discovered in cells, the formation of what is now known as the **nucleolus.**

Theodor Schwann (1810–1882), a German professor of anatomy, began work with Schleiden's cell formation theory, which he accepted in its entirety, and expanded it into a general theory of the basis and origin of all life phenomena. He applied Schleiden's discoveries in plants to animals and invented the term **cell theory** in 1838; however, the term is generally attributed to Schleiden and Schwann and dated 1839. Schwann refined this theory in the following way:
1. The cell is the smallest building element of a multicellular organism and as a unit is itself an elementary organism.
2. Each cell in a multicellular organism has a specific task to accomplish and represents a working unit.
3. A cell can only be produced from another cell by cell division.

This concept of the cell as a general unit of life and as a common basis for the vital phenomena in both the animal and vegetable kingdoms was immediately and universally accepted.

Rudolf Ludwig Carl Virchow (1821–1902), a German professor of pathological anatomy, was a figure of great importance in the intellectual, social, and political history of nineteenth-century Europe. He confirmed the principle that cells arise only from preexisting cells, which is also referred to as the **theory of cell lineage.** Diseases and their causes were the chief objects of Virchow's studies, and this led him to the realization of the cells as basic constituents of the organism both in health and sickness. In 1858 he published his *Cellular Pathology,* a theory identifying the cells as the true causes of disease. His observations gave Schleiden and

Schwann's **cell theory** its impact in terms of heredity and development, for if present cells have come from preexisting cells, then all cells trace their ancestry back to the first cells in an unbroken line of descent.

Gregor Mendel (1822–1884), prelate of an Austrian monastery, developed the fundamental principles of heredity. In 1866 he published his famous paper on *Experiments in Plant Hybridization* describing his garden pea experiments. This paper was the first to come from many years' work, and it testified to a keen observation of nature and a thorough grounding in mathematical thought. His findings were obscured for many years but finally were rediscovered in 1900.

Peas have many advantages for research. They are easy to pollinate and to protect from foreign pollen. Mendel noted the points of resemblance and difference between certain varieties and convinced himself of the constancy of several pairs of characteristics such as the round or wrinkled shape of the seed, its yellow or green color, the different colors of the seed pods, and the tallness or dwarfness of the plants. He then studied these characteristics through several daughter generations of the hybrids. He did not merely note the development of the characteristics or their failure to appear in the hybrids, but determined the frequency of their appearance in the progeny resulting from various crosses. His counts put the phenomena of inheritance for the first time on a numerical basis, and the new principles which he eventually revealed became known as Mendel's Laws, which are:

1. **Law of Segregation:** There are pairs of factors within the sexual organism, and one factor of each pair goes into each mature germ cell. Therefore, each member of a pair segregates from the other in the parent and reunites in the offspring. This law deals with one pair of genes and discusses the behavior of genes or alleles at the same locus.
2. **Law of Independent Assortment:** Each of the genes segregates from the other and pairs again in an independent fashion, thus giving rise to new combinations of characteristics. This law deals with two or more pairs of genes and discusses the behavior of genes or non-alleles at separate loci.

Francis Galton (1822–1911), an English explorer and Fellow of the Royal Society, in 1869 published *Hereditary Genius* and so founded the scientific study of human heredity. This book also introduced the statistical method into the study of heredity. In 1876 he reported on studies of human twins in which he tried to separate the effect of heredity and environment. He developed the **concept of regression,** a measurement of the degree of resemblance of relatives. He also developed the concept of quantitative analysis of continuously variable or polygenic traits such as diabetes and height.

To the science that he placed highest of all he gave the name **eugenics,** a name that has become universally accepted. According to Stern (1960), eugenics is the study of agencies under social control that may improve or impair the hereditary physical or mental qualities of future generations of men.

Fredrick Miescher (1844–1895), a Swiss chemist, in 1871 reported that he had isolated nucleic acid and nucleoprotein. By that time it was clear from its universal occurrence that the nucleus was a peculiarly important cell constituent. However, before any chemical analysis of the nucleus was possible, it was first necessary to separate, in quantity, nuclei from the remainder of the cell. Miescher decided to attempt this and made, what seems at first sight, the bizarre choice of pus cells. He found that it was possible to get the cells in suspension and furthermore, by treatment with dilute hydrochloric acid, pepsin, and ether, to separate the nuclei from everthing else. From those nuclei he prepared a substance of remarkable properties, namely nucleic acid, which later became known as DNA. It was distinguished by a high content of phosphorus, an element then rarely found in organic substances of physiological origin. So remarkable did Miescher's results then appear, that the publisher Hoppe-Seyler was reluctant to print them until he had himself confirmed Miescher's conclusion by experiment.

Then Miescher worked on Rhine winter salmon sperm. In the isolated heads of the sperm, he not only found nucleic acid but also a highly basic nitrogenous substance to which he gave the name protamine. When protamine is combined with nucleic acid, the compound is referred to as nucleoprotein.

Wilhelm August Oscar Hertwig (1849–1922), a professor of anatomy, in 1876 and 1877 studied reproduction in the sea urchin, *Paracentrotus lividus,* and concluded that fertilization involves the union of sperm and egg. This study initiated the period of experimental cytology.

When he was younger, Oscar, along with his brother Richard, was supposed to take over their father's factory, but a high school teacher recognized their natural scientific abilities and prevailed upon their father to allow his boys to study chemistry. At the University, however, the professor of chemistry turned out to be so dull that the brothers soon changed to medicine, which also included zoology.

When he became a professor of anatomy, Oscar Hertwig chose the eggs of sea urchins for his first investigations of reproduction. This proved to be a good choice because many favorable conditions are met in the sea urchin, which is still unsurpassed today for many experimental purposes. It is particularly favorable for observations of living cells. The discovery of a new favorable object of investigation often opens an entire new area of research. The eggs of the sea urchin are transparent, small (0.1 mm in diameter) and favorable for studies under high magnification. Hertwig was able to watch the sperm nucleus pass through the translucent cytoplasm of the egg and realized that it would unite with the ovum nucleus. Until Hertwig's investigations, only unfavorable objects like the nontransparent eggs of frogs were used, eggs that were fertilized inside the mother and could not be observed easily.

Walter Flemming (1843–1915), an Austrian cytologist, in 1882 proposed the term **mitosis.** He showed that the chromosomes split longitudinally during nuclear division and the formation of daughter nuclei. He also applied the name **chromatin** to the stainable portion of the nucleus. He was a distinguished observer, technician, and teacher.

In an important paper (1879) he described mitosis in living and fixed cells of the salamander. An essential contribution was his development of improved fixing and staining methods to make visible cytological details. His monograph (1879) included a remarkably foresighted treatment of the problems of cell division, some of which are still active research problems today. In 1882 the studies on the human chromosome complement began with Flemming's demonstration of cell division in the corneal epithelium of humans.

August Weismann (1834–1914), a German biologist, in his essays of 1883 and 1885 put forth his **germplasm theory,** which was an alternative explanation to Lamarck's theory of acquired characteristics. Weismann's theory was based on the early separation, in the animal embryo, of the germplasm from the somatoplasm. It emphasized the remarkable stability of the hereditary material. Conceivably, little, if any, environmental influence could affect the genes, even though environmental modifications of external characteristics occurred. Reproduction in animals was accomplished not by body cells or somatoplasm but by the germplasm, which was transmitted essentially unchanged from generation to generation.

The fundamental premise is well established, although some details of the germplasm theory have been modified. Weismann speculated that the chromosomes of the sex cells were the carriers of his germplasm, but he erred in assuming that each chromosome could contain all hereditary material. He also postulated that a periodic reduction in chromosome number must occur in all sexual organisms and that during fertilization a new combination of chromosomes and hereditary factors takes place. His theory was that the alternation of reduction and fertilization is necessary for maintaining constant chromosome numbers for sexual reproduction. At that time this process had not been observed under the microscrope, and its mechanism was a matter of speculation.

Wilhelm Roux (1850–1924), a German zoologist, in 1883 proposed that it was the chromosomes that contain the units of heredity. He speculated on the question of how the hereditary units could behave in such a way that each daughter cell receives all that is in the parent cell and becomes a complete cell and not half a cell or only part of a parent cell.

The only mechanism he could devise to test his speculation was to line up objects in a row and duplicate them exactly. He therefore suggested that the significance of cell division lies in the fact that nuclei have strings of bead-like structures that line up and duplicate themselves. If nuclei really have such structures, he reasoned, it might be possible to explain the mechanics of hereditary transmission from cell to cell. The most likely constituents of the nucleus to fill these requirements were the chromosomes. His hypothesis was that not only the chromosomes but individual parts of each chromosome were important in determining the individual's development, physiology, and morphology. Proof of this hypothesis was not given until later. This was in direct contrast to Weismann's idea, that each chromosome could contain all hereditary material.

Edouard van Beneden (1845–1910), a highly reputed Belgian zoologist, in 1883 showed that in the round worm, *Ascaris megalocephala,* the number of chromosomes in the gametes is half the number that is in the body cells, and that in fertilization, the chromosome contributions of egg and sperm to the zygote are numerically equal. Through this observation he confirmed Weismann's theory on reduction and fertilization.

Ascaris was a favorable object for his studies. This species has only 2n = 8 chromosomes and in a particularly favorable race only 2n = 4. Also the size of its chromosomes make them ideal for observation. Basic to the clarification of the role of the chromosomes as the physical agents of Mendelian phenomena was the discovery of their behavior in meiosis. In his monograph, Beneden (1883) traced the spermatogenesis and oogenesis of the round worm and proved Weismann's theory of reduction and fertilization. Confirmation of these discoveries soon followed, but the exact mechanism of reduction by synapsis and the formation of bivalent pairs of chromosomes was not understood until the 1890s.

Carl Wilhelm von Nägeli (1817–1891), a Swiss botanist, though certainly best known as a cytologist, in 1884 made his major contribution with his theory of the **idioplasm.** According to this theory, the idioplasm is the sum of all hereditary determinants of an organism, which in its fine structure ought to contain the **life formula** of that organism. With this theory he provided the chief stimulus for the view that there is a single hereditary substance. This theory led Beneden, Strasburger, Oscar Hertwig, and Weismann to the belief that the idioplasm is located in the substance of the chromosome. Thus, nuclear division and chromosomes were shifted into the center of a uniform and large field of biological studies. In 1842 Nägeli published the first drawings of chromosomes.

Eduard Strasburger (1844–1912), a German cytologist, in 1884 described fertilization in angiosperms. In the field of plant cytology, he was preeminent in his time. Equally distinguished as a research worker and a teacher, he attracted a large number of students from many countries to his institute. He was a leading writer of textbooks, and his scientific productivity included his epoch making cell studies (1884, 1904, 1905). Strasburger demonstrated that the principles of fertilization developed by Oscar Hertwig for animals held also for plants.

Hertwig and Strasburger are considered the discoverers of fertilization. Before Mendel's work was discovered, interest had already developed in locating the source of hereditary transmission. Since sex cells, that is, eggs and sperm, were known to be involved in fertilization and both parents were known to transmit their characteristics to the progeny, the first problem was to determine which part of the cell was involved in hereditary transmission. Strasburger observed that the egg carried more cytoplasm than the sperm. Just like Kölreuter 100 years before him, Strasburger made reciprocal crosses between different plant species and found that the results were similar. Since the egg and sperm were unequal with respect to size and amount of cytoplasm carried, he suggested that the cytoplasm was not responsible for hereditary differences between species. Consequently, he came to the conclusion that the nucleus and its chromosomes are the material basis of heredity and, at the same time, the material governing development. Strasburger stated that molecular stimuli are passed from the nucleus to the cytoplasm that surrounds it, controlling the process of metabolism in the cell and giving a specific characteristic to its growth.

It is an important historical fact that after the German zoologist Otto Bütschli had discovered and understood mitotic division, the botanist, Strasburger, called on him, studied his preparations, and realized that what he had seen in plant cells was exactly the same thing. Since then cytology has not made a serious distinction between animals and plants. All the basic facts of chromosomal structure and behavior, mitosis, fertilization, sex chromosomes, cytoplasmic inclusions, and cell physiology are identical. Thus, cytology developed into an independent science, drawing its discoveries from animals, plants, and humans.

Ernst Abbé (1840–1908), a German physicist, by 1886 had produced oil immersion objectives with a resolution of 0.25 μm. This advanced the resolving power of the light microscope to the absolute limit set by the wave length of light. A further advantage of this system was that the performance of these lenses was independent of the thickness of the coverslip. Abbé was head of the German Zeiss corporation, which was the leading microscope manufacturer.

Early in the 1880's Abbé joined with Otto Schott, a glass manufacturer, in experiments on adding various chemical elements such as boron and phosphorus to the silicate base of glass. By 1886 they had produced their Jena glass, which had novel characteristics. The improved lenses that these new materials made possible were called **apochromatic**, because they eliminated the residual chromatic aberration, the secondary spectrum, of the achromat. Cytologists such as Hertwig and Flemming were using apochromatic objectives within a few years of their introduction.

Theodor Boveri (1862–1915), a professor at Würzburg, Germany, and a student of the brothers Hertwig in his celebrated *Zellstudien* (1887, 1888) together with Oscar Hertwig (1890) discovered the real nature of reduction division. In 1892 he described meiosis and, particularly, synapsis in *Ascaris*. He also explored the question of the source of hereditary transmission in animals, which Strasburger had studied in plants.

By shaking sea urchin eggs at a critical time in their development, he produced some eggs without nuclei and some with nuclei as usual. Each of these kinds of eggs were fertilized by a normal sperm from another species of sea urchin. Eggs lacking a nucleus produced larvae resembling the species from which the sperms were obtained, but those with nuclei developed into hybrids, showing the characteristics of both species. The cytoplasm in the two kinds of eggs had not been altered and it was therefore presumed that the nucleus and not the cytoplasm was responsible for the transmission of hereditary traits.

With his experiments on the double fertilization of sea urchin eggs, *Toxopneustes* (1902, 1904, 1907), Boveri also contributed to the formulation of the **chromosome theory of inheritance**, which will be discussed later. He found eggs that had been fertilized by two spermatozoa. Since each sperm introduced a centrosome into the egg, and each centrosome divided in anticipation of the first cleavage division, the initial metaphases and anaphases were often characterized by a **tetraster**, which is a spindle with four poles. Since the dividing nucleus was triploid, the distribution of the chromosomes to four poles in anaphase was irregular. Boveri isolated many of the first-division blastomeres from these dispermic eggs and demonstrated that most were abnormal in development, but that all were not alike in their abnormalities. He concluded that abnormal development resulted from the irregular distribution of chromosomes brought on by the multipolar division. Each chromosome must consequently have possessed a certain individual quality that expressed itself in development.

Hermann Henking (1858–1942), a German zoologist, in 1891 described in the hemipteran insect, *Pyrrhocoris,* chromatin elements that he labeled X and that now are known to be the sex- or X-chromosomes. He found a peculiar chromatin element that in the second spermatocyte division first lagged behind the separating anaphase chromosomes and then passed undivided to one pole while all the other eleven chromosomes were equally divided. From this it followed that the sperms were of two numerically equal classes distinguished by the presence or absence of this chromosome element. This element had to have a close relationship for the determination of sex. If the egg was fertilized by one class, a male was formed, if by the other, a female. All the essential features of Henking's description were subsequently confirmed in other animals by other observers. This mechanism is now called the XO system of sex determination.

Edmund Beecher Wilson (1856–1939), an American biologist and Professor at Columbia University, was a superb synthesist as well as a stimulating teacher and investigator. By 1896 he had been able to organize the cytological and embryological knowledge of his day in the first edition of his classic *The Cell in Development and Inheritance.* Mendel's principles of genetics were still to be rediscovered, but the beginning of cytogenetics and of the **chromosome theory of inheritance** were clearly outlined by Wilson's statement that the visible chromomeres on the chromosomes were in all probability much larger than the **ultimate dividing units** and that these units must be capable of assimilation, growth, and division without loss of their specific characteristics. Wilson brought the past in relation to the future. Four principles were laid down by Wilson as the foundation of the chromosome theory:

1. The exact lengthwise division of the chromosomes at mitosis allows for the equal distribution of linearly arranged particles to the daughter cells.
2. The assumed material existence of the chromosomes in the nucleus between mitoses gives the genetic continuity necessary for the organs of heredity.
3. The fact that the nucleus goes where things are happening shows its governing position in the work of the cell.
4. The equality of the chromosomes of the fusing germ cells corresponds to the equality of male and female in heredity.

These arguments had long been known but were still widely disputed or misunderstood at this time.

Carl Franz Joseph Correns (1864–1933), a German botanist, in 1900 along with Hugo de Vries and Erich von Tschermak was one of the three rediscoverers of the fundamental principles of heredity, first developed by Mendel in 1866. He had carried out extensive hybridization experiments on maize, stocks *(Matthiola)*, beans, peas, and lilies at the University of Tübingen during the 1890s. In 1899 he had data from several generations of garden pea and maize and had arrived at conclusions similar to those of Mendel. He studied Mendel's paper, because he had read a statement that Mendel believed he had found constant numerical relationships in his experiments. Correns compared his own and Mendel's data, and in 1900 he reported that he had observed the same kind of results with maize. He disagreed with de Vries in that he thought there were cases that did not conform to the Mendelian scheme.

Hugo de Vries (1848–1935), a Dutch biologist and rediscoverer of Mendel's laws, was also known for his mutation theory and studies on the evening primrose and maize. De Vries published three papers on Mendelism in 1900, one of which, for the most part, has been overlooked. He later stated that he had worked out the Mendelian scheme for himself and was later led to Mendel's paper.

In the 1880s de Vries, a keen observer and objective scientist, saw striking variations in the plant called Lamarck's evening primrose, *Oenothera lamarckiana*, which had been introduced from America and had grown wild in Europe. He collected seeds from plants that differed from the standard type and raised them in his botanical garden at Hilversum, a few miles east of Amsterdam. On careful observation, many differences in growth form were seen among the different plants. One type of *Oenothera* called *gigas* was much larger than the average and no intermediate gradations were observed between it and smaller types. It seemed to represent a distinct and discontinuous change from the usual size. On one occasion, a gigas plant was found alone in a bed of plants of the standard size. The new plant produced only giants like itself. A dwarf type called *nanella*, which gave rise only to dwarfs, was also observed. Other abrupt changes that affected the color and shape of various parts of the plant were studied and the variations seemed to breed true.

De Vries visualized these changes as a source of variation in evolution, in contrast to the gradual process suggested by Lamarck and Darwin. The word **mutation,** implying change, was used to describe such alterations. In 1901 de Vries published his accumulated data in a book entitled *The Mutation Theory*. Mutations were considered to be rare in nature but capable of providing variations by which races and species were distinguished. De Vries was careful to make a distinction between hereditary and environmental variation, but his mutations are now known to include changes in chromosome structure and number. The term mutation today is used in a more restricted sense to specify only gene changes or point mutations and not visible chromosome changes.

Erich von Tschermak (1871–1962), an Austrian botanist, is also considered to be one of the rediscoverers of Mendel's laws. His interests were in practical plant breeding, and this led to studies of the effects of crossing and inbreeding on vegetative vigor in peas. He published two papers on the subject in 1900. He later wrote that the three rediscoverers were fully aware of the fact that the independent discovery of the laws of heredity was far from being the accomplishment it had been in Mendel's time since it was made considerably easier by the work that appeared in the interval, especially the cytological researches of Hertwig and Strasburger. Consequently, the three were less interested in being celebrated as rediscoverers of rules they themselves designated *Mendel's laws* than in the successful utilization of these laws for the development of their various fields—de Vries for the mutation theory, Correns for fundamental research in inheritance, and Tschermak for practical plant breeding (Tschermak-Seysenegg, 1951).

Walter S. Sutton (1876–1916), a young American graduate student, in 1902 and 1903 showed the significance of reduction division and proposed the **chromosome theory of heredity.** He independently recognized a parallelism between the behavior of chromosomes and the Mendelian segregation of genes.
The first paper (1902) contained the earliest detailed demonstration that the somatic chromosomes of the lubber grasshopper, *Brachystola magna,* occur in definite distinguishably different pairs of like chromosomes. [He knew of Boveri's first paper (1902) on dispermic eggs]. His 1903 paper contains a full elaboration of his hypothesis, including the view that the different chromosome pairs orient at random on the meiotic spindles, thus accounting for the independent segregation of separate pairs of genes seen by Mendel. This cytological basis for genetic theory is also often called the *Sutton–Boveri theory of chromosomal inheritance.* With the second paper by Sutton (1903), this phase of the separate histories of cytology and genetics was finished. The conclusions were not immediately generally accepted, but from then on cytology and genetics began to have strong effects on each other, and this is generally considered the birth of cytogenetics.

William Bateson (1861–1926) a British biologist at Cambridge University, immediately became interested in Mendel's work after its rediscovery in 1900, and in cooperation with R. C. Punnett investigated nine plant genera and four animal genera, but sweet peas and poultry most intensely. During the course of his work he introduced the terms **genetics** (1906), **allelomorph** (= allele), F_1 and F_2 for daughter generations (1902), **homozygote** and **heterozygote** (1902), and **epistasis** (1907). He worked and fought hardest to prove the universality of the Mendelian theory in plants and animals. He was engaged in experimental breeding with poultry. These animals breed rapidly and large numbers of progeny were obtained in a short time.
In 1906 Bateson and Punnet reported the first case of **linkage,** which they discovered in sweet peas. Since the number of hereditary factors seemed large in comparison with the number of pairs of chromosomes, it was to be expected that several factors should be associated with one chromosome, and therefore linked together. In sweet peas certain factors for color and pollen shape were always inherited together. The bearing of this discovery on the chromosome theory of inheritance was pointed out later.

Wilhelm Ludvig Johannsen (1857–1927), a Danish geneticist and plant physiologist, in 1905 coined the terms **gene, genotype,** and **phenotype** and stressed the importance of making a clear distinction between genes and characteristics (1909).

He can be closely identified with the development of genetics as a science although his early scientific work was in the field of plant physiology. His first genetic paper, *On Heredity and Variation,* appeared in 1896, and in 1898 he began the investigations that have since become classics on barley and beans.

In his paper on *Pure Lines* (1903) he showed a difference in the effects of selection when applied to populations of ordinary cross-fertilizing organisms as compared with self-fertilizing ones. Self-fertilization was found to produce homozygosity or pure lines. In cross-breeding populations, selection was found to be effective in altering the proportion of different types. When plants were self-fertilized over long periods, selection was no longer effective. The plants had become completely homozygous and no genetic variation was left for selection to act on. All variation in a pure line is environmental.

Frans Alfons Janssens' (1863–1924) name is generally associated with the partial **chiasmatype theory,** which he advanced in 1909. In 1905 he described the configurations of the bivalent pairs in the spermatogenesis of Amphibia that showed chiasma-like configurations. He indicated that chiasmata are produced by exchange between chromatids of nonhomologous chromosomes and later suggested the possible genetic significance of these crossed segments of the chromosomes. The partial chiasma-type theory postulates that true chiasmata are the direct result of crossing-over, being formed at precisely the points where the exchange of segments between non-sister chromatids took place. This theory is now the most satisfactory account of the relationship between the cytologically visible chiasmata and genetic crossing over.

Thomas Hunt Morgan (1866–1945) in 1910 discovered the mutant *white eye* and consequently sex linkage in *Drosophila.* With this discovery *Drosophila* genetics had its beginning.

Drosophila melanogaster, the fruit fly, proved to be one of the most ideal laboratory animals for cytogenetic studies. It is a tiny organism, about 6 mm long. Completing its life cycle from egg to fly in about 10 days, this insect supplies as many as thirty generations a year, an enormous advantage compared to the relative slowness of the usual laboratory animals. It is easily bred, fertile, and with a life span that can reach ninety days. Thousands of these flies can be handled in a few milk bottles, while the cost of feeding and upkeep is negligible. The giant chromosomes in the salivary glands are several hundred times larger than normal somatic chromosomes, and the bands reveal the necessary detail for cytogenetic study. The low chromosome number of $n = 4$ also is ideal.

Within a short time, Morgan's **fly room** at Columbia University in New York became a very popular place. A steady stream of American and foreign students, both doctoral and post-doctoral, passed through his laboratory. Morgan was concerned about the exceptions to Mendel's second law of independent assortment. This law implies that an organism cannot possess more gene pairs than the number of chromosomes in a haploid set, if it is granted that the genes are borne on chromosomes. Within the first decade after the rediscovery of Mendelism, this logical consequence of the theory was sharply contradicted by experience.

Obviously, some extension or revision of the theory was necessary. Morgan's **alternative linkage theory** supposed that genes are organized in a definite linear order within the chromosome. These genes are expected to exhibit linkage if they lie within the same chromosome, but should they lie in nonhomologous chromosomes, they would be transmitted according to the principle of independent assortment. The possibilities of recombination for linked genes were, thus, envisioned to depend on the breakage of chromosomes and their rejoining in such a way as to result in the exchange of equal segments without disturbance of the basic linear sequence.

It must be recalled at this point that Morgan's theory, unlike the purely formalistic approach of Bateson and Punnet (1906), rested solidly on a body of accumulated cytological evidence concerning the intimate details of chromosome behavior during the prophase of the first meiotic divisions. The first test of the validity of these assumptions was provided by Morgan in 1911 when he showed that several sex-linked mutants in *Drosophila* were associated with the behavior of the heteromorphic sex chromosomes. During the following decades, thousands of experiments in a wealth of diverse biological forms, have confirmed the universality of Morgan's interpretation of linkage. In 1933 Morgan was the first to receive the Nobel Prize in medicine and physiology for accomplishments in the field of genetics for his development of the theory of the gene.

Ralph A. Emerson (1873–1947), together with E. M. East, in 1913 published a paper on maize in which they reported that the F_2 was much more variable than the F_1. They interpreted this as being due to the segregation of several pairs of genes. Their joint paper is a classic in the field of genetics and marks the bringing of the **inheritance of quantitative characters** into the general scheme of Mendelism. In 1914 Emerson discovered the first mutant in maize, *blotched leaf (bl.* Chrom. 2. Emerson et. al., 1935).

Emerson, the son of a long line of American farmers, including the celebrated Adams family, was teaching horticulture at the University of Nebraska. Later he came to Cornell University where he trained a group of workers in genetics, who are now spread all over America, and initiated a remarkable organization for cooperative work. The maize work which started independently of the *Drosophila* work took fresh impetus from the publication of the first important *Drosophila* papers. The disadvantage of maize was that, as mentioned before, the fruit fly would produce 30 generations before the maize plant completes one. But the maize workers had one great advantage over the fly workers; the chromosomes in meiotic maize cells are more easily studied under the microscope, and this cytological simplicity made their mapping less difficult. Most of the delicate cytogenetic work in maize was later done by Barbara McClintock, one of the most skillful and persevering of the younger genetics workers in America at that time.

In 1914 Emerson also suggested that some genes might not be completely stable. He studied a variegated variety of maize that had a white pericarp with numerous red spots of varying size. Genetically, these plants were homozygous for the recessive gene for *white*. Emerson concluded that this gene must be unstable, that it can mutate spontaneously into the dominant allele for *red*, and that each red spot on the kernel is made up of cells that came from one cell in which such a mutation arose. This was a very revolutionary idea at that time. Such unstable genes are now referred to as **mutable genes.**

Albert Francis Blakeslee (1874–1954), American botanist and geneticist, in 1921 discovered **trisomics** in the Jimson weed, *Datura stramonium,* a plant species from which the drug belladonna is obtained. He worked at the Cold Spring Harbor Station for Experimental Evolution for the Carnegie Institution of Washington. In 1937, together with Oswald T. Avery, he discovered that doubling of the chromosome set in plants may be induced by use of the alkaloid colchicine. After Blakeslee discovered trisomics, he initiated the first critical study of aneuploid plants with Belling (1924). The trisomics differed morphologically from the wild-type plants in several specific ways. Conspicuous deviations from the normal were observed in the shape and spine characteristics of the seed capsules. These traits were associated with the extra chromosome, which gave the plant an extra dose of all those genes contained in the extra chromosome. It was thus possible to identify some genes with their chromosomes, giving further evidence that certain genes are located in particular chromosomes.

Calvin Blackman Bridges (1889–1938), a research associate of the Carnegie Institution of Washington, in 1923 was the first to discover **duplications, deficiencies,** and **translocations** in *Drosophila* chromosomes. He also observed triploid intersexes in *Drosophila*. Bridges joined T. S. Painter's investigations of the giant salivary gland chromosomes for further refinement of technique and fuller and more salient details. He stretched the chromosomes of the salivary gland cells until they were more than 150 times longer than those of the egg cells. He made preparations from larvae that had been raised to their maximum size by supplying them with an extra diet of yeast.

Bridges kept working, studying deficiencies and duplications, in an effort to revamp his chromosome maps. In 1935 he published the first complete map of all four *Drosophila melanogaster* chromosomes. Several revised maps of these chromosomes were published later. The genetic linkage map was superimposed on the cytological map. He also devised a special numbering system that is still used today to identify particular bands illustrated. In 1936 his son, Philip N. Bridges, came to his help. Calvin Bridges, a brilliant, simple, and unaffected worker, never finished this study. He had driven himself so hard during this work that he died of a heart attack in Los Angeles in 1938.

Robert Joachim Feulgen (1884–1955), German biochemist, in 1924 together with H. Rossenbeck described a test for the presence of DNA. This specific staining reaction is now called **Feulgen reaction.** Through this reaction it was proven that DNA is located in the chromosomes of interphase cells. In 1914 Feulgen showed that the unstable carbohydrate of the thymus type of nucleic acid was not a hexose, as it had been regarded until then, but a pentose. On gentle hydrolysis this pentose liberated aldehyde, which could be detected by the usual reagent for this class of substance, the dye fuchsin decolorised by sulphurous acid. Ten years later this test was applied to sections of tissue under the microscope. Feulgen and Rossenbeck were then greatly surprised to find that the nuclei of the wheat germ gave a strong reaction to this test, for this result showed that a nucleic acid of the thymus type could be found in plant cells. The thymus type of nucleic acid, of course, is now known as DNA. The full significance and potentiality of the Feulgen reaction, applied as histochemical method, only slowly became understood. Cytologists did not begin to employ it until the late 1920s. This method is still the safest one to distinguish DNA or chromatin from cytoplasm and nucleoli or RNA.

Herman Joseph Muller (1890–1967) in 1912 joined Morgan's fly group at Columbia University with an assistantship in zoology. In 1915 he completed his Ph.D study with a thesis on fruit flies called *The Mechanics of Crossing Over,* showing how genes are exchanged between chromosomes. Columbia University was an exciting place during Muller's graduate years; the young science of genetics was getting greater impetus in America.

In 1914, while working for his doctorate, he came across a new fly with a bent wing. The usual routine for a new mutant was followed in an attempt to find its linkage group. After an elaborate series of selected breeding experiments, the new character refused to associate itself with any of those in the three demonstrated linkage groups of chromosomes. The obvious conclusion was that it belonged to the small fourth chromosome that all this while had been floating apparently uselessly in the nucleus, waiting for a mutant character with which to be associated. What for a moment had appeared as an obstacle to the acceptance of the validity of the linkage theory was converted into additional evidence of its plausibility.

The discovery of crossing over and the elucidation of its cytological basis necessarily occasioned a major revision of the gene concept that was to include another fundamental property. In crossing over, genes behaved as units between which, but not through which, exchanges occurred. Once a mutant was detected it was a relatively simple matter to discover both the chromosome with which it was associated as well as its specific genetic or cytological locus on that chromosome. But the study of gene mutation was seriously hampered by the low rate of spontaneous mutations. This serious limitation to direct study of the gene was removed by the epochal discovery of Muller in 1927 that the mutation rate could be increased several thousand percent through the action of x-rays. For this discovery Muller received the Nobel Prize in 1946. In 1928, one year after Muller's paper, L. J. Stadler verified the increase of the mutation rate by x-rays in plants.

Hitoshi Kihara (b. 1893), former Japanese Professor at Kyoto University, in 1930 formulated a method he called **genome analysis.** This method is designed to determine the diploid ancestors of allopolyploid species. Several years earlier Kihara together with Ono in 1926 had introduced the terms **auto-** and **allopolyploidy** in order to better distinguish between these two important classes of ploidy. The method of genome analysis was first used in such important plant genera as *Triticum, Aegilops, Nicotiana, Raphanus, Brassica,* and *Rosa.* The method consists of the analysis of meiotic chromosome pairing in the hybrids between polyploids and diploids. If the diploid has at one time or another contributed to the formation of the polyploid, chromosome pairing should occur between two sets of homologous chromosomes in the hybrid. This method has subsequently contributed to the knowledge of systematic relationships between many cultivated and wild polyploid species.

Curt Stern (b. 1902), a German born American geneticist, in 1931 presented cytological proof of crossing over in *Drosophila*. This was done independently of McClintock's and Creighton's demonstration in maize during the same year. Stern's study, which involved the sex chromosomes, provided considerably larger. populations in which genetic crossing over was more precisely localized. More specifically, Stern used an X-chromosome with an arm of a Y-chromosome attached to its right end, and an X-IV translocation. In both cases, two marker genes between the cytologically identifiable regions were available, and it was demonstrated that recombination between the marker genes was regularly accompanied by recombination between the cytological markers. These papers gave the final cytological proof that genetic crossing over is accompanied by an exchange of parts between chromosomes. Beyond any question these were some of the truly great experiments of modern biology.

G. K. Chrustschoff et al. in 1931 published a first attempt to study human chromosomes using cultures of leucocytes from peripheral blood. In 1935 Chrustschoff and Berlin published details of culture techniques for human leucocytes. This important paper has until recently passed relatively unnoticed, and no effort was made at the time to adapt this technique to the chromosome analysis of humans.

J. Belling (1866–1933) in 1931 developed a new classical model of crossing over. He studied plants of the lily and related families. The cytological study of the meiotic processes was actively investigated at about this period to see what really happened at crossing over. His model was based on the assumption of random breaking of the thin, paired chromosome strands with reunion of the broken ends, which could lead to interchanges between homologues if two breaks happened to occur at the same level. He related the phenomenon to the production of new daughter chromatids, an idea that has been involved in many of the more recent interpretations.

Cyril Dean Darlington (b. 1903), professor of botany at the University of Oxford, in an attempt to explain meiosis advanced the **precocity theory.** He published his opinions in a long series of papers and first developed the general scheme in *Recent Advances in Cytology* (1932). The scheme was very generally accepted and for a time was considered the very backbone of cytogenetics. He assumed that the chromosomes have a tendency to be in a paired state at all times. In mitosis this condition is met in that the chromosomes entering prophase are already double. According to this theory meiotic prophase is assumed to start precociously with chromosomes that have not yet split, and this is held responsible for chromosome pairing.

Darlington said that the chromosomes are in an unsatisfied, or unsaturated, state electrostatically. To become saturated they must pair homologously. When the chromosomes become double in late pachytene, the satisfied state is between sister chromatids instead of homologous chromosomes. The paired homologues consequently fall apart and diplotene is initiated. This theory was logically beautiful in superficially explaining the genetic implications of meiosis. But since both DNA and protein synthesis have now been shown to be completed before meiotic chromosome pairing occurs, the precocity theory appears no longer to be valid.

This theory is only one of the many thought provoking ideas that Darlington developed during this period. These ideas were challenging and stimulating and initiated a wealth of research all over the world. In 1929, for instance, Darlington coined the term **chiasma terminalization** in order to explain the progressive shift between diplotene and metaphase I in the distribution of chiasmata along the arms of paired chromosomes from their points of origin to more distal positions.

Ernst August Friedrich Ruska (b. 1906), Director of the Max Planck Institute for Electron Microscopy in Berlin, Germany, together with Knoll published a description of one of the first electron microscopes in 1932. Their instrument consisted of an electron source and two magnifying lenses. A condenser lens was not used. The resolution obtained with this instrument was below that attained with the light microscope. Nevertheless, they obtained the first electron micrographs of an illuminated specimen.

In 1934 Ruska described an improved version of this electron microscope to which a condenser lens was added. The micrographs obtained indicated that the potential existed to surpass the resolving power of the light microscope. The uniquely high resolving power of a microscope using electrons as the illuminating beam is explained quite simply from the fact that electrons have an associated wavelength smaller than that of any other radiation practicable for use in a light microscope system. By 1940 commercial instruments with limiting resolving power of 2.5 nm were manufactured in Germany and America. By 1946, improvement in technique and design made it possible to demonstrate resolutions from 0.85 nm to 1.5 nm. Because the electromagnetic lenses that must be used for focusing the electron beam cannot be corrected for spherical and chromatic aberrations, the resolution limit of 0.8 nm to 1 nm is still the experimental and theoretical limit of the microscope, despite the approximate 0.005 nm wavelength of electrons.

Most of the advances possible with the electron microscope have involved the cytoplasm of the cell since the chromosomes are notable for their lack of the membranous and granular structures so prominent in the cytoplasm. Techniques of spreading interphase and metaphase chromosomes for electron microscopy are very recent developments that can contribute to new ideas of chromosome ultrastructure.

Emil Heitz (b. 1892), German geneticist, professor at the University of Tübingen and associated with the Max Planck Society, in 1933 together with Bauer discovered the importance of the **giant chromosomes** in the salivary gland cells of dipteran insect species as important objects in cytogenetic research. These structures had been discovered prior to this in 1881, but had not been identified as chromosomes. They represent bundles of chromosome subunits or chromatids. Since they are not spiralized, they are about 100 to 150 times longer than ordinary mitotic chromosomes. This unusual length and their banding pattern make them very suitable for chromosome identification and gene localization.

In 1928 and 1929 Heitz was the first to distinguish two types of chromatin, which he named **euchromatin** and **heterochromatin.** Euchromatin stains lightly or not at all in interphase and prophase, while heterochromatin stains darkly in these stages. Heterochromatin is an extremely helpful marker for chromosome mapping in the pachytene stage of meiotic prophase. In 1931 Heitz showed a correlation between the number of nucleoli in the interphase nucleus and the number of a particular type of chromosome, now called the **nucleolus organizer chromosome.** A study of these chromosomes indicated that the nucleolus is organized at a specific site on the chromosome.

Tobjörn Oskar Caspersson (b. 1910), head of the Medical Cell Research and Genetics Department of the Karolinska Institute of Stockholm, Sweden, in 1936 began to develop ultraviolet photomicrography for the study of nucleic acids within the nucleus. These substances absorb ultraviolet light very strongly in a most characteristic and selective fashion. The method has the great advantage of being able to use unstained material as the object, and, thus, the contrast in the resulting photomicrographs was indirectly due to the components themselves and not to their affinity for a stain, the depth of which is largely dependent on the conditions of use. So the density of nucleic acids could readily be compared from one tissue to another. It was found that wherever cells of tissues are growing rapidly, the density of nucleic acids within them was relatively high. Evidence of this kind thus pointed to the conclusion that nucleic acids have some biological function in the process of synthesis within the cell. It also helped to pinpoint the time period during which such nucleic acid synthesis does take place (Caspersson, 1947).

George Wells Beadle (b. 1903), later president of the University of Chicago, with Edward L. Tatum in 1941 and in a monograph in 1945 developed the **one gene-one enzyme concept.** In 1958, together with Lederberg, they received the Nobel Prize in medicine and physiology for the discovery that genes regulate certain definite chemical processes. Their study on the boichemical genetics of the pink bread mold, *Neurospora crassa,* is now a classic, and it marked a significant turning point in the analysis of the general problem of genetic control in metabolism and development.

Instead of attempting to work out the chemical basis for known genetic characters, they deliberately reversed the procedure and set out to determine if and how the gene controlled known biochemical reactions. The wild strain of bread mold could be made to grow on a medium containing sugar and inorganic salts. The salts included nitrogen compounds out of which the mold was able to manufacture for itself all the necessary amino acids. Not one amino acid had to be added to the medium. Beadle and Tatum then subjected the spores of the mold to x-rays. Occasionally, an irradiated spore would refuse to grow on the medium, but it would grow if a certain amino acid such as lysine were added. Apparently, the irradiated spore had lost the capacity to manufacture its own lysine out of the inorganic nitrogen compounds. Without lysine it could not grow. If lysine were added to the medium, it could grow. It seemed evident that an enzyme that normally would have catalyzed one of the reactions that led to lysine was not formed by the spore. Supposedly a particular gene had been mutated by the x-rays.

According to their one gene-one enzyme hypothesis, the gene controls the synthesis or the activity of but a single protein or enzyme with catalytic activity. Since its formulation this concept has been verified in principle even though it was controversial when first announced.

Oswald T. Avery (1877–1955), a member of the staff of the Rockefeller Institute Hospital, New York, together with MacLeod and McCarty, in 1944 showed the significance of DNA as the hereditary material through studies of transformation in bacteria. The phenomenon that they called "transformation" involves a transfer of genetic information by means of naked extracellular DNA. They showed that purified DNA preparations extracted from a particular smooth strain of *Pneumococcus* bacteria can confer an inheritable smoothness on bacteria that were formerly rough. The experiments also showed that the preparations most active in bringing about transformation were those purest and most free of protein. This fact effectively cast doubt on the wide-spread and commonly accepted belief that proteins were the mediators of biological specificity and cellular inheritance. To the chemists at that time and earlier, the problem of nucleoprotein was first of all the problem of protein. The structure of the nonprotein portions of nucleoprotein appeared too simple to them. It was the protein portion that counted. Avery's discovery set the stage for the rapidly ensuing elaboration of the structure, function, and importance of DNA, which ten years later led to the development of the Watson-Crick model for the DNA molecule.

Murray Llewellyn Barr (b. 1908), a Canadian physician, together with Bertram in 1949 unexpectedly discovered a small, stainable body in the nondividing nuclei of females and its absence in those of males. This body is now called sex-chromatin or **Barr body,** after its senior discoverer. It can be seen in many tissues of females including the epidermis and the oral mucosa and also in the amniotic fluid surrounding female fetuses. Researchers often wondered whether developmental sex deviants had XX or XY constitution. Before the discovery of the Barr-body, no knowledge was available on this topic. We now know that the Barr-body is a heterochromatic X-chromosome that during interphase is completely, or for the most part, positively heteropycnotic and condensed. The discovery a few years later of the sex chromatin and the correct human chromosome number was rapidly followed by the discovery of the first human chromosome abnormalities in the late 1950s. These discoveries were followed by a world-wide outburst of research on human chromosomes.

Barbara McClintock (b. 1902) in 1950 discovered the **Activator–Dissociation system** in maize. She is a Distinguished Service member of the Genetics Research Unit of the Carnegie Institution at Cold Spring Harbor, New York. She was mentioned earlier for her skillful and persevering cytogenetic work on maize in her earlier years and for her presentation of cytological proof of crossing over in 1931.

In her studies of the Activator-Dissociation system, McClintock demonstrated that genic expression is intimately related to chromosomal organization. In a variable and mutable strain of maize, two loci were shown to be in control of genic action in the course of development. One of the loci, called *Activator (Ac),* seemed to be a master locus in that the second locus, called *Dissociation (Ds),* was unable to function in its absence. Both loci were believed to be blocks of heterochromatin. The presence of both loci in the same nucleus gave rise to an increase in spontaneous chromosome breaks and unstable and mutable genic loci. But *Ds* also had a second function in the presence of *Ac.* It affected genes lying adjacent to it in that they mutated to the recessive condition. McClintock discovered several different such gene controlling systems in maize. The full significance of McClintock's findings is still not appreciated. Similar systems were later found in *Drosophila* and mouse. But most important of all, her findings led to the epoch-making discoveries in bacteria ten years later that revealed an entirely new class of regulatory genes.

John Albert Levan (b. 1905), a professor of cytology at the University of Lund, Sweden, together with the Indonesian born American, Joe Hin Tjio, in 1950 showed favorable results with their oxyquinoline squash technique in 40 plant species. Together they first worked out the importance of this chemical agent for chromosome analysis. The metaphase chromosomes were contracted, the spindle was destroyed and did not interfere with the spreading of the chromosomes at squashing, and many cells were arrested in metaphase, which increased the chance of finding good preparations.

Later, the two scientists applied the squash technique to human tissue. In 1956 they published a paper on the chromosome number of man giving 46 as the 2n number. Their counts were made from tissue culture preparations of lung tissue from four different human embryos. With previously used techniques, it had been extremely difficult to make counts in human material. Until then, the human chromosome number was assumed to be 2n = 48. Tjio and Levan's demonstration was soon verified by several other research workers. As a matter of fact, two English investigators a few months later reported 46 chromosomes in testicular preparations of three adults. This represented the basis for cytogenetic research in man, and vertebrates, a field of investigation that has developed with an avalanche-like rapidity.

In the following years, the Turner's, Klinefelter's and Down's syndromes were linked to chromosome aneuploidy. Levan's special scientific interest is chromosomes in relation to cancer. In 1956 he reported 70 to 80 chromosomes in two highly malignant effusions of lung and stomach carcinoma of man. Recently (1966) he studied the nonrandom representation of chromosomes in tumor stem cell lines from 40 human cancers and concluded that chromosomes of the C group were over-represented while those of D and G groups were under-represented.

Sir Francis Harry Compton Crick (b. 1916), British biophysicist and geneticist and Kieckhefer Research Professor at the Salk Institute at San Diego, California, in 1953 with the American, James Dewey Watson, published a paper in which they proposed a model for the molecular structure of DNA. The model they proposed is now widely known as the Watson–Crick model. For the discovery of the molecular structure of DNA and its significance for the transfer of information in living material, they received the 1962 Nobel Prize in medicine and physiology together with Maurice Hugh Frederick Wilkins from New Zealand.

The discovery of the double helix structures of DNA was based on the achievements of Wilkins and his colleagues at the Kings College in London. They had taken good x-ray diffraction pictures and had analyzed and interpreted the photographs. Watson and Crick had made the brilliant deductions that revealed the structure of the molecule. This model of DNA proved immediately fruitful. Its structure and the theory of its replication was so clear and uncomplicated that geneticists at once accepted it. All investigations since this discovery supported it. Enzymatic synthesis of RNA and DNA followed in the 1950s, and by 1961 Crick and his coworkers in an ingenious experiment furnished evidence of the triplet nature of the **codon,** the smallest combination of bases in a polynucleotide, which determines that a specific amino acid shall be inserted at a specific position into a polypeptide chain.

The discovery of the triplet genetic code is based on the work of Crick and Nirenberg as an answer to the problem of designating 20 amino acids by a nucleotide code consisting of only four characters. In 1966 Crick advanced the **wobble hypothesis,** which was proposed to provide rules for the pairing of a codon in messenger RNA and for an anticodon in transfer RNA of the third position of the codon; degeneracy of the code was explained by this hypothesis. The first 2 positions of the triplet codon on messenger RNA pair precisely with the anticodon on transfer RNA, but pairing of the third position may be wobbly, and independent of the nucleotide that is present at the third position.

Francois Jacob (b. 1920), Chief of the Department of Microbial Genetics at the Pasteur Institute in Paris, France, together with Jacques Monod in 1961 published a classic paper on the regulation of protein synthesis through which they introduced many new concepts into genetics and established others that had been debated for several years by researchers. Together with André Lwoff they received the 1965 Nobel Prize in medicine and physiology for their discovery of a previously unknown class of genes whose function is to regulate the activity of other genes.

In their concept there is an interplay between three kinds of genes, **structural genes, operator genes,** and **regulator genes.** The structural gene corresponds to the classical gene, possessing the ability to synthesize a specific protein or enzyme that has a special task during the life and development of the individual. It would, however, be inconvenient if this enzyme production occurred all the time. It would be advantageous if it were stopped and started again when necessary; this activity is controlled by the operator and regulator genes. The operator gene apparently is located in immediate proximity to the structural gene and represents something like a switch mechanism that either turns on or shuts off the activity of the structural gene. The operator gene, however, does not know when this should be done, but receives orders from a specific regulator gene. This gene may not necessarily be located close to the operator gene, but may be located in another part of the chromosome. The regulator gene gives its orders via a repressor product that interacts with the operator gene to shut it off under conditions in which the structural gene products are not needed.

Mary Francis Lyon, Head of the Genetics Section of the Medical Research Council Radiobiology Unit at Berkshire, England, in 1961, independently from L. B. Russel's work during the same year, developed the **single active X hypothesis** of dosage compensation in man and mammals known as the *Lyon Hypothesis.* Lyon worked with mice. She provided evidence suggesting that one X-chromosome is inactivated in some early embryonic cells and their descendants, that the other is inactivated in the rest, and that females are consequently X-chromosome mosaics. This is a specific manifestation of a much wider biological phenomenon, the inactivity of whole chromosome sets, specific chromosomes, or specific chromosomal regions. Lyon's genetic findings verified the cytological discovery by Barr of small stainable bodies in the female nuclei of nondividing nerve cells in cats. This stainable body is one of the two X-chromosomes that is genetically inactivated by heterochromatinization.

Wolfgang Beermann (b. 1921), Director of the Max Planck Institute of Biology at Tübingen, Germany, in 1961 demonstrated that a puffing locus on a polytene chromosome of a dipteran insect, *Chironomus,* is the site of a gene. These puffs arise at different points on these chromosomes and many are found only in specific tissues but vary within a tissue at different times. The present view is that the puff signifies RNA synthesis, and this view is supported by experiments that stain RNA differentially and show its localization in the puff.

Sol Spiegelman (b. 1914), professor of microbiology at the University of Illinois, in 1961, together with B. D. Hall, demonstrated that hybrid molecules can be formed containing one single stranded DNA and one RNA molecule that are complementary in base sequence. This technique opened the way to the isolation and characterization of different kinds of RNA. **Nucleic acid hybridization** has since been exploited to study the cell's mechanism for manufacturing proteins. One strand of nucleic acid will combine with another wherever the subunits of the two strands are complementary. Artificial hybrid combinations clarify the flow of information in the living cell. It is now known that the actual synthesis of protein molecules is accomplished with the help of ribosomes, which serve as workbenches of protein synthesis in the cytoplasm and evidently hold the translatable RNA in position while the message is being read. In 1965 Spiegelman together with Ritossa showed that the genes producing the ribosomal RNA of *Drosophila* are located in the nucleolus organizer regions of the chromosomes. It appears now that the precursor material or ribosomal RNA is manufactured by the nucleolus organizer, and is then transferred to the nucleolus for final assembly into ribosomes. These findings are in line with recent research that indicates that living organisms cannot exist without nucleolar organizer chromosomes.

James Bonner (b. 1910), a professor of biology at the California Institute of Technology, in 1962, together with R. C. Huang, studied the protein components of chromosomes and found that in some cases the rates at which messenger-RNA was produced could be increased by removing histones. This pointed strongly to the involvement of histones in regulating gene action. If such a mechanism exists, some proteins could serve as locks inhibiting the action of certain nucleic acid molecules. Every cell, regardless of its level of specialization, could still contain all genes, but each might possess its own kinds of regulating proteins that would block out certain genes in certain cells.

Margit M. K. Nass (b. 1931) and **Sylvan Nass** (b. 1929), a research couple from the Department of Therapeutic Research, University of Pennsylvania, School of Medicine, and of the Department of Molecular Biology at the Eastern Pennsylvania Psychiatric Institute, Philadelphia, in 1962 and 1963 furnished one of the earliest reliable reports of mitochondrial DNA.

Under the electron microscope they observed fibrous DNase-sensitive regions in thin-sectioned chick mitochondria. In the same year (1962) Ris and Plaut demonstrated by electron microscope and cytochemical methods that chloroplasts in the plant *Chlamydomonas moewusii* contained DNA. Other cytochemical tests had preceded these observations in mitochondria and chloroplasts, but they had not been as conclusive as the ones mentioned. The demonstrations under the electron microscope initiated a search in many laboratories that confirmed these findings. When viewed with the electron microscope, these so-called **extranuclear chromosomes** differ from nuclear chromosomes of the same cells by their closer resemblance to pure DNA. In general, they tend to carry much less protein, are believed to lack histone, and in these and other respects are similar in organization to bacterial or viral chromosomes. Even the total amounts of DNA per mitochondrion or per chloroplast are similar to the amount per cell in bacteria such as *Escherichia coli*. The study of extrachromosomal or cytoplasmic inheritance has entered a new phase through these new and interesting discoveries.

Henry Harris (b. 1925), a professor of pathology and Head of the Department of Cell Biology at the University of Oxford, England, together with Watkins in 1965 developed a technique that uses appropriate viruses to cause somatic cells of very different origins, such as from species of different genera, to fuse into one binucleate cell. The method is now generally referred to as **somatic cell hybridization** or **cell fusion.** Hybridization between somatic cells in vitro promises to provide the basis for cytogenetic analysis of somatic cells in culture. The assignment of genes to specific chromosomes is perhaps the simplest and most immediately achievable goal to the analysis of hybrid lines formed by fusing different cells. These pioneering studies that are now being conducted by several laboratories around the world have led to rapid advances in the knowledge of the human chromosome map. Harris and his colleagues are now heavily engaged in the analysis of malignancy by cell fusion.

Ernest Joseph DuPraw (b. 1931), a professor of anatomy, who was engaged in clinical training and research at Stanford University School of Medicine, in 1965 and 1966 published techniques of spreading interphase and metaphase chromosomes for electron microscopy and contributed new ideas on chromosome structure. His method of **whole mount electron microscopy** involves growing leucocytes or other cells in culture. They are blocked at metaphase by using colchicine to disaggregate the mitotic apparatus. The blocked cells are spread on an air-water interface that bursts them and releases the chromosomes. Finally, the intact chromosomes and nuclei are picked up on electron microscope grids, washed or treated with analytical reagents, and dried from liquid CO_2 by the critical-point method.

DuPraw studied honey-bee and human chromosomes by this method. Honey-bee chromosomes are well suited for this approach because they are extremely small and rod-shaped. Whole mount electron microscopy has an advantage over thin-sectioning because thin sectioned chromosomes are much more difficult to interpret with respect to fiber configurations and dimensions, and there has been wide disagreement among published estimates of fiber diameters. DuPraw's interpretation of chromosomal organization is called the **folded fiber model.** Before interphase replication, each chromosome is thought to consist of a **unit chromatid,** that is a single, long 20 nm to 50 nm fiber, that contains a DNA double helix in supercoiled configuration. DuPraw was only one of many workers who were trying to solve the puzzle of the molecular structure of the chromosome.

Tobjörn Oskar Caspersson (b. 1910), Swedish cell biologist, who was mentioned earlier for his pioneering studies of nucleic acids, (see p. 21) in 1968, along with his colleagues, was the first to demonstrate that when metaphase chromosomes are stained with quinacrine mustard or related substances and examined by fluorescence microscopy, each pair stains in a specific pattern of dark and light bands called **Q-bands.** This revived a whole new search for methods that permit distinction between individual metaphase chromosomes and chromosome segments.

One of the more prominent and simpler, new methods developed in the beginning of the 1970s is the **Giemsa staining method.** When chromosomes are treated with a denaturing agent such as trypsin and then stained with Giemsa stain, they take up stain patterns of dark and light bands, called **G-bands,** very similar but not identical to Q-bands. Giemsa banding is simpler and less expensive than fluorescent banding and provides much the same information, so it will probably become more widely used. Chromosome banding techniques have greatly broadened the usefulness of chromosome analysis in cytogenetics. Now that metaphase chromosomes can be individually identified, chromosomal rearrangements can be more easily recognized, and the chromosomes involved can be specifically identified. As a consequence, mapping of genes on chromosomes is facilitated.

Daniel Nathans (b. 1928), a professor of microbiology at Johns Hopkins University, Baltimore, Maryland, and his coworkers in 1973 published a paper on the use of **restriction enzymes** for chromosome mapping (Danna et al., 1973). They set the stage for an explosion of research on **restriction maps, transcript maps,** and **nucleotide sequencing** of isolated restriction fragments. The so-called **restriction effect** was first discovered by Dussoix and Arber in 1962. They demonstrated that restriction enzymes act like chemical knives that cut DNA strands into defined fragments. In 1970 Hamilton O. Smith and coworkers reported the purification and characterization of a **specific restriction endonuclease** of the type II (Kelly and Smith, 1970).

Restriction enzymes belong to two different types according to their restriction products. Type I cuts at unique DNA sites resulting in specific fragments with unique terminal sequences. The cuts are within sequences that show twofold symmetry around a given reference point $\left(\text{e.g.,} \dfrac{\text{AAG CTT}}{\text{TTC GAA}} \right)$. Type II restriction endonucleases are smaller and simpler in subunit composition than type I and are more specific in their cleavage sites. Nathans, Smith, and Arber were the winners of the 1978 Nobel Prize for Medicine. Arber was credited with having first predicted the existence of restriction enzymes, Smith with having isolated the first such enzyme, and Nathans with having first applied these enzymes to the study of genetics. Restriction enzymes have become a very important tool in the study of DNA. They allow the isolation of sufficiently short DNA fragments and the sequencing of their nucleotides.

Stanley Norman Cohen (b. 1917), Department of Medicine, Stanford University School of Medicine, in Stanford, California, in 1973, together with A. C. Y. Chang, developed the technique of **DNA cloning** by which DNA molecules from prokaryotic and eukaryotic sources can be spliced together via plasmid vehicles (Cohen et al., 1973). They isolated DNA pieces from the bacterium *Staphalococcus* and spliced them into nonconjugal plasmids. Such resulting **recombinant plasmids** in turn were then introduced into *Escherichia coli*. Once the isolated DNA segment is incorporated in the *E. coli* bacterium, it can be reproduced therein to provide researchers with enough recombinant DNA to determine the exact sequence of the nucleotides.

The potential usefulness of such genetic manipulation lies in the fact that in bypassing the sexual cycle, a new genetic combination of inherited properties is established. Large amounts of a particular gene or combination of several genes can be obtained for study by this method. Cloning individual eukaryotic genes with their adjoining control elements could reveal the process of **gene expression** in eukaryotes, which has been very difficult to study because of the enormous complexity of the eukaryotic genome. For instance, a λ phage-mouse β-hemoglobin chromosome was constructed, and it was discovered that so-called **intervening sequences** in the mouse chromosome about 550 nucleotides in length do not code for β-globin at all (Leder et al., 1977). Intervening sequences have also been found in rabbit globin genes, in genes corresponding to *Drosophila* 28S ribosomal RNA, adenovirus, Simian virus 40, mouse immunoglobin, yeast tRNA, and chicken ovalbumin and appear to be a common occurrence of eukaryotic gene organization (Leder et al., 1978). Cloning is considered by some geneticists to be of potential major medical and agricultural benefit. Insulin genes have been spliced into bacteria (Villa-Komarov et al., 1978), and work that will introduce nitrogen-fixing genes *(nif)* into genomes of crop plants is in progress (Streicher and Valentine, 1977).

Charles Allen Thomas Jr. (b. 1927), Department of Cellular Biology, Scrips Clinical and Research Foundation, La Jolla, California, in 1974, together with D. A. Wilson discovered the widespread occurrence of the so-called **palindromes**, hairpin-like structures resulting from inverted **repetitious DNA** which is located at intervals along the chromatids of eukaryotic chromosomes. The name "palin-dromes" applies because these sequences read the same both backward and for-ward $\left(\text{e.g., } \frac{\text{ATCTA}}{\text{TAGAT}} \text{ t*} \right)$. Palindromes may be miniature "handles" that could be useful in the dissection of chromosomes. Boyer (1974) reported that many sites recognized by restriction endonucleases prove to be palindromes. Many palin-dromes, particularly those recognized by restriction enzymes, are only 3 to 10 base pairs long. Longer ones are hundreds of base pairs in size.

This short history has shown some of the trends in cytology, genetics, and cytoge-netics during the last four hundred years. It demonstrates the close interdepen-dence of tool development, the imagination of people, and the art of integrating bits of information into a framework of facts and working hypotheses. From the early microscope builders, who saw the first cells and discovered some of the first principles of life, to the sophisticated researchers of the 20th century, who have the most advanced technology at their disposal, it is a story of fascinating development that can be read from the lives and ambitions of many devoted scientists.

---

*t stands for **turn-around region,** the point where the single linear DNA chain folds back.

Part II
Structure of Chromosomes

Chapter 2
Gross Morphology of Chromosomes

This chapter emphasizes the aspects of gross morphology of chromosomes that are visible under the light microscope. In Chapter 3, aspects of fine structure will be discussed.

There are several stages at which chromosomes can be studied, and each stage has advantages and disadvantages. The stage of the cell cycle in which the chromosomes are most easily identified and distinguished is during **mitotic metaphase** when they are usually most condensed or coiled. In the past, methods for preparing mitotic metaphase chromosomes did not reveal many morphological characteristics that could be used to distinguish them within the complement. Only a few criteria could be employed to describe them. Due to the lack of simple and reproducible differential staining procedures for such ordinary metaphase chromosomes, cytologists turned their major attention to special chromosome types such as the **giant salivary gland chromosomes** of insects and some other organisms that exist in the prophase stage. Because of their polyteny—an increase in lateral multiplicity—they reveal much detail that usually cannot be studied in ordinary prophases. Other advantages of the study of prophase are (1) the possibility to distinguish between eu- and heterochromatin, (2) the visibility of chromomeres, and (3) the presence of nucleoli that are associated with specific chromosomes and that mark them as **nucleolus organizer chromosomes.** For these reasons many species have been subject to **pachytene analysis.** But there are disadvantages to the morphological study of the pachytene chromosomes of meiosis. Because of their considerable length, they are not usually all visible in squash preparations. The higher the **n-number** of chromosomes, the more difficult is a pachytene analysis.

However, not every organism can be analyzed in this manner. Those scientists who worked on the majority of species, including man, had to rely on the ordinary metaphase chromosome analysis. But a recent major breakthrough in cytogenetic technology has suddenly changed this situation (see Chapter 1). Several reliable methods are now available that reveal unique **banding patterns** in mitotic metaphase chromosomes.

In 1971 an ad hoc committee meeting on the Standardization of Human Chromosomes was held to revise the nomenclature system in light of new techniques and new findings (Paris Conference, 1971). This system of cytogenetic human nomenclature was again revised in 1978 (International System, 1978). The Paris Conference describes four different chromosome banding methods now known as

C-banding, G-banding, Q-banding, and R-banding. In this chapter we will consider the different applications for studying the gross morphology of chromosomes.

2.1 Mitotic Metaphase Chromosomes

Because of the recentness of the discovery of banding patterns in mitotic metaphase chromosomes, the majority of metaphase chromosome analyses have been carried out with the aid of other methods. It is therefore important for the student of chromosome morphology to familiarize himself with earlier approaches. Mitotic metaphase chromosomes usually range in sizes from about 0.5 μm to 30 μm in length and from 0.2 μm to 3 μm in diameter. Plants and animals alike can have very small chromosomes, but on the average, plants have larger chromosomes than animals.

2.1.1 Total Length of Chromosomes

The morphology of a chromosome in mitotic metaphase is described by two major factors: its total length and the position of the centromere. In order to demonstrate these characteristics, cytologists construct **idiograms** of the **karyotypes** of species. The karyotype as described by Battaglia (1952) is the particular chromosome complement of an individual or a related group of individuals, as defined by chromosome size, morphology, and number. An idiogram is a diagrammatic representation of the gametic chromosome set (n) of a given species and is used to compare the karyotype of one species with those of other species. Figure 2.1 shows an idiogram of *Agropyron orientale* (Schulz-Schaeffer and Jurasits, 1962). There exist karyotypes with chromosomes essentially similar in size and others with chromosomes differing greatly in size. The average size of chromosomes is 6 μm. The longest chromosomes exist in the plant genus *Trillium* and are longer than 30 μm. The shortest chromosomes are less than 1 μm in length and occur in fungi, rushes, sedges, and in some animals. In many species we find two distinct sizes of chromosomes, large ones and small ones. Such karyotypes occur in the plant genera *Yucca* and *Haemanthus* (Fig. 2.2) and in birds and lizards. In polyploid plant species, groups of chromosomes in different size classes give clues of parental origin. For instance, in the grass genus *Bromus*, the North American octoploids

A. ORIENTALE   2n=28

Fig. 2.1. Idiogram of *Agropyron orientale* (2n = 28). The satellite chromosomes are placed at the beginning of the idiogram and are arranged according to the length of their satellites. The rest of the chromosomes are arranged according to the length of their short arms. One unit of the scale to the left of the idiogram equals 0.72 μm. (From Schulz-Schaeffer and Jurasits, 1962. Reprinted by permission of McClure Newspapers, Inc., Burlington, Vermont).

Fig. 2.2. Mitotic metaphase chromosomes of the plant *Haemanthus katharinae* (2n = 18). × 2000. (Courtesy of Dr. A. H. Sparrow, Biology Department, Brookhaven National Laboratory, Upton, New York).

(AABBCCLL) have 6 **basic genomes*** (6x) of medium size chromosomes and 2 basic genomes (2x) of long chromosomes. According to genome analysis by Stebbins (Stebbins and Tobgy, 1944; Stebbins, 1947a), the medium size chromosomes are homologous with the chromosomes of the hexaploid species of section Ceratochloa confined to South America (AABBCC) while the long chromosomes (LL) are homologous with those of the North American diploids of section Bromopsis. Similar homologies exist between the genomes of Old World and New World cottons (Skovsted, 1934).

2.1.2   The Centromere

Centromeres could be classified as follows:
1. Localized centromeres
2. Neocentromeres
3. Nonlocalized centromeres
   a. Polycentromeres
   b. Holocentromeres

The **localized centromere** constitutes the normal condition in which a chromosome possesses a permanently localized region to which the spindle fiber attaches during chromosome movement. **Neocentromeres** form under certain conditions in which the centromere region is replaced by a secondary center of movement. These are exceptional cases in which the chromosome ends move first during anaphase of

---

*Basic genome: A group of chromosomes that are thought to have been present in the gametes of the diploid ancestors of polyploids and those groups that are present in the gametes of the still existing diploids of a genus. The number of chromosomes in a basic genome is represented by the **basic chromosome number** or x−**number** (x = 7 in *Bromus*).

meiosis (Rhoades, 1952). **Nonlocalized centromeres** are those in which the spindle attachment is not confined to a strictly localized chromosome area. In the case of **polycentromeres,** each chromosome is attached by many spindle fibers. Here many centromeres are separated by noncentric segments. Examples of this type of centromere are some ascarid nematodes. **Holocentromeres** (Hughes-Schrader and Ris, 1941) are diffuse in nature where every point along the chromosome shows centromeric activity. Such centromeres have been observed in Hemiptera, Homoptera, Protista, and the higher plant genus *Luzula*.

The site of the localized centromere is often referred to as the primary constriction or kinetochore. Its location on the chromosome is probably the most important character in determining the morphology of the chromosome. The centromere is observed as a constriction in the metaphase chromosome and is stained lighter than other parts of the chromosome. This constriction can be located toward the end of the chromosome, in the center, or in between. According to its position, it will subdivide the chromosome into 2 equally or unequally sized arms. Chromosomes are categorized according to the position of the centromere as **telocentric, subtelocentric, submetacentric,** and **metacentric** chromosomes.

Chromosome arms formed by the location of the centromere can be measured and their lengths expressed in different ways. A very popular nomenclature for expressing these measurements is the one used by the Human Chromosome Study Group (Chicago Conference, 1966). This nomenclature designates the short arm with the letter "p" (abbreviation for *petit,* French for *short*) and the long arm with the letter "q." The ratio between the arms is often calculated as the arm ratio:

$$A = \frac{p}{q}$$

or as the centromeric index (Chicago Conference, 1966):

$$C = \frac{p \times 100}{p + q}$$

Other indices or formulas have been used also.

Telocentric chromosomes are those with a terminally located centromere. Telocentric chromosomes may arise by centromere misdivision or breakage induced within the centromere region. Telocentrics are generally considered to be unstable, since fracturing of the centromere is usually involved. The instability of telocentrics is considered to be the reason for their rarity in nature.

Most chromosomes are **monocentric,** having only one centromere per chromosome. Chromosomes with two centromeres are called **dicentric.** Such dicentric chromosomes are usually the product of structural changes. Dicentric chromosomes may pass through cell divisions without difficulty. But it may happen that the two centromeres pass to opposite poles, which causes bridge formation. If such bridges break, each daughter nucleus will contain two broken chromatid ends. Freshly broken chromatid ends have the tendency to fuse, which in this case will result in newly formed dicentrics. A similar cycle may happen during the next division. McClintock (1938a, 1938b, 1941a, 1941b, 1941c, 1942, 1944) studied this chromosome behavior in maize and called it the **breakage-fusion-bridge cycle** (see Chapter 11). The irregular behavior of dicentric chromosomes must be the reason that they normally do not occur in nature and cannot generally be maintained in laboratory or field stocks of most organisms. However, a transmissible dicentric chromosome was discovered by Sears and Camara (1952) in wheat, *Triticum aestivum* L. The reason for the transmissibility must be the partial inactivity of the second centromere. As long as one of the two centromeres has stronger centromeric activity than the other, it can cause the chromosome to be pulled to one pole in its entirety without tearing. Other forms of centromeric activity in chromosomes are of the neocentric and diffuse kind. But since they are not localized, they do not contribute to the morphology of the chromosome. They therefore will be discussed later.

2.1.3   The Nucleolus Organizer Region

In addition to a primary constriction formed by the centromere, certain chromosomes reveal a region that is called **secondary constriction.** This region is responsible for the formation of the nucleolus during telophase and is associated with this structure during interphase and prophase; it is therefore called the **nucleolus organizer region** (McClintock, 1934). It is also the chromosomal site of ribosomal RNA synthesis as mentioned in Chapter 1 (see Spiegelman, p. 25). However, in metaphase the nucleolus is generally not visible. This region in its true sense is really not a constriction like the centromere since its diameter is mostly as great as the remainder of the chromosome. But the region is usually very strikingly marked since it is negatively heteropycnotic to such a degree that the remaining portion of the chromosome, the so-called **satellite,** seems to be removed from the rest of the chromosome like a chromosome fragment. Each species usually possesses at least one homologous pair of **nucleolus organizer chromosomes,** a pair that has a nucleolus organizer region. Very often, each basic genome (x) has such a pair of nucleolus organizer or **satellite chromosomes.** In the genus *Bromus,* 31 species were investigated that represented 166 basic genomes. The number of satellite chromosomes found were 160, almost coinciding with the number of basic genomes (Schulz-Schaeffer, 1960). In this way, satellite chromosomes can serve as marker chromosomes for specific basic genomes, and they are, therefore, valuable for cytotaxonomic studies. The size of the satellite can vary considerably. Satellites are generally attached to the short arm of a nucleolus organizer chromosome. In the human chromosome complement, the so-called **acrocentric** (subtelocentric) chromosomes of the D and G groups all have tiny satellites that are so small they are often not manifested in every cell (Fig. 2.3). Larger satellites can possess a separate constriction and are then called **tandem satellites** (Taylor, 1926). Satellite type XIV (Fig. 2.4) of *Bromus sitchensis* Trin., *B. haenkeanus* (Pressl.) Kunth, *B. coloratus* Steud., and *B. valdivianus* R. A. Phil., of the section

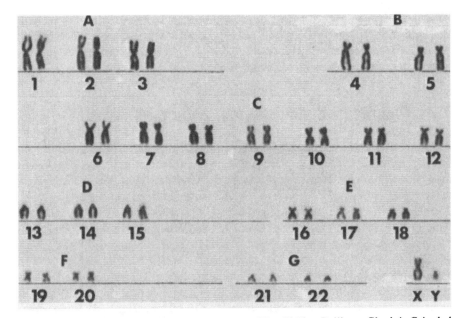

Fig. 2.3. Karyotype of human male. (Courtesy of Dr. Philipp Pallister, Shodair Crippled Children Hospital, Helena, Montana).

Ceratochloa of *Bromus* has such a tandem satellite (Schulz-Schaeffer, 1960). Satellites also may show considerable variation in size. This may be correlated to the fact that satellites are mostly believed to be **heterochromatic.** Heterochromatin is generally considered to be void of genetic activity found in **euchromatin.** Consequently, it is relatively dispensable to the genome. However, Phillips et al. (1977) demonstrated that the gene for *polymitotic (po)* in maize is located in the satellite of chromosome 6. Giant satellites in man were first reported by Tjio et al. in 1960. Other so-called secondary constrictions have been detected that are not connected with nucleolus organization. The author at one time proposed to call these **tertiary**

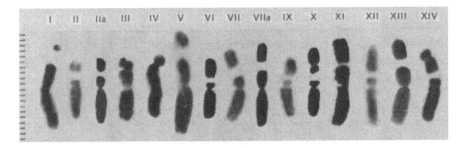

Fig. 2.4. Satellite chromosomes of the genus *Bromus* arranged according to the length of the satellites. One scale unit equals 0.5 μm. (From Schulz-Schaeffer, 1960. Reprinted by permission of the American Genetic Association, Washington, D.C.).

constrictions in order to avoid confusion with nucleolus organizer regions (Schulz-Schaeffer and Haun, 1961). Such tertiary constrictions have been observed in plants and animals and may represent regions of differential spiralization, nucleic acid content, or weakness (Kaufmann, 1948).

2.2   Meiotic Pachytene Chromosomes

As mentioned in the introduction to this chapter, the pachytene stage of meiosis in many respects is ideal for the study of chromosome morphology. **Pachytene analysis** has been made by several investigators in a variety of species. Such studies were carried out in *Zea mays* L. (McClintock, 1929a; Longley, 1938, 1939), *Euchlaena* (Longley, 1937), Solanaceae (Gottschalk, 1954; Gottschalk and Peters, 1955, 1956; Peters, 1954), *Oryza sativa* (Shastry et al., 1960b), *Sorghum* (Magoon and Shambulingappa, 1960, 1961), *Hordeum* (Sarvella et al., 1958; MacDonald, 1961), *Melilotus* (Shastry et al., 1960a; Rao and Shastry, 1961), *Brassica* (Röbbelen, 1960), *Aquilegia* (Linnert, 1961a, 1961b), *Luzula* (Kusanagi and Tanaka, 1960) and human beings (Schultz and St. Lawrence, 1949; Yerganian, 1957; Eberle, 1963; Ferguson-Smith, 1964; Hungerford and Hungerford, 1978). Pachytene analysis is particularly rewarding in species with short chromosomes. The relative length of the pachytene stage of meiosis guarantees a high number of cells for analysis. Valuable morphological criteria in pachytene are the centromere, the chromomeres, the telomeres, the nucleolus organizer region with the nucleolus attached, the knobs, and the possibility to distinguish between heterochromatin and euchromatin.

2.2.1   Heterochromatin vs. Euchromatin

In order to profitably discuss the morphology of pachytene chromosomes, something should be mentioned here about the nature of **euchromatin** and **heterochromatin.** The two structures were first discovered on the cytological level and are different forms of chromatin (Flemming, 1882). Even the name **heterochromatin** was not coined until later. The structure was already recognized early during the century. Rosenberg in 1904 described **prochromosomes** as heterochromatic blocks observed in the interphase nucleus. They were also referred to as **chromocenters.** They represent large heterochromatic segments that really are regions that do not undergo despiralization and decondensation at the end of each cell division. Instead, they remain tightly coiled at a time when the rest of the chromosomes and chromosome segments are in a relatively uncoiled condition. Montgomery (1904, 1906) and Gutherz (1907) described the concept of **heteropycnosis** in relation to chromosomes or chromosome regions that during interphase or prophase are out of phase in respect to their coiling cycle and staining properties. Heitz (1928, 1929) finally coined the word **heterochromatin.** This material is often found proximal to the centromere or in the distal parts of the chromosomes. At metaphase these regions are usually indistinguishable from euchromatin. For this reason, pachytene is an ideal stage during which to detect these differences.

2.2.2   The Chromomeres

The **chromomeres** are bead-like projections, along the entire length of a pachytene chromosome, that are heavier stained than the **interchromomeric regions** (Fig. 2.5B). They are typical for mitotic and meiotic prophase alike. The now almost generally accepted interpretation of chromomeres is that they are structures resulting from local coiling of a continuous DNA thread. They probably represent units of DNA replication, RNA synthesis, and RNA processing (Rieger et al., 1976). The heterochromatic chromomeres stain darker than the euchromatic chromomeres. They also seem to be larger than the euchromatic chromomeres and have, therefore, been referred to as **macrochromomeres** (Gottschalk, 1954) as opposed to **microchromomeres.** Chromomeres also vary in size within these artificial size classes. For instance, the chromomeres next to the centromere are large and become progressively smaller toward the chromosome ends. Lima-de-Faria (1952) detected a **chromomere size gradient** that described this progression in diminishing chromomere size. He concluded that a detailed pachytene chromosome analysis includes a study of the number, size, and disposition of the chromomeres of each chromosome, thus permitting construction of a map of each chromosome type (Fig. 2.5A). The number of chromomeres within a pachytene chromosome seems to be reasonably constant and can serve as a reliable morphological characteristic. Different methods of constructing pachytene chromosome maps have been applied. One method (Gottschalk, 1954) uses a schematic illustration of the chromosome in which the heterochromatin is depicted as a dark bar and the euchromatic portions as a thin line. The numbers of the chromomeres in

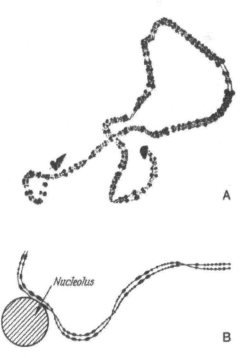

Fig. 2.5. (*A*) Pachytene bivalent depicting chromomere size gradient indicating that chromomeres next to centromeres are large and become progressively smaller toward chromosome ends. (From Lima-de-Faria, 1949. Redrawn by permission of the Mendelian Society, Lund, Sweden). (*B*) Pachytene bivalent showing bead-like chromosomes (from Rieger et al., 1968).

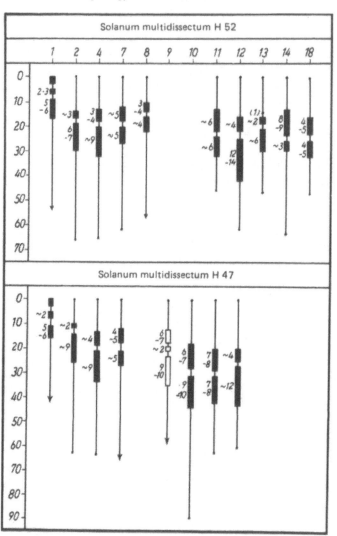

Fig. 2.6. Schematic representation of pachytene chromosomes of 2 cultivars of *Solanum multidissectum*. The numbers are macrochromomeres in heterochromatin (dark bars). (From Gottschalk and Peters, 1956. Redrawn by permission of Verlag Paul Parey, Hamburg, Germany).

the heterochromatin (macrochromomeres) are given next to the dark bars (Fig. 2.6).

2.2.3   The Centromere

Some essential characteristics of the centromere have been already given in Section 2.1.2. Centromeres in pachytene often show up characteristically different from those in mitotic metaphase. In many animal and plant species, the pachytene centromere consists of one to three chromomere pairs of different sizes, that are connected to the chromosome arms by thin fibers. Here again, Lima-de-Faria (1949, 1954) made a careful study of this structure. He suggested that the centromere is a compound structure that could be fractured with each broken part still functioning as a separate centromere.

2.2.4   The Telomeres

Telomeres are the enlarged terminal chromomeres of chromosomes. They seem to be an integral part of chromosomes, just like the centromere in that the chromosomes do not function normally when the telomeres are missing. They seem to seal off the ends of normal chromosomes so that they cannot join with other broken chromosome ends. In special instances, telomeres can have centromeric activity and are then called neocentromeres (Rhoades and Kerr, 1949). Lima-de-Faria and Sarvella (1958) studied the compound structure of the telomere in several plant species and stated it consists of two separately distinctive regions: the **protelomere** and the **eutelomere.** According to their observation, the protelomere is a terminal deep-staining structure with sharp limits, usually consisting of one to three dark staining large chromomeres. The eutelomere is a weakly staining subterminal segment adjacent to the protelomere. One compound telomere may consist of as many as eight different chromomeres. Parts of such a structure may break off without loss of genetic function of such a structure. According to electron-microscopic investigations, the telomere consists of irregularly folded chromatin fibers that rarely terminate at the chromosome ends, but loop back into the chromatid (Rieger et al., 1976).

2.2.5   The Nucleolus Organizer Region

The identification of the **nucleolus organizer region** (NOR) is simplified in the pachytene stage since the nucleolus is in immediate contact with this region at this time of meiotic division. Specific chromomeres are recognizable in this region during prophase, and they are called **nucleolus organizer bodies.** In maize, McClintock (1934) demonstrated a heteropycnotic knob in this region on chromosome 6 and showed that this knob is the organizer of the nucleolus. As in the case of the centromere and the telomere, this knob is a compound structure. After breakage, both fractured portions are capable of forming nucleoli. For some reason the smaller portion is able to form the larger of the two nucleoli. When only one of the two broken parts of the nucleolus organizer body are present in a cell, it is capable of collecting the entire mass of the nucleolar material. More recent studies show the role of this organizer in nucleolus formation (Givens, 1974; Givens and Phillips, 1976). Brown and Gurdon (1964) first demonstrated in the toad, *Xenopus laevis,* that the nucleolus is not only the site where ribosomal RNA is accumulated and stored but also the site where it is synthesized. They detected a deletion in the nucleolus organizer region of chromosome 12, called *O-nu.* Normal *Xenopus* cells $(+/+)$ contained 2 nucleoli but heterozygous cells $(+/O\text{-}nu)$ only one. Heterozygous toads were viable. When two heterozygous toads were hybridized, approximately 25% of their offspring died during the swimming larva stage. These obviously were the ones with the homozygous deletion *(O-nu/O-nu)*. The embryo tail tips of these homozygotes were investigated and showed lack of nucleoli in their cells. A particularly large nucleolus organizer body (Fig. 2.7) was recently found in the dryland leguminous forage plant Sainfoin *(Onobrychis viciifolia)* (Schulz-Schaeffer, unpublished).

Ohno et al. (1961) demonstrated that all 10 satellited chromosomes in man never form more than 6 nucleoli in any given interphase cell. The Ag-As ammoniacal silver staining method stains transcriptionally active nucleolus organizer regions. In humans, Miller et al. (1977) found that each person has a characteristic number of Ag-stained chromosomes, always fewer than 10. The frequency of Ag-stained chromosomes was correlated with the number of satellited chromosomes.

2.2.6   The Knobs

Another valuable landmark on the pachytene chromosomes of some species, particularly maize but also alfalfa (Buss and Cleveland, 1968), are the **knobs.** They are darkly stained bodies reminiscent of heterochromatin. Their position and number are constant for a particular race, but they vary between different races of the same species (McClintock, 1930; Longley, 1939; Brown, 1949). Their position is most frequently terminal or subterminal and less frequently interstitial (Fig. 2.8). McClintock (1930) could demonstrate a reciprocal translocation between chromosomes 9 and 10 of maize by using the terminal knob on the short arm of chromosome 9 as a marker (Fig. 2.9).

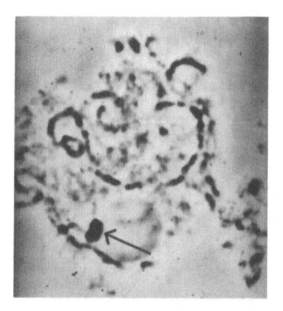

Fig. 2.7. Pachytene cell of *Onobrychis viciifolia* (2n = 28). The arrow indicates a large heterochromatic nucleolar organizer body. (Schulz-Schaeffer, unpublished).

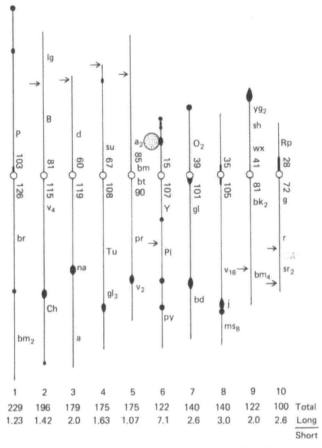

Fig. 2.8. Pachytene morphology of the chromosomes of *Zea mays* (n = 10). The lengths are relative, based on a length of 100 for chromosome 10. Arrows indicate additional knobs as found by McClintock. (From Burnham, 1962).

2.3 Banding Patterns in Mitotic Metaphase Chromosomes

As mentioned in the introduction of this chapter and in Chapter 1, new methods for the identification of mitotic metaphase chromosomes have been developed since 1970. They are making it possible to distinguish between chromosomes from many different species. With the modern squash, smear, and air drying techniques for chromosome preparation, mitotic metaphase is one of the most easily accessible stages for the study of chromosome morphology. The new banding techniques will explore the advantages of this stage with the added detail for morphological study. The flood of research on the subject of banding since 1970 verifies the great importance of the new discoveries. In the following sections a description of the four most popular banding techniques, of a group of other miscellaneous banding methods,

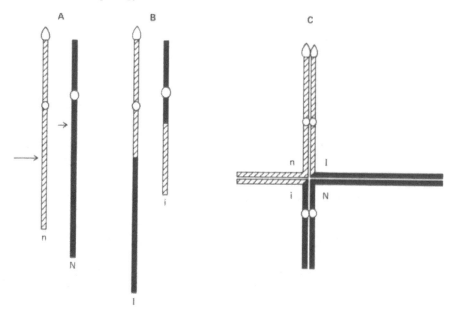

Fig. 2.9. (*A*) Diagram of chromosomes 9 (smaller) and 8 (larger) of *Zea mays* L. Chromosome 9 terminates in an enlarged deeply staining knob. The arrows indicate the places at which the translocation occurred to produce situation shown in *B*. (*B*) The 2 translocated chromosomes. (*C*) The type of synaptic configuration in pachytene obtained by combining a normal chromosome complement with a translocated one through crossing. n: normal chromosome 9. N: normal chromosome 8. i: translocated chromosome 8. I: translocated chromosome 9. (From McClintock, 1930. Redrawn by permission of National Academy of Science, Washington, D.C.).

and the first results with these methods will be presented. The popular banding techniques are presented in alphabetical order: C, G, Q, and R. A band is defined as a part of a chromosome that is clearly distinguishable from its adjacent segments by appearing darker or lighter.

2.3.1   C-Bands

The method to obtain **C-bands** is called **C-staining method** and it demonstrates constitutive heterochromatin. This method and all the Giemsa methods had their beginning with a paper by Pardue and Gall (1970). Constitutive heterochromatin is the common form of heterochromatin that usually does not change its nature, it is redundant and present in the proximity of centromeres and telomeres and in the nucleolus organizer region (Brown, 1966). It is so designated in order to distinguish it from **facultative heterochromatin,** which in short is euchromatin that has been heterochromatinized. This heterochromatin is not redundant, and it may or may not be condensed in interphase. A typical example for facultative heterochromatin is the chromatin of the Barr body X-chromosome. As mentioned in Section 2.2.1, heterochromatin (= constitutive heterochromatin) is naturally dif-

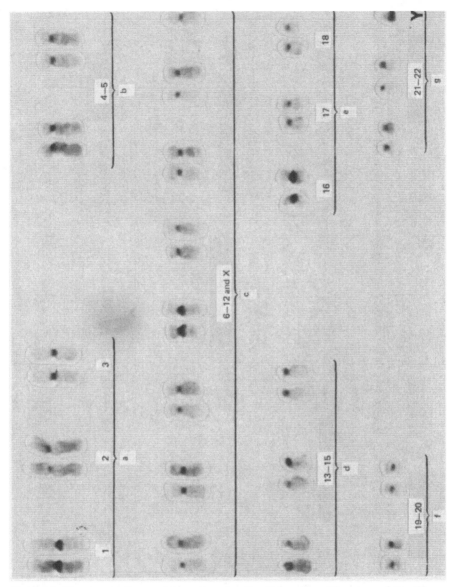

Fig. 2.10. Karyotype of normal human male after C-banding. (Courtesy of Dr. Cheng Wou Yu, Louisiana State University, Shreveport).

ferentiated by darker staining during interphase and prophase. In metaphase it usually does not show up. Therefore, it takes special methods to make hetero-chromatin visible in metaphase chromosomes. These methods usually involve treatment with acid, alkali, or elevated temperature. It is presumed that cellular DNA is denatured by these treatments. An overnight incubation at 60°C in saline-citrate solution presumably renatures the DNA. Therefore, it seems rea-sonable that highly repetitious DNA such as constitutive heterochromatin rena-

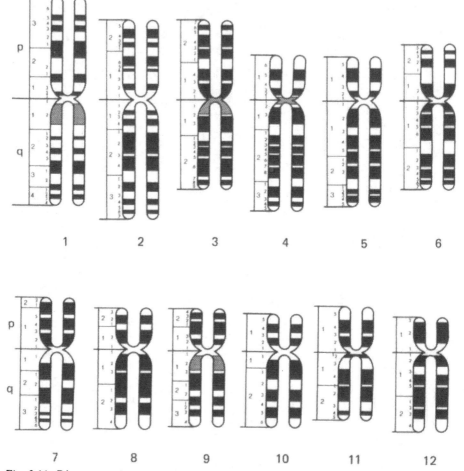

Fig. 2.11. Diagrammatic representation of human chromosome bands as observed with the Q-, G-, and R-banding methods. (From Paris Conference, 1971. Reprinted by permission of S. Karger A.G., New York).

tures under prescribed conditions while low repetitious DNA and unique DNA do not, thereby resulting in the differential staining reaction (Hsu, 1973). In most of the mammalian species, constitutive heterochromatin is located in the proximity of the centromere. The amount of it in each chromosome seems to be characteristic, but in man, polymorphism does occur (Craig-Holmes and Shaw, 1971; Craig-Holmes et al., 1973). Nevertheless, the Human Chromosome Study Group has started to use C-banding for the characterization of chromosomes. Banding patterns obtained with this method do not permit individual identification of each human chromosome but are helpful in the process (Fig. 2.10). The C-band technique is very useful for identifying the Y chromosome of mammals, which is often entirely heterochromatic particularly in species where its length extends 2 μm (Hsu, 1973). C-banding has been also applied to plant material. Vosa and Marchi

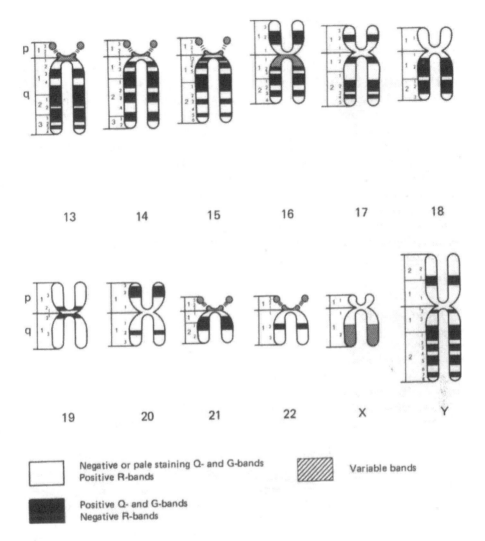

Negative or pale staining Q- and G-bands
Positive R-bands

Variable bands

Positive Q- and G-bands
Negative R-bands

(1972) demonstrated that plant heterochromatin may show up even more dramatically than animal heterochromatin. For instance, Linde-Laursen (1978) tried to explore the extent of **band heteromorphy** in barley by Giemsa C-banding in order to evaluate the use of the bands as markers in cytogenetic investigations.

2.3.2 G-Bands

The G-banding technique provides more detail than C-banding. Along with Q- and R-banding, it is ideal for the cytogeneticist since almost every chromosome within a complement can be identified and variation in chromosome structure from a standard type can be detected. This method uses Giemsa (= G) staining and usually pretreatment with a diluted trypsin solution, urea or protease (Berger, 1971; Dutrillaux et al., 1971; Seabright, 1972). Along with Q- and R-banding, it has been particularly promoted by the Human Chromosome Study Group (Paris Conference, 1971). They published a diagrammatic representation of chromosome bands as observed with G-, Q-, and R-staining methods (Fig. 2.11). The first reports of banding human chromosomes with Giemsa staining appeared in 1971 (Sumner et al.; Drets and Shaw; Patil et al.; Schnedl). Since then, reports on many other animal species have accumulated. The nature of G-bands is a subject of much discussion. Recent light microscope findings have demonstrated that they reflect stronger chromatin condensation (Ross and Gormley, 1973; McKay, 1973; Yunis and Sanchez, 1973; Ruzika and Schwarzacher, 1974). However, Giemsa banding could also be interpreted as a result of alteration of the histones and other proteins of the chromosomes. This would imply that such proteins are distributed along the chromosomes in clusters. A third possibility is that both DNA and proteins are involved in the cytochemical reactions following G-banding procedures (Hsu, 1973). Figure 2.12 shows the human G-band karyotype.

2.3.3   Q-Bands

As mentioned in Chapter 1, Q-banding was demonstrated by Caspersson et al. starting in 1968 and revived the search for dependable morphological characteristics for the differentiation of mitotic metaphase chromosomes. In their first three papers (1968, 1969a, 1969b), they demonstrated in several plant species that chromosomes stained with quinacrine mustard show bright and dark zones under ultraviolet light. Later they applied the procedure to human chromosomes and found that every chromosome pair could be recognized by fluorescent characteristics (Caspersson et al., 1970a, 1970b, 1970c). Figure 2.13 shows the human Q-band karyotype. Mouse genetics has progressed considerably through the discovery of Q-banding. The mouse chromosomes were hard to distinguish before the possibility of banding occurred. Now a nomenclature system has been established on the basis of Q-banding (Committee on standardized genetic nomenclature for mice, 1972). There are an enormous amount of genetic information and a large number of translocation stocks available in mice that now can be immediately utilized to correlate genetic and cytological discoveries (Hsu, 1973).

2.3.4   R-Bands

R-Bands show a pattern that is *reverse* to G-bands in that lightly stained G-bands become darkly stained when they are treated for R-banding. The method was first demonstrated by Dutrillaux and Lejeune (1971). Sehested (1974) suggested the use of a low pH (4–4.5) as a prerequisite for obtaining R-bands at heat treatment of 88 °C with incubation in NaH_2PO_4 (1M). This banding is reversible to G-band-

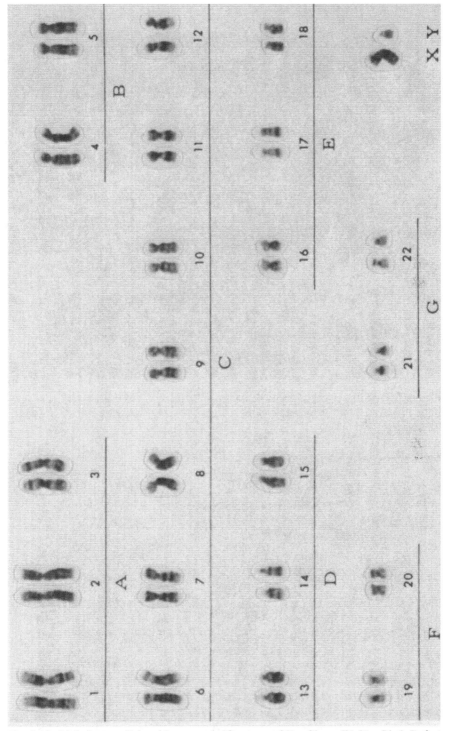

Fig. 2.12. Male human G-band karyotype. (Courtesy of Dr. Cheng W. Yu, Birth Defect Center, Department of Pediatrics, Lousiana State University Medical Center, Shreveport).

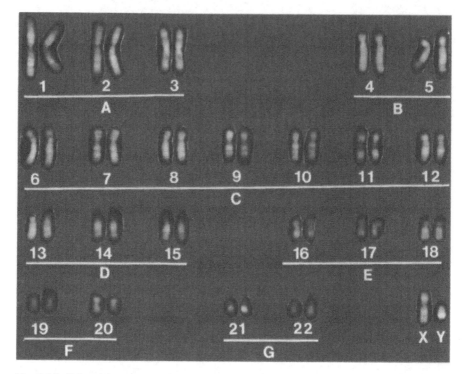

Fig. 2.13. Q-band karyotype of the human male. (Courtesy of Dr. Cheng W. Yu, Birth Defect Center, Department of Pediatrics, Lousiana State University Medical Center, Shreveport).

ing if the pH is readjusted to 5.5–5.6. While the number of published G-band methods has been steadily increasing, the number of R-band methods available has remained low. This may be due to the fact that R-banding has not been perfected to the same degree as G-banding. R-banding may be an important tool in deciphering chromosome organization in the future, as it may represent the other side of the same coin (Hsu, 1973).

2.3.5 Miscellaneous Bands

The classical findings in mitotic metaphase chromosome banding were obtained in the genus *Trillium* (Darlingtom and LaCour, 1938, 1940; Callan, 1942; Wilson and Boothroyd, 1941, 1944). This kind of band is now called H-band. They were originally obtained by prolonged cold treatments (0°C). But Yamasaki (1956) found that they also can be revealed by acetic orcein–HCl treatment without prior chilling of plants.

T-banding (T = terminal band) was developed by Dutrillaux (1973) and uses high temperatures (e.g., 87°C) at a pH of 6.7 and Giemsa staining. It shows especially a staining of some terminal regions of chromosomes. The application of this technique to translocations allows the precise location of juxta-telomeric breakpoints. N-banding was developed by Matsui and Sasaki (1973). It originally was used to demonstrate the nucleolus organizer. Along with C-banding, it has proven to be a superior method for plant material (Gerlach, 1977; Jewell, 1979).

The popular G-banding method has refused to reveal G-banding patterns in plant chromosomes. The absence of G-bands in plants was explained by Greilhuber (1977). He states that plant chromosomes contain much more DNA in metaphase than vertebrate chromosomes of the same length. For simple optical reasons vertebrate chromosomes would not show G-bands either at such a high degree of contraction.

Chapter 3
Fine Structure of Chromosomes

In 1976 Watson wrote that "even today" our fundamental knowledge of the molecular structure of chromosomes is very incomplete. This is particularly relevant for the more complex chromosomes of higher plants and animals. The main chromosome component of bacteria and viruses is **deoxyribonucleic acid** (DNA). However, up to 50% of the chromosomes of higher organisms is **protein.** Information on the ultrastructure of chromosomes has been obtained by various techniques including x-ray diffraction, chemical analysis, electron microscopy, and autoradiography.

3.1   The Structure of DNA

Deoxyribonucleic acid, the genetic material of all cells, is a polymer of deoxyribonucleotides. Its primary building block is called the nucleotide and consists of 3 types of simple molecules: a phosphate, a pentose sugar deoxyribose, and one of four nitrogenous bases. The sugar molecules are linked together by the phosphates, and each sugar molecule is attached to a single base. The bases are either purine (adenine and guanine) or pyrimidine (cytosine and thymine) bases. Nucleotides linked together by phosphatediester bonds form a polynucleotide. The secondary structure of DNA has been successfully described by several authors. Wilkins and Randall in 1953 concluded from x-ray diffraction studies in sperm heads of the cuttlefish, *Sepia,* that the polynucleotide chains of DNA are helical and not extended. Watson and Crick (1953a, 1953b), Franklin and Gosling (1953), and Wilkins et al. (1953) all came to the conclusion that two helices are present in the DNA molecule. As was mentioned in Chapter 1, Watson and Crick (1953a, 1953b) made the brilliant deductions that showed how the two helices fit together. They are linked together by hydrogen bonding of the base pairs (thymine-adenine, cytosine-guanine) so that each base pair forms a link between the sugar molecule on one helix and the opposite sugar molecule at the same level on the other helix (Watson-Crick model) (Fig. 3.1). The two right-handed helices are coiled in an interlocked form (plectonemically) about the same axis. Each turn or pitch of the so-called double helix includes 10 base pairs (Fig. 3.2). When DNA in crystalline form is studied by x-ray diffraction, the double helix makes one full turn every 3.4 nm.

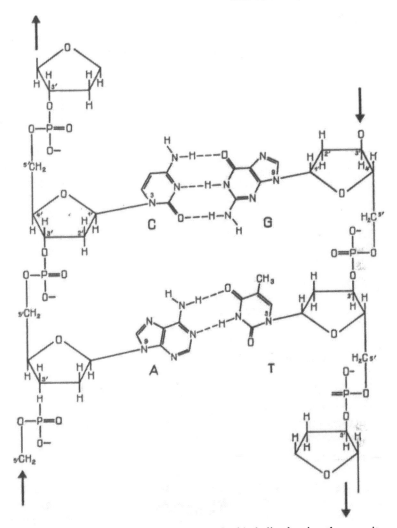

Fig. 3.1. A two-dimensional representation of a DNA double helix showing the opposite polarities of the sugar-phosphate linkages in the two strands. (From Herskowitz, 1967).

3.2   The Structure of RNA

Closely related in structure and function to DNA is ribonucleic acid or RNA. DNA and RNA differ in the composition of their pentose. The RNA pentose sugar is a ribose instead of a deoxyribose. Further, RNA contains no thymine but rather the closely related pyrimidine uracil.

In contrast to DNA, the RNA molecules are usually single stranded. In connection with the chromosome structure, RNA is important since it is the primary carrier of genetic information in some viruses. In these viruses DNA is replaced by RNA. The major function of RNA in the cell is to serve as a template substance. The

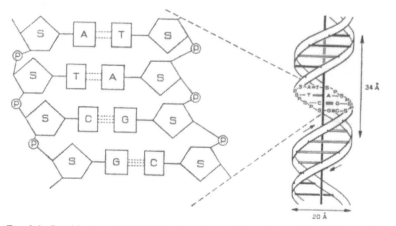

Fig. 3.2. Double stranded DNA helix with the dimensions of the helices indicated. (From Herskowitz, 1967).

template RNAs are mostly called messenger RNA or mRNA. Other RNAs in the cell are ribosomal or rRNA and transfer or tRNA.

3.3   Nucleoproteins

As mentioned, up to 50% of the chromosomes of higher organisms is protein. Proteins associated with DNA in the nucleus are basic proteins such as protamine and histone. They are of low molecular weight: the protamines between 1000 and 5000; the histones between 10,000 and 20,000. Protamines are a component of animal sperm chromosomes. Several theories have been advanced about the special relationships of DNA and proteins in the chromosome. One earlier theory postulated that a histone α-helix fits into the grooves of the DNA double helix (Zubay and Doty, 1959). More recently it is believed that the eukaryotic DNA is tightly complexed to proteins and comprises the nucleoprotein fibers called chromatin (Watson, 1976). According to electron micrographs this chromatin has a beaded structure and the components of this structure are spheroid chromatin units called **ν-bodies** or **nucleosomes** 6.0–8.0 nm in diameter (Olins and Olins, 1974; Oudet et al., 1975). Olins and Olins and Oudet et al. used this name because of the **new** discovery of these bodies. In spite of their different magnitude, such nucleosomes are very reminiscent of the chromomeres visible under the light microscope in leptotene and pachytene, which have been known since at least 1896 when Wilson described them in the first edition of his book *The Cell in Development and Heredity* (see Chapter 1). However, nucleosomes and chromomeres should not be mistaken for one another. Each of the nucleosomes, or repeating units, is believed to have 140 DNA base pairs and eight histone molecules made up of the four main types of histone: H2a, H2b, H3, and H4 (Kornberg, 1974). The structure of the ν-bodies has not yet been entirely revealed, but the eight histone molecules are believed to fill the central part of the nucleosome. The histones in this structure are believed to be involved in the process of chromosomal contraction. Chromosome condensation may be a function of cyclic chemical changes of these histones as they are being phosphorilated, methylated, and acetylated (Watson, 1976).

Vengerov et al. (1978) proposed a model for a **nucleosome package** that is a 20 nm globule formed by six nucleosomes (4 to 8 estimated by Müller et al., 1978) of a **nucleosomal fiber** with internucleosomal DNA segments being wound around the nucleosomes in which part of the DNA is covered by histone H1. The diameter of the internucleosomal DNA linkers is 2 nm (Fig. 3.3). Such DNA linkers vary in length from about 30 to 70 base pairs (Elgin and Weintraub, 1975). Shelton et al. (1978) estimated the size of the circular Simian virus 40 chromosome as being 5,224 base pairs, with a nucleosome size of 187 base pairs, the number of nucleosomes being 22 and the DNA linkers varying in size from 0 to 172 with 23% of them 20 base pairs in size. Nucleosomes are actually too small in size to be in the magnitude of actual genes. The present idea of gene size is in the order of approximately 1000 DNA base pairs (Goodenough, 1978).

3.4   Models of Chromosome Ultrastructure

Models of the ultrastructure of chromosomes reflect theories based on available data and serve as the basis for further experimentation. Data that serve for such models are collected from a wide range of approaches. The broad scope of attacking this problem is reflected in a recent symposium on *Chromosome Structure and Function* (Cold Spring Harbor Symposia on Quantitative Biology, 1974). Almost a hundred different ways of *dissecting* the chromosome are represented,

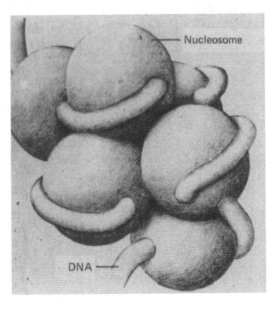

Fig. 3.3. Model of the nucleosome package. (From Vengerov et al., 1978. Reprinted with permission of VEB Georg Thieme, Leipzig).

ranging from many different biochemical approaches to different ways of cytologically analyzing the ultrastructure.

The classical approach in the study of chromosome structure is, of course, the cytological analysis. With the increase of resolution obtained by the discovery of the electron microscope, it was hoped that the structure of the chromosome could be studied in detail. But because of the lack of good fixation methods and sufficient contrast, there still does not exist a convincing picture of the ultrastructure of chromosomes (see Chapter 1). Very fine fibrils, 2 nm to 4 nm in diameter, that appear like the double helix strands of the Watson-Crick DNA model have been observed in sectioned material. The dimensions of the chromatid pairs at the light microscope level are at the order of 200 nm. Since both phenomena—double helix and chromatids—are believed to have analogous linear arrangement of genes that are involved in crossing over, chiasma formation, and mutational events, the question of how the molecular structure can be built into the visible chromosomes arises. The present prevailing concept is that one long thread of DNA is arranged in the chromatid in some coiled or folded manner.

3.4.1   The Folded Fiber Model

The **folded fiber model** of chromosome ultrastructure is based on the method of whole mount electron microscopy described in Chapter 1. This model was developed by DuPraw (1965b, 1968) in order to integrate the large body of experimental genetic, cytological, and biochemical data with the new morphological discoveries. In this model each unreplicated **monad chromosome** or **unit chromatid** is loosely packed in irregularly transverse and longitudinally folded spirals of a single 20 nm to 50 nm elementary fiber, which contains one extremely long single DNA double helix in supercoiled configuration held together by protein molecules. Replication of the chromosome occurs at several sites along the length of this fiber, where DNA polymerase catalyzes DNA synthesis at fork configurations. The late replicating segments of the fiber at the centromere and elsewhere serve to hold together the sister chromatids (Fig. 3.4).

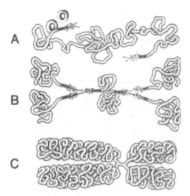

Fig. 3.4A–C. Diagram illustrating a folded fiber model of chromosome structure. (*A*) Each unreplicated chromosome (unit chromatid) is essentially a single 20-50 nm fiber, that contains a DNA double helix in supercoiled configuration. (*B*) Replication of the chromosome occurs at several sites along the length of the fiber where DNA polymerase catalyzes DNA synthesis at fork configurations. (*C*) The late replicating segments of the fiber at the centromere and elsewhere serve to hold together the sister chromatids. (From DuPraw, 1965b).

3.4.2   The Molecular Chromosome Model

DuPraw's folded fiber model is in line with the **molecular model of chromosomes** as formulated by Taylor et al. (1957), Taylor (1958, 1963), Freese (1958), and Schwarz (1960). The most important conclusion was that the chromosome is composed of only one DNA strand probably consisting of several molecules of DNA associated linearly. This model is based on the assumption that the chromosomal DNA follows a semiconservative mode of replication. The semiconservative mode was confirmed by tritiated thymidine labeling experiments that demonstrated that the DNA double helix separates into two separate polynucleotide strands as the hydrogen bonds between the nucleotide pairs break. As the two strands unwind, each synthesizes a new complementary copy of DNA nucleotides, leading to the formation of the two double stranded DNA molecules, half derived from the parent molecule and half newly synthesized (Fig. 3.5).

One of the not so essential features of the molecular chromosome model is that the chromosome is composed of a number of subunit DNA helices linked end to end by a series of protein molecules (Fig. 3.6). Coiling and uncoiling of the chromosome may be explained by the interaction of such protein molecules with each other (Freese, 1958). Recent data seem to give strong evidence against the presence of such protein linkers. The integrity of DNA has been found to be sensitive only to deoxyribonuclease and not to ribonuclease or proteolytic enzymes.

3.4.3   The Multistranded Chromosome Model

One of the oldest and most seriously discussed arguments of chromosome fine structure has been the question of **polynemy** vs. **uninemy.** In spite of a recently prevailing consensus of uninemy or single strandedness of DNA in the chromosome, some aspects of chromosome structure are still not readily explained by a

Fig. 3.5. Diagram showing the separating double helix at the beginning of replication and each of the strands producing a new strand. (Modified after Bollum, 1963).

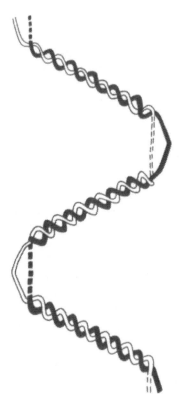

Fig. 3.6. The molecular chromosome model consisting of a number of linked double helical subunits. (After Taylor, 1963. Redrawn with permission of Academic Press, Inc., New York).

single DNA strand. Such phenomena are the visible doubleness of chromosomes at anaphase (Nebel, 1941; Sparvoli et al., 1965; Bajer, 1965; Wolff, 1969) and isolabeling (Peacock, 1965). In isolabeling, part of a second division chromosome has both chromatids radioactively labeled (Fig. 3.7). This has been explained by doublestranded chromosomes in telophase II.

A typical polyneme chromosome model is that proposed by Ris (Ris, 1961; Ris and Chandler, 1963) which suggests that each 25 nm fiber consists of two 10 nm fibers twisted around each other. Each 10 nm fiber consists of two 4 nm fibers that contain a DNA double helix surrounded by protein. Thus, a prophase chromosome would consist of 8 continuous DNA threads.

Fig. 3.7. Second division chromosome with part of both chromatids radioactively labeled as in isolabeling.

Fig. 3.8. The "General Chromosome Model" according to Crick (1971). According to this model, the globular control DNA is in the bands, and the fibrous coding DNA is in the interbands. (Redrawn with permission of Macmillan Journals LTD., London).

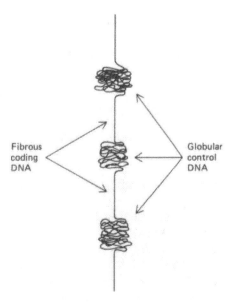

3.4.4   General Chromosome Model

In 1971 Crick proposed a general model for the chromosomes of higher organisms. In contrast to the folded fiber model, this model is strongly based on findings in molecular genetics. The model suggested that chromosomal DNA falls into two classes: (1) globular control DNA in the bands and (2) fibrous coding DNA in interbands (Fig. 3.8). This model also suggests that the DNA in a chromatid is a

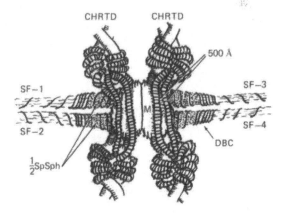

Fig. 3.9. The Hoskins (1969) model of the centromere. The chromatid arms (CHRTD) are extending upward and downward, the spindle fibres (SF-1, -2, -3, and -4) extend from the centromere to the left and to the right. The chromonemata (50 nm) extend across the centromere from one arm of one chromatid to its other arm. (M–matrix of the centromeres.) (Courtesy of Dr. Godfrey Hoskins, Hoskins Pathology Laboratories, Dallas, Texas. Reprinted by permission of Caryologia, Firenze, Italy).

very long monomere (Prescott, 1970; Laird, 1971) that probably runs continuously from one end of the chromatid to the other.

This model has 3 basic features:

1. The coding sequences of DNA are postulated to be mainly in the interbands (Vogel, 1964).
2. The recognition sites needed for control purposes in higher organisms are mainly unpaired single stretches of double stranded DNA (Gierer, 1966).
3. The forces and energy needed to unpair the recognition stretches of DNA are provided by the combination of DNA with chromosomal proteins—probably histones.

3.5. Ultrastructure of the Centromere

A well-documented model of the fine structure of the centromere is based on electron microscopic analysis of Hoskins (1969). Hoskins used a method of micromanipulation by which he pulled the centromeres out of the cells for detailed study. The model (Fig. 3.9) shows two chromatids of a metaphase chromosome held together in the centromere regions by two hemispheric or valentine-shaped matrices (M), associated base to base. The two arms of each chromatid are interconnected across the matrix by chromonemata (50 nm in diameter) that are continuous with the chromonemata of the chromatids. This model seems to tend to a polyneme concept of the chromosome. Also attached to the matrix are the spindle fiber bundles (SF), two to each matrix (a total of 4 to each centromere). In the area of attachment to the matrix, there is a swelling of the bundles which is called the **spindle spherule.**

Part III
Function of Chromosomes

Chapter 4
Function of Autosomes

Part III is divided into "Function of Autosomes" and "Function of Sex Chromosomes." All chromosomes that are not sex chromosomes are called **autosomes.**
After the discussion of chromosome structure, it is important to learn some basic facts about chromosome function. Structure and function are necessarily closely related. Many structural features are the basis for the function of chromosomes. The genetic units of the chromosomes are the genes. Genes seem to function as groups and also seem to be transferred as such from generation to generation. As mentioned in Chapter 1, the first one to recognize such gene complexes was Morgan (1910b). He called them linkage groups (1911). A linkage group is the association of certain genes that are located on the same chromosome.

4.1 Linkage

When it was realized that organisms had more genes than chromosomes, it was concluded that each chromosome must carry more than a single gene. Out of this arose the concept of **chromosome stability.** This concept, of course, excludes the possibility of chromosome breakage of any form. Genes that tend to be inherited as groups rather than individually are established as linkage groups. In a number of plants and animals such linkage groups have been worked out. The most complete ones are in *Drosophila melanogaster* (Fig. 4.7), *Zea mays* (Fig. 4.8), and *Neurospora.*
But linkage is obviously not complete because of the existence of chromosome or chromatid breakage. This fact has been used to determine the relative position of genes in the linkage groups. The result of this research are linkage maps that show the position of genes on chromosome maps.
Linkage of genes is only seldom complete. Linked genes are separated from one another through the process of crossing over. The higher the incident of crossing over, the further apart are the genes on the linkage map.
Linkage data are obtained from crossing results. If the parents used for crossing differ in respect to two linked pairs of alleles *(AB/AB* and *ab/ab),* four classes of gametes are expected: *AB, ab, Ab, aB. AB* and *ab* are the **parental classes** and *Ab* and *aB* the **recombinant classes** that are the result of crossing over. The larger the recombinant classes, the further away *a* and *b* are located on the linkage map.

4.2   The Mechanism of Crossing Over

The mechanism of crossing over is still not understood and is closely related to the unsolved problem of chromosome ultrastructure. Consequently, several classical and newer hypotheses of crossing over exist. Only some will be discussed here.

4.2.1   The Partial Chiasmatype Theory

As mentioned in Chapter 1, Janssens in 1909 advanced the **partial chiasmatype theory,** which is now accepted as the most reasonable explanation of the relationship between cytologically observable chiasmata and experimentally demonstrated genetic crossing over. According to this theory, chiasmata are the direct result of crossing over and are formed exactly at the points where the breakage or exchange of non-sister chromatids occurs (Fig. 4.1). Crossing over occurs only between 2 of the 4 chromatids present at any given point, but three- and four-strand crossing over is possible in any given region.

4.2.2   The Belling Hypothesis

Belling's hypothesis (1931a, 1931b, 1933) correlates crossing over with the reproduction of new chromatids. The historical importance of this work was briefly mentioned in Chapter 1. Belling's hypothesis is the basis for some of the more recent theories and merits discussion in more detail. This hypothesis requires some kind of **relational coiling** (Section 6.3) between homologues at the time of chromatid reproduction (Fig. 4.2). The theory further postulates that new chromomeres are formed alongside their respective sister chromomeres without the formation of interconnecting fibers (Fig. 4.2A). The next step in this scheme is the formation of the connecting fibers between the newly synthesized chromomeres (Fig. 4.2B). During this process, sections of nonsister chromatids get intercon-

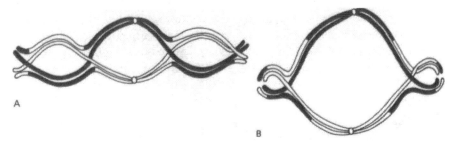

Fig. 4.1A and B. Schematic drawing of the cytological manifestation of crossing over in diplotene according to the partial chiasmatypy theory. (*A*) Four cytologically visible chiasmata occur at exactly the points where genetic crossing over has occurred in pachytene previously. Black chromatid segments may symbolize paternal chromosome origin and white ones maternal. (*B*) Advanced stage of chiasma terminalization in diakinesis. Chiasmata have started to move toward chromosome ends. Crossover points and chiasmata do not coincide anymore. Two chiasmata have become end-chiasmata. (From Swanson, 1957. After Darlington, 1930. Redrawn by permission of Prentice-Hall, Englewood Cliffs, N.J.).

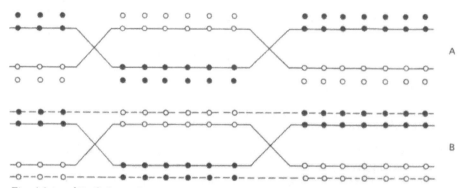

Fig. 4.2A and B. Schematic representation of the Belling hypothesis of crossing over. (*A*) The chromosomes have coiled relationally and the chromomeres have replicated (black chromomeres paternal, white maternal). No interchromomeric fibers have formed as yet. (*B*) Interchromomeric fibers have been formed. Non-sister chromatid segments (black and white) have linked forming the new chromatids. (From Swanson, 1957. Redrawn by permission of Prentice-Hall, Englewood Cliffs, N.J.).

nected, and crossing over is being accomplished in this fashion. The Belling system restricts crossing over to the newly formed chromatids and seemingly rules out three-strand and four-strand double crossing over. But if sister-strand crossing over is considered to be an independent event occurring at a different time, it allows for four-strand crossing over even in the Belling system. In sister-strand crossing over the two chromatids involved in the exchange belong to the same chromosome.

4.2.3   The Copy-Choice Hypothesis

The **copy-choice hypothesis** is considered to be a revival of the classical Belling hypothesis. It was formulated by Lederberg in 1955 following the discovery that crossing over could occur within a single gene, which threw doubt on the possibility of breakage and reunion as postulated in the chiasmatype theory. The copy-choice hypothesis relates recombination to DNA replication. During replication the new DNA strand is synthesized along the maternal chromosome and at a given point switches templates whereupon it is copying the paternal chromosome (Fig. 4.3). Assuming that the complementary copy of the paternal DNA switches templates at the same point, the result will be the two reciprocal recombinant strands.

4.2.4   The Polaron Hybrid DNA Model of Crossing Over

This model was developed by Whitehouse (Whitehouse, 1963, 1965; Whitehouse and Hastings, 1965). It has been suggested that the bulk of DNA synthesis has been completed at the time of crossing over and that only little DNA synthesis is occurring during the time of crossing over. In order to account for this apparent fact, Whitehouse developed his crossing over model. The **polaron model** also explains both reciprocal recombination (crossing over) and nonreciprocal recombination (**conversion**). A basic assumption of this model is that the chromatid consists of a single DNA double helix at the time of recombination. In Fig. 4.4, only

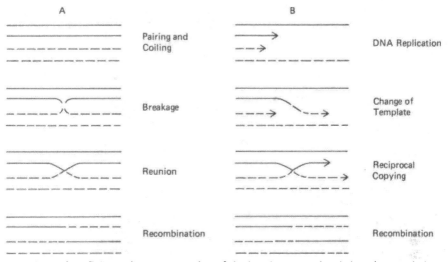

Fig. 4.3A and B. Schematic representation of the breakage-reunion (*A*) and copy-choice (*B*) hypotheses of crossing over. (From Hamerton, 1971a. Redrawn by permission of Academic Press, New York).

two of the four chromatids present during crossing over are shown. The figure (A) shows two non-sister chromatids each consisting of a DNA double helix. The horizontal lines represent polynucleotide chains. The arrows indicate the direction of the sugar-phosphate backbone and also delimit the extend of the **polaron.** The polaron is the unit of the chromosome by which it is subdivided in terms of **linkage points** where crossing over can be initiated. The short vertical lines depict the hydrogen bonds between the bases of the complementary nucleotide chains. The figure (B) also shows two opposite non-sister chromatid nucleotide chains of opposite polarity breaking off enzymatically at one end of the polaron. The broken nucleotide chains separate from their complementary chains over the main part of the polaron length. New chains (broken lines in C) are synthesized along the polaron where the old ones were broken off. After the new chains have been synthesized, they in turn also break off from their complementary nucleotide templates (D). The old and newly synthesized break products are now pairing up as **hybrid duplexes** to form new complementary DNA double helices (E). Any still existing gaps are now being filled with complementary nucleotide pieces (F). The old unpaired nucleotide chains are now breaking down by digestion and are thus eliminated (G, H). Crossing-over is now completed.

4.3   The Cytological Basis of Crossing Over

Chromatid exchanges in meiosis are observable under the microscope by the formation of unmistakable structures between homologous chromosomes (homologues) called chiasmata (Fig. 4.5). As can be seen clearly, a chiasma involves only one chromatid, but each of the two homologues are involved in its formation.
The events taking place in meiosis, described later (Chapter 7), give us further

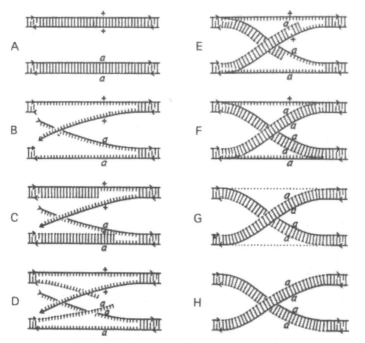

Fig. 4.4. Diagram of the polaron hybrid DNA model of crossing over. (From Rieger et al., 1976. After Whitehouse, 1965).

insight into the physical basis of crossing over. It may be only mentioned here that homologous chromosomes are brought together in a snug union during early prophase and then separate during diplotene. By this time the chromatid exchanges have taken place, and at the exchange points the chiasmata begin to show up cytologically. These, then, are naturally the points of so-called "crossing over," a phys-

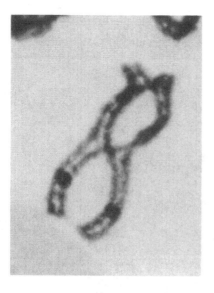

Fig. 4.5. Photograph of a late diplotene bivalent in a spermatocyte of the Costa Rican plethodontid salamander, *Oedipina poelzi.* (Courtesy of Dr. James Kezer, Department of Biology, University of Oregon, Eugene).

Recombination Diagram

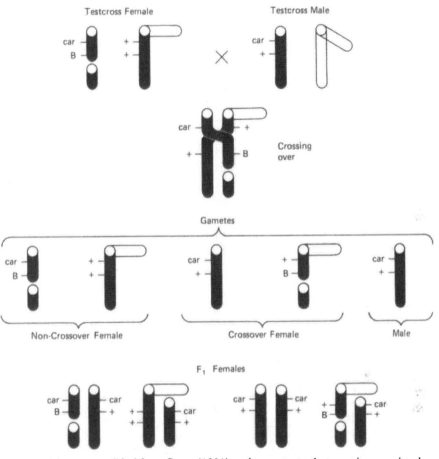

Fig. 4.6. Diagram modified from Stern (1931) to demonstrate that crossing over involves an exchange of chromatin between homologous chromosomes. Detail in text. (From Swanson, 1957. Drawn by permission of Prentice-Hall, Inc., Englewood Cliffs, New Jersey).

ical embodiment of a genetic term that originally designated a cytological or physical happening. Crossing over therefore represents the exchange of chromatid material between homologous chromosomes.

The actual demonstration of this exchange was accomplished by Stern (1931) and Creighton and McClintock (1931), as mentioned in Chapter 1. In Stern's experiment with *Drosophila*, a female carrier of two heteromorphic X-chromosomes was used in a test cross. The two X-chromosomes are explained in Fig. 4.6. The broken X-chromosome carried the two marker genes, the recessive eye color *carnation* (*car*: 62.5) and the dominant eye shape *Bar* (*B*: 57.0). The elongated X-chromosome carried the two wild type genes *(+)*. The testcross males carried an X-chromosome with the two recessive genes (*car*, +). The female offspring of the testcross formed four classes: (1) *carnation* and *Bar* with the parental broken X-chromosome, (2) normal eye color and shape with the parental elongated X-chro-

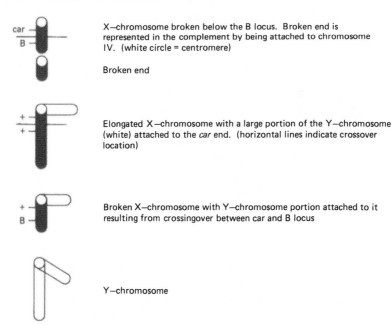

car

B

X—chromosome broken below the B locus. Broken end is represented in the complement by being attached to chromosome IV. (white circle = centromere)

Broken end

+

+

Elongated X—chromosome with a large portion of the Y—chromosome (white) attached to the *car* end. (horizontal lines indicate crossover location)

+

B

Broken X—chromosome with Y—chromosome portion attached to it resulting from crossingover between car and B locus

Y—chromosome

Fig. 4.6 (contd.) Explanation of chromosomes used in Stern's (1931) experiment.

mosome, (3) *carnation* and normal shape, a crossover type and (4) normal eye color and *Bar*-shaped eye with a crossover type broken and elongated chromosome. Stern made a genetic and cytological analysis of 364 non-crossover and crossover F_1 females, and in almost all of them there was exact agreement between the genetic and cytological data. This demonstrated that genetic recombination was accompanied by a reciprocal exchange of chromatid material between the two homologous chromosomes.

The number of chiasmata per pair of homologues is limited by the length of the chromosomes. Adjacent crossovers or chiasmata do not occur independently. A chiasma in a given chromosome region suppresses a chiasma in the adjacent region. This has been called **chromosome interference** (Muller, 1916), **chiasma interference** (Mather, 1933), or **crossover interference** (Whitehouse, 1965). Interference increases with decreasing distance between successive genes and decreases with increasing distance.

4.4   Locating Genes on Chromosomes and Genetic Mapping

One of the main functions of the chromosome is the block transfer of genes. Cytogeneticists have been interested in first assigning genes to specific chromosomes and, if possible, locating their position on the chromosome.

Several methods have been used to assign genes to their respective chromosome or linkage groups. Some of them will be only mentioned here and described in detail at the appropriate place in this book. Species with high numbers of chromosomes are difficult to work with for gene location. The discovery of **monosomy** and the establishment of **monosomic series** has helped in the assignment of genes to specific chromosomes or linkage groups. The first report of monosomics was in tobacco (*Nicotiana tabacum*, 2n = 48) by Clausen and Goodspeed (1926) who at that time reported about plants with 47 chromosomes $(4x - 1)$. Locating genes by the use of monosomics is discussed in Chapter 16.

The most extensive studies involving gene location have been carried out with trisomics (e.g., $2x + 1$). In Chapter 1 the discovery of trisomics by Blakeslee (1921) was described. By the detection of trisomic ratios, genes can be associated with specific chromosomes. A more detailed description of trisomic gene analysis will also be given in Chapter 16. The use of translocations in mapping genes will be shown in Chapter 14.

Genetic mapping involves the assignment of genes to specific linkage groups and the determination of the relative distance of these genes to other known genes in that linkage group. This process assumes that genes are arranged in a linear order along the chromosome as first postulated by Morgan. The sequence of the genes on the chromosome can be determined from the **three-point cross** devised by Sturtevant (1915). The following example is taken from Swanson (1957). If one assumes that the correct sequential arrangement of three genes is *abc*, then a testcross of the heterozygote $+ + +/abc$ to the triple recessive *abc/abc* would result in the following genotypes:

$$\left. \begin{array}{l} + + + \ /abc \\ a \ b \ c \ /abc \end{array} \right\} \quad \text{noncrossover parental individuals}$$

$$\left. \begin{array}{l} + \ b \ c \ /abc \\ a \ + \ + \ /abc \end{array} \right\} \quad \text{single crossover recombinants, class 1}$$

$$\left. \begin{array}{l} + \ + \ c \ /abc \\ a \ b \ + \ /abc \end{array} \right\} \quad \text{single crossover recombinants, class 2}$$

$$\left. \begin{array}{l} + \ b \ + \ /abc \\ a \ + \ c \ /abc \end{array} \right\} \quad \text{double crossover individuals}$$

As we know from incomplete linkage studies, the number of recombinant individuals is less than the number of parental combinations. Consequently, the number of the noncrossover parental individuals would be the highest. The frequencies in the two single crossover recombinant classes will depend on the distance between genes *a* and *b*, and between *b* and *c*. The number of individuals in the double crossover class will be the smallest. The double crossover class gives information concerning the linear arrangement of the genes along the chromosome. In the double crossover class ($+b+$ and $a+c$), gene *b* has shifted its position with its dominant allele with respect to genes *a* and *c*. The order of the three genes must therefore be *abc*. An actual example of the three-point cross is shown in Chapter 14 (see Table 14.1).

Many such genetic data lead to the construction of genetic maps. Classical examples of such genetic maps are those of *Drosophila melanogaster* (Fig. 4.7) and of maize, *Zea mays* (Fig. 4.8). Each gene is shown on the genetic map as a point on

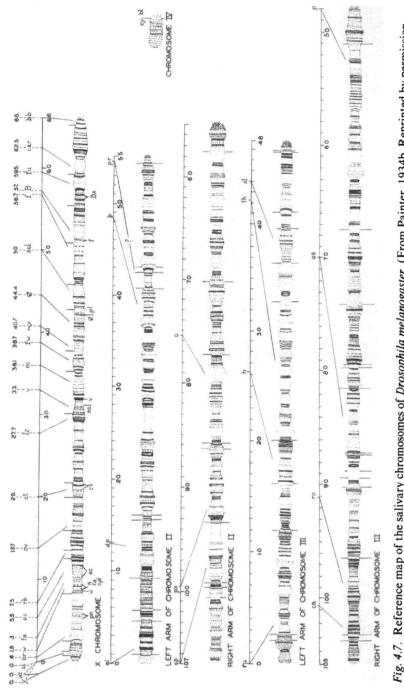

Fig. 4.7. Reference map of the salivary chromosomes of *Drosophila melanogaster.* (From Painter, 1934b. Reprinted by permission of the American Genetic Association, Washington, D.C.).

the linear chromosome. The distance between any two genes is a function of the recombinant frequencies.

Genetic maps do not reveal the *actual distance* of the genes because crossing over does not occur at the same frequency at different sections of the cytological or chromosome map. In order to get actual distances, one has to resort to **deletion mapping** or **cytogenetic mapping.** This mapping seeks to determine the locus of a specific gene on the chromosome map. Such locus may be detected for a specific arm, for a fraction of such an arm, or for a minute deleted segment of a chromosome. To carry out such mapping, a specific mutation can, for instance, be pinpointed to a deficiency (Chapter 11) in the corresponding chromosome.

When an organism has a recessive, nonlethal mutation in a chromosome and is heterozygous for a deficient segment on its homologous chromosome, the recessive mutation will be expressed in the phenotype. This phenomenon has been referred to as **pseudodominance.**

The best results with deletion mapping have been achieved in the localization of genes on the giant salivary gland chromosomes of *Drosophila.* These chromosomes are extended to such a size and have so much detail that small deletions can be traced with an excellent degree of accuracy.

Mackensen in 1935 and Slizynska in 1938 used deletions to locate genes on the *Drosophila* chromosome map. Slizynska's study is shown in Figure 4.9. The black areas on the diagram show the deficient regions of 14 different deficiency mutants all of which produce the *white-Notch* phenotype. These areas are correlated to the numbering system of Bridges (1935) that divides parts of chromosomes with Arabic numbers, uses capital letters for subdivisions, and gives Arabic numbers to the bands within the subdivisions. As an example, band 3C7 is deleted in deletion mutants N8 Mohr, 264-38, 264-36, 264-30, 264-31, 264-32, 264-33, 264-37, 264-39, 264-2, and 264-19 (Fig. 4.9). Bridges numbering system in turn is correlated to the bands on the chromosome map in Figure 4.9.

In order to be able to appreciate the phenotypic expression of such mutated genes, a picture of a fly with *notched wing (N)* is shown in Figure 4.10.

Cytogenetic mapping also has been carried out in humans. By 1977 over 110 gene loci had been assigned to specific human autosomes, and about 100 more to the X-chromosome (McKusik and Ruddle, 1977). The total number has since climbed to 347 (see Fig. 4.11). The use of somatic cell hybridization for cytogenetic studies has been mentioned in Chapter 1 (Harris and Watkins, 1965). The assignment of genes to specific chromosomes is a possible outcome of such studies. One such approach is the **synteny test** by which one can investigate if two genetic loci are linked to the same chromosome depending on their correlated loss or retention in hybrid cells. The first such successful test was performed by Nabholtz et al. (1969) who demonstrated that the loci for *HGPRT* (hypoxanthine-guanine phosphoribosyltransferase) and *G6PD* (glucose-6-phosphate dehydrogenase) are both located on the X-chromosome.

Another approach is the **assignment test** where the location of a particular gene on a specific human chromosome is demonstrated by the concordance between the presence or absence of this chromosome and a specific phenotype in many hybrid clones (McKusick and Ruddle, 1977).

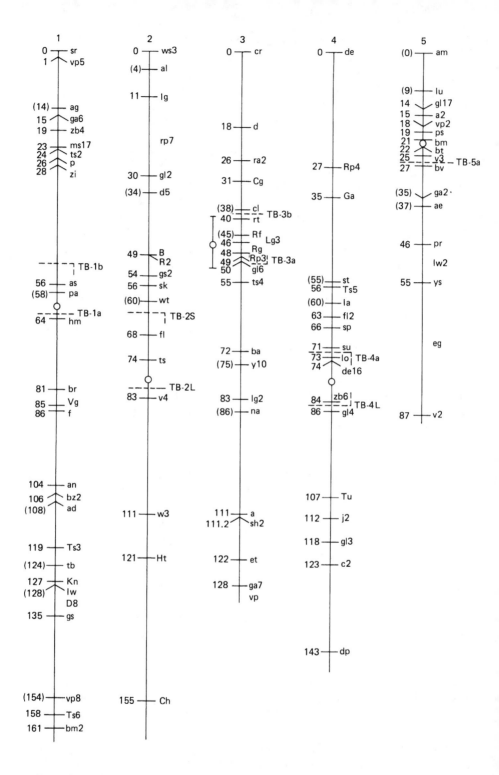

Fig. 4.8. Genetic linkage map of maize. (Revised from Neuffer et al., 1968. Reprinted from Maize Genetics Cooperation News Letter 51, 1977).

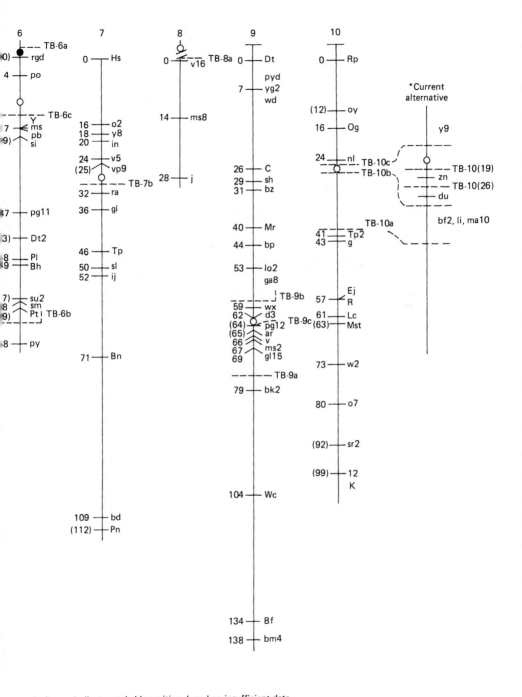

( ) Indicates probable position, based on insufficient data
O Indicates centromere position
● Indicates organizer (NOR)
TB Indicates translocation involving A and B chromosomes, with A
    break point at broken line or in direction indicated. TB-2S, TB-2L
    and TB-4L are short designations for TB-2S, 3L(6270), TB-2L,1S(4464)
    and TB-4L, 9S(6222) from Rakha and Robertson (1970). TB-10(19)
    and TB-10(26) are as designated by Lin (1974).
* Lin (1974); Beckett (Personal communication)

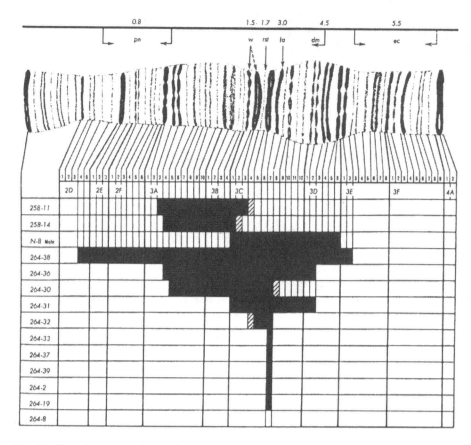

Fig. 4.9. Deletions in the left end of the X-chromosome that have been used to locate genes on the *Drosophila* chromosome map. Black segments shown below indicate bands removed by each deletion. (From Slizynska, 1938. Redrawn by permission of Prentice-Hall, Inc., Englewood Cliffs, N.J.).

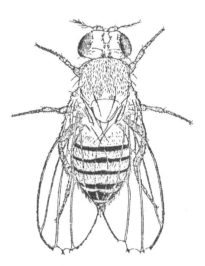

Fig. 4.10. Drosophila fly with *notched* wing (N^8). (From Mohr, 1924).

Regional mapping is used to assign a human gene locus to a specific chromosome segment. Here, a human cell with a particular chromosomal rearrangement is used as parent in a cell hybrid. Such rearrangement would be a translocation. The translocation breakpoint divides the chromosome under investigation into two independently segregating units that can be separately studied in the assignment test. A **human mutant cell bank** (Coriell, 1973) has helped to store, distribute and analyze such chromosomal translocations and deletions. The first regional mapping assignment through the use of a chromosomal translocation was that of *G6PD, PGK,* (phosphoglycerate kinase) and *HGPRT* to the long arm of the X chromosome (Ricciuti and Ruddle, 1973).

In situ hybridization of RNA and DNA has been used as an approach to gene mapping in humans. Such nucleic acid hybridization can be performed in the test tube (Spiegelman, Chapter 1) as well as inside the cell. If dissociated and denatured DNA is left in place inside the cell nucleus, it can be subjected to hybridization with isolated RNA. If the RNA is radioactively labeled, the chromosomal location of the specific genes from which the RNA is normally transcribed can be actually identified under the microscope. Henderson et al. (1972) as well as Guanti and Petrinelli (1974) used this method to localize genes for 18S and 28S ribosomal RNA on the short arms of the satellite chromosomes 13, 14, 15, 21, and 22.

Deletion mapping, discussed in Chapter 11, has also been applied in human chromosome mapping. The first assignment by deletion mapping of a human gene not previously located by any other method was that of *ACP-1* (acid phosphatase) to the distal end of the short arm of chromosome 2 (Ferguson-Smith et al., 1973).

Duplication mapping, like deletion mapping, is a gene dosage method. A person trisomic for part or all of a chromosome has about 50 percent more of a particular gene product. The location of five human genes has been confirmed by duplication mapping. These are the assignment of *antiviral protein* (Tan et al., 1974) and *SOD-1* (superoxide dismutase-1) (Sinet et al., 1975) to chromosome 21, *If-1* (Galactose-1-phosphate uridyltransferase) to chromosome 3 (Allderdice and Tedesco, 1975), *ACP-1* (red cell acid phosphatase-1) to chromosome 2 (Magenis et al., 1975) and *glutathione reductase* to chromosome 8 (Chapelle et al., 1976). A genetic map of human chromosomes is shown in Figure 4.11.

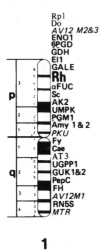

1

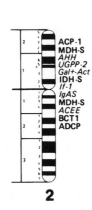

2

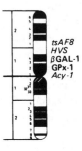

3

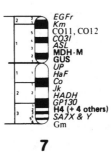

6

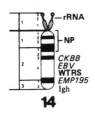

7

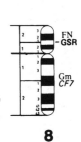

8

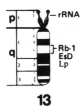

13

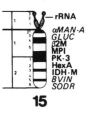

14

15

19

20

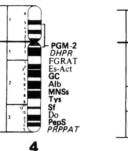
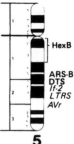
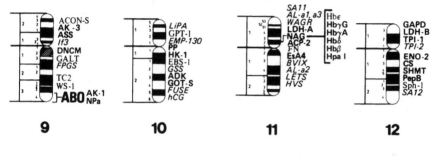
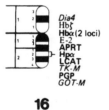
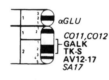

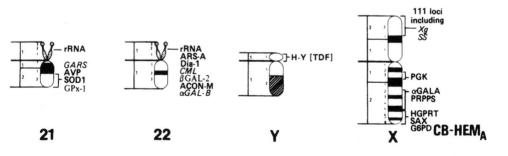

Fig. 4.11. Genetic map of the human chromosomes. (Courtesy of Dr. V. A. McKusick, Department of Medicine, Johns Hopkins Hospital, Baltimore, Maryland).

Chapter 5
Function of Sex-Chromosomes

By now it may be obvious that the author emphasizes the historical importance of the discoveries that made cytogenetics the science it is today. That is why the first chapter was written in such detail and the student is often referred back to it in order to freshen his recollection. The first studies of chromosomes that determine sex were undertaken at the end of the last century. As mentioned in Chapter 1, Henking in 1891 for the first time described what is now known as the X chromosome. Half of the sperm of the insect *Pyrrhocoris apterus* received this chromosome and half did not. This system is now known as the X-O system. A much more common system is the X-Y system, which will be discussed first.

5.1 The X-Y System

The X-Y system is a basic form of sex determination in animals and in some plants. This involves a structural difference between the sex chromosomes that can be observed cytologically. A homologous pair of sex chromosomes may be unequal in size and shape in one sex (**heteromorphic**) but equal in size in the other (**homomorphic**). The sex that harbors the heteromorphic sex chromosome pair was called by Wilson (1911) the **heterogametic sex** because during meiosis it produces two types of gametes, one male determining and one female determining. He called the sex with the homomorphic chromosome pair the **homogametic sex** since it produces only one type of gametes. In most vertebrate and in many insect species, the heterogametic sex (XY) is the male, and the homogametic sex (XX) is the female. However, in birds, in some moths, and in fishes, amphibians, and reptiles, for instance, the opposite relation is true in that the male is the homogametic sex (XX). The homologies of the X and Y chromosomes can vary from species to species. The sex chromosomes (X and Y) may have a long **homologous region** and a very short **differential region** or a short homologous region and a long differential region. In some species such regions have been mapped according to their size. In *Melandrium album,* for instance, the homologous or pairing region is very short, proportionally speaking. In Fig. 5.1 a diagram of the X and Y chromosomes of *Melandrium* is presented.

As shown in Fig. 5.1, there are male and female supressor regions as well as male and female promoting regions on the sex chromosomes, depending on the role of

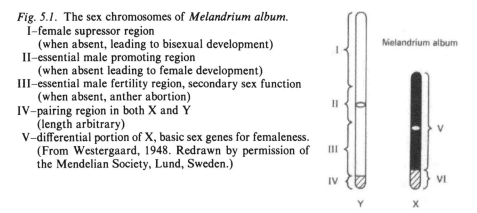

Fig. 5.1. The sex chromosomes of *Melandrium album.*
  I–female supressor region
    (when absent, leading to bisexual development)
  II–essential male promoting region
    (when absent leading to female development)
  III–essential male fertility region, secondary sex function
    (when absent, anther abortion)
  IV–pairing region in both X and Y
    (length arbitrary)
  V–differential portion of X, basic sex genes for femaleness.
    (From Westergaard, 1948. Redrawn by permission of
    the Mendelian Society, Lund, Sweden.)

these chromosomes in sex determination. Several theories on sex determination have been developed by different investigators. Naturally, these theories depended greatly on the organisms that were investigated by these different researchers.

5.1.1 Bridges' Balance Theory

In 1932 Bridges developed the **balance theory.** He studied *Drosophila,* which normally has 2n = 8 chromosomes, but for his studies he used individuals that had 2n = 12 chromosomes called **triploids** (Chapter 16). Bridges crossed these individuals with normal diploid males and received several different chromosome combinations in the offspring. Depending on the balance between the X chromosomes and the autosomes (A), Bridges observed different degrees of maleness or femaleness. He expressed this balance in an X/A ratio between X chromosomes and autosomes. The Y chromosome did not seem to have any effect on sex determination in *Drosophila.* The results of this study are shown in Table 5.1. As seen

Table 5.1. Chromsome constitution and sex in *Drosophila* (A = one set of autosomes)

Chromosome constitution	Sex	X/A ratio
2A XXX	Superfemale	1.5
2A XX 2A XXY 3A XXX 4A XXXX	Female	1.0
3A XX 3A XXY	Intersex	0.67
4A XXX	Intersex	0.75
2A X 2A XY 2A XYY 4A XX	Male	0.50
3A X	Supermale	0.33

in a normal female, the balance between two sets of autosomes and two X chromosomes produces femaleness ($X/A = 1$), while in a normal male, the autosomes outweigh the X chromosomes ($X/A = 0.5$) resulting in maleness. The X chromosomes seem to influence female development, while the autosomes seem to influence male development. Other imbalances produce superfemales, supermales, and intersexes.

5.1.2   Goldschmidt's Theory

Goldschmidt's work (1934) was carried out on the gypsy moth, *Lymantria dispar*. In this organism the male is homogametic (XX), and the female is heterogametic (XY). The gypsy moth was chosen for these studies because intersexes (gynandromorphs) are common in this species. Goldschmidt concluded that in *Lymantria* the sex was determined by the relative strength or balance of a female-determining factor (F), which is inherited from the mother, and male determining factors (M), which are located on the X chromosomes. At different times Goldschmidt thought that the F factor was either carried on the Y chromosome or in the cytoplasm. Figure 5.2 depicts the second contention. In this scheme the female formula is F/M (F>M), and the male formula is F/MM (MM>F). F- and M-determiners differed in potency in different races, their relative strengths were approximately the same in every race. There were strong and weak F's and strong and weak M's. Intersexes occurred regularly among the offspring of crosses between different geographical races. If a female with strong F and weak M was crossed with races that had M's of different strength, the XX offspring, even though being genetically males (chromosomal sex), were not necessarily phenotypical males. If the introduced M-determiners were strong, the offspring was male; if the introduced M-determiners were weak, the phenotypical expression of the offspring was female. Intersexes occurred when male- and female-determiners were in balance. Goldschmidt concluded further that the different strengths of M depended on 13 genes, the different strengths of F on 8 genes or cytoplasmic dosage factors. Goldschmidt (1915, 1916, 1920, 1922, 1923, 1929, 1934) and his coworkers also

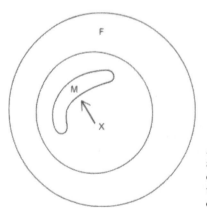

Fig. 5.2. Sex expression in *Lymantria dispar* according to Goldschmidt (1934). The female-determining factor is carried in the cytoplasm, the male-determining factor in the X-chromosome.

proposed a mechanism by which these sex factors influenced the development of the sex phenotype. He called this mechanism the **time law.** He assumed that the intersexes begin their development as females or as males (chromosomal sex) and develop as such to a certain point, called the **turning point,** after which they developed into the opposite sex. The degree of intersexuality is determined by the timing of the switch-over in differentiation.

5.1.3 Pipkin's Theory

Pipkin (1940, 1942, 1947, 1960) concluded from her work that the sex in *Drosophila melanogaster* was decided by a balance between male-determining factors in the second and third and female-determining factors in the X chromosomes (Fig. 5.3). Her work was based on a method developed by Dobzhansky and Schultz (1934) that added or subtracted broken pieces of an X to the normal 2X of triploid intersexes (see Table 5.1). The feminizing effect of the X portion could be measured by the degree of intersexuality in the flies. No single female sex genes could be located through these very thorough studies. The feminizing effect of the extra X sections was proportional to the size of these sections. It was concluded that many female sex genes were spread over the X chromosome. The conclusion that the male-determining factors were located in the third chromosome came from the following process of elimination: Bridges (1922) originally assumed that the male-determiners must be carried in the autosomes. Obviously, the fourth chromosome must be disregarded as a carrier because haplo-IV and triplo-IV *Drosophila* individuals do not affect the intersexes. Pipkin (1947) carefully checked the second chromosome with the translocation and triploid method and did not find any sex-determiners in it. She likewise did not find sex-determiners if the second chromosome only was involved (Pipkin, 1960). Her suggestion was that both the second and the third chromosomes are responsible for the shift toward maleness found in ordinary 2x3A triploid intersexes (see Table 5.1).

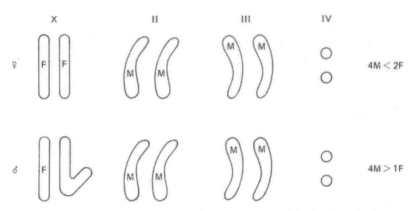

Fig. 5.3. Pipkin's (1947) theory on sex determination in *Drosophila.* Explanation in the text.

5.2   The Function of the Y Chromosome

The function of the Y chromosome varies according to the organism. An excellent review on this subject is published by Dronamraju (1965). This chromosome may vary from bearing some functional units of great importance to being completely lost as in the already mentioned X-O system. The Y chromosome mostly has a high proportion of **heterochromatin,** which is generally considered as having a high degree of genic inertness. Historically, the Y chromosome was considered to contain degenerate genes or no genes at all (Muller, 1914a, 1914b). This idea was based on some of the early discoveries. In *Drosophila,* flies without Y chromosomes (XO) were viable, but flies without X chromosomes (YO, YY) were inviable. The Y chromosome consequently was not necessary for survival. Females with additional Y chromosomes (XXY) were indistinguishable from normal flies (Muller, 1914a, 1914b). Males with two Y's (XYY) also did not show any different morphology. As shown in Table 5.1, the Y did not seem to have any appreciable effect on the sex expression of *Drosophila.*

The sudden discovery in 1959 (Ford et al.; Jacobs and Strong) that the Y in man is strongly male determining drastically changed the earlier conclusions. Later it was found that even XXXXY individuals and mosaics of the type XXXY/XXXXY/XXXXXY are phenotypically male in man (Anders et al., 1960). Work with mice, cats, and other mammals also indicated that the Y is male-determining in these animals.

In the flowering plant wild campion of the pink family, *Melandrium diocium,* as in man and mammals, the Y chromosome is also strongly male-determining. This is the only plant species in which the function of the Y chromosome has been intensively investigated (Warmke, 1946; Westergaard, 1958). It is interesting that this study predates the findings in man by 14 years, but it did not have the same impact as the discovery in man. In *Melandrium* the XYY, XXY, and XXXY types are all male, but the XXXXY is hermaphrodite. The Y chromosome is larger than the X chromosome in this species (Fig. 5.1). In many species the Y chromosome is much smaller than the X, which also has been used as an argument of its relative inertness.

Very few genes have been located on the Y chromosome. They are referred to as **holandric genes** (Enriques, 1922). The first Y-linked gene in any species was the one for a black pigment spot in the fish, *Lebistes reticulatus* (Schmidt, 1920). Only two Y-linked characteristics are presently listed for the human gene map (McKusick and Ruddle, 1977). They are the *histocompatibility gene* (*H-Y*) and the *testis determining factor* (*TDF*). From studies of chromosome aberrations it was concluded that these two genes may be at the same locus on the short arm of chromosome Y close to the centromere (Wachtel et al., 1976). *H-Y* regulates immunological properties of histocompatibility antigens. Histocompatibility antigens determined by the Y chromosome have also been reported for mice, rats, and guinea pigs (Wachtel et al., 1974). Other characteristics were located on the human Y chromosome at various times but firm evidence is lacking.

5.3   Dosage Compensation

The term **dosage compensation** was coined by Muller et al. (1931) in order to account for the fact that in *Drosophila* there must exist mechanisms that equalize the effective dosage of sex linked genes in the male (XY, XO) and female (XX) organisms. For instance, in *Drosophila* the X chromosome carries many sex linked genes that are not responsible for sex expression. Such genes do not have corresponding alleles on the Y chromosome and are consequently present in a **hemizygous** condition in the males. However, males and females are morphologically and physiologically so similar in expression that it seems that one gene dosage is as effective as two. We will see later that the effect of the dosage has an appreciable effect on gene expression in individuals that have missing or additional chromosomes (Chapter 17). In man even the monosomic or hemizygous condition of the smallest chromosome (one-fourth of the size of the X) is lethal. The fact that the hemizygous (XY, XO) and disomic (XX) conditions of the X chromosome have similar phenotypic expression has been explained with some kind of dosage compensation mechanism.

In *Drosophila,* dosage compensation has been explained to be the result of the action of modifier genes, so-called **dosage compensation genes,** on the X chromosome that cancel the effect of different doses of a given gene (Muller, 1947). The dosage compensation mechanism of *Drosophila* appears to operate by forcing a given X-linked gene in the XY-male to work harder, while restraining the activity of the same X-linked gene on each of the two X's of the female (Ohno, 1967).

5.3.1   The Single Active X Hypothesis

In man and mammals another mechanism seems to provide for the inactivation of the second X chromosome as a means of dosage compensation. This mechanism is called the **single active X hypothesis** or **Lyon hypothesis** (Lyon, 1961, 1962a, 1962b, 1963, 1970, 1971, 1972) mentioned in Chapter 1. The Lyon hypothesis makes the following conclusions:

1. In XY-males, the single X chromosome is active in all cells, while in each cell of the female (XX) one of the two X chromosomes becomes inactivated.
2. Paternal and maternal X chromosomes have an equal chance of being inactivated.
3. Inactivation occurs early in the life of the female embryo. This implies that in XX organisms both X chromosomes are euchromatic and active in RNA synthesis during early embryonic development.
4. Once it has been decided which X chromosome is inactivated in a cell, the same X chromosome will always be inactivated in the descendants of that cell.
5. The inactive X chromosome becomes heterochromatinized and forms the sex chromatin found in interphase, which is believed to be the late replicating X chromosome (Ohno and Hauschka, 1960).

Lyon's hypothesis is derived from her studies of the X-linked coat color genes in mice. These dominant genes produce different phenotypes in males and females. If males, for instance, carry the gene for mottled (*Mo*), the mice will have a uniform coat color (*Mo/Y*). If females are heterozygous for mottled (*Mo/+*), they have a

Fig. 5.4. A female mouse heterozygous for the X-linked gene, *dappled* ($Mo^{dp}/+$), having a variegated coat with particles of mutant and wildtype color. (After Lyon, 1966. Reprinted by permission of Paul Elek Limited, London).

variegated coat with patches of mutant or wild-type color. The same pattern is produced in females carrying the gene for dappled (Mo^{dp}) (Fig. 5.4). The pigmented wild-type patches ($+$) descended from cells in which the X chromosome carrying the mutant gene (Mo) was inactivated. The mutant patches (Mo) originated from cells in which the X chromosome carrying the wild-type gene ($+$) was inactivated.

Application of the Lyon hypothesis to human beings has come from the study of cultured skin fibroblasts (Beutler et al., 1962; Davidson et al., 1963). They found electrophoretic variants of glucose-6-phosphate dehydrogenase (*G6PD*). Clones from *G6PD* heterozygous females were 50% normal and 50% deficient.

5.3.2   Sex Chromatin and Drumsticks

The discovery of sex chromatin or **Barr bodies** by Barr and Bertram (1949) in cats is closely related to the phenomenon of dosage compensation. Barr's discovery was based on some earlier findings. During the first decade of this century, Montgomery (1904, 1906) discovered the **heteropycnotic** behavior of the X chromosome in the male germ line of the hemipteran insect *Pyrrhocoris*. Heteropycnotic chromosomes are those that are out of phase (**allocycly**) if they are compared with the **coiling cycle** in which the rest of the chromosomes of the set are engaged. They are also out of phase in respect to their staining properties. **Positive heteropycnosis** is the condition of the chromosome when it is densely coiled; **negative heteropycnosis** is the condition when it is less spiralized. These terms correspond to the expressions **heterochromatin** and **euchromatin,** mentioned in Chapters 1 and 2, that designate the staining properties of chromosomes. Densely coiled, positively heteropycnotic material is considered to be heterochromatic or darkly staining and less spiralized. Negatively heteropycnotic material is considered to be euchromatin; it has normal and less darkly staining properties in interphase and prophase.

In order to understand the nature of sex chromatin, one should distinguish between **facultative** and **constitutive** heterochromatin (Brown, 1966). Facultative heterochromatin is euchromatin that can be heterochromatinized during the cell cycle. Constitutive heterochromatin is always heterochromatinized, and it is the usual form. It is the chromatin that is found in the centromere regions, near the telomeres, in the satellites and in the nucleolus organizer region. Constitutive het-

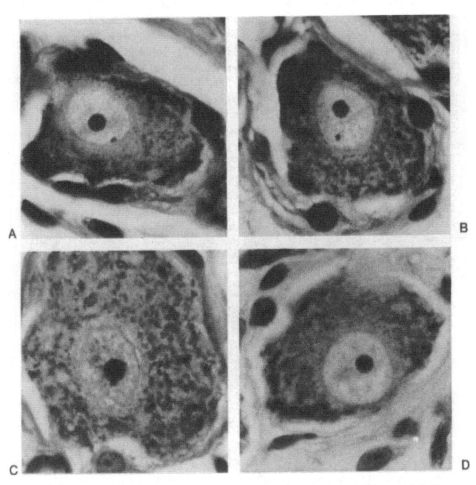

Fig. 5.5A–D. Different locations of the sex chromatin in the neurons of the cat. (*A*) Close to the inner surface of the nuclear envelope. (*B*) Free in the nucleus. (*C*) close to the nucleolus. (*D*) no sex chromatin present ($\times$ 1600). (Courtesy of Professor M. L. Barr. Reprinted by permission of Academic Press, New York).

erochromatin has also been termed **satellite DNA** (Yunis and Yasmineh, 1970, 1971). It is thought that satellite DNA is composed of relatively short, repeated polynucleotide sequences (Walker and McClaren, 1965; Britten and Kohne, 1968). This heterochromatin is also referred to as **redundant** or **repetitive.** It occurs commonly among eukaryotes. It comprises some 10% of the genomes of higher organisms. Both chromatins have the formation of interphase chromocenters and late DNA replication in common.

The discussion of sex chromatin is mainly related to the facultative heterochromatin. Sex chromatin is found in 20% to 96% of the nuclei of all females in humans and many mammalian species, but it is absent or rarely found in the nuclei of males of the same species. Sex chromatin is generally located at the periphery of the interphase nucleus away from the main chromatin mass just inside and close to or flattened against the inner surface of the nuclear envelope (Fig. 5.5). But it also can be located at other sites of the nucleus. The size of this body is about 0.8x1.1 μm. As mentioned, the first suggestion of the possible relationship between the sex chromatin and one of the two X chromosomes present in females was made in 1959. The basis for this assumption was the fact that whenever two Xs were found in the karyotype of an organism, a Barr body was detected in interphase nuclei. A minimum of two X chromosomes is the prerequisite for the presence of a Barr body, and if more than two X chromosomes are present, more than one Barr body can be expected in at least some of the interphase nuclei. The ease with which sex chromatin now can be detected as in scrapings from the oral mucosa (Marberger et al., 1955; Moore and Barr, 1955) makes it a most valuable tool for sex diagnosis.

Another method for sex diagnosis is the determination of the presence or absence of the so-called **drumsticks.** These were first discovered by Davidson and Smith (1954) in the circulating polymorphonuclear neutrophil leucocytes of human blood. These are drumstick-like chromatin appendices that are attached by a fine chromatin thread to one lobe of the polymorph nucleus. Drumsticks do not vary with

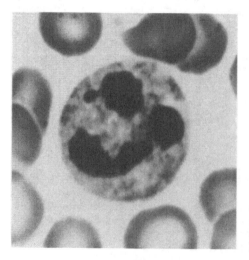

Fig. 5.6. Chromatin appendixes called "drumsticks" of the neutrophil leucocytes of human female. Drumsticks do not vary with age. (× 2666). (Courtesy of Carolina Biological Supply Co., Burlington, N.C.).

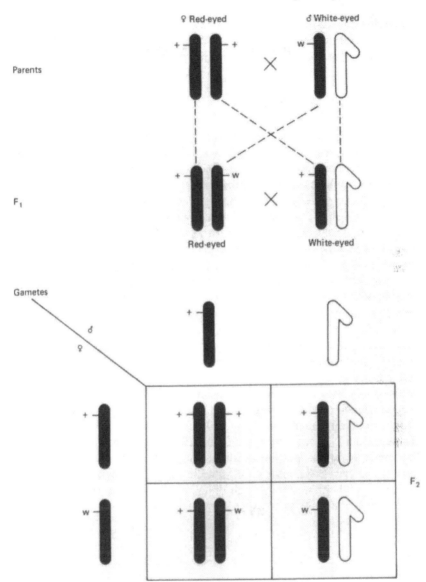

Fig. 5.7. Diagram of a cross between a red-eyed female (+/+) and a white-eyed male (w/y), the resulting F₁ (+/w, +/y) and the F₂ from a cross between the F₁ individuals. This diagram demonstrates sex linkage. (After Morgan, 1910b).

age (Fig. 5.6). Their size is 1.4 μm to 1.6 μm in diameter. Drumsticks are found in about one out of 40 leucocytes of normal females and in less than one in 500 cells of normal males. It has been hypothesized that drumsticks correspond to sex chromatin in that they represent the heteropycnotic region of the X chromosome. It is, however, doubtful if such a conclusion can be made at this time.

Sex chromatin and drumsticks are both valid structures for sex diagnosis although drumsticks are much less useful and interpretation is subject to error. Therefore, sex chromatin is more widely used in clinical tests than drumsticks and requires less skill in observation and interpretation (Hamerton, 1971a).

5.4   Sex Linkage

As was mentioned before, only a small number of loci on the sex chromosomes are probably responsible for sex inheritance. The majority of sex-linked loci are responsible for other morphological and physiological expressions, and most of those are located on the X chromosome. Consequently, the function of the sex chromosomes in general, and of the X chromosome specifically, is similar to that of the autosomes with the exception that the inheritance pattern is different.

The first X-linked character was reported by Morgan (1910b) in *Drosophila*. He discovered a white-eyed mutant male individual (w/Y) in a culture of normal red-eyed flies ($+/+$). By mating this abnormal fly to a red-eyed, he found a new mode of inheritance, which initiated the study of sex linkage. All F_1 flies had the normal wild-type red-eye characteristic (Fig. 5.7). When the F_1 females ($+/w$) were mated with the F_1 males ($+/Y$), the offspring (F_2) did not yield the normally expected 3:1 ratio for males and females. Instead, all the F_2 females were red-eyed and half of the F_1 males were red-eyed and half white-eyed. Morgan concluded that the mutation must have occurred on the X chromosome instead of on one of the autosomes.

Other X-linked characteristics were determined following this discovery. Some X-linked genes are shown in Figure 4.7.

The underlying principle of X-inheritance is the absence of father-to-son gene transmission. The reason for this is that the X chromosome of the male is not transmitted to any of his sons but to all his daughters. This kind of inheritance is also called the **crisscross pattern of inheritance.** Characteristics occurring in the fathers are inherited through their daughters who inherit them to the grandsons who express them and so on.

Part IV
Movement of Chromosomes

Chapter 6
Chromosomes During Mitosis

In this part on the movement of chromosomes, the so-called **normal behavior** of the chromosomes during the processes of cell division and cell union will be discussed. These two processes guarantee the continuation of species from one generation to the next.

In cell division, there exist two major phases, **karyokinesis** (or **mitosis**) and **cytokinesis.** The term **mitosis** is usually preferred over **karyokinesis.** During mitosis the hereditary information that is contained in the chromosomes is passed on to the daughter nuclei. During cytokinesis, which usually follows mitosis, the cytoplasm and its inclusions are divided, finalizing cell reproduction. Since the chromosomes, as the carriers of the hereditary units, are the major topic of discussion, only mitosis is treated in this chapter. The principal stages of mitosis are **prophase, metakinesis, metaphase, anaphase,** and **telophase.** However, in order to understand the entire cell cycle, **interphase** is included in this discussion. Figure 6.1 depicts the different mitotic stages in an animal cell.

6.1   Interphase

In the interphase nucleus, the chromatin, the substance that contains the genetic material, appears to be dispersed throughout the entire nucleoplasm. The nucleolus and the chromocenters (Chapter 2) stand out by their dense staining properties.

In terms of the cell cycle, the interphase takes up a relatively long time interval. The reason for this long period is that during interphase some very important functions of the cell cycle are taken care of, such as metabolism and synthesis. The interphase period is generally subdivided into three phases—the G_1 period, the S period (synthesis), and the G_2 period (Howard and Pelc, 1953). The length of these periods varies according to species and probably also according to different tissues of the same species. In Fig. 6.2, the relative duration of these interphases is indicated along with the estimated period of mitosis for cells such as cultured cells of man and hamster, root tip cells of the broad bean *(Vicia faba),* and spermatogonia and spermatocytes of grasshoppers. The duration of the cell cycle in these organisms is 18 to 19 hours, while the S period lasts 6 to 8 hours. During the S period, some chromosomes are early replicating and others are late. Heterochromatic chromosomes such as the allocyclic X chromosome mentioned in Chapter 5 are usually

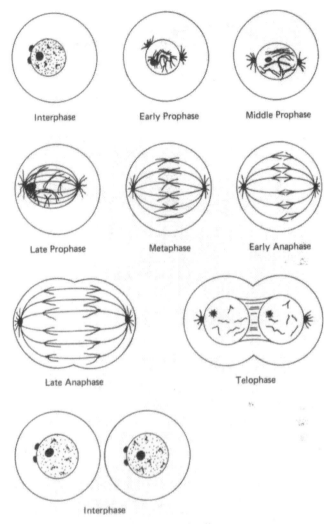

Fig. 6.1. Mitotic stages in an animal cell.

late replicating. The S period is usually preceded by a presynthetic gap period called G_1. The physiological condition of the cell determines the length of the G_1 period. Some cells such as inactively growing yeast may not have a G_1 period. During the G_1 period the chromosomes are not reduplicated. This period can be considered as a preparatory period for DNA synthesis. The chromosomes are released from their condensed condition, which they had assumed during mitosis. It has not been possible as yet to ascribe with certainty any specific biochemical events to either the G_1 period or the G_2 period (John and Lewis, 1969). Both RNA and protein synthesis are evident during the G periods as well as during the S period. Near the end of the G_1 period, the beginning of the S period is triggered by a yet unknown event. This is a period of active DNA synthesis during which the chromosomes replicate. The S period can be traced by the use of labeled DNA

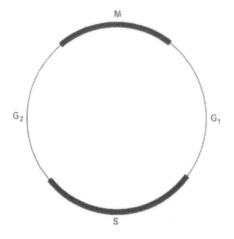

Fig. 6.2. Life cycle of dividing cells such as those of humans, hamster, root tips of the broad bean, and spermatogonia and spermatocytes of grasshoppers. Phases indicated are mitotic division (M), synthesis (S), gap between M and S (G_1), and gap between S and M (G_2). (After Swanson et al., 1967. Redrawn by permission of Prentice-Hall, Inc., Englewood Cliffs, N.J.).

precursors such as thymidine. Several enzymes catalyze the process of DNA replication. The most critical one is *DNA polymerase*. This enzyme copies each strand of the DNA double helix into a complementary strand. Since the DNA strand in a human chromosome has an estimated length of 30,000 μm and DNA replication occurs at a calculated speed of 0.5 μm per minute, DNA replication would take much too long to account for a typical S period of 6 to 8 hours, assuming replication proceeds sequentially from one end to the other. But it was discovered that DNA replication occurs at many different places along the chromosome at once. During synthesis short transient DNA fragments were observed that are called **Okazaki pieces** (Okazaki et al., 1968; Huberman and Riggs, 1968), which correspond to the earlier hypothesized **replicons** (Jacob and Brenner, 1963). Such Okazaki pieces are autonomous DNA units with an average estimated length of about 1000 nucleotides or 30 μm that are eventually joined together by the action of DNA ligases forming the final product, the complete DNA daughter strand.

The post-synthetic G_2 period is a gap between synthesis and mitosis. During this period the chromosomes are in a reduplicated state. Irradiation experiments have verified this. When irradiated during G_1, the chromosomes yield chromosme aberrations and when irradiated during G_2, chromatid aberrations. During the interphase directly preceding meiosis the G_2 period is either very short or completely missing. Duration of the interphase periods G_1, S, and G_2 in relation to the mitotic period have been determined for many species. A recent summary for higher plants was published by Van't Hof (1974).

6.2   Preparation for Mitosis

One of the preparatory phenomena of the cell for mitosis is cellular growth (Mazia, 1961). A product of a mitotic division such as a late telophase daughter cell usually almost doubles its volume by the end of interphase before it divides again. This applies particularly to cells in mitotically active tissues. It demonstrates that mitosis is a cellular process and not entirely limited to the nucleus. The preparation

Fig. 6.3. Cross section of a centriole from a human lymphocyte at the interphase stage. The 9 triplet fibers consist of 3
microtubules each. (× 316,000). (From Sitte, 1965).

for mitosis is going on continuously throughout the life of the cell. Preparation for
the next cell division already has started during the course of the previous division.
There are many preparations for a given division which all have to be completed
before the mitotic apparatus becomes functional. Bradbury et al. (1974a, 1974b)
proposed that during the G_2 period, the initiation of mitotic cell division is controlled by the level of the growth-associated enzyme F1 histone phosphokinase
(**HKG**) but direct proof is not yet available. Another preparatory process for
mitosis is the replication and poleward movement of the **centrioles** of the **centrosome** (Boveri, 1888).

6.2.1   The Centrosome

The centrosome, which contains the centrioles, is a region of clear cytoplasm adjacent to the outer side of the nuclear membrane of cells in many animals and in
some lower plants (Boveri, 1895). It was first described by Beneden in the 1870's
and by Boveri in 1888 in his famous *Cell Studies* mentioned in Chapter 1.
Centrioles have been studied in detail with the electron microscope (Fig. 6.3). They
are shaped like a short hollow cylinder about 300 nm to 800 nm long and 160 nm
to 250 nm in diameter. The wall of the cylinder contains 9 triplet fibers that consist
of 3 microtubules each. The microtubules are about 15 nm to 20 nm in diameter.
The three microtubules of each triplet fiber are arranged in a line tilted about 30°
to 40° to the tangent of the circumference of the centriole.
During the G_1 period of interphase, usually two centrioles are observed. They are
generally replicated during the S period. The ultrastructure of centriole replication
was first described by Bernhard and DeHarven (1960) and Gall (1961).
The new centriole arises as smaller **procentriole** at an angle greater than 90° to
the mother centriole forming an L-shaped angle with the mother (Fig. 6.4). The
diameter of the procentriole is almost equal to that of the mother, but the length
is only about 70 nm when it first becomes visible under the electron microscope.
The procentriole gradually increases in size until it reaches the dimensions of the
mother centriole. In most organisms, centriole replication is finished by the end of
interphase (G_2 period). At that time two pairs of centrioles or two **centriole
duplexes** are visible at one side of the nuclear envelope (Fig. 6.5).
At the beginning of prophase, one centriole duplex starts to move away from the
other one and migrates around the periphery of the nucleus. The other centriole

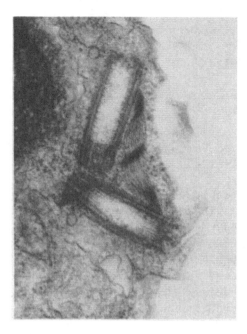

Fig. 6.4. Longitudinal sections of centrioles from the alga *Nitella*. The new centriole arises at an angle greater than 90° to the mother centriole forming an L-shaped angle with the mother. × 52,200. (Micrograph from Turner, 1968. Reprinted by permission of the Rockefeller University Press, New York).

duplex remains in its previous position. During this time small spindle fragments appear between the separating centriole pairs.

At the beginning of metaphase the migrating centriole pair has obtained a position opposite to its two partners. A complete spindle fiber system formed by the **asters** now stretches from centriole pair to centriole pair. During anaphase the centriole pairs seem to be pushed farther apart by the continuous spindle fibers.

There may be some exceptions to this general centriole behavior described above.

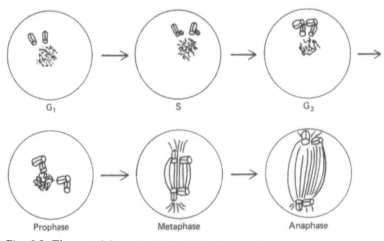

Fig. 6.5. The centriole cycle during interphase and cell division. (Modified after *Elements of Cytology*, Second edition by Norman S. Cohn, 1969, by Harcourt Brace Javanovich, Inc. Redrawn by permission of the publisher).

The centriole duplex separation may be delayed beyond the beginning of prophase (Fig. 6.5). Both centriole duplexes may migrate around the cell as a unit with little change in their center to center spacing prior to separation. Duplex separation may occur at any point within the mid-prophase-prometaphase period (Rattner and Berns, 1976).

The function of centrioles seems to be connected with the formation of the spindle fiber mechanism and the detecting of chromosome migration during mitosis. The centriole is the place where the **tubulin** (Borisy and Taylor, 1967) is assembled and serves as the center of the spindle microtubules (spindle fiber) organization. Tubulin is the subunit protein of the microtubules.

6.3   Prophase

Prophase is the beginning stage of mitosis. During this stage the chromosomes become visible as thin threads. This is accomplished by progressive coiling and folding. Each prophase chromosome now consists of two adjacent chromosome threads called **chromatids,** which are the result of chromosome reduplication during the S period of interphase. The coiling and folding transform the largely extended metabolic chromosomes into a shape suitable for transport. A diagrammatic representation of the coiling cycle is shown in Fig. 6.6. During early prophase the two chromatids are twisted about each other in **relational coils** (No. 2, Fig. 6.6). The coils interlock in such a manner that the chromatids cannot be separated without unwinding the coil. This kind of twisting is also called **plectonemic coiling** (Fig. 6.7A). Such chromatid association is different from the one occurring in meiotic prophase, which is called **paranemic coiling** (Fig. 6.7B) where the chromatids are easily separated laterally (Sparrow et al., 1941). As prophase proceeds and the chromosomes become shorter, the relational coiling disappears, the chromatids disengage themselves and lie side by side (No. 4, of Fig. 6.6). Later, during the coiling cycle (metaphase and anaphase), two levels of coiling can be seen (No. 6, Fig. 6.6). The large coils are called **somatic coils,** the small ones that are imposed upon the large ones are called **minor coils.** As seen in the diagram, the initially small somatic coils decrease in number with progressing prophase and at the same time increase in diameter. This causes an apparent thickening of the chromosomes that is often referred to as contraction.

During prophase, the nucleolus of most species breaks down and disappears. Electron microscopic studies have revealed that the component parts of the nucleolus disperse throughout the nucleus during this stage. In some lower forms of life the nucleolus persists through metaphase and anaphase and divides into two halves that are distributed to the daughter cells.

At the end of prophase, the nuclear envelope breaks down into fragments. This allows the chromosomes to spread over the greater part of the cell and gives them a better chance to separate as chromatids during poleward movement. Electron microscopic investigation seems to prove that pieces of the nuclear envelope disperse into the cytoplasm and become part of the **endoplasmic reticulum.** It is possible that the nuclear envelope originates from the endoplasmic reticulum (Bern-

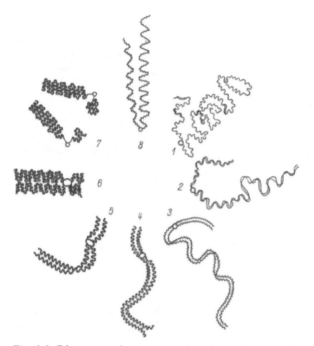

Fig. 6.6. Diagrammatic representation of the mitotic coiling cycle of a chromosome. Centromeres are shown as circles—(1) Interphase (2, 3, and 4) Prophase (5) Prometaphase (6) Metaphase: metaphase chromatids show major and minor coils (7) Anaphase (8) Telophase. (Modified after De Robertis et al., 1965).

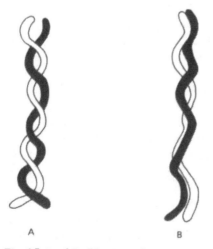

Fig. 6.7 A and B. Diagram of two possible types of coiling between chromosomal subunits. (*A*) Plectonemic coiling. (*B*) Paranemic coiling.

hard, 1959; Whaley et al., 1960; Porter, 1961). Protozoa and fungi comprise an exception in that the nuclear envelope remains intact throughout the entire mitotic division.

Right after the disappearance of the nuclear envelope, the spindle fiber apparatus appears.

6.4 Metakinesis

The term **metakinesis** was first used by Wassermann in 1926 and was brought into popular use by Mazia (1961). Many textbooks do not distinguish between metakinesis and metaphase but include the discussion of both of these stages under metaphase. But many of the important features of metakinesis are omitted from the discussion if it is not considered separately. Metakinesis is often also referred to as **prometaphase** (Lawrence, 1931).

Darlington (1937) divided the movement of the chromosomes during the metakinetic stage into three substages:

1. chromosome congression
2. centromere orientation
3. chromosome distribution

6.4.1 Chromosome Congression

During **chromosome congression** the chromosomes move to the equatorial plate half way between the two poles of the spindle where the centriole pairs are located. The chromosomes reach a position of equilibrium at the equatorial plate. In general, this movement is coordinated except in instances where small chromosomes perform their movement out of step with the large ones in the same complement. Detailed cinematographic studies by Bajer and Molè-Bajer (1954, 1956) show that individual chromosomes may move toward the pole at first, then make a turn and finally arrive at the equator. The movement toward the equator may be very abrupt. The chromosomes are now freely floating in the cytoplasm, unrestricted by the nuclear envelope. In the grasshopper neuroblast, this metakinetic movement lasts only four minutes out of a total duration of mitosis of three hours (Carlson and Hollaender, 1948). It is possible that the spindle fibers are required for the movement of the chromosome toward the equatorial plate since they are not able to migrate when the spindle has been destroyed with the spindle fiber poison colchicine (O'Mara, 1939; Eigsti, 1942; Berger and Witkus, 1943; Allen et al., 1950; Hadder and Wilson, 1958; Malawista et al., 1968).

6.4.2 Centromere Orientation

This metakinetic movement was described in detail by Darlington (1936). It deals with the orientation of the **kinetic sites** of the chromosomes toward opposite poles through movements that lead toward their orderly arrangement in the equator (coorientation). Each metaphase chromosome consists of two chromatids and each

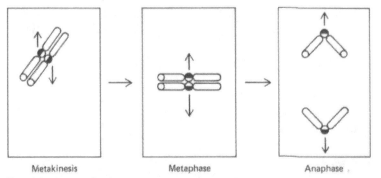

Metakinesis   Metaphase   Anaphase

Fig. 6.8. Auto-orientation of mitotic chromosomes. (From Rieger and Michaelis, 1958).

of these chromatids has a kinetic site and an **akinetic site** (Fig. 6.8). The kinetic sites are oriented toward the poles through forces that originate from the poles and that very well could be the spindle fibers. While the chromosomes previously were located in the cell at random, centromere orientation places them into a stable equilibrium at the equator. According to Darlington, the same forces that accomplish centromere orientation are responsible for moving the chromosomes toward the poles at anaphase (Fig. 6.8).

6.4.3   Chromosome Distribution

The third process of metakinetic movement is **chromosome distribution.** The centromeres come to be oriented in such a way that their corresponding chromosomes are more or less evenly distributed on the equatorial plate. Darlington thinks that this even distribution of the chromosomes is caused by some kind of body repulsion. The chromosomes are not always distributed at random on the equatorial plate, but they may be subject to specific arrangement. This was already known by such prominent cytologists as Wilson (1925) and Schrader (1953). The metaphase chromosomes of many insects are arranged in such a way that the larger chromosomes lie on the periphery of the equatorial plate while the smaller chromosomes lie in the middle. On the other hand, there is also the reverse tendency such as observed in human cells were the largest chromosomes, numbers 1 and 2, were found near the middle while the smaller chromosomes, Y and numbers 13, 17, 18, and 21, lay near the periphery (Miller et al., 1963).

6.5   Metaphase

At metaphase the chromosomes are at their highest level of coiling and therefore appear to be shorter and thicker than in any other stage. This makes them ideal for cytotaxonomic studies because they are most sharply defined during this stage (Chapter 2, Section 2.1). There is no longer much relational coiling present, and, consequently, the chromatids are no longer twisted about each other but lie side

Fig. 6.9A and B. C-pairs in mitotic metaphase of *Allium cepa* (2n = 16). Four hours of treatment with 0.2% colchicine solution. (*A*) Uncoiling of colchicine treated chromosomes has reduced the number of turns in each chromosome arm producing figure-8 and forceps-types. (*B*) Cross-type c-pairs are only connected at the centromere. (× 1258). (Schulz-Schaeffer, unpublished).

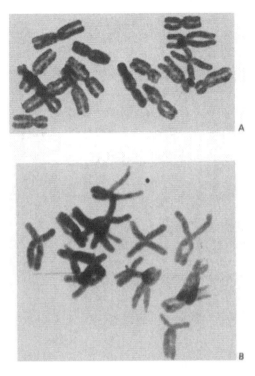

by side (No. 6, Fig. 6.6). Proof of this conclusion is the ease of separation of the chromosome arms as a result of colchicine treatment, which leaves the chromosomes only attached at the undivided centromeres. Such colchicine influenced chromatid associations in metaphase are called **c-pairs** (Fig. 6.9) and have a cross-shaped appearance (Levan, 1938).

Metaphase is much shorter than prophase but on the average somewhat longer than anaphase. Table 6.1 gives a comparison of the length of mitotic stages of different tissues of a number of animal and plant species.

The end of metaphase is signaled by an almost simultaneous splitting of the centromeres and separation of all sister chromatids at the centromeres. Brown (1972) writes that somehow all the chromosomes *know* when to separate and start anaphase and that they all do this at the same time even if most of them have to wait for one **laggard** to arrive late on the metaphase plate. The controlling mechanism for this separation remains to be discovered.

6.6   Anaphase

Anaphase is a stage of active and rapid movement and is the shortest of all mitotic stages (Table 6.1). During this stage the spindle elongates and the centriole duplexes—if present—move closer to the cell periphery (Fig. 6.5). As the centro-

Table 6.1. Comparative duration of the mitotic phases, representing direct observations (or cine records) of living dividing cells (Mazia, 1961)

| Cell | Minutes | | | | References |
	Prophase	Metaphase	Anaphase	Telophase	
Yoshida sarcoma (35°C)	14	31	4	21	Makino and Nakahara (1953)
MTK-sarcoma I (35°C)	10	44	5	18	—
Mouse spleen in culture	20–35	6–15	8–14	9–26	Hughes (1952)
Triton liver fibroblast (26°C)	18 or more	17–38	14–26	28	Hughes and Preston (1949)
Chortophaga (grasshopper) neuroblast (38°C)	102	13	9	57	Carlson and Hollaender (1948)
Pea endosperm	40	20	12	110	Bajer and Molè-Bajer (1954)
Iris endosperm	40–65	10–30	12–22	40–75	Bajer and Molè-Bajer (1954)
Micrasterias rotata (desmid)	60	21–24	6–12	3–45	Waris (1950)

meres become functionally double at the end of metaphase, the chromatids immediately move toward opposite poles. The newly formed daughter centromeres lead in this poleward movement while the two chromatid arms drag passively behind. Exceptions to this will be discussed later. Two types of spindle fibers have been observed (Schrader, 1944), the **continuous spindle fibers** and the **chromosomal spindle fibers** (Fig. 6.10). The continuous spindle fibers connect the two polar

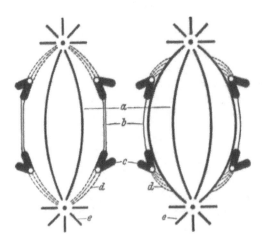

Fig. 6.10. The two main types of spindle fibres. a–continuous spindle fibres that connect the two polar regions with each other and persist from early prometaphase to early prophase; d–chromosomal spindle fibres that are directly attached to the kinetic sites of the chromatids; b–interzonal connections; c–chromosomes; e–spindle poles. (From Schrader, 1944).

regions with each other and persist from early prometaphase to early prophase. The chromosomal spindle fibers are directly attached to the kinetic sites of the chromatids. The anaphase movement of the chromosome is observed as a shortening of the chromosomal spindle fibers. There are several theories on the chromosome movement at anaphase but the mechanism is not known (Mazia, 1961). Depending on the position of the centromere on the chromatid (daughter chromosome), the lagging chromosome arms describe V shapes, J shapes, or rods. If an abnormal chromosome has two centromeres, called **dicentric,** the two centromeres usually move toward opposite poles forming an **anaphase bridge** that eventually tears apart somewhere between the two centromeres. If one of the two centromeres has weaker properties than the other, both centromeres may be included in one daughter nucleus. Acentric fragments may move passively along in the poleward current toward the poles but often lag behind in the equatorial plate and are not included in either **daughter nucleus.**

6.7 Telophase

Telophase usually is considered to start when the chromosomes reach the opposite poles. At that time the nuclear envelope reconstitutes around the two daughter nuclei, the nucleoli form at the distinct site of the nucleolus organizer chromosomes, and the chromosomes fuse into an indistinguishable mass of chromatin. Hydration and uncoiling of the chromatin threads aid in this process of reforming an interphase nucleus where the chromosomes lose their density and stainability. Also, the chromocenters reappear at this time. Cytokinesis and **cleavage** in animal cells complete cell division. In plants cytokinesis is completed with the formation of a **cell plate** that eventually will form a cell wall that cuts the cell into two parts. The telophase stage concludes the mitotic cycle and ushers in a new interphase period.

Chapter 7
Chromosomes During Meiosis

As mentioned in Chapter 1, the essential facts of meiosis and fertilization in animals and plants were demonstrated by Beneden (1883), Strasburger (1884), Boveri (1890) and Oscar Hertwig (1890). These investigators found that the most important result of fertilization was the fusion of gametes of maternal and paternal origin. Since the nuclei of a particular species maintain their constant chromosome number (2n) from generation to generation, they concluded that a mechanism had to operate that would compensate for the increase of chromosome number during fertilization. This mechanism was found to be a reduction of chromosomes before fertilization, as it occurs in **meiosis** of higher plants and animals. Other essential characteristics of meiosis are the pairing of chromosomes, which makes reduction possible, and crossing over as discussed in Chapter 4, which provides for recombination.

Meiosis includes two nuclear divisions that generally succeed each other rapidly and during which the chromosomes divide only once. In spite of modified types of meiosis having been observed, the details of the basic process are very similar for humans and for the majority of higher animals and plants. These two divisions have been called different names according to the different functions carried out during these divisions by the chromosomes. Names like **heterotypic** vs. **homeotypic** as well as **reductional** vs. **equational** division are some more familiar terms. The first of the two divisions has been called *heterotypic* because it is the more unusual one, while the second was called *homeotypic* because it is more similar to a normal mitotic division. The terms "reductional" for the first and "equational" for the second meiotic division are misnomers and would be correct if crossing over would not occur. In the presence of crossing over, however, portions of the chromosomes divide reductionally during the first meiotic division (**prereductional separation**), while other portions divide reductionally during the second meiotic division (**postreductional separation**). During reductional separation, homologous segments of non-sister chromatids disassociate, while during equational separation, homologous segments of sister chromatids separate. Logically, there is also **preequational** and **postequational separation.** These phenomena are illustrated in Fig. 7.1. During the first meiotic division (MI) of this illustration, a predominantly paternal chromosome (black) moves away from a predominantly maternal chromosome (white) in a reductional fashion, while portions that have been exchanged by crossing over separate from one another in an equational fashion (black from black and white

Fig. 7.1. Prereduction and postreduction in meiosis. Chromosome separation is predominantly reductional in meiosis I of this illustration (white segments separating from black ones). Only small segments separate equationally (white from white and black from black). Chromatid separation in meiosis II is predominantly equational in this illustration.

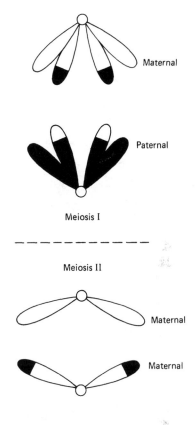

Maternal

Paternal

Meiosis I

Meiosis II

Maternal

Maternal

from white). During the second division a predominantly maternal chromatid (white) moves away from a predominantly maternal chromatid (white) in an equational fashion while small portions (black vs. white) divide reductionally. In reality, none of the two meiotic divisions are truly reductional or equational. The real condition depends on the crossover situation of a given meiosis.

The most commonly used nomenclature for the two divisions are the terms **meiosis I** (MI) and **meiosis II** (MII) or first and second meiotic divisions. In order to understand meiosis, some other terminology should be introduced here. The prophase of the first meiotic division usually is of long duration since **homologous chromosomes** (homologues) are pairing during this period in **synapsis.** Since each partner of such a pair of sister chromatids (reduplication occurred in the preceding synthesis period as described for mitosis), there are now four chromatids present immediately after synapsis. Such a group of four chromatids is generally referred to as a **tetrad chromosome** (Nemec, 1910) (Fig. 7.2). A tetrad chromosome is also called a **bivalent** referring to the two homologous chromosomes belonging to it. During anaphase I the tetrad chromosome separates into two **dyad chromosomes** (Nemec, 1910), each of which consists of two chromatids. Such a dyad chromosome can also be called a **univalent.** During anaphase II the dyad chromosomes divide into **monad chromosomes** or unit chromatids. This completes the entire cycle of two meiotic cell divisions (Fig. 7.2).

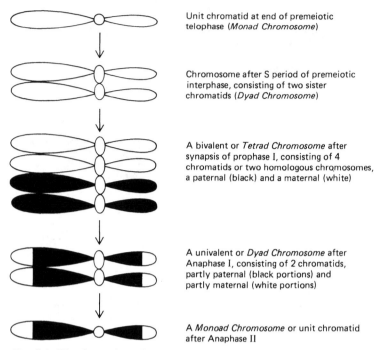

Unit chromatid at end of premeiotic telophase (*Monad Chromosome*)

Chromosome after S period of premeiotic interphase, consisting of two sister chromatids (*Dyad Chromosome*)

A bivalent or *Tetrad Chromosome* after synapsis of prophase I, consisting of 4 chromatids or two homologous chromosomes, a paternal (black) and a maternal (white)

A univalent or *Dyad Chromosome* after Anaphase I, consisting of 2 chromatids, partly paternal (black portions) and partly maternal (white portions)

A *Monoad Chromosome* or unit chromatid after Anaphase II

Fig. 7.2. The formation of dyad chromosomes, tetrad chromosomes, and monad chromosomes during a regular meiotic cycle.

Other relationships during meiosis and mitosis can be expressed in terms of total DNA present during these different stages. For this purpose the **C-value** has been employed. If C represents the amount of DNA in a haploid gamete before fertilization, then a gamete will have an amount of 1C and a zygote of 2C (Fig. 7.3). During the S period of premeiotic mitosis, the DNA content in the cell will rise to 4C. Mitotic anaphase will bring it back down to 2C. During the S period immediately preceding meiosis, it will rise back to 4C. Anaphase I will reduce the DNA content to 2C and anaphase II to the original 1C value.

Meiosis and gametogenesis (Chapter 8) occur only in special tissues of organisms that during development are differentiated and are set aside as gamete forming tissues. Weismann in 1883 and 1885 in his germ plasm theory (see Chapter 1) called this special tissue the **germ line.**

Meiosis like mitosis has been divided into stages and substages. They are called:

Premeiotic Interphase	Anaphase I
Prophase I	Telophase I
Leptotene	Interkinesis
Zygotene	Prophase II
Pachytene	Metaphase II
Diplotene	Anaphase II
Diakinesis	Telophase II
Metaphase I	

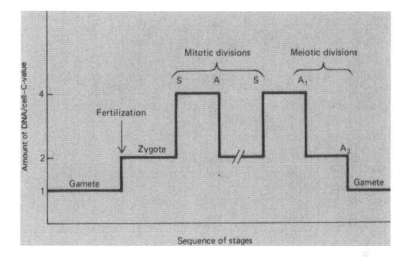

Fig. 7.3. Increase and decrease of the C-value during mitotic and meiotic cycles of animals and plants. S–synthesis; A–anaphase. (From Swanson et al., 1967. Redrawn by permission of Prentice-Hall, Inc., Englewood Cliffs, N.J.).

A schematic drawing of a homologous pair of chromosomes passing through these stages is represented in Fig. 7.4.

7.1   Premeiotic Interphase

Premeiotic interphase is very similar to premitotic interphase in that during the S period the chromosomes are reduplicated. Compared with the mitotic S period, the meiotic S period is longer.

In cells of anthers of lily *(Lilium longiflorum)* and tulip *(Tulipa gesneriana)*, a unique histone was found that was absent or nearly so from the somatic tissues of these plants (Sheridan and Stern, 1967). This histone was therefore called **meiotic histone.** It is synthesized during the S period of premeiotic interphase and persists through meiosis, microsporogenesis, and pollen maturation. The possible function of a meiotic histone is not known. But histones are thought to be involved in some way or another in the regulation of genetic activity. The discoveries of Huang and Bonner (1962) have been discussed in this connection in Chapter 1. We know that the entire process of cell division in general and of meiosis specifically is under rigorous control of certain genes or groups of genes. It is not hard to follow that a specific meiotic histone would have a certain function in the regulation of the rather specific phenomena of meiosis. Very few analyses of meiotic histones have been reported so far, and future research in this area seems promising.

The duration of the different stages of meiosis varies from species to species. A summary of data from some higher plants was recently published by Van't Hof (1974) and is shown in Table 7.1. If these data are compared with Table 6.1, it can be seen that meiosis in general, and its first division in particular, lasts considerably longer than mitosis.

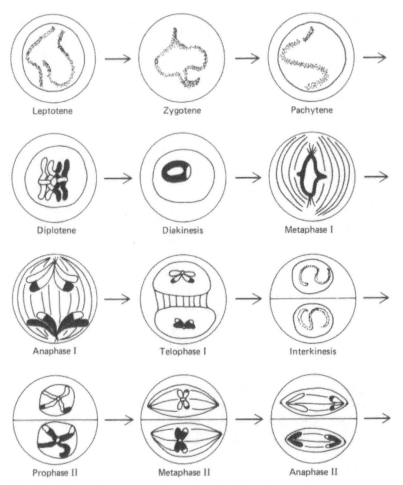

Fig. 7.4. Schematic representation of a pair of homologous chromosomes passing through the stages of meiosis.

7.2   Prophase I

An important feature of Prophase I is the great increase in volume of the nucleus. This increase is greater than that during mitosis and is partly due to an increase in hydration which is several times greater than in mitosis. According to Beasley (1938), the volume of the meiotic prophase nuclei in plants and animals is 3 to 4 times that of mitotic prophase nuclei. Prophase I of meiosis is also of extremely long duration if it is compared with mitotic prophase (Table 6.1 and Table 7.1). The chromosomes have to perform specific functions during this period that they do not have to carry out during mitotic prophase. Such functions are chromosome pairing, chromatid exchange, repulsion, and terminalization. According to these functions, prophase I is divided into 5 substages: leptotene, zygotene, pachytene, diplotene, and diakinesis. These stages are not rigid entities but rather arbitrary

Table 7.1. The duration of stages of meiosis of 14 plants[a]

Species	Leptotene	Zygotene	Pachytene	Diplotene	Diakinesis	Metaphase I	Anaphase I	Telophase I	Dyads	Metaphase II	Anaphase II	Telophase II	Total meiotic duration	Ploidy level
Antirrhinum majus	6.0	3.0	3.0						9.0				24.0	
Allium cepa		72.0							24.0				96.0	
Hordeum vulgare	12.0	9.0	8.8	2.2	0.6	1.6	0.5	0.5	2.0	1.2	0.5	0.5	39.4	2x
Lilium candidum	40.0	40.0	40.0						24.0				168.0	2x
Lilium longiflorum	40.0	36.0	40.0						48.0				192.0	2x
Rhoeo discolor		24.0							24.0				48.0	2x
Secale cereale	20.0	11.4	8.0	1.0	0.6	2.0	1.0	1.0	2.5	1.7	1.0	1.0	51.2	2x
Secale cereale	13.0	9.0	6.4	1.0	0.6	1.8	0.7	0.7	2.0	1.4	0.7	0.7	38.0	4x
Tradescantia paludosa	48.0	24.0	24.0						13.0				126.0	2x
Tradescantia reflexa		111.0							33.0				144.0	4x
Trillium erectum	70.0	70.0	50.0						64.0				274.0	2x
Triticum aestivum	10.4	3.4	2.2	0.6	0.4	1.6	0.5	0.5	2.0	1.4	0.5	0.5	24.0	6x
Triticale	7.5	3.0	2.3	1.0	0.5	1.8	0.5	0.5	1.5	1.3	0.5	0.5	20.8	8x
Tulbaghia violacea		102.0							28.0				130.0	2x

Duration of meiotic stages, hr

[a]Data from Bennett and Smith (1972); Bennett (1971,1972).

divisions of a continuum. They serve mainly to help the student to better understand the functions.

7.2.1   Leptotene

Leptotene does not differ very much from early prophase in mitosis, with the exception that meiotic prophase cells are larger than mitotic ones. Leptotene chromosomes are longer and thinner than those in early mitotic prophase. Also, during leptotene, bead-like structures, called chromomeres (Section 2.2.2), are appearing along the entire length of the chromosomes. Early observers, such as Belling (1928), considered these structures to be the visible manifestations of the genes, but other studies seem to substantiate that they are merely regions of the chromatin threads (chromonemata) that are more tightly coiled than the interchromomeric regions (Ris, 1945). In Crick's (1971) chromosome model (see Fig. 3.6), he interprets the gene regions (coding DNA) to be located in the interchromomeric regions (interbands). Other studies, such as the very thorough investigation of the *zeste-white (zw)* chromosome region of the *Drosophila* X chromosome (Judd et al., 1972), seem to indicated that there is a correspondence between the presence of an amplified chromomere (polytene band) and the presence of a gene. In other cases, several genes were located in a region of a polytene chromosome where only one band is present.

In some instances, particularly in animals, a certain type of polarization has been observed in this stage in which the ends of the chromosomes seem to be attached to the nuclear envelope (Moens, 1969b) at the site where the centrosome is located in animal cells. This has been referred to as the **bouquet stage** (Eisen, 1900), which can be observed in both leptotene and pachytene. A barley cell in pachytene suggesting such arrangement is shown in Fig. 7.5. It has been speculated that bouquet formation may aid in the union of homologous chromosomes during synapsis in the next stage, zygotene. A similar phenomenon has been described in plants where the chromosomes are densely clumped to one side leaving the rest of the nucleus clear. Such clumping into a more or less dense knot has been referred to as **synizesis** (McClung, 1905) or **synizetic knot** (Fig. 7.6). It has been claimed by some that synizesis may be due to a fixation artifact. An increase of size of the nucleolus during leptotene has been related to RNA and protein synthesis.

7.2.2   Zygotene

As in mitosis, the chromosomes during meiosis gradually become shorter in length and wider in diameter as a result of progressive coiling. Coiling mechanisms were described in the last chapter. Swanson (1957) states that in meiosis the coiling picture is comparable to mitosis but is more complicated and that this is due to the occurrence of synapsis and chiasma formation. A greater degree of contraction is attained by the chromosomes in meiosis. This greater contraction is partly accomplished by the **major coils** in metaphase I and anaphase I that are larger in diameter but fewer in number than the somatic coils (Chapter 6) in mitosis.

Zygotene is primarily the stage of pairing of homologues. This pairing is envisioned

Fig. 7.5. A barley cell in pachytene suggesting bouquet arrangement of the chromosomes. (Courtesy of Mrs. Christine E. Fastnaught-McGriff, Department of Plant and Soil Science, Montana State University).

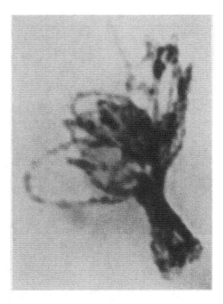

as being in a zipper-like fashion starting at any or even at several "contact points" along the chromosomes and proceeding until all homologous segments are in a pairing equilibrium. Pairing is not always completely finished. Exceptions are also those chromosomes that have nonhomologous sections such as the sex chromosomes (see Fig. 5.1). But in general, meiotic pairing of homologous chromosomes is remarkably precise and specific and gene by gene. Exceptions have been observed and summarized (Riley and Law, 1965).

The phenomenon of chromosome pairing is called **synapsis** and was apparently first observed by Moore in 1895. Synapsis is a prerequisite for an orderly separation of homologous chromosomes during first meiotic anaphase. Electron microscope studies have reveiled some remarkable insight into synapsis in recent years.

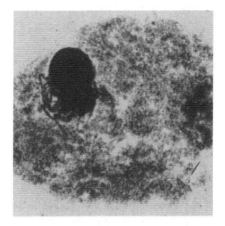

Fig. 7.6. Maize cell in leptotene. Chromosomes are clumped into a dense synizetic knot. ($\times$ 632). (Schulz-Schaeffer, unpublished).

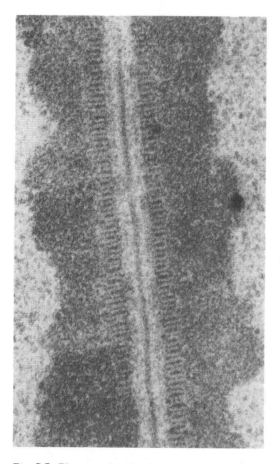

Fig. 7.7. Electronmicrograph of the ultrastructure of the synaptonemal complex of the ascomycete *Neottiella*. (From Westergaard and Wettstein, 1970. Reprinted by permission of Carlsberg Laboratory, Copenhagen)

Moses in 1956 first discovered a tripartate ribbon at the site of synapsis in crayfish called the **synaptonemal complex.**

7.2.2.1 The Synaptonemal Complex (SC). This complex is one of the few configurations in which the 10 nm fibers of the chromosomes are arranged into a superstructure that can be viewed under the electron microscope. It has now been established that this complex occurs in all animals and plant nuclei undergoing synapsis. It is composed of three parallel, electron dense elements that are separated by less dense areas (Fig. 7.7). The two **lateral elements** seem to be composed of fibers that are slightly wider than 10 nm (**synaptomeres**). They vary in structure between different stages of meiotic prophase I within a species. The **central element** is a ladder-like configuration in the center of the SC. It is more pronounced in some species than in others. The **transverse elements** are electron-dense filaments that interconnect the central element with the lateral elements. The lateral elements may be spaced from 20 nm to 30 nm to as much as 100 nm to 125 nm. Cytochemical studies have demonstrated that the lateral elements are rich in DNA, RNA, and proteins (histones included) but that the central element contains mainly RNA and protein and none or little DNA. The presence of RNA in the complex is questionable (Moses, 1968).

Carpenter (1975) described certain SC modifications at the crossover sites. She called these **recombination nodules** (RN) in order to indicate the correlation of their frequency and distribution with the crossover sites in female *Drosophila*. Similar RN's have been observed in the ascomycetous fungi, *Neurospora crassa* (Gilles, 1972), *Saccharomyces cerevisiae* (Byers and Goetsch, 1975) and *Sordaria macrospora* (Zickler, 1977). They also were reported in *Chlamydomonas* (Storms and Hastings, 1977), *Ascaris* (Bogdanov, 1977), and *Maize* (Mogensen, 1977). Holliday (1977) in his **model of crossover position interference** proposed that a crossover between naked DNA molecules is initially weak in structure and must subsequently be stabilized into a mechanically strong chiasma. He believes that such stabilization is accomplished by DNA binding protein that aggregates with DNA and forms a recombination nodule. The depletion of DNA binding protein in the neighborhood of a crossover prohibits the formation of a second crossover adjacent to it (Section 4.3).

7.2.2.2 The Synaptomere-Zygosome Hypothesis of Synaptonemal Complex Formation.   According to this hypothesis (King, 1970), the **synaptomeres** are coiled polynucleotide segments scattered along the length of a synapsed chromosome lying in close proximity to one another. Each synaptomere is composed of three segments: A, B, and C (Fig. 7.8C). The lateral segments of the synaptomeres (A and C) pair with the respective segments of the adjacent synaptomeres. The B segments are directed toward the central element. The B segments are the sites where the so-called **zygosomes** are attached. These are rod-shaped subunits that are assembled in the nucleoplasm and are each visualized as protein molecules having a folded *head* end by which they can attach to the central segment (B) of the synaptomere (Figs. 7.8B, 7.8C). The *tail* ends of the zygosomes contain charge sites that are represented by four dots in Fig. 7.8C. These charges allow the zygosomes to bind laterally with adjacent zygosomes in a ladder-like fashion as indicated in Figs. 7.8B and 7.8C. Fig. 7.8A shows a possible mechanism of synapsis according to this hypothesis. The left and right telomeres of a pair of homologues (T_L and T_R) attach to specific adjacent sectors of the inner membrane of the nuclear envelope (ne). These specific sectors of the nuclear envelope often may be void of nuclear pores and may be thickened. Such attachment may be prepared by the phenomenon of polarization that has been observed during leptotene. The chromosomes are shorter and thicker because of the folding that results from the pairing of synaptomeres that are distributed along the chromosomes. Then follows the attachment of the zygosomes described above (Fig. 7.8B). They seem to be coiled structures that uncoil as they attach to the synaptomeres. When each synaptomere possesses a zygosome, a peg extends from the base of each chromosomal fold. Interdigitation of the pegs can be compared to the zipper-like action mentioned before.

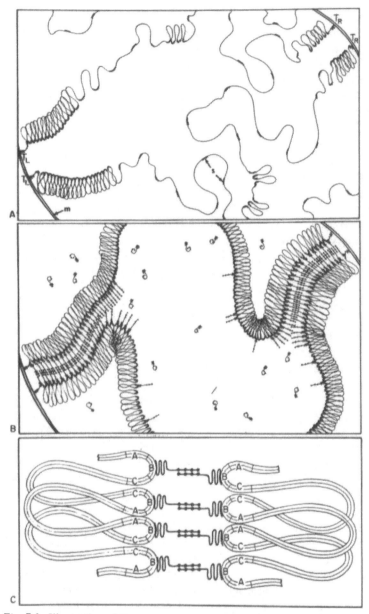

Fig. 7.8. Illustration of the synaptomere-zygosome hypothesis of synaptonemal complex formation. (From King, 1970. Redrawn by permission of Academic Press, New York.).

The exact nature of events that lead to synapsis is still not clear and is vigorously debated in the literature. A strong group of investigators believes that homologous chromosomes are prepared for synaptic pairing by the attachment of their telomeres to so-called "attachment sites" on the nuclear envelope as mentioned in the above hypothesis (Wettstein and Sotelo, 1967; Woollam et al., 1967; Moens,

1973), Comings (1968) believes that these attachments are already evident in the interphase nucleus and that some of the attachment sites may correspond to the **points of initiation** of DNA synthesis in each replicon. This viewpoint is called by Maguire (1977) the **nuclear envelope homologue attachement site model.** As an alternative to such a hypothesis, Maguire proposes an **elastic connector model,** according to which homologous chromosome pairing may be accomplished by chance meeting of homologous chromosome segments followed by the establishment of elastic connectors at congression in premeiotic mitosis (Maguire, 1974). Similar connectors have been hypothesized by Holliday (1968) and Bennet et al. (1974). Homologous chromosome segments may be connected in such a fashion in the stages intervening between premeiotic metaphase and zygotene. **Premeiotic mitotic pairing** has been observed by quite a few investigators (Comings, 1968; Grell, 1969; Dover and Riley, 1973).

A special protein has been identified that may be involved in meiotic and mitotic chromosome pairing. An increase of this protein coincides with the leptotene-to-pachytene-period of meiosis. Hotta and Stern (1971) called this substance **colchicine binding protein.**

7.2.3 Pachytene

Pachytene seems to be a stable stage in that the pairing of homologues is completed. The homologues are closely appressed and form bivalents. In pachytene the chromosomes are shorter than during early prophase and in well flattened cells can be distinguished as separate entities. This made possible pachytene analysis as described earlier (Chapter 2). Figure 7.9 shows an exceptionally well spread pachytene cell of the grass *Bromus secalinus* (2n = 28). In Fig. 7.10 a pachytene cell of a *Triticum durum* x *Agropyron intermedium* backcross derivative plant is shown. This is the more typical situation in pachytene where individual chromosomes are hard to discern. The arrow shows the paired nature of the bivalent threads. At the middle of a pachytene, a longitudinal cleavage becomes apparent in each homologue. This demonstrates that each pachytene bivalent consists of 4 chromatids forming a tetrade chromosome (Fig. 7.2). The nucleoli are particularly evident during pachytene. In many species they have already all united into

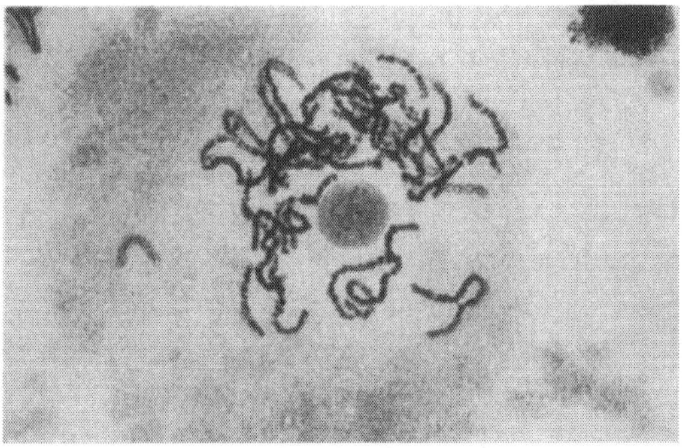

Fig. 7.9. Pachytene cell of *Bromus secalinus* (2n = 28) (× 2111). (From Schulz-Schaeffer, 1956. Reprinted by permission of Verlag Paul Parey, Berlin.).

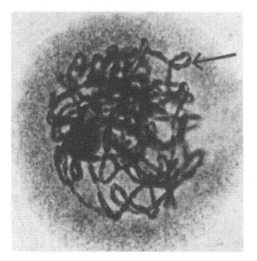

Fig. 7.10. Cell of *Triticum durum* × *Agropyron intermedium* back-cross derivative. (Arrow: Paired nature of bivalent thread) (× 1338). (Schulz-Schaeffer, unpublished).

one big nucleolus by pachytene that is attached to the nucleolus organizer chromosomes. There seems to be evidence that DNA synthesis can extend into meiotic prophase and that it overlaps with synapsis and extends beyond it (Hotta et al., 1966). Such late DNA replication apparently has not been observed during mitosis, and the question arises if such replication may be related to a chromosome break-repair mechanism that may be connected with crossing over (see polaron hybrid DNA model, Chapter 4). The major function of the chromosomes during late zygotene and pachytene is the phenomenon of crossing over, which has already been discussed (Chapter 4). This function has been closely related to the structural discovery of the synaptonemal complex (King, 1970; Wolfe, 1972). Wolfe concluded that the concept of the synaptonemal complex can be conveniently fitted into the phenomena of initial chromosome pairing, close pairing, recombination, and chiasma formation and can even be related to the features of both classical and intragenic recombination. That the complex is involved in at least some of these processes seems certain. Whether all of the listed mechanisms take place with the direct intervention of the complex remains to be seen (Wolfe, 1972).

7.2.4 Diplotene

During diplotene the chromosomes further contract and thicken (Fig. 7.11). This is also the stage where the **chiasmata** (Chapter 4) become apparent as visible evidence of crossing over. Figure 7.12 shows a diplotene cell of a *Triticum* x *Agropyron* derivative with evidence of chiasmata. The synaptic attraction of the chromosomes suddenly comes to an end, the homologues move apart in **repulsion** and are only held together at exchange points that are the result of crossing over. Only two of the four chromatids are involved in the exchange at any given exchange point. More than two chromatids can be involved over larger regions. In organisms

Fig. 7.11. Diplotene cell of barley
(n = 7). (Courtesy of Mrs. Christene
F. Fastnaught-McGriff, Depart-
ment of Plant and Soil Science,
Montana State University,
Bozeman).

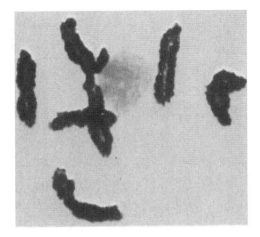

with large chromosomes, the two chromatids involved in such a chiasma are seen
to cross reciprocally from one homologue to the other (see Fig. 4.5). The number
of chiasmata per homologous chromosome pair seems to depend on the species and
on the length of the chromosomes. The longer the chromosome the more chiasmata
are present. Up to 12 chiasmata have been observed in the long chromosomes of
the broad bean *(Vicia faba)* (Swanson, 1957). As diplotene progresses, the chiasmata seem to move away from the centromere and diminish in number. This process is called **chiasma terminalization** (Darlington, 1929a). The terminalization
process may be complete, partial, or altogether absent. In the last instance, the
chiasmata are called **localized chiasmata**. Especially in large chromosomes, terminalization may not be complete. One of the main forces during chiasma terminalization seems to be a strong repulsion force at the centromeres (Darlington,
1937). As the centromeres move apart, the chiasmata slide over the crossover

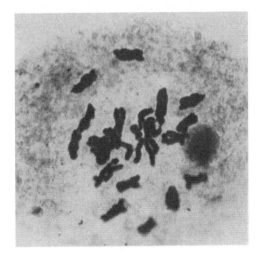

Fig. 7.12. Diplotene cell of a *Triticum* × *Agropyron* derivative with
evidence of chiasmata (× 1585).
(Schulz-Schaeffer, unpublished).

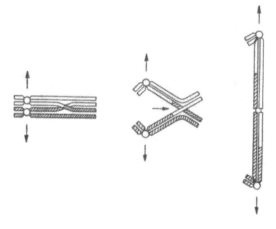

Fig. 7.13. Diagrammatic illustration of three successive stages of chiasma terminalization (Rieger et al., 1976).

points along the chromatids that are involved in the exchange and toward their distal portions (Fig. 7.13). As the chiasmata move toward the chromosome ends, they seem to become arrested there, forming **endchiasmata.** As a consequence, an endchiasma is the result of the terminalization of one or more **interstitial chiasmata.** It is as if they become locked in at the telomeres without being able to slide over these end structures. The purpose of this mechanism is to hold the homologues together until metaphase orientation is completed and extreme tension is exerted on the endchiasmata, at which time a special trigger mechanism separates all chromosomes and at the same time ushers in anaphase I. The forces that hold the chromosomes together at the end of terminalization are not known as yet.

There are three main hypotheses to explain the mechanism of chiasma terminalization.

1. Electrostatic hypothesis (Darlington and Dark, 1932, Darlington, 1937).
2. Coiling hypothesis (Swanson, 1942, 1957).
3. Elastic chromosome repulsion hypothesis (Östergren, 1943).

Darlington's **electrostatic hypothesis** states that two separate repulsion forces are responsible for movement of the chiasmata to the chromosome ends. One of these forces, **localized repulsion,** is supposed to be the dominating one that repels the centromeres. The second force, **generalized repulsion,** is supposed to be evenly distributed over the entire surface of the chromosomes and tends to force the chromosomes apart. The result is a movement of the chiasmata in a distal direction, as described before, because the greater force exists between the centromeres. Darlington based this hypothesis on the observation that in some organisms there is a greater stretching of the centric loop between the two most interstitial chiasmata.

Swanson's **coiling hypothesis** questions the reality of Darlington's forces and explains terminalization as being affected by despiralization of the chromosomes. During the coiling cycle (see Fig. 6.5), the initial coils are small and numerous and as prophase proceeds the coils decrease in number but the gyres of the coils increase in diameter. This theory postulates that the coiling and associated short-ening of the chromosomes develops mechanical tension that will force the chias-mata to slide along the chromosomes. The advantage of the coiling hypothesis over the electrostatic hypothesis is that the coiling hypothesis can be tested by varying the degree of coiling. This has been accomplished by exposing the biva-lents to different temperatures. Support for the coiling hypothesis came from a comparison of normal and mutant types in the plant *Matthiola incana* (Lesley and Frost, 1927). In the mutant the chromosomes failed to attain their normal state of contraction, and the chiasmata remained at the original interstitial posi-tions. In the normal plants the chromosomes shortened, and the chiasmata were terminal.

The **elastic chromosome repulsion hypothesis** of Östergren is based on the idea that chiasma movement eliminates the tension created by the chiasmata them-selves. A chiasma forces the chromosomes out of shape by preventing repulsion, which is effective in the adjacent areas of the bivalent. The repulsion force tends to push the chiasmata distally since this is the only direction in which relief from tension can be gained.

During diplotene the bivalents are generally observed more distinctly because of a widening gap between them, which suggests a repulsion force. This is even more evident in the next substage, **diakinesis.**

In some specialized tissues, diplotene can be very much prolonged and can last a year or much longer, not only an hour or two as indicated in Table 7.1. The long duration of this stage in these instances is associated with a specialized function of the cells involved. Such prolonged diplotene stages are found in the **primary oocytes** of some vertebrates such as fishes (sharks), amphibians, reptiles, birds, mice, in human beings, and in the **primary spermatocytes** of some insects such as *Drosophila*. In these cells the chromosomes acquire a very characteristic appearance. They become very diffuse by forming thin threads or loops that are transverse to the main axis of the chromosomes. This despiralization makes these chromosomes look like old-fashioned, oil-lamp chimney brushes, and they are therefore called **lamp brush chromosomes** (Section 9.6). These chromosomes also increase enormously in length. The purpose of this increase in surface and length is to provide for increased metabolic activity of these chromosomes. The loops are believed to be active genetic material such as DNA, which synthesizes messenger RNA that is responsible for protein synthesis in the cells' cytoplasm. This causes an enormous growth of the oocytes. Oocytes of the frog, *Rana pipiens,* increase in size by a factor of 27,000 over a period of three years (Balinsky, 1970). In chicken that factor is 200 (last rapid growth), and in mice 43, and in both cases the growth proceeds at a much faster rate and takes a shorter period for comple-tion. In female human beings the diplotene oocytes are already formed by the fifth month of prenatal life. Here they remain in diplotene for a period of 12 to 50 years, from the age of sexual maturity of humans to the age when the last eggs are ovulated. The functional significance of such a very prolonged diplotene stage is unknown (Swanson et al., 1967). This prolonged diplotene condition is referred to as **dictyotene.**

7.2.5   Diakinesis

If chromosome counts are desirable, then diakinesis may be one of the most ideal stages for this purpose. The only disadvantage is the shortness of this stage (see Table 7.1). But the extreme coiling and the seeming repulsion between the bivalents space them all over the cell in squash preparations. The restricting nuclear envelope can be ruptured under the pressure of squashing. In this respect diakinesis also has the advantage over metaphase I in that the spindle fiber apparatus is not yet attached to the chromosomes. This apparatus generally prevents an even spread of chromosomes and keeps them in a bunch.

The bivalents have their greatest degree of terminalization and contraction in diakinesis. In some cases they become almost spherical configurations (Fig. 7.14). If interstitial chiasmata, located close to the centromeres, remain localized, then the bivalents appear like crosses. Similar configurations can be formed by **collochores** (Cooper, 1941). Collochores are small conjunctive segments in the regions adjacent to the centromere that are responsible for meiotic chromosome pairing without chiasma formation and for the coherence of such pairing associations until the beginning of anaphase I. **Cross bivalents** observed in certain *Triticum* x *Agropyron* hybrids and their backcross derivatives (Schulz-Schaeffer et al., 1971; Schulz-Schaeffer and McNeal, 1977) were interpreted as being formed by collochores (Fig. 7.15). There is also good evidence from research in mantids (White, 1938; Hughes-Schrader, 1943a, 1943b), lepidoptera (Bauer, 1939), mites (Cooper, 1939), scorpions (Piza, 1939), bugs (Schrader, 1940a, 1941), and flies (Cooper, 1944) that chromosomes can join and hold together during meiosis by mechanisms other than synapsis and chiasmata. Further electron-micrographic studies like those that led to the discovery of the synaptonemal complex may shed light on the possible formation of collochores.

Other possible chromosome associations in diakinesis are **rod bivalents** and **open ring bivalents.** Rieger and Michaelis (1958) defined rod bivalents as pairing associations that had chiasma formation and terminalization in only one chromosome arm in each of the two homologous chromosomes involved (Fig. 7.16D). This def-

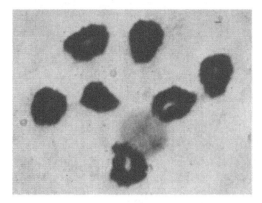

Fig. 7.14. Diakinesis cell of barley (n=7). (× 3436). (Schulz-Schaeffer, unpublished).

Fig. 7.15. Photomicrograph of a diakinesis cell of a *Triticum* × *Agropyron* derivative with 25 cross bivalents and 1 univalent (2n = 51). (× 885). (From Schulz-Schaeffer et al., 1971. Reprinted by permission of Verlag Paul Parey, Berlin)

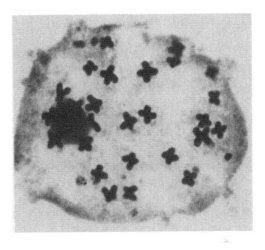

inition could also apply to the more frequent, fairly normally occurring phenomenon of **open ring bivalents** (Fig. 7.16E). There has to be a reason for the obvious difference in shape between rod bivalents and open ring bivalents. A possible explanation would seem to be that rod bivalents have previously synapsed over only a minute distal portion of one arm of each of the two chromosomes involved or that these chromosomes join by organelles similar to the collochores. In the case of open ring bivalent formation, synapsis and the repulsion following usually cause the bow shape that is also typical for closed ring bivalents and is not expressed in rod bivalents. A schematic representation of several possible chromosome configurations in diakinesis is presented in Figure 7.16.

The nucleoli usually fuse toward the end of prophase I to form one large nucleolus. At the end of diakinesis, the nucleolus begins to disappear.

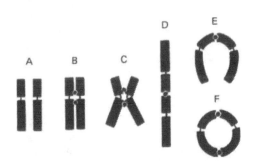

Fig. 7.16 A-F. Schematic representation of chromosome configurations in the diakinesis of *Triticum* × *Agropyron* derivatives. Configurations are arranged in a meaningful way to imply tendency for progressive pairing from complete asynapsis (*A*) to normal closed ring bivalents (*F*). (*A*)–Two homologous chromosomes in asynapsis. (*B*)–H-type cross bivalent. (*C*)–Standard-type cross bivalent. (*D*)–Rod bivalent. (*E*)–Open-ring bivalent. (*F*)–Closed-ring bivalent. (●–collochores and endchiasmata (|–centromeres) (From Schulz-Schaeffer et al., 1971. Reprinted by permission of Verlag Paul Parey, Berlin).

7.3   Metaphase I

Similarly, as in mitosis, the end of prophase I is also marked by the disappearance of the nuclear envelope and the nucleolus as well as by the division of the centrosome and formation of the spindle. Bivalents assemble at the equatorial plate and become oriented with their centromeres poleward. Figure 7.17 shows a metaphase I cell of barley in which each of the seven bivalents is clearly visible in the equatorial plate. If more than 14 chromosomes are involved, they are not as easily identifiable as in barley. An example is shown in Figure 7.18 in which the 2n = 108 chromosomes of the hexaploid native American rubber plant guayule *(Parthinium argentatum)* are shown in metaphase I.

There is a pronounced difference between mitotic metaphase and metaphase I of meiosis (Fig. 7.19). In mitosis, the sister chromatids are held together by functionally undivided centromeres (C), which are located on the equatorial plate exactly halfway between the poles. In meiosis, the two centromeres (C) of the homologues are not located on the equatorial plate but are oriented in the long axis of the spindle equidistant from the equator, while the endchiasmata (E-C) are located in the equatorial plate. The coordinated arrangement of all tetrad chromosomes on the equatorial plate in the described manner is called **coorientation** (Darlington, 1937). An equilibrium is established at the equatorial plate after all tetrad chromosomes (bivalents) have attained this position, until the chromosomes yield to the tension exerted on them originating from the two opposite poles via the spindle fibers.

This final arrangement of the bivalents on the equatorial plate also has some genetic consequences. Just as crossing over during zygotene and pachytene provides for recombination of paternal and maternal genes on the chromosomes as discussed, so does the coorientation of the bivalent on the equatorial plate provide for recombination of paternal and maternal chromosomes during metaphase I. In general, the position of each chromosome of a bivalent with respect to the poles seems to be at random (Fig. 7.20, **random assortment**). There may be exceptions in which preferential segregation is involved (Section 17.5). The random orientation of the bivalents on the equatorial plate determines the **meiotic segregation** and distribution of the paternal and maternal chromosomes to the daughter cells of the first

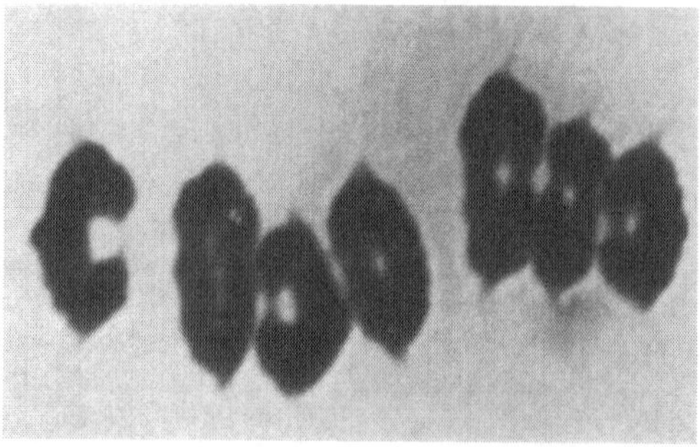

Fig. 7.17. Metaphase I in barley. (n = 7). (Courtesy of Mrs. Christine E. Fastnaught-McGriff, Department of Plant and Soil Science, Montana State University, Bozeman).

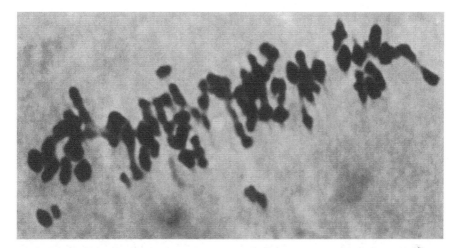

Fig. 7.18. Metaphase I of hexaploid guayule *(Parthenium argentatum).* (n = 54). (×
1438). (Courtesy of Dr. Duane Johnson, Department of Plant Science, University of
Arizona, Tucson).

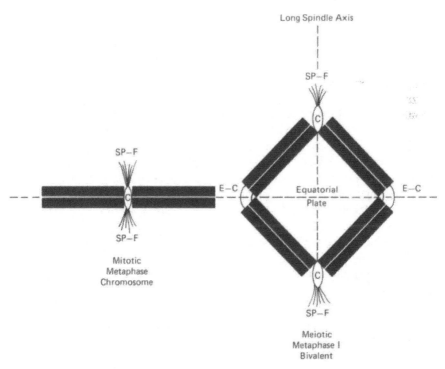

Fig. 7.19. Illustration of the differences in chromosome orientation between mitotic
metaphase and metaphase I of meiosis. SP-F = spindle fiber attachment. E-C = endchiasma. C = Centromeres.

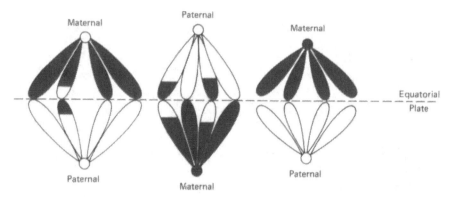

Fig. 7.20. Diagrammatic representation of random assortment of homologous chromosomes at the equatorial plate of a metaphase I nucleus.

meiotic division. Meiotic segregation is the substance of Mendel's second law (See Chapter 1), the law of independent assortment. Mendel's law made statements concerning gene segregation, while meiotic segregation deals with gene blocks. Mendel's law agrees with the assumption that two factor pairs under consideration are located on different bivalents. The discovery of linkage brought cytological and genetical discoveries into alignment in that gene blocks or linkage groups were now accounted for.

7.4 Anaphase I

During anaphase I the tetrad chromosomes separate into dyad chromosomes (Fig. 7.21), as the two coorienented centromeres move toward opposite poles (Fig. 7.2). The difference between anaphase I (A I) and mitotic anaphase (or A II) is best exemplified in Fig. 7.1. A typical meiotic anaphase I always has four chromosome arms dangling behind the centromere, while a mitotic anaphase has only two such arms showing. The reasons for this difference is the fact that an anaphase I dyad chromosome consists of two chromatids, while a mitotic anaphase chromosome is really a single chromatid. The four arms of an anaphase I dyad chromosome do not stick closely together but diverge as if they are mutually repelling each other. Anaphase I is shorter in duration than metaphase I (see Table 7.1). The only real function of this stage is to evenly distribute the partners of homologous chromosome pairs to the daughter nuclei, with the result of a reduction by half the number in each resulting nucleus. The original **somatic chromosome number** (2n) is reduced to a **gametic chromosome number** (n). These two symbols are very often used in reports of chromosome numbers. If a species is investigated in somatic tissue (mitosis), the chromosome count is reported as 2n (e.g., 2n = 28). If a count is made in gametogenesis (meiosis) the report is made with an n-number (e.g., n = 14). This understanding enables the cytologist to quickly identify the nature a particular chromosome report. Very often, ploidy levels are erroneously reported with n-numbers. But the number reserved for ploidy levels is the x-number or

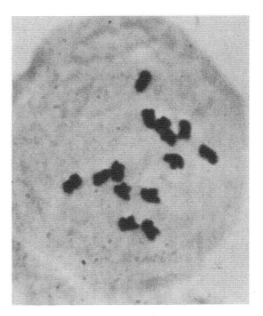

Fig. 7.21. Anaphase I of barley (n = 7). (× 1453). (Schulz-Schaeffer, unpublished).

basic genome number (x, 2x, 4x, 6x, etc.) (see also Chapter 2). Well-known summaries of chromosome numbers have been published for animals by Makino (1951) and for plants by Darlington and Fedorov (Darlington and Janaki Ammal, 1945; Darlington and Wylie, 1955; Fedorov, 1974). Makino's report has data on some 2,800 animal species, and Fedorov's has data on 35,000 plant species.

Each anaphase I dyad chromosome has two chromatids that remain joined at the centromere until anaphase II.

7.5   Telophase I and Interkinesis

Telophase I is similar to mitotic telophase in that the chromosomes assemble at the poles (Fig. 7.22). But since the following interphase (called **interkinesis**) is different from normal interphase, telophase I can also be different in several respects, depending on the organism. During interkinesis, which is a short stage, the chromosomes do not synthesize new DNA and consequently there is no reduplication. The chromosomes are already prepared for the second division in that each of them consists of two chromatids only held together by a centromere. Therefore, despiralization, uncoiling, and hydration of chromosomes are not necessary. As a matter of fact, in some species following the disappearance of the spindle, the chromosomes orient themselves at the poles and pass directly to the equatorial plate of the second division (M II). This has been reported for *Trillium* (Swanson, 1957) and certain members of the Odonata (Cohn, 1969). In this case the coiling of the chromosomes is retained throughout interkinesis. In other instances, the chromosomes become partially uncoiled during interkinesis, and nuclear envelopes form. This

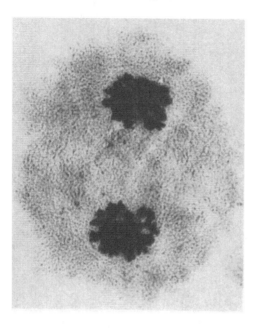

Fig. 7.22. Telophase I of *Bromus inermis* (n = 28). (× 1077). (Schulz-Schaeffer, unpublished).

kind of interkinesis has been observed in *Tradescantia, Zea mays,* and grasshopper (Swanson, 1957). Figure 7.23 shows an interkinesis cell of *Agropyron intermedium.*

In other organisms there is no cytokinesis after the first meiotic division as reported for *Paeonia* where no walls are formed at interkinesis (Swanson, 1957). Cytokinesis is postponed until after the second meiotic division, and this process is referred to as **quadripartitioning.** In contrast, the normal process as found in many other plants, where a cell plate forms between the telophase nuclei of the first division, is called **bipartitioning.**

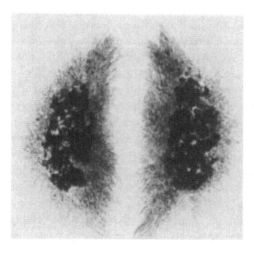

Fig. 7.23. Interkinesis of *Agropyron intermedium* (n = 21) (× 1239). (Schulz-Schaeffer, unpublished).

Fig. 7.24. Prophase II in barley (n = 7). (× 1093). Insert: single prophase II cell showing nucleolus. (Schulz-Schaeffer, unpublished).

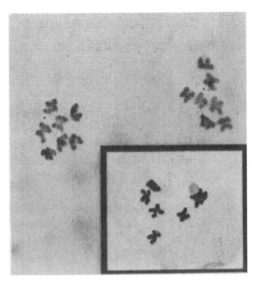

7.6   Prophase II

The second meiotic division in many respects is very similar to a mitotic division. Second prophase differs in appearance from first prophase in that the sister chromatids of each dyad chromosome show a very striking repulsion so that the chromatid arms are widely separated from each other (Fig. 7.24). This makes the dyad chromosomes look like crosses. The shortness of the only partially uncoiled chromosomes makes it possible to view each prophase II chromosome individually, and chromosome counts can sometimes be made.

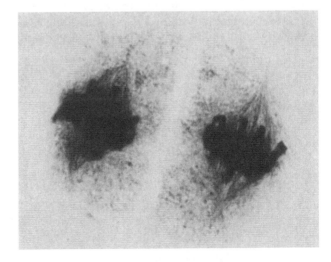

Fig. 7.25. Metaphase II cell of *Agropyron intermedium* (n = 21). (× 1333). (Schulz-Schaeffer, unpublished).

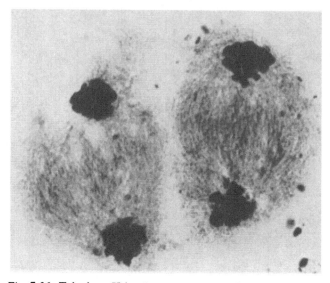

Fig. 7.26. Telophase II in *Agropyron intermedium* (n = 21). (× 1554). (Schulz-Schaeffer, unpublished).

The genetic constitution of the two sister chromatids in the second meiotic division depends on the extent of crossing over during the prophase of the first division. In this respect the second meiotic division differs from a mitotic division that is strictly equational. Figure 7.2 illustrates that prophase II dyads can be partly maternal and partly paternal in genetic makeup. In this respect the second meiotic division really completes the process of **genetic recombination** that was started with crossing over in prophase I. Crossing over of chromatid segments and random distribution of maternal and paternal chromosome segments during the first and second meiotic divisions are together the agents of genetic recombination in meiosis.

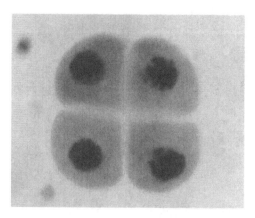

Fig. 7.27. Radial quartet in barley. (Schulz-Schaeffer, unpublished).

Fig. 7.28. Microspore of *Triticum* × *Agropyron* derivative. (Schulz-Schaeffer, unpublished).

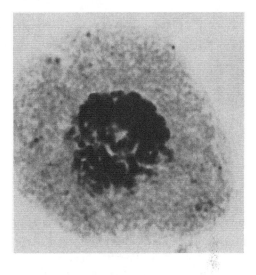

7.7   Metaphase, Anaphase, and Telophase II

The centromeres of the dyad chromosomes situate on the equatorial plate as in an ordinary somatic division. In higher plants, the two adjacent cells, separated only by a cell wall, generally go through a synchronized procedure of chromosome congression, orientation, and distribution (see Section 6.4). The two equatorial plates are lined up across the separating cell wall. Figure 7.25 shows a metaphase II cell of *Agropyron intermedium*. As the centromeres become functionally double, the **monad chromosomes** (Fig. 7.2) move toward the four poles of the **duet cells** of **microsporogenesis** and form a telphase II cell (Fig. 7.26). As the cell walls form at the end of telophase II, the so-called **radial quartet cells** form. The quartets are the four adherent cells resulting from the two meiotic divisions in microsporogenesis (Fig. 7.27). The quartets then differentiate into four haploid microspores (Fig. 7.28), which are the endproducts of microsporogenesis as described in the next chapter. The two meiotic divisions in **spermatogenesis** also lead to four daughter cells (**spermatids**) that subsequently differentiate into four sperms. **Oogenesis** and megasporogenesis are different (unequal) in that only one large haploid egg and three small **polar bodies** in animals and only one large megaspore in plants are produced. This completes the meiotic cycle in plants and animals.

Chapter 8
Chromosomes During Sexual Reproduction

In this chapter special attention is paid to the importance of chromatin and chromosomes in reproduction. However, in order to understand what happens at the chromosomal level, the whole cell is considered carefully. In the last chapter we stated that the processes of genetic recombination in meiosis are: (1) crossing over of chromatid segments during prophase I and (2) random distribution of maternal and paternal chromosome segments during the first and second meiotic divisions. But meiosis is only one part of sexual reproduction. During the life cycles of **haplontic, diplontic,** and **diplo-haplontic** organisms, there is a regular alternation between meiosis and fertilization. The second part of genetic recombination is the union of paternal and maternal gametes during syngamy.

In order to understand the differences in the life cycles of various plants and animals, the newly introduced terms **haplontic, diplontic,** and **diplo-haplontic** should now be explained (Cook, 1965). **Haplonts** are common in most unicellular or filamentous algae and protozoa. These are organisms in which the **haplophase** is more prominent than the **diplophase** (Renner, 1916). The haplophase is the **functionally haploid** period (n) during a particular life cycle lasting from meiosis to fertilization, while the diplophase is the **functionally diploid** period (2n) of a life cycle that spans from fertilization to the beginning of meiosis. The term **diploid** is used here in a wider sense since it really designates only organisms with two basic genomes (2x) (Section 2.1.1). This usage of **functional diploidy** is further explained in Section 15.2. In the haplonts, only the zygote is diploid (Fig. 8.1), and in some species it becomes a resistant spore, which guarantees the survival of the species under difficult conditions. The mature individuals in this life cycle, which in multicellular organisms differentiate by mitotic divisions, are functionally haploid (n).

Diplonts are found in humans, higher animals, and in some algae. They are organisms in which the diplophase is more prominent than the haplophase. In diplonts the products of meiosis function directly as gametes. This is the regular life cycle of all multicellular animals. Only the gametes are haploid in this type of life cycle (Fig. 8.2). The mature individuals are produced by mitotic division and differentiation and are functionally diploid (2n). In contrast to diplo-haplontic organisms, the diplonts do not have alternation of generations but only alternation of nuclear phases (n and 2n).

Diplo-haplonts are typical in higher plants and in many algae and fungi. As early as 1851, the concept of **alternating generations** for such organisms had been devel-

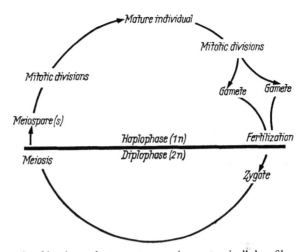

Fig. 8.1. Diagram of the life cycle of haplonts that are common in most unicellular, filamentous algae and protozoa. (From Cook, 1965).

oped by Hofmeister. He demonstrated that the life cycle of a typical plant consists of two unique generations: a spore-bearing (**sporophyte,** 2n) and a gamete-bearing (**gametophyte,** n) (Fig. 8.3). The sporophyte of higher plants makes up the more prominent generation. Meiosis in diplo-haplonts does not immediately produce gametes as in higher animals; instead, a parasitic structure, which in turn produces the gametes, is inserted as an alternating generation in higher plants. This parasitic structure is called the gametophyte. The relationship between the sporophytic and gametophytic generations varies depending on plant groups. In the **Spermatophyta** (seed plants) the sporophyte is dominant and independent. Its life duration is from one to several years. The gametophyte is extremely small and

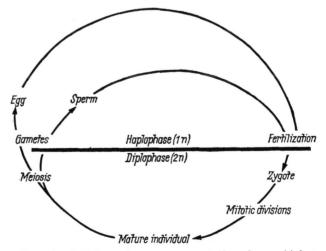

Fig. 8.2. Diagram of the life cycle of diplonts that are representative of man, higher animals, and some algae. (From Cook, 1965).

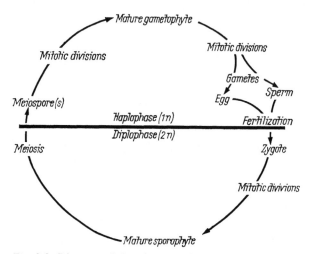

Fig. 8.3. Diagram of the life cycle of diplo-haplonts that are typical of higher plants and many algae and fungi. (From Cook, 1965).

parasitic on the sporophyte, as mentioned, and it lives only from a few days to a few weeks. In the **Pteridophyta** (ferns and related plants), the sporophyte is dominant, vegetatively independent, and often perennial. The gametophyte, though small, is independent of the sporophyte and lives a few weeks or longer. In the **Bryophyta** (mosses and liverworts), the sporophyte is partially parasitic on the gametophyte and lives a few weeks. A gametophyte is dominant, vegetatively independent, and lives one to several years. The diagram in Fig. 8.4 illustrates these relationships between Spermatophyta, Pteridophyta, and Bryophyta. Single lines represent the functionally haploid (n) gametophyte generation and double lines represent the functionally diploid (2n) sporophyte generation. The lengths of the lines generalize the relative prominence and length of the two generations in the life cycle. Generations represented by unbroken lines are independent and

Fig. 8.4. Diagrammatic representation of the relationsbip between gametophyte and sporophyte generations in plants:
Single lines–haploid gametophyte generation.
Double lines–diploid sporophyte generation.
Length of lines–approximate relative prominence of gametophyte and sporophyte generations.
Unbroken lines–independent generations.
Broken lines–parasitic on other generation. (From Alexander, 1954.
Redrawn by permission of Barnes and Noble, Inc., New York).

those shown by broken lines are parasitic on the other generation in the life cycle
(Alexander, 1954).

8.1   Sexual Reproduction in Plants

Reproduction in plants can be subdivided into sporogenesis, gametogenesis, and
syngamy. The example for reproduction in plants will be taken from the Spermatophyta, which make up the dominant part of our vegetation. As mentioned
above, they exemplify a diplo-haplontic life cycle. The diplophase is called the
sporophyte. The sporophyte differentiates two kinds of tissues that later will
develop the **meiospores**. The tissue that leads to **microspores** is called the **microsporangium** and is located in the **anthers**. The tissue that eventually will produce
megaspores is called the **megasporangium** and is located in the **ovary**.

8.1.1   Microsporogenesis and Spermatogenesis

An illustration of microsporogenesis and spermatogenesis in angiosperms is shown
in Fig. 8.5. **Microsporogenesis** is the process of microspore formation that leads
to the first cell of the male **gametophyte** generation. Microsporogenesis takes
place in the microsporangium. A typical anther of a phanerogam (Spermatophyta) has four elongated microsporangia that develop into **microsporocytes** or
pollen mother cells (PMC's). These cells enlarge and go through meiosis, as
described in Chapter 7. A microsporocyte that goes through meiosis is also
referred to as a **meiocyte**. The end product of microsporogenesis is a **radial quartet cell**. The radial quartet cells each separate into four haploid **microspores** with
one nucleus each. Each of these microspores may develop into a pollen grain
(**spermatogenesis**).

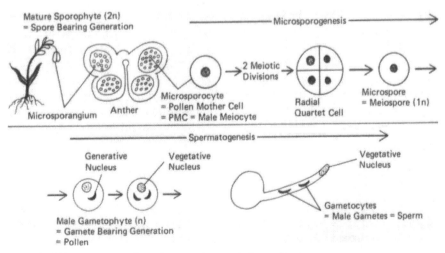

Fig. 8.5. Microsporogenesis and spermatogenesis in angiosperms.

Between the microspore stage and the first **pollen mitosis** there is a rest period that varies from a few hours to several months (Dahlgren, 1915; Finn, 1937). The developing pollen grain or male gametophyte enlarges during this period either because of an increase in the amount of cytoplasm (Sax and Edmonds, 1933), or because of the formation of vacuoles in the cytoplasm (Steffen, 1963), or because of both of these reasons together. Naturally, there is also an increase in the amount of DNA before pollen mitosis (Bryan, 1951) because during synthesis the chromosomes replicate.

Pollen mitosis is an ideal stage for chromosome analysis. The chromosomes are reduced by half (n), which facilitates the count and the spread of the chromosomes in organisms with high chromosome numbers. They are also less contracted than during the two meiotic divisions, which makes it easier to discern their centromere positions.

The nuclei that result from the first pollen mitosis differentiate into a **generative** nucleus and a **vegetative** or **tube nucleus**. The male gametophyte now has two nuclei that were produced by **karyokinesis** (nuclear division) without the event of **cytokinesis** (cytoplasmic division).

The two nuclei of the male gametophyte usually differ in shape. The generative nucleus is densely compacted and often is crescent shaped, while the vegetative nucleus is larger and spherical and less densely stained. These nuclei are often considered to be **protoplasts** or cells, but they definitely lack cell walls. Electron microscope studies have shown a clearly defined double membrane around the cytoplasm of the generative nucleus but no wall as earlier reported from light microscope studies (Bopp-Hassenkamp, 1960). The DNA content of the two morphologically different nuclei seems to be similar (Swift, 1950; Bryan, 1951; Ogur et al., 1951).

A **second pollen** mitosis leads to the production of a **mature male gametophyte** or **pollen**. This mitosis occurs in the generative nucleus of the pollen and leads to the formation of the two **male gametes** or sperms. These are also referred to as the **male gametocytes**. The time of the second pollen division varies according to species. In many grasses the three nuclei are present before pollen tube formation. In lily, second pollen mitosis happens in the pollen tube when the sperm passes through it down the style on its way to the micropylar opening of the ovule. If pollen mitosis takes place in the pollen tube, the metaphase chromosomes form rows parallel to the axis of the pollen tube (Fig. 8.6). The vegetative nucleus seems to be responsible for the growth of the pollen tube and is the one in the tip of the pollen tube trailed by the two sperms (Fig. 8.5). The vegetative nucleus and the sperms are believed to be passively transported through the streaming cytoplasm in the pollen tube (Navashin et al., 1959). The sperms are now ready for syngamy.

8.1.2   Megasporogenesis and Syngamy

An illustration of megasporogenesis and syngamy is shown in Fig. 8.7. **Megasporogenesis** is the process of megaspore formation in the **megasporangium** or **ovule**. The megasporangium consists of the **nucellus** and of one or two **integuments**. Enclosed in the nucellus is the **megasporocyte** or **embryo sac mother cell** (EMC).

Fig. 8.6. Pollen mitosis in the pollen tube of colchicine treated tetraploid *Polygonatum commutatum.* (n = 2x = 20). (Redrawn by permission from Colchicine—in Agriculture, Medicine, Biology and Chemistry by O. J. Eigsti and P. Dustin, Jr. © 1955 by The Iowa University Press, Iowa 50010).

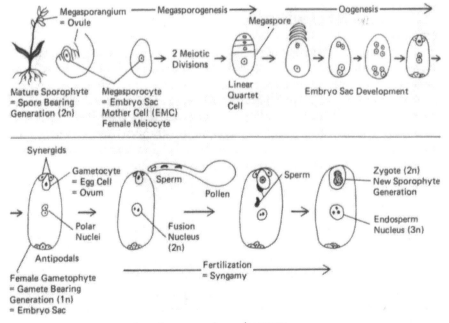

Fig. 8.7. Megasporogenesis and syngamy in angiosperms.

Each ovule contains only one megasporocyte. The megasporocyte divides by meiosis to form a **linear quartet cell,** which is a row of four cells, each of which is a potential megaspore. Three of these cells degenerate, but the fourth enlarges and forms the large **megaspore.** The megaspore typically develops into the **embryo sac** through three mitotic divisions. Eight nuclei are formed by these divisions that organize into the **egg apparatus,** two **polar nuclei,** and three **antipodals** (Fig. 8.7). The embryo sac thus becomes the **female gametophyte** (n) or the gamete bearing generation. The egg apparatus consists of the two outer **synergids** and the egg nucleus in the center. The egg nucleus becomes the **egg cell, female gametocyte,** or **female gamete.** The two polar nuclei often fuse and become a diploid **fusion nucleus** (2n).

Syngamy is preceded by the landing of the pollen grain on the **stigma** of the female portion of the flower, which consists of stigma, **style,** and **ovary (pistil).** The pollen grain germinates on the stigma. The time for germination is variable. For instance, germination takes three minutes in *Reseda* (Eigsti, 1937) and five minutes in *Zea mays* (Randolph, 1936). The pollen forms a tube that passes down the style and reaches the opening of the ovule, called **micropyle.** After the two sperm enter at the mycropyle, one fuses with the egg cell to form the zygote (2n), and the other fuses with the fusion nucleus to form the **endosperm nucleus** (3n) in what is often referred to as **double fertilization.** The time from pollen germination to fertilization varies according to species. In general, it takes 12 to 48 hours (Maheshwari, 1949). The zygote is the first cell of the new sporophyte (2n), which by mitotic cell division develops and differentiates into the mature sporophyte. This completes the life cycle of a typical diplo-haplont with its characteristic alternating generations (2n, 1n).

8.2   Sexual Reproduction in Animals

In higher animals, sexual reproduction consists of **gametogenesis** and **syngamy.** Higher animals are diplonts in which the diplophase is more prominent than the haplophase. The fertilized diploid ovum or zygote divides and differentiates by regular mitosis to form the adult mature animal body, a portion of which differentiates into the **germ line** (Weismann, 1885). The germ line is a group of cells that early during the development of an animal organism are separated from somatic cells as potential gamete forming cells. Gamete formation or gametogenesis takes place in the testes of the male and in the ovary of the female.

8.2.1   Spermatogenesis

Gametogenesis in the male animal is called **spermatogenesis.** The testes contains the immature potential germ cells called **primary spermatogonia.** By rapid mitotic multiplication, they produce the so-called **secondary spermatogonia.**

The chromosomes in spermatogonial mitosis differ in shape and size from normal mitotic chromosomes. They show a spiral configuration and are very much contracted (Fig. 8.8). Sasaki and Makino (1965) observed that they were extremely fragile with a tendency to break and scatter. A fairly high degree of **polyploidy** has been observed in spermatogonia of humans (Sasaki, 1964; Sasaki and Makino,

Fig. 8.8. Human chromosomes in spermatogonial mitosis. (× 1600). (From Sasaki and Makino, 1965).

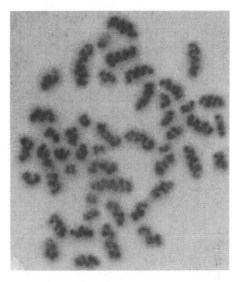

1965; Kjéssler, 1966; McIlree et al., 1966). McIlree et al. found from 0% to 25% polyploid cells in different individuals, and Sasaki reported an average of 7% to 8.7%. Kjéssler's (1970) observations are in agreement with the above percentages. In the vertebrates, the spermatogonia are found next to the basal membranes of the seminiferous tubules (Fig. 8.9). While part of the spermatogonia remain in this

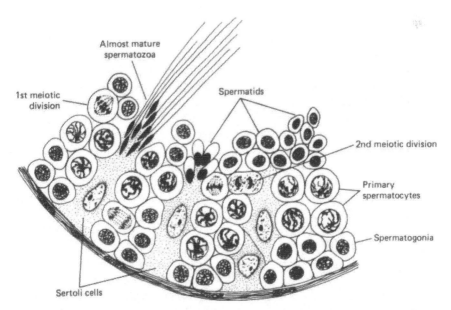

Fig. 8.9. Diagram of part of a seminiferous tubule in a mammal. The spermatogonia are found next to the basal membranes of the seminiferous tubules. (From Balinsky, 1970. Redrawn by permission of W. B. Saunders Co., Philadelphia).

condition and form a source of new sex cells throughout the reproductive life of the animal, some of the cells produced move toward the lumen of the tubule, and enter into the growth phase of spermatogenesis and are then called **primary spermatocytes.** The growth of the spermatocytes is limited, but as a result they become about twice as large in volume as the spermatogonia (Balinsky, 1970). The chromosomes also associate in pairs during this period. The primary spermatocytes undergo the first meiotic division and become **secondary spermatocytes.** After the second meiotic division, the **spermatids** are formed. The spermatids are haploid sex cells but they are still not capable of functioning as male gametes, which are the so-called **spermatozoa** or **sperm.**

8.2.2 Spermiogenesis

The process of differentiation from an immature spermatid to a mature spermatozoon is called **spermiogenesis.** Spermiogenesis is a striking metamorphosis and involves a very radical change. After the second meiotic division, the nucleus of the spermatid goes into a typical interphase forming dispersed chromatin. The cytoplasm of a spermatid contains all the inclusions that a normal cell usually has. Among them are the mitochondria, the centrioles, and the **Golgi apparatus.** These organelles are instrumental during the development from a spermatid to a spermatozoon. One of the most important changes in the morphology during spermiogenesis is the reduction in cytoplasmic material and the condensation and elongation of the nucleus.

It seems that the cell tries to eliminate all extra material that is not of importance in motility. The main function of the mature spermatozoon is the transportation of the male genetic material to the female egg. The spermatozoon differentiates the **acrosome** so that it can penetrate the egg membrane and enter the egg's cytoplasm. The acrosome is derived from the Golgi apparatus. In the early spermatid, the Golgi apparatus consists of the typical flattened membrane-bound sacs, called **cisternae.** As the development of the spermatid proceeds, vacuoles develop in the Golgi apparatus. Within the vacuoles, small dense bodies appear that are called **proacrosomal granules** (Fig. 8.10). The smaller vacuoles coalesce into a larger one, and the granules they contain fuse into one. The vacuole and the Golgi apparatus now approach the tip of the elongating nucleus. As the Golgi apparatus moves toward the nucleus, the vacuole containing the granule actually moves toward the edge of the Golgi apparatus and attaches to the nuclear envelope (Fig. 8.10C). The proacrosomal granule now increases in size as small vacuoles continue to arise from the Golgi apparatus and coalesce with the large vacuole or vesicle, thus adding more proacrosomal material. The granule thus becomes the **acrosomal granule.** As the development proceeds further, the vesicle loses its liquid content and becomes completely filled with granule substance (Burgos and Fawcett, 1955).

The next step in the development of the spermatozoon is the formation of its **middle piece** (Fig. 8.11B). This part eventually contains the base of the flagellum and the mitochondria, which will serve as a power plant supplying the flagellum with energy. This formation of the middle piece begins with the movement of the two centrioles to a place just behind the nucleus opposite the acrosome. One of them, the **proximal centriole,** becomes located in a depression of the nucleus. The other

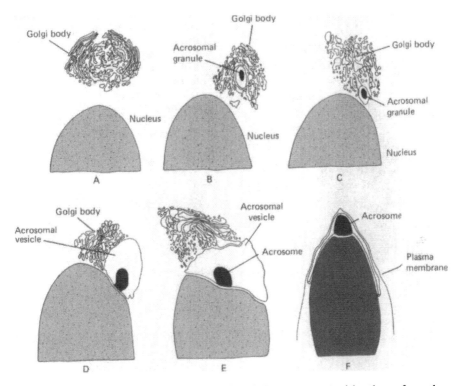

Fig. 8.10. Sequence of stages in the formation of the acrosome and head cap from the Golgi complex during spermatogenesis in the cat. (From Burgos and Fawcett, 1955. Redrawn by permission of the Rockefeller University Press, New York).

one, the **distal centriole,** lines up right behind it at a right angle and coincides with the longitudinal axis of the spermatozoon (Fig. 8.11A). The distal centriole is responsible for the development of the **axial filament** of the flagellum. Mitochondria aggregate around the axial filament. In mammals the mitochondria, which become concentrated in the middle piece from other parts of the cell, lose their individuality and form a **mitochondrial spiral** that winds around the axial filament. At the end of the middle piece, between the middle piece and the tail, is the dense so-called **ring centriole** with unknown function. The name *centriole* is misleading since its fine structure does not resemble a centriole.

The tail or **flagellum** (Fig. 8.11B) is usually the longest part of the spermatozoon. It enables the sperm to swim. Its main part is the axial filament that continues into it from the middle piece. Its fine structure reveals ten pairs of longitudinal fibers, one in the middle and nine surrounding it in a ring.

The final structure of the mature spermatozoon, as shown in Fig. 8.11, contains **head, middle piece,** and **tail.** The head consists of the acrosome and an elongated nucleus. The middle piece is composed of the proximal centriole (the distal centriole often disintegrates), the axial filament, and the mitochondrial spiral. The tail is composed mainly of the axial filament.

As seen in Fig. 8.11, the sperm has lost most of the cytoplasm during maturation.

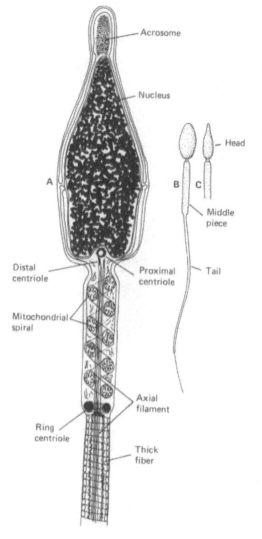

Fig. 8.11A–C. Diagram of a mammalian spermatozoon. (*A*) Detailed diagram as seen under the electron microscope. (*B, C*) Diagram of spermatozoon as seen in the light microscope. The sperm head is seen from the flattened side in *B* and from the narrow side in *C*. (From Balinsky, 1970. Redrawn by permission of W. B. Saunders, Co., Philadelphia).

Only the plasma membrane remains as a sheath around the mature sperm. This means that very little male cytoplasm is transferred to the egg during fertilization. This may be the reason for so-called **maternal effects** that can be caused by extrachromsomal genetic factors in the female cytoplasm. They are thought to be transmitted through the egg but not to be controlled by the genes of the developing embryo (see Chapter 20).

As pointed out at the beginning, the main concern in this chapter is the focus on the chromosomes or the chromatin (DNA) during sexual reproduction. But in order to fully understand what happens in the nucleus, some surrounding structures had to be considered also.

DuPraw (1970) writes that the sperm of animals and some plant cells are highly specialized as motile carriers of the species' haploid DNA component. Associated with this role is the development of various unique chromosome configurations characterized by tight packing of DNA, a low percentage of DNA-linked proteins, and complete cessation of RNA and DNA synthesis. Inoué and Sato (1966) studied the head of an intact cave cricket sperm with polarized light. They developed a model of chromosome configurations in this sperm (Fig. 8.12). Their studies revealed that the DNA base pairs lie in zigzags, with gaps that seem to indicate spaces between chromosomes. They postulated that the chromosomal DNA is a supercoiled supercoil arrangement.

8.2.3   Oogenesis and Syngamy

Oogenesis is the female's equivalent of gametogenesis. The first stages of gamete development are similar to those in the male. The ovary contains the immature germ cells called **primary oogonia.** These multiply rapidly like the primary spermatogonia by mitotic division to become the **secondary oogonia.** After several oogonial mitoses, the germ cells enter meioses and are called **primary oocytes,** which are enlarged cells. Because the egg contributes the greater share of developmental substances, growth is much more prominent and important in oogenesis than in spermatogenesis. The mature human egg, for instance, has a volume of 2,000,000 μm^3, while the volume of the mature human sperm is 30 μm^3. Thus, the volume of the egg is about 85,000 $\times$ that of the sperm (Fig. 8.13). The growth factor of the egg was mentioned in Chapter 7. It varies from 43 $\times$ in the mouse to 27,000 $\times$ in the frog, *Rana pipiens,* from a young oocyte to a mature egg (Balinsky, 1970).

Not only do the oocytes enlarge during this period, but also do their nuclei. They are often referred to at this stage as **germinal vesicles** (Purkinje, 1825). The size increase is due to the production of large quantities of nuclear sap. The duration of the primary oocyte stage is very long and may last from one to many years. The cells enter meiotic prophase. The homologous chromosomes pair but do not proceed further than diplotene. As mentioned in Chapter 7, during this time the chromosomes enter into a diffuse state by loosening up and extending into thin threads and loops. Because of their appearance, they are called **lamp brush chromosomes.** During this period great amounts of RNA are transcribed on the chromosomes. mRNA synthesis occurs in the loops of the lamp brush chromosomes. There seems to be also a very active rRNA synthesis in the nucleoli during this period. The nucleolus of the oocyte increases very much in size.

In amphibians many smaller nucleoli are formed during this stage. Approximately 1000 nucleoli line up on the periphery of the nucleus close to the nuclear envelope. Each of these nucleoli contains a circular DNA molecule up to 200 μm in circumference (Fig. 8.14) (Miller, 1964, 1966). At certain stages these DNA circles open up. They can be seen with the light microscope because they are heavily coated with RNA and protein. Callan (1966) called them **ring nucleoli.** Gall and Pardue (1969) tried to explain the origin of these ring nucleoli. They hypothesized that the nucleolar DNA's are synthesized on the nucleolus organizer region of the oocytes' lamp brush chromosomes and that this replication generates circular DNA molecules (**nucleolus chromosomes**) that eventually take their described position as vis-

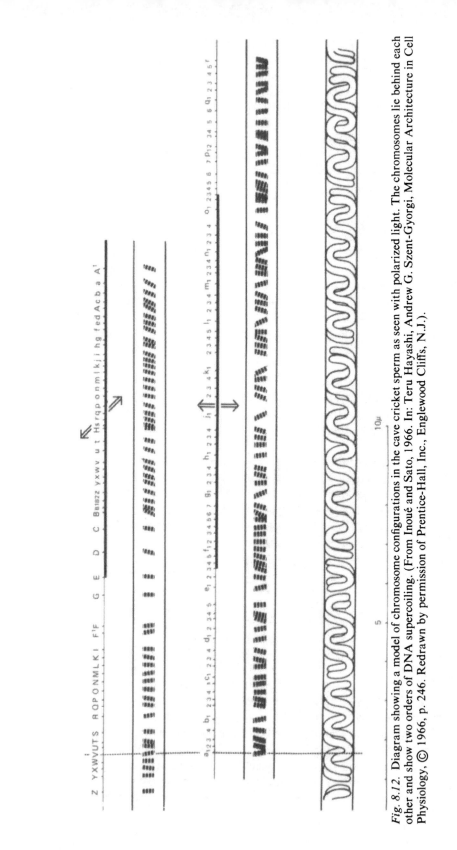

Fig. 8.12. Diagram showing a model of chromosome configurations in the cave cricket sperm as seen with polarized light. (From Inoué and Sato, 1966. In: Teru Hayashi, Andrew G. Szent-Gyorgi, Molecular Architecture in Cell Physiology, © 1966, p. 246. Redrawn by permission of Prentice-Hall, Inc., Englewood Cliffs, N.J.).

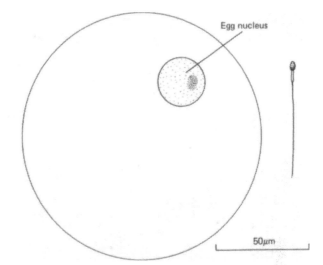

Fig. 8.13. Mammalian egg and spermatozoon drawn to the same scale. (From Mittwoch, 1973. Redrawn by permission of Academic Press, New York).

ible ring nucleoli just inside the nuclear envelope. DuPraw (1970) speculated that such metabolic DNA synthesis may eventually prove to be of widespread occurrence, possibly accounting for the origin of extranuclear DNA's such as mitochondrial or chloroplast chromosomes (Chapter 20). There is evidence that probably all ribosomal components (including proteins) are synthesized in the nucleolus (Flamm and Birnstiel, 1964; Birnstiel and Flamm, 1964; Birnstiel et al., 1964). The close aggregation of the nucleoli on the inside of the nuclear envelope suggests some transport to the cytoplasm via the envelope. Many reports suggest such a passage from the nucleus to the cytoplasm during the growth of the oocyte. Bretschneider and Raven (1951) and Logachev (1956) claimed to have observed nucleolar material passing through gaps in the nuclear membrane. More recent electron microscopic studies have confirmed these observations (Balinsky and Davis, 1963).

A large part of the growth of the oocyte is due to the accumulation and storage of nutrients such as lipids, polysaccharides, and proteins. A particularly great accumulation of such nutrients takes place in **oviparous animals,** which lay eggs that hatch after they are separated from the mother. Such animals need a large nutritive supply for the developing embryro, which cannot receive them directly from the mother at this time.

Most of the proteins and lipids are stored in large granules called **yolk platelets,** which are oval and flattened in one plane. It may be of interest in this connection that 80% of the cytoplasmic DNA found in amphibian oocytes is bound to these yolk platelets and only 20% to mitochrondria (Tyler, 1967; Brachet, 1969).

While part of this material is accumulated from internal synthesis, some of the growth of the oocyte also results from food substances secreted by special nurse cells that surround the oocytes (Fig. 8.15). These cells are called **follicle cells** and in mammals they originate from the germinal epithelium of the ovary. During the maturation process of the egg, the number of follicle cell layers increases.

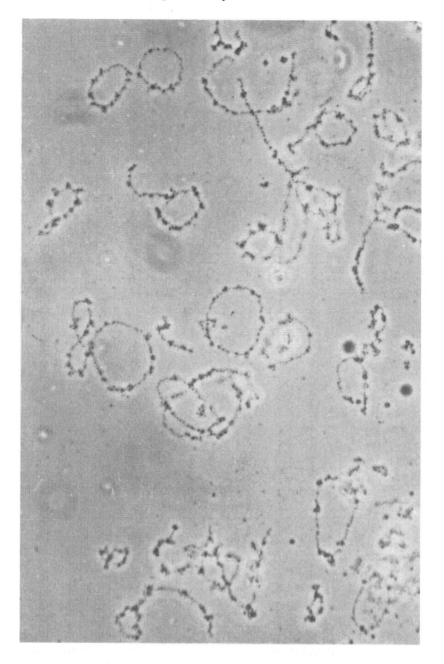

Fig. 8.14. Light microphotograph of circular, DNA-containing ring nucleoli of amphibian oocytes. ($\times$ 424). (From Miller, 1966. Reprinted with permission of the National Cancer Institute, Bethesda, Maryland).

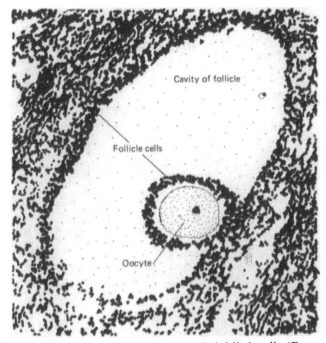

Cavity of follicle

Follicle cells

Oocyte

Fig. 8.15. Nurse cells that surround the mammalian oocytes, called *follicle cells.* (From Balinsky, 1970. Redrawn by permission of the W. B. Saunders Co., Philadelphia).

The primary oocyte eventually resumes meiotic division. At the end of meiosis I, two haploid cells are present, a large one, the **secondary oocyte,** and one small abortive cell, the so-called **first polar body** or **polocyte** (Fig. 8.16). The polocyte remains attached to the oocyte. Meiosis II also produces two cells of unequal size, a large functional egg called **ootid** or **ovotid** and a small abortive **second polar body.** During meiosis II the first polar body either disintegrates, remains undivided, or undergoes division to form a **third polar body.** In most species the three polocytes eventually disintegrate. Figure 8.16 shows the actual process of the two meiotic divisions. As the nuclear membrane breaks down at the end of prophase I, the chromosomes move from the center of the oocyte toward the periphery. The division then actually appears as a bulging and pinching off of the small polar bodies from the large oocyte. Two major purposes of this pinching off seem to be the elimination of half of the chromosomes by discarding them in the abortive primary polar body and a further growth of the oocyte that essentially receives all nutrient material because of unequal cell division. This process completes the maturation of the oocyte into a mature egg.

Fertilization or syngamy in animals is the process of sperm penetration (Fig. 8.17) into the egg and the union of the paternal and maternal gametes resulting in **zygote** formation. Thus, the homologous chromosomes that lost their partners during the process of male and female meiosis are now going to be matched again as homologous pairs. This is the final step in the recombination of genes from different sources as mentioned before. It makes possible the sharing of favorable genes

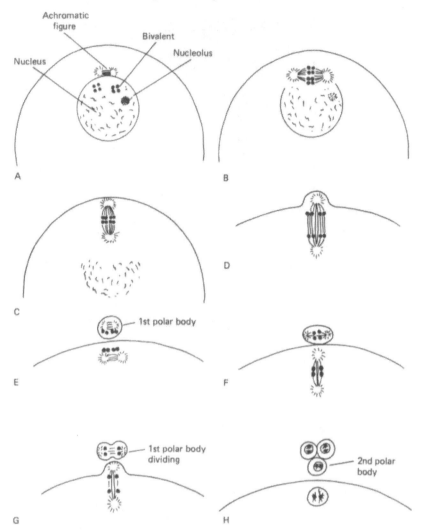

Fig. 8.16. The two meiotic divisions of the primary oocyte resulting in the maturation of the oocyte into a mature egg. (From Balinsky, 1970. Redrawn by permission of the W. B. Saunders Co., Philadelphia).

throughout a group of cross-fertile individuals such as a species, where before such genes were available only to one individual.

Fertilization in animals usually starts before the completion of oogenesis. Penetration of the sperm into the oocyte seems to be necessary for the maturation of the egg in a number of animal species. In some mammals, amphibians and insects, the oocyte is still in the prophase I stage at the time of sperm entry. The entrance of the sperm into the egg seems to activate the egg. If the egg is not fertilized at all, it may eventually degenerate. If it is fertilized, it goes into action. If meiosis has not been completed then sperm entrance will bring it to completion.

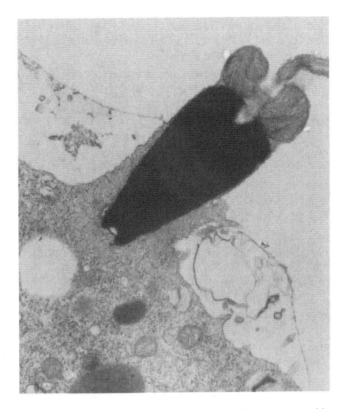

Fig. 8.17. Electronmicrograph of the process of sperm penetration in the sea urchin
Arbacia punctulata. (Courtesy of Dr. Everett Anderson, Department of Anatomy, Harvard Medical School, Boston, Massachusetts).

The process of fertilization can be divided into three substages:
1. The penetration of the oocyte membrane by the sperm,
2. **plasmogamy,** or the fusion of the cytoplasms of the two gametes,
3. **karyogamy,** or the fusion of the two pronuclei.

The penetration of the oocyte membrane has been studied in detail by light and
electron microscopy (Colwin and Colwin, 1967), but it will not be discussed here
in detail. The mechanism of the sperm penetration seems to be spurred by a chemical in that the acrosome produces enzymes known as **sperm lysine** that dissolve
the egg membrane locally (Tyler, 1948; Colwin and Colwin, 1961). Thus, the
acrosome seems to be a vital part of sperm penetration as it moves in front of the
sperm with the nucleus, centriole, middle piece, and tail trailing behind. Often,
the tail breaks off as the sperm enters the egg cytoplasm. Soon after entry the
nucleus usually turns around 180 degrees so that the centriole is now in a forward
position with the nucleus following behind. Any other parts of the former sperm
have now disconnected and disintegrate. The nucleus and the centrosome of the
sperm now both change their appearance. The nucleus, which was closely packed
in the sperm, becomes dispersed, granular, and enlarged (Longo and Anderson,
1968). The centrosome forms an **aster.** Both the sperm and egg nuclei become

similar in appearance and are now referred to as male and female **pronuclei** (Beneden, 1875). In higher animals the pronuclei increase progressively in volume until they are about 20 times their original size (Austin, 1969).

The female pronucleus also has to change position. It is located at the periphery of the egg where it completes its meiotic division (Fig. 8.16). As division is completed, the female pronucleus also migrates toward the male pronucleus in preparation of the fusion of both. This fusion generally takes place near the center of the egg. Fusion of the two pronuclei varies from species to species. In the sea urchin, for instance, fusion is complete at the onset. In higher animals, the two pronuclei suddenly diminish in volume and finally fade out altogether giving place to two chromosome groups. These chromosome groups move together and form a single group, which represents the prophase of the first cleavage division (Austin, 1969). But such complete fusion does not occur in all animal groups. In *Ascaris,* some molluscs and in annelids the chromosomes of the male and female pronuclei attach to the zygote spindle and remain in two separate groups until completion of the first cleavage division. Only then do the paternal and maternal chromosomes become enclosed by a common nuclear envelope. In the fresh water crustacean, *Cyclops,* the paternal and maternal chromosomes remain in two separate groups until after the gastrulation stage of embryonic development. The two groups actually can be observed forming bilobed nuclei. Each group forms its own nuclear envelope. But fusion of paternal and maternal chromosomes eventually occurs in all animal species.

Part V
Variation in Chromosome Types

Chapter 9
Polyteny and Lampbrush Chromosomes

Beginning with this part a new phase of this book is introduced. The preceding parts of the book dealt with the theme: structure (Part II), function (Part III), and movement (Part IV) of the chromosomes. The following parts of the book deal with the variations on the theme: variations in chromosome types (Part V), variations of chromosome structure (Part VI), variation of chromosome number (Part VII), and variation of chromosome function and movement (Part VIII).

In the discussion of chromosome structure in Chapters 2 and 3, *normal* chromosomes usually seen during mitotic and meiotic cell divisions were described. However, in specialized tissues or in certain species or species groups, nonstandard or unusual chromosomes exist that serve as valuable tools for cytogenetic research. The first group of such chromosomes to be discussed distinguish themselves by their unusual size. These are the **polytene** and **lampbrush chromosomes.** These chromosomes reveal structural details that cannot be seen in ordinary somatic chromosomes.

9.1 Polyteny vs. Endopolyploidy

Polyteny (Koltzoff, 1934) is found in salivary gland nuclei, nurse cells, and other larval tissues of dipterous flies and in intestinal cells of larval mosquitoes. The phenomenon was discovered by Balbiani in 1881, but its cytogenetic significance and importance were not revealed until Kostoff (1930), Heitz and Bauer (1933), and Painter (1933, 1934) made their important contributions. Heitz and Bauer's research was briefly mentioned in the historical treatise (Chapter 1).

Polyteny and **endopolyploidy** both are results of **endomitosis** (Geitler, 1937). Endomitosis is the process of chromosome duplication without karyokinesis and cytokinesis. In endomitosis the DNA content of the cell is doubled or at least increased. In polyteny the duplicated chromosomes do not separate into sister chromatids, while in endopolyploidy they separate, resulting in chromosome doubling. Figure 9.1 demonstrates this difference. If a diploid cell with 4 chromosomes ($2n = 4$) goes through the process of endomitosis, it can form a polytenic cell in which the chromosome number does not change ($2n = 4$) but the chromatid number changes from 8 to 16, doubling the amount of DNA. Each chromosome then has 4 chromatids instead of 2. White (1935) called such chromosomes

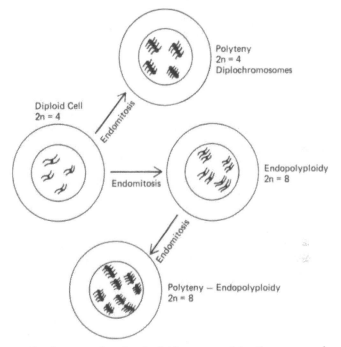

Fig. 9.1. The phenomena of polyteny and endopolyploidy as caused by the process of endomitosis.

diplochromosomes. At this point they have gone through two duplications without centromere division since they were exposed to the last effective mitosis. If they go through three such duplications without centromere division, they will consist of eight chromatids each and are then called **quadruplochromosomes.** If the diploid cell (Fig. 9.1), by the process of endomitosis, forms an endopolyploid cell, the chromosome number doubles (2n = 8) as well as the chromatids and the DNA. A third possibility is indicated in Fig. 9.1 where endopolyploidy and polyteny coexist in the same cell. In this example, a diploid cell (2n = 4), which by endomitosis formed an endoploid cell (2n = 8), in turn experiences a second endomitosis that results in polyteny. Such instances have been observed by White (1946, 1948), Matuszewski (1964, 1965), and Henderson (1967a, 1967b) in the salivary gland nuclei of various Cecidomyiidae. In the salivary glands of the gall midge, *Lestodiplosis,* White found one super-giant cell at the ascending portion of the duct that had the customary polytene chromosomes but in a polyploid number, in this instance 32-ploid.

The prefix "endo" in endomitosis and endopolyploidy indicates that the process takes place within the nucleus of the cell without a breakdown of the nuclear membrane. The chromosomes condensate, but there is no spindle formation or orientation of the chromosomes on a metaphase plate. In the case of endopolyploidy, the chromosomes seem to fall apart. There is no passing to any poles. At the end of such an incomplete mitosis, the chromosomes decondensate and may become diffuse.

Endopolyploidy is apparently a much more extensively occurring phenomenon than it was originally visualized. Apart from the pathological occurrence of endomitosis in malignant tissues of mice (Levan and Hauschka, 1953), endopolyploidy is a normal process in plant and animal cells. A review on the subject was published by Tschermak-Woess (1963). In some diploid plants, for instance, the vascular tissues of the roots are polyploid (Jacobj, 1925). Such an increase in chromosome number is also characteristic of the mother cells from which latex tubes, vessel members, collenchema cells, and other specialized cells such as idioblast cells arise (Tschermak-Woess and Hasitschka, 1954). Müntzing (1961) reports endopolyploidy in the roots of spinach. Wipf and Cooper (1938) found that in legumes like red clover, common vetch, and garden peas, the root nodules that are involved in nitrogen fixation have plant cells with twice the chromosome number observed in the rest of the plant. In the vetch in addition to cells with the normal chromosome number 2n = 12, cells with 2n = 24, 48, and even 96 chromosomes have been reported. Endopolyploidy in animals is best known in insects. Geitler (1937, 1939, 1941) showed that in the salivary glands of the water insect *Gerris lateralis* (2n = 21, XO type), many cells were 512-ploid, some 1,024-ploid, or even 2,048-ploid.

9.2   Morphological Characteristics of Polytene Chromosomes

The **polytene chromosomes** have been most thoroughly studied in the salivary glands of *Drosophila*. Their discovery has been briefly mentioned in Chapter 1. Bridges was one of the most diligent researchers in mapping these **giant chromosomes.** The length of these chromosomes in the late larval stage of *Drosophila* is about 100 times the length of somatic chromosomes. They are believed to be in the interphase stage, and their enormous length could be explained by molecular unfolding. According to the **polyteny hypothesis** of Bauer (1935), these chromosomes originate from repeated endomitotic cycles forming bundles of numerous chromonemata (chromatids) that are held together by **somatic pairing** and consequently they are represented in the haploid number. Each so-called **polytene chromosome** really represents a very close union of two homologous chromosomes that simulate the appearance of only one single chromosome.

In a typical salivary gland cell of *Drosophila melanogaster,* the four chromosomes are all united in the chromocenter by their centromere regions (Fig. 9.2). This gives the cell the appearance of five long and one short strand radiating out of the chromocenter. The strands represent the arm of the telocentric X chromosome, two arms of chromosome 2 (2R, 2L), two arms of chromosome 3 (3R, 3L), and the tiny arm of the small telocentric chromosome 4. Along these arms appear the distinct darkly stained **bands** that divide the chromosomes into band and **interband** regions. These bands are of different size and staining capacity, and some of them appear as **doublets.** According to Lewis (1945), doublets represent **one-band-tandem-repeats.** The bands are visualized as chromomeres that associate closely at the same level of the interphase chromonemata. Each band really is a disk that

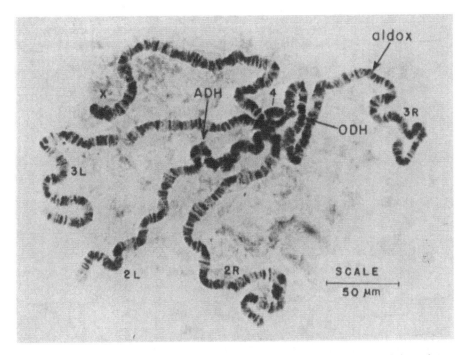

Fig. 9.2. Giant polytene chromosomes from the salivary glands of *Drosophila melano-gaster*. The chromosomes are closely connected at their centromere regions in a compact chromocenter from which the chromosome arms protrude. The positions of the structural genes for the three enzymes aldehyde oxidase (aldox), alcohol dehydrogenase (ADH), and octanol dehydrogenase (ODH) are indicated. X–X chromosome; 2L and 2R–left and right arms of chromosome 2; 3L and 3R–left and right arms of chromosome 3; 4–chromosome 4. (From Ursprung et al., 1968. Copyright 1968 by the American Association for the Advancement of Science, Washington, D.C.).

extends throughout the thickness of the chromosome, and it is composed of tightly packed chromomeres. These chromomeres are represented in a number that corresponds to that of the chromonemata. Since chromomeres are actually areas of tightly folded and coiled chromatids, the bands, consequently, are also tightly packed areas on the chromosomes. DuPraw's (1970) interpretation of such bands is shown in his model in Fig. 9.3. The high degree of visibility of these bands is caused by the multistrandedness of the polytene chromosomes. In DuPraw's simplified model, only 4 strands are visible in the interbands (Fig. 9.3). Most recent estimates envision the polytene chromosome as a cable-like structure formed by intertwining of about 1000 to 4000 identical threads with diameters from about 20 nm to 30 nm. In *Chironomus* more than 16,000 threads may be present resulting from up to 14 replications. Each thread is thought to be a unit chromatid containing a single DNA molecule 2 nm in diameter, coated with protein and tightly packed into the fiber by supercoiling (DuPraw, 1970). An unusually good photograph of a highly stretched section (division 101) of chromosome 4 of *D. melanogaster* is shown in Fig. 9.4.

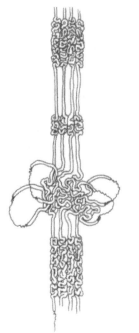

Fig. 9.3. DuPraw's model of a polytene chromosome consisting of multiple, side-by-side strands, each corresponding to an unfolded metaphase unit chromatid. Concentrated bands of DNA arise by tight folding in all the unit chromatids at specific sites. At some sites DNA may unpack for RNA synthesis, leading to the formation of a puff. (From DuPraw and Rae, 1966. Redrawn by permission of Macmillan Journals, LTD., London).

Polytene chromosomes have also been reported in plants. They were detected in certain cell types of the embryo by Tschermak-Woess (1956), Hasitschka (1956), and Nagl (1962, 1965, 1969a). Nagl (1967, 1969b, 1970) and Avanzi et al. (1970) also found polytene chromosomes in the suspensor cells of two species of *Phaseolus* and they studied the structure and function of these chromosomes in detail.

9.3   Puffing

Since the early discovery of the salivary gland chromosomes by Balbiani, it has been known that they may have a number of small to large swellings on them that were later called **Balbiani rings** and **puffs**. Balbiani rings are restricted to the Chironimidae while puffs occur in all Diptera. Balbiani rings are extremely large puffs. There are also some structural differences between Balbiani rings and puffs.

Puffing seems to throw light on the metabolic function of the giant interphase chromosomes that seemed to puzzle cytologists for a long time. The first one to observe this phenomenon was Bridges (1935). Puffing involves an unfolding of DNA in the band regions (Fig. 9.5). Such bands originally have very distinct outlines, but with progressive puffing they become more diffuse until they disappear completely. This process is reversed as puffs disappear. Puffing also can spread to other adjacent bands (Pelling, 1966). Apparently, there is a difference between **RNA-puffs** and **DNA-puffs.** Both kinds of puffs are unusually active in RNA synthesis but in the DNA-puffs, additional DNA is produced during the time of puffing. DNA-puffs

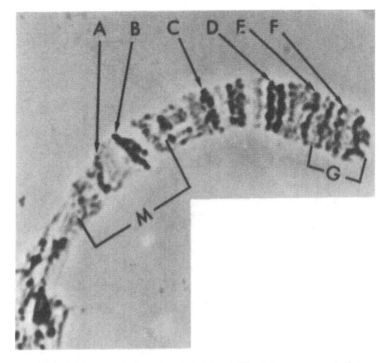

Fig. 9.4. Photograph of highly stretched section (Division 101) of chromosome 4 of *Drosophila melanogaster*. (From Hochman, 1974. Reprinted by permission of Cold Spring Harbor Laboratory, Cold Spring Harbor, New York).

have been observed in the midges of the family Sciaridae (Breuer and Pavan, 1955; Rudkin and Corlette, 1957; Swift, 1962; Crouse and Keyl, 1968; Pavan and DaCunha, 1969). It has been proposed that such DNA-puffs are sites of specific **gene amplification** and that they are responsible for differentiation and development. This theory is supported by the fact that puffs develop at different sites during different stages of larval development (Beermann, 1952; Pavan and Breuer, 1952; Breuer and Pavan, 1955).

Not all bands are visibly involved in the puffing phenomenon. Only 15% of the estimated 2000 bands in *Chironomus tentans* have been associated with puffing (Pelling, 1964, 1966). Puffs vary in size from just barely swollen bands to very extensive swelling. A possibility exists that minor puffs cannot be detected microscopically. The phenomenon of puffing is closely associated with the different theories on gene location. Bridges (1938), Berger (1940), Welshons (1965), and Lefevre (1974), for instance, claim that the bands are the sites of the genes. These investigators are proponents of the so-called **one band-one gene** concept. They lean toward the idea that puffing would be an indication of genic activity. Koswig and Shengun (1947) and Fujita and Takamoto (1963) are in favor of the idea that the genes are located in the interbands. This hypothesis would be in line with Crick's

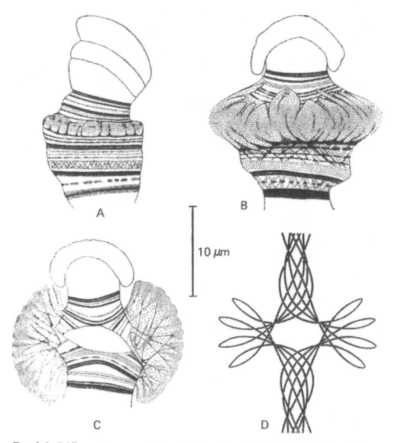

Fig. 9.5. Different degrees (*A* to *C*) of puffing of the Balbiani ring in the salivary gland chromosome IV of *Chironumus tentans*. The interpretation of the course of the chromosomal fibers in the region of the Balbiani ring is given in *D*. (After Beerman, 1952, from Clever, 1964).

(1971) **general chromosome model** that suggests that the fibrous coding DNA is in the interbands and that the globular DNA in the bands is control DNA (see Section 3.4.4).

9.4   Super Chromosomes

The extent of polyteny can be increased by infection with microsporidian protozoan parasites (particularly of the genus *Thelophania*). Such parasites cause the salivary gland cells to grow very large and to produce **super chromosomes.** Such chromosomes have been studied by Diaz and Pavan (1965), Pavan and Basile (1966), Pavan (1967), Pavan and DaCunha (1968), and Diaz et al. (1969). When super chromosomes are stained, they are visible to the naked eye. They may have from

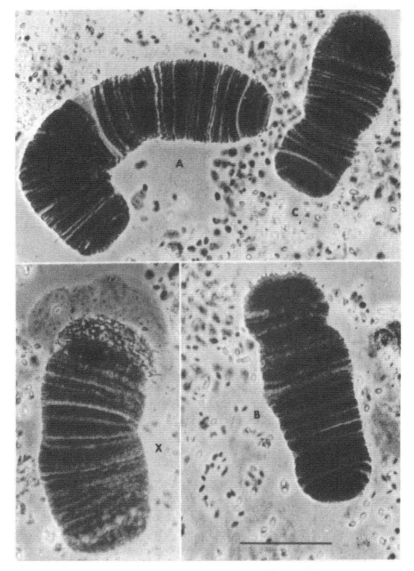

Fig. 9.6. The four (X, A, B, C) polytene super chromosomes of an infected salivary gland cell of *Rhynchosciara angelae.* Scale: 50μm. (Courtesy of Dr. C. Pavan, University of Campinas, Sao Paulo, Brasil. Reprinted by permission of the C. V. Mosby Company, Saint Louis).

250,000 to 1,000,000 chromonemata. Roberts et al. (1967) estimated that nuclei infected by parasites may have 2, 4, 8, 16, or 32 times the DNA content of normal salivary gland nuclei. This indicates that up to five extra cycles of replication may have occurred. Puffing was inhibited in larvae of *Rhynchosciara* that were infected by parasites. In Fig. 9.6 super chromosomes of an infected salivary gland cell of *Rhynchosciara angelae* can be seen.

9.5   Somatic Synapsis

Synapsis is generally considered to be a meiotic process. However, as we have seen in this chapter, it is also occurring in other types of cells. Depending on the closeness of the pairing association of the homologous chromosomes in such pairing configurations, the phenomenon is either called **somatic pairing** or **somatic synapsis.** Somatic pairing has been known for a long time in plants (Strasburger, 1904, 1905; Sykes, 1908; Digby, 1910; Nemec, 1910) and animals (Montgomery, 1906; Stevens, 1908). The most striking type of somatic pairing exists in the Diptera where the homologous chromosomes lie next to one another and closely parallel throughout interphase. Somatic pairing is less close at metaphase. The most intimate pairing is observed in the salivary gland chromosomes, where it is usually associated with synapsis. There is, however, an important difference between meiotic and somatic synapsis. Meiotic synapsis is always two-by-two. If more than two homologous chromosomes are present during meiotic synapsis (in trisomics or triploids, Chapters 15 and 16), only two-by-two pairing occurs at any given section of the chromosomes. In the somatic synapsis of salivary gland chromosomes, pairing occurs three-by three in triploids (Fig. 9.7).

9.6   Lampbrush Chromosomes

As already mentioned in Chapters 7 and 8, the so-called **lampbrush chromosomes** are another type of giant chromosomes that occur in the diplotene stage of primary oocyte nuclei in vertebrates and invertebrates and also in the Y chromosome of *Drosophila* spermatocytes. Lampbrush chromosomes can be even larger than the polytene giant chromosomes of the Diptera, but their diameter is much less. The longest such chromosomes of about 1 mm have been found in urodele amphibia (Rieger et al., 1976). These chromosomes are characterized by a typical diplotene bivalent appearance showing chiasmata, and from the darkly stained chromosomes thousands of thin chromatin loops are extending laterally at rectangles (Fig. 9.8).

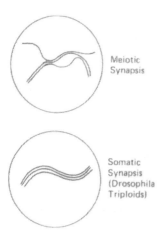

Meiotic
Synapsis

Somatic
Synapsis
(Drosophila
Triploids)

Fig. 9.7. Comparison of chromosome pairing in meiotic and somatic synapsis.

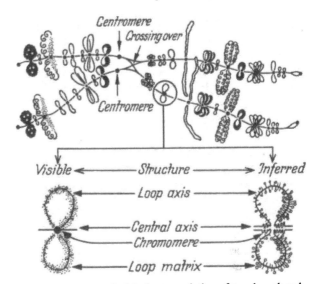

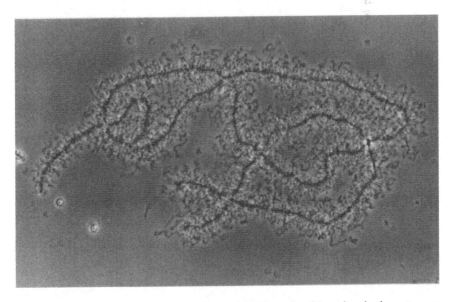

Fig. 9.8. Diagram showing typical appearance of a bivalent consisting of two lampbrush chromosomes. (From Lewis and John, 1963).

The first lampbrush chromosomes were discovered by Rückert (1882) in shark. Almost all of the recent research has been with the urodeles, particularly of the genus *Triturus*. The most improved techniques in studying lampbrush chromosomes were applied by Gall and Callan (1962). A phase-contrast photograph of a portion of an isolated lampbrush chromosome is shown in Fig. 9.9.

Fig. 9.9. Phase contrast photograph of a portion of a pair of lampbrush chromosomes isolated from an oocyte of the newt *Triturus viredescens* ($\times$ 337). (From Gall, 1966).

Beermann (1952) first realized that the lampbrush chromosome loops are similar in function to the puffs of the polytene chromosomes. But certain major differences are apparent between the puffs and the lampbrush loops. A puff consists of thousands of identical DNA loops, each arising from a chromomere at the puff producing locus. While in the lampbrush chromosome, all or almost all chromomeres have formed loops, in the polytene puff only a few unfold at any one stage of development or in any particular tissue (White, 1973).

DuPraw (1970) mentions that each pair of loops along a giant lampbrush chromosome has its own specific morphology, and often one side of the loop is thicker than the other, as if more RNA had accumulated there. This suggests that the loops are units of genetic activity. Apparently, along each loop mRNA transcription occurs.

Chapter 10
Ring-Chromosomes, Telocentric Chromosomes, Isochromosomes, and B Chromosomes

This chapter is a continuation of the discussion of unusual chromosome types. As mentioned before, the term "unusual" in this connection is a relative term. We very often have a certain concept of things, and whatever deviates from this concept we call "unusual". Because **ring-chromosomes, telocentric chromosomes, isochromosomes, or B chromosomes** differ from the majority of chromosomes in humans, animals and plants, they are considered unusual. But apparently, in some instances, such chromosomes fulfill a need that cannot be met by any other chromosome type. There is no obvious connection between these four chromosome types except that they all deviate in some way or another from the prototype as described in Chapters 2 and 3.

10.1   Ring-Chromosomes

Standard chromosomes of higher organisms (**eukaryotes**) usually have two ends and do not form a continuous ring. However, the chromosomes of lower organisms such as **prokaryotes** (e.g., bacteria like *Escherichia coli* and some viruses) normally have ring-shaped chromosomes. Often such chromosomes are referred to as **genophores** (Ris, 1961) in order to emphasize the difference in structure. A linkage map of *E. coli* is shown in Fig. 10.1. It demonstrates the circular structure of the bacterial genophore. Such genophores are more than 1 mm in length and consist of a single DNA molecule that is tightly packed into a **nucleoid** (bacterial nucleus without a nuclear envelope) of only 1 μm in length. The DNA in such genophores is considered to be naked or pure, seemingly lacking any kind of histones that are generally associated with DNA in the chromosomes of eukaryotes and carrying less protein. The diameter of a genophore of *E. coli* is reported to be 4 nm, which is twice the diameter of a Watson-Crick double helix (Miller et al., 1970). A possible **unidirectional model** for DNA replication of circular chromosomes has been developed by Cairns (1963). This model is shown in Fig. 10.2. Replication starts at a fixed point and always proceeds in the same direction. Because the complementary strands of the DNA molecule are twisted in a double helix, the circular DNA must rotate around its own axis as the old and new strands separate. Another more recent model indicates that DNA replication of the circular *E. coli* chromo-

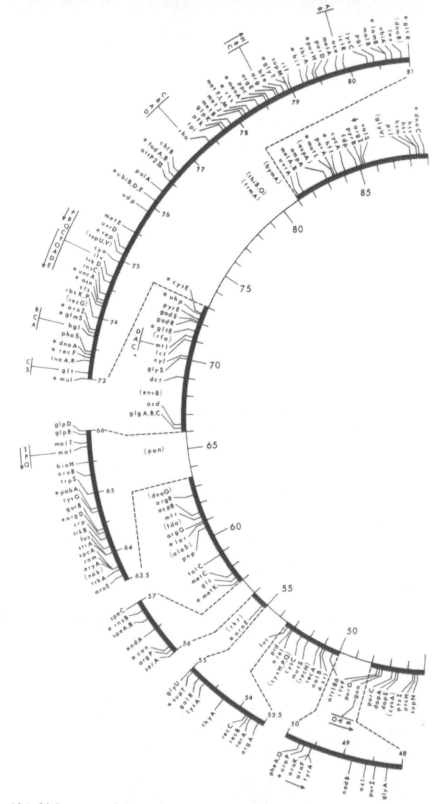

Fig. 10.1. Linkage map of the *Escherichia coli* genophore. The numbers represent minutes required to transfer the character during conjugation. Some portions of the map are expanded to show particular regions in greater detail. (From A. L. Taylor and C. D.

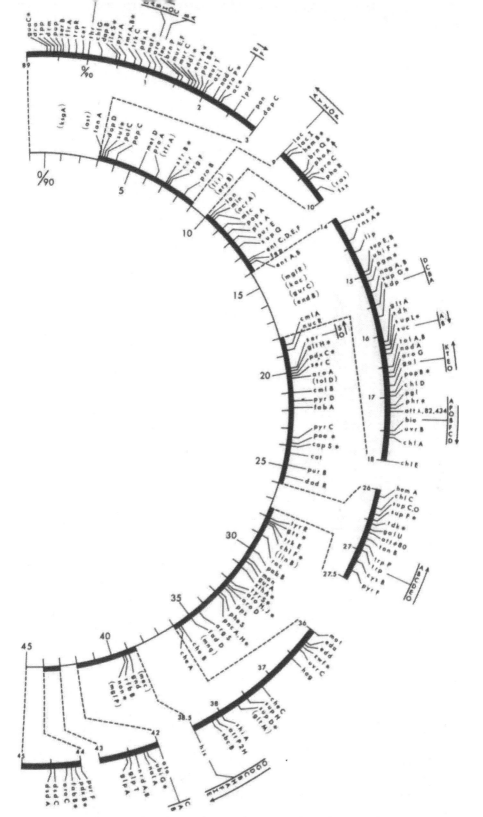

Trotter, 1972. Reprinted with permission of the American Society of Microbiology, Washington, D.C.).

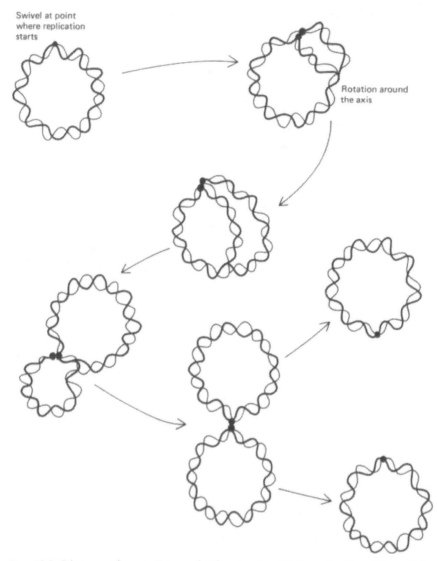

Fig. 10.2. Diagram of a possible model for circular DNA replication. (From Cairns, 1963. Redrawn by permission of the Cold Spring Harbor Biological Laboratory, Cold Spring Harbor, New York).

some proceeds in a **bidirectional fashion** from a fixed point (Masters and Broda, 1971).

Apart from such normally occurring ring-chromosomes in prokaryotes, such chromosomes frequently form in eukaryotes as a result of structural chromosome changes. Chromosomes in higher organisms are not naturally ring-shaped. Ring-chromosomes have been detected in humans, *Drosophila,* and certain plant species. They were most thoroughly studied in maize by McClintock (1938b, 1941a, 1941b, 1944). In maize, ring-chromosomes are likely to form dicentric double sized rings

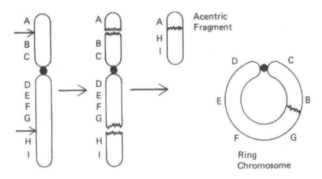

Fig. 10.3. Diagrammatic presentation of the formation of a ring chromosome.

that break at anaphase. This very often leads to instability of the rings. The details of this behavior will be discussed in Chapter 12. Normal chromosomes do not form rings because they are believed to have telomeres on each end (Chapter 3). Telomeres seem to prevent the union of chromosome arms into ring formation. A chromosome can form a ring-chromosome by fusion of the raw ends (Fig. 10.3) only if it has two terminal deletions (Chapter 12) producing a centric segment with two raw ends and two acentric fragments. As seen in the illustration, the ring-chromosome (BC.DEFG) inherits the centromere, and the terminally deleted material can unite into an acentric fragment (AHI) that eventually gets lost from the nucleus. A ring-chromosome lacks the genetic information that was carried by the terminally deleted fragments. After the occurrence of such a deletion, an organism or tissue will be heterozygous for the deletion having a normal standard chromosome (ABC.DEFGHI) and a deleted ring-chromosome (BC.DEFG). Ring-chromosomes generally are meiotically unstable. During mitosis they can produce two daughter rings of equal size that are regularly distributed to the daughter cells. Such ring-chromosomes can be somatically stable.

Ring-chromosomes have been observed in several human syndromes. According to Borgoankar (1975), they involve all human chromosomes except 2, 10, 11, and 12. Since 1975 a ring-chromosome 2 was found in a newborn (Vigfusson, unpublished). A karyotype of the ring-chromosome 2 individual is shown in Fig. 10.4. Figure 10.5 shows an enlarged photograph of homologous chromosomes 2, one of them being the ring-chromosome.

Eight children with heterozygous ring-chromosome conditions in the E group (Er) (see Fig. 2.3) had generalized mental and developmental retardation and a variety of congenital malformations (De Grouchy et al., 1968; Hamerton, 1971b). Such clinical features could be expected since the ring-chromosome necessarily involves two terminal deletions. In these instances the ring-chromosomes were unstable. Another eleven patients were observed with 2n = 46 chromosomes and with a ring-chromosome replacing one of the standard chromosomes in the D group (Lejeune et al., 1968a; Hamerton, 1971b). The majority of these had developmental and mental retardation and variable congenital malformations. It could not be established if any of these ring-chromosomes involved the same D chromosome. In six of the eleven D ring-chromosome (Dr) patients, the rings seemed to be relatively

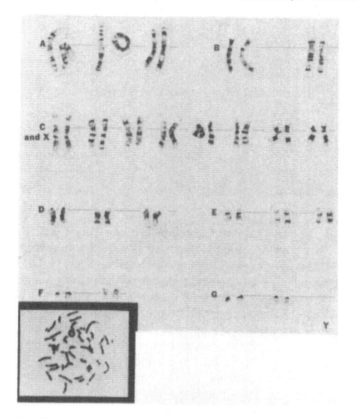

Fig. 10.4. Human G-banding karyotype of a newborn with a ring chromosome 2. Insert: Original cell. (Courtesy of Dr. N. V. Vigfusson, Department of Biology, Eastern Washington University, Cheney, Washington).

stable. Consequently, the D rings seem to be more stable than the E rings. C ring stability was studied by Shaw and Krooth (1966) over about 70 cell generations. The proportion of the cells containing ring-chromosomes diminished with time and they disappeared completely in two cell lines. A case of ring-chromosome **mixoploidy** was observed in a female infant with apnoeic spells (partial suspension

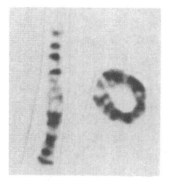

Fig. 10.5. Normal and ring-chromosome 2 of same patient as in Fig. 10.4. (Courtesy of Dr. N. V. Vigfusson).

of breath), abnoral ears, and hypoplastic (below normal size) nails. Mixoploids (Nemec, 1910) are chimeras in which the cells vary according to their chromosome numbers. In this case the blood cultures revealed two different cell lines, one normal (46, XX) and one trisomic (47, XX r+). The trisomic cell line had a ring-chromosome in addition to the normal cell complement. The ring-chromosome varied in size from a G group to a D group chromosome (see Fig. 2.3). The proportion of cells containing the ring-chromosome diminished from 20% in early blood cultures to 8% in blood cultures taken at the age of 13 months (Hamerton, 1971b). A probable case of an A ring-chromosome (Ar) was recorded by Cooke and Gordon (1965).

Other ring-chromosome syndromes in humans causing congenital malformations have been recorded by Smith-White et al. (1963), Aula et al. (1967), and Gripenberg (1967). In some of these the X chromosome formed the ring.

Ising and Levan (1957) observed one or more ring-chromosomes in some cells of lung and stomach carcinoma with chromosome numbers between $2n = 70$ to $2n = 80$. Ring-chromosomes form various kinds of syndromes in men that are expressed more or less severely according to the size of the deletions involved. If only a small amount of genetic information is missing, the patient may be affected little or not at all.

Reasonably stable stocks of *Drosophila* having a ring-shaped X chromosome have been maintained in the laboratory (Morgan, 1933; Schultz and Catcheside, 1937; Swanson, 1957). Brown et al. (1962) discovered that such stocks showed mosaicism (mixoploidy) as a result of somatic crossing over and because of ring elimination. A ring Y-chromosome in *Drosophila hydei* was described by Beck et al., 1979.

10.2   Telocentric Chromosomes

Telocentric chromosomes (Darlington, 1939a) or **telocentrics** are those that have a terminal centromere. They are generally not considered to exist in nature but are formed by centromere misdivision. If at mitosis the centromere divides transversely instead of longitudinally (Fig. 10.6), the result is two telocentric chromosomes each of which inherits a part of the original centromere. Since apparently the entire centromere is required for normal centromere function, such telocentric chromosomes are usually eliminated after a few cell divisions. If, however, **isochromosomes** (Section 10.3) are formed from telocentrics, such chromosomes seem to be maintained and stabilized. Isochromosomes can form if the telocentric arm reduplicates in interphase and the two chromatid arms do not become completely separated during the next mitotic interphase but stay united at the half-centromere. The telocentric then becomes a metacentric isochromosome (Fig. 10.6). Such chromosomes have two identical arms that are, genetically speaking, homologous.

There is some reason for the fact that a centromere is not completely functioning if it does not have chromosome arms on both sides. It is possible that the centromeric properties reside in the chromomeres flanking the unstainable centromere on either side (White, 1973).

Telocentrics originating from centromere misdivision have been reported in maize,

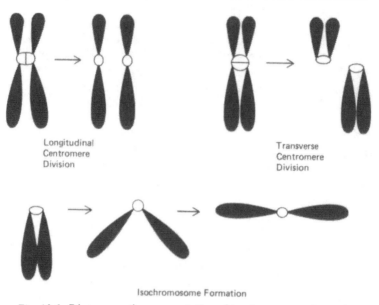

Longitudinal
Centromere
Division

Transverse
Centromere
Division

Isochromosome Formation

Fig. 10.6. Diagrammatic representation of isochromosome formation.

wheat, cotton, and tomato. Maize telocentrics were reported by McClintock (1932) and Rhoades (1936, 1938, 1940). They are mitotically unstable. The telocentrics of wheat (Sears, 1952a, 1952b; Morris and Sears, 1967; Sears, 1969) are not as unstable as those of maize and can be called semipermanent. Brown (1972) suspects that the centromere fractions in wheat telocentrics are more than just half-centromeres. Sears (1962, 1966, 1969) reports the successful use of wheat telocentrics in gene mapping. Telocentrics are now available for most of the 42 chromosome arms in wheat. Most of them are maintained as ditelosomic lines. A ditelosomic line in wheat $(2n = 42)$ is one that has two homologous telocentric chromosomes in addition to 20 normal chromosome pairs $(20^{II} + 2^{t})$. The absence of a whole chromosome arm in a telocentric allows positioning of a gene in that arm as well as determining the distance of that gene from the centromere. One advantage of mapping genes with telocentrics is the fact that most telocentrics are transmitted poorly through the pollen. Endrizzi and Kohel (1966) mapped three chromosomes in cotton by the use of telocentrics. Khush and Rick (1968c) have used telotrisomics to map genes in tomato. A telotrisomic is a normal disomic with an extra telocentric chromosome in addition $(2n + t)$. So-called "natural telocentrics" have been reported for Protozoae by Cleveland (1949), for Crustaceae by Melander (1950a, 1950b), for mouse by Tjio and Levan (1954), for cattle by Melander (Melander and Knutsen, 1953; Melander, 1959), for grasshoppers by John (John and Hewitt, 1966, 1968; John and Lewis, 1968), and for fish by McGregor (1970). Such observations may really involve **acrocentrics,** which are often mistaken for telocentrics. Acrocentric chromosomes (White, 1945) are those in which the centromere is very close to one end of the chromosome. In line with what was stated before, White (1957, 1973; White et al., 1967) thinks that all naturally occurring chromosomes cannot be telocentrics but are acrocentrics. He believes that there is always a minute second arm even if it cannot always be seen by conventional techniques.

10.3   Isochromosomes

Isochromosomes (Darlington, 1940) are metacentric chromosomes with two homologous arms. Such chromosomes really are a reverse duplication of the constitution ABC.CBA (see Fig. 10.7).

Their origin by centromere misdivision and reduplication of the telocentric fragments has been explained in Section 10.2 (Fig. 10.6). At meiosis isochromosomes can act in three different ways (Sen, 1952; Elliott, 1958):

1. internal pairing
2. fraternal pairing
3. normal pairing

In **internal pairing** the two arms of the isochromosome pair with each other in pachytene (**autosynapsis**) and after terminalization form a **ring univalent** at the end of first prophase in diakinesis (Fig. 10.7A). In **fraternal pairing** one or both of the arms of the isochromosome pair with a homologous arm of another chromosome (Fig. 10.7B); this can happen if the carrier is a secondary trisomic (Chapter 16) in which the isochromosome exists as an extra chromosome ($2n+i$). In **normal pairing** the isochromosome pairs with another one just like it (Fig. 10.7C).

The attached X chromosome of *Drosophila* is a classic example of an isochromosome (Morgan, 1922). This chromosome consists of two normally acrocentric X chromosomes attached at their centromeric regions and possessing a single centromere. The origin of this chromosome is not known. This attached X causes a

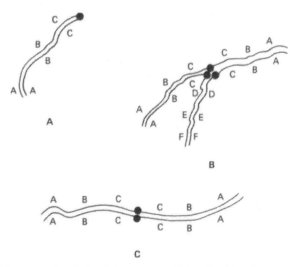

Fig. 10.7A–C. Different possible ways of meiotic chromosome pairing if an isochromosome is involved: (*A*) Internal pairing. (*B*) Fraternal pairing. (*C*) Normal pairing.

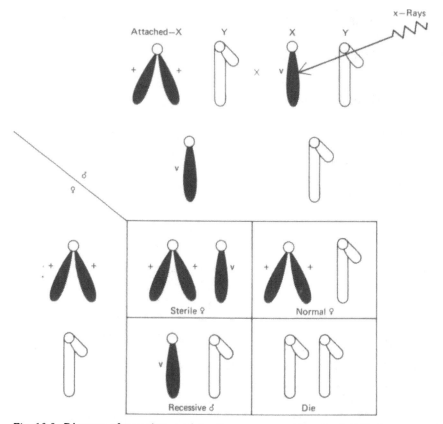

Fig. 10.8. Diagram of noncrisscross inheritance as caused by attached X chromosome in *Drosophila melanogaster* showing transmission of *vermilion* eye color (*v*). (After Morgan, 1922. Redrawn by permission of the Marine Biological Laboratory, Woods Hole, Massachusetts).

noncrisscross inheritance. It can be used for detecting sex-linked induced mutations (Fig. 10.8). If XXY females with an attached X are mated with males in which a recessive mutation such as *vermilion* eye color (v) has occurred by x-ray treatment, then all the males of the next generation express this recessive character.

There apparently is a strong case for the occurrence of attached X chromosomes in humans. The isochromosome in this case is formed by the two long arms of the X chromosome. In an individual heterozygous for such an isochromosome (XX qi) the long arm of X is represented three times and the short arm only once (Fig. 10.9). Such condition causes the phenotypic expression of an XO type or Turner syndrome (Chapter 15). Such **isochromosome Turner syndromes** can be diagnosed by unusually large Barr bodies that are formed by the iso-X (XX) (Brown, 1972). Most humans with an isochromosome are mixoploids (45, X/46, XX qi) (Hamerton, 1971a). The isochromosome is shut off and the functioning X behaves like an XO Turner syndrome.

Other isochromosomes in humans involve the Y chromosome, the D group, and the

Fig. 10.9. Diagrammatic representation of the hetero-zygous X isochromosome condition (XX qi) in humans.

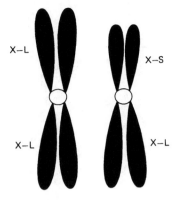

G group. Jacobs and Ross (1966) found two females with one X chromosome and a possible isochromosome formed by two long arms of the Y (46, XY qi). The clinical characteristics of these individuals were ovarian dysgenesis (failure of men-struation), streak gonads, and primary amenorrhea (defect of the ovary). The fact that both cases expressed femaleness may indicate that the male determining fac-tors of the Y are located on the short arm. Another kind of isochromosome has been presumed for the long arm of a D group chromosome by Therman et al. (1963) and by Giannelli (1965a). Therman et al. found mixoploidy of two cell lines, one having a short arm deficiency in a D group chromosome, the other hav-ing a D group isochromosome (46, XX, Dp-/46, XX Dqi). They presumed that the Dqi line arose from the Dp- line during early cleavage. Hamerton (1962) and Polani et al. (1965) found an isochromosome that they believed involved the long arm of a G group chromosome. Hsu (1969) found isochromosome heterozygosity in six out of seven investigated rats of the species *Sigmodon minimus* from New Mexico. The isochromosome consisted of the fusion of the long arms of the two homologues of an acrocentric chromosome.

Isochromosomes have been used to test the effect of colchicine on chiasma fre-quency (Driscoll and Darvey, 1970). Colchicine was applied to wheat after the last premeiotic mitosis until metaphase I with the result that chiasma frequency was reduced to about 50% of the normal level, except in an isochromosome. This seems to prove that the homologous regions involved in crossing over are subject to some physical forces that prearrange proximity of homologous segments prior to synap-sis. The two homologous portions of the isochromosome were held together by the centromere and were not affected by the colchicine.

Also mentioned here are the **pseudoisochromosomes** that were obtained by x-radiation (Caldecott and Smith, 1952). These chromosomes are similar in their genetic constitution to isochromosomes in that the ends of their chromosome arms are homologous, but the chromosome segments next to the centromere (**interstitial chromosome segments**) are nonhomologous. Pseudoisochromosomes are the result of reciprocal translocation between end segments of opposite arms of chromosomes of the same homologous pair (Fig. 10.10). Internal pairing at meiosis of such chro-mosomes is like that shown for isochromosomes.

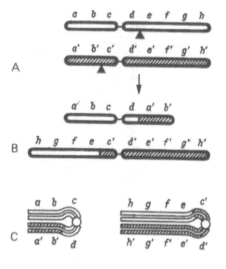

Fig. 10.10A–C. Diagram illustrating the origin of pseudoisochromosomes. (*A*) Indication of breakpoints in two homologous chromosomes. (*B*) The formation of pseudoisochromosomes after reciprocal translocation at the breakpoints. (*C*) Meiotic chromosome pairing of these pseudoisochromosomes. (From Rieger et al., 1976).

10.4 B Chromosomes

The term **B chromosomes** was given by Randolph (1928) to a type of chromosome that is present in many plant and animal species and differs in many respects from normal chromosomes, which he termed **A chromosomes.** Other terms for this chromosome type have occurred in the literature such as **accessories, supernumerary,** and **extra chromosomes.** In the present discussion, the term *B chromosome* is preferred over the other terms since it restricts this type to a more well-defined group of chromosomes. Since B chromosomes were first discovered in maize (Kuwada, 1925; Longley, 1927), the maize-type of accessory chromosome should be the one that delimits the definition of B chromosomes. In maize these chromosomes are distinguishable from normal chromosomes (A chromosomes) according to the following characteristics:

1. structure
2. genetic constitution
3. numerical variability
4. meiotic behavior
5. mitotic behavior

B chromosomes in maize are noticeably smaller in size than the normal chromosome set. They are about ⅔ of the size of the smallest maize chromosome. The centromere of the maize B chromosomes is terminal (Fig. 10.11) (Rhoades, 1955). These chromosomes are largely heterochromatic. They are also genetically ineffective in that they do not noticeably influence the phenotype of the plant. Maize B chromosomes are present in excess to the normal 2n chromosome number of this

Fig. 10.11. Diagram of B chromosome of maize in pachytene. (From Rhoades, 1955. Redrawn by permission of Academic Press, New York).

species. They vary in number between different cells, tissues, individuals, populations, and generations. Such B chromosomes do not pair with any of the A chromosomes in meiosis, and they do not pair as regularly among themselves as A chromosomes. They have abnormal postmeiotic behavior in that they undergo **nondisjunction** at the second pollen grain division. In nondisjunction the two B chromatids do not separate and go to opposite poles but rather stick together and move to the same pole. This, in combination with **preferential fertilization,** causes an increase in the number of B chromosomes in the next generation and, thus, in the population. Preferential fertilization results in combinations having nonrandom frequencies. In this case the male gametes with B's unite more often with the eggs than do those without B's. Maize B chromosomes usually are maintained and not lost in the mitotic tissue (Blackwood, 1956) but **mitotic elimination** has been recorded for other plants.

The foregoing characteristics shall be the guiding criteria for the classification of B chromosomes in this discussion. Deviation in one or the other point will make the classification as B chromosomes more or less questionable. If, for example, chromosomes are in excess of the normal somatic chromosome number of a certain species, they definitely should not be classified as B chromosomes. Such chromosomes easily fit the description **supernumeraries** or **accessories**.

10.4.1   B Chromosome Structure and Genetic Constitution

Since a partial requirement for B chromosomes is their heterochromatin content, many recorded accessories should probably not be included in this category. The high content of heterochromatin in B chromosomes must be connected with their genetical ineffectiveness and with the fact that they can accumulate up to a certain limit, at which point they become deleterious. Heterochromatin is generally considered to be less genetically active than euchromatin.

B chromosomes are generally shorter than A chromosomes and thus susceptible to nondisjunction. Since they are considered to be nonessential chromosomes, they seem to undergo morphological changes that do not have any genetic consequences. In maize, Randolph (1928) distinguished between B, C, D, E, and F chromosomes, thus designating deviation from the B or standard type. Such **polymorphism** has also been observed in other plants. The most extreme case of B polymorphism has been observed by Matsuda (1970) in *Aster ageratoides* in which 24 morphological deviations from the standard type occur. The most characteristic B chromosome shape is the telocentric or acrocentric. This is certainly the case in the most thoroughly described types in maize and rye. The normal rye chromosome complement (A chromosomes) consists of metacentric and submetacentric chromosomes. Thus, the rye B chromosomes are very easily distinguishable by their morphology (Fig. 10.12).

In maize, Randolph (1941b) tested 46 linked genes distributed among 17 of the 20 arms of the A chromosomes. None of these genes showed disturbed ratios in combination with the B chromosomes. This demonstrates that there is complete absence of any known major genes on the B chromosomes.

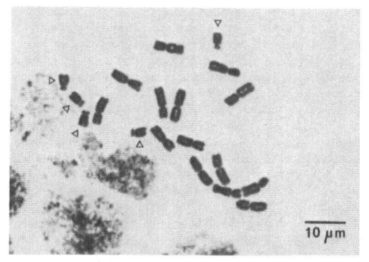

Fig. 10.12. Root tip cell of rye with 14 normal A chromosomes and 5 extra B chromosomes. Arrows indicate B chromosomes. (From Jones, 1975. Reprinted by permission of the Academic Press, New York).

10.4.2   Numerical Distribution, Variability, and Effects of B Chromosomes

In general, if a chromosome or a group of chromosomes occur in an even number in all individuals of a species, they are not considered to be B chromosomes. The number of B chromosomes in maize normally ranges only from 1 to 3. But up to 34 B's have been observed (Rhoades, 1955). When the B's ranged from 10 to 15 in number, plant vigor was not affected. There was a direct correlation between increasing number of B's and vigor, seed set, and fertility when the B's ranged in number from 15 to 25. Plants with 30 to 34 B's were very low in vigor and entirely sterile. In diploid rye (2n = 14), a maximum number of 10 B chromosomes and in tetraploid rye (2n = 28), a maximum number of 12 B chromosomes have been recorded (Müntzing, 1963).

As mentioned, B chromosomes can vary in number among populations, among individuals, among different tissues of the same individual, and between cells of the same tissue. Battaglia (1964) summarized this subject. In the grass *Poa alpina,* for instance, Müntzing (1948b) found that B's are eliminated from the leaves and adventitious roots but that they are present in the primary roots, the central part, and the germ cells. In *Sorghum purpureosericeum,* Darlington and Thomas (1941) reported that the B's are lost in the roots during seed development, that in the growing inflorescence the B's are eliminated from tissues that are not going to produce germ cells, but that the pollen mother cells contain a constant number of B's. The anther walls and ovaries are intermediate in the amount of loss. In the flatworm, *Polycelis tenuis,* Melander (1950b) found that B chromosomes tend to be lost from the somatic cells of fully grown animals but are retained in the ovarial tissue.

On the other hand, there are many different species in which the number of B's is very constant within an individual such as in the grasses *Agrostis, Alopecurus, Anthoxanthum, Briza, Dactylis, Festuca, Holcus, Phleum,* and *Secale* (Battaglia, 1964).

In plants, B chromosomes seem to be limited to the angiosperms in which they have been reported in more than 475 species of 163 genera in 42 families (Brown and Bertke, 1969). In animals, B chromosomes have been reported in flatworms (Melander, 1950b), snails (Evans, 1960), Isopoda (Rocchi, 1967), grasshoppers, scale insects, Heteroptera, Lepidoptera, beetles, and some Diptera (White, 1973). B's seem to be very rare in vertebrates but have been reported in Urodela, Reptilia, Anura, and Mammalia. White (1973) published a table with species of animals in which B chromosomes have been recorded.

Kodani (1957a, 1957b, 1958a, 1958b) suggested the presence of one or two supernumerary chromosomes in Japanese human populations, but Makino and Sasaki (1961, Sasaki and Makino, 1965) were not able to confirm this finding.

10.4.3   Meiotic and Postmeiotic Behavior of B Chromosomes

B chromosomes are not in any way homologous with A chromosomes but pair with each other. However, their pairing efficiency is not as high as it is among A chromosomes. If B chromosomes are unpaired, they can divide at anaphase I, at anaphase II, or at either one of these divisions, depending on the species. In most species with B chromosomes, their meiotic transfer is normal. However, in the grasshopper, *Locusta migratoria,* Rees and Jamieson (1954) observed that the univalent B's lag in the first anaphase spindle and divide tardily, causing up to a 20% loss in meiosis. Mendelson and Zohary (1972) detected a similar meiotic loss of B's in *Aegilops speltoides.* The B remains lagging outside the equatorial plate in anaphase I and then undergoes a precocious division. It then fails to be included in the daughter nuclei at the end of the first meiotic division. At the end of meiosis, the B appears as a micronucleus in 80% to 85% of the pollen mother cells.

A search for the cause of B chromosome accumulation in populations of plants and animals has led to the discovery of a **nondisjunction** mechanism. Battaglia (1964) distinguished between three major types of nondisjunction in the B chromosomes of plants and a fourth is added here:

1. *Secale* type
2. *Sorghum* type
3. *Zea* type
4. *Lilium* type

The **Secale type** of B chromosome nondisjunction results in a postmeiotic preferential distribution of B's. It was first discovered by Hasegawa in 1934 and later carefully studied by Müntzing and coworkers for microsporogenesis (Müntzing, 1946, 1948a; Müntzing and Lima-de-Faria, 1949, 1952, 1953; Lima-de-Faria, 1953) and by Håkansson (1948) for megasporogenesis. At the first postmeiotic division, the centromere of the B's divides normally but the two chromatids remain closely attached to each other in the regions close to the centromere. This has been explained as a stickiness of the heterochromatin of these regions. The two B's then are preferentially directed toward the pole that is responsible for sperm or egg formation. *Secale cereale* (rye) is the only species in which such preferential segregation is recorded for micro- and megasporogenesis alike. Secale type nondisjunction has also been reported for the male line only in the grasses *Anthoxanthum aristatum* (Östergren, 1947), *Festuca arundinacea, Festuca pratensis, Phleum phleoides, Alopecurus pratensis, Briza media, Holcus lanatus* (Bosemark, 1957a, 1957b), *Dactylis glomorata* (Puteyevsky and Zohary, 1970), *Deschampsia bottnica, D. caespitosa,* and *D. wibeliana* (Albers, 1972).

In the **Sorghum type** of B chromosome nondisjunction (Darlington and Thomas, 1941), the first pollen grain division is regular, producing a vegetative and a generative nucleus. The vegetative nucleus undergoes one or more hastened divisions (called **extra divisions** or **polymitosis,** Beadle, 1933a) giving rise to supernumerary (above the normal two) generative nuclei. The results of these divisions is a sterilization of the pollen. At the first of such extra divisions, the B's pass undivided to the generative pole. Apparently, this division takes place so rapidly that the B's are incapable of dividing.

The **Zea type** of B chromosome nondisjunction occurs at the second pollen grain division (Roman, 1947). As already mentioned, this type is coupled with **preferential fertilization.** The generative nucleus possessing the B's unites more frequently (60%) with the egg than does the generative nucleus without the B's (Roman, 1948; Blackwood, 1956).

In the **Lilium type** of B chromosome nondisjunction (Kayano, 1957), the preferential distribution of B's takes place during the first meiotic division of megasporogenesis. The nondisjoined B chromosome preferentially passes to the anaphase I pole of the megaspore so that the two B's are present in 75% to 85% of the eggs rather than in 50%. This type of nondisjunction also occurs in *Trillium grandiflorum* (Rutishauser, 1956), *Tradescantia virginiana* (Vosa, 1962), *Plantago serraria* (Fröst, 1959), *Phleum nodosum* (Fröst, 1969), and *Cochlearia pyrenaica* (Gill, 1971).

Jones (1975) postulated that most of the systems of B nondisjunction in animals are premeiotic. Ehrendorfer (1961) also proposed such a system for the plant *Achillea.*

Part VI
Variation in Chromosome Structure

Chapter 11
Chromosome Deletions

Four different classes of structural chromosome changes are being considered in Part VI (Chapters 11–14): deletions, duplications, inversions, and translocations. These four classes can be grouped as follows: In deletions and inversions the chromosome breaks are confined to one pair of chromosomes only, whereas in duplications and translocations more than one chromosome pair can be involved in chromosome breakage.

11.1 Breakage-Reunion and Exchange Hypotheses

All structural changes of chromosomes must be connected in some way or another to chromosome damage and breakage. The interpretation of the ultrastructure of such chromosome damage and breakage is very limited by the still persisting ignorance of chromosome fine structure (Chapter 3). A summary by Brinkley and Hittelman (1975) on the ultrastructure of mammalian chromosome aberrations shows the reality of this dilemma. They conclude that the actual mechanism involved in the formation of a break or exchange is still in the realm of postulation.

Structural chromosome changes are generally considered to depend on **breakage** of chromosomes and on **reunion** of chromosome segments. Chromosome breakage results in injured chromosome ends, which differ from natural chromosome ends or telomeres by being **sticky** and having the tendency for reunion with other such injured ends.

Most structural chromosome or chromatid changes involve both breakage and reunion. Thus, the so-called **breakage-reunion hypothesis** was formulated and put forward by Stadler (1931, 1932), Sax (1938, 1941), Muller (1932, 1938, 1940a, 1940b; Muller and Herskowitz, 1954), Wolff (1961), and by Evans (1962).

According to this hypothesis, breaks occur spontaneously or as a result of mutagens and usually rejoin in the original order by repair processes. This phenomenon is called **restitution** (Darlington and Upcott, 1941). If restitution to the original structure does not take place, the chromosomes may undergo structural changes through the phenomenon of **reunion** (Darlington and Upcott, 1941), where the broken ends of the chromosomes or chromatids reunite in a new arrangement. If only a single break occurs, the centric fragment may undergo **sister-strand reunion** between the two chromatids. Such reunion leads to dicentric chromosomes in the next division with breakage and further complications in the following cell divisions until loss of the chromosome or death of the cells involved occurs (see discussion of **breakage-fusion-bridge cycle,** Section 11.5).

Obviously, single chromosome breaks involving loss of larger chromosome segments do not generally produce viable cytogenetic changes, but exceptions may occur. For instance, McClintock (1941a, 1941b) observed in maize, that freshly broken ends (1) seemed to "heal" in the sporophyte, (2) did not fuse with other broken ends, and (3) were not subject to sister-strand reunion. Since such healing did not occur in the gametophyte, this process is not perpetuated into the next generation. Generally, a minimum of two breaks must occur to effect change in the karyotype. An alternative to the breakage-reunion hypothesis, is the newer **exchange hypothesis** of Revell (Revell, 1955, 1959, 1960, 1963, 1966; Evans, 1962; Rieger, 1966; Brewen and Brock, 1968). According to this hypothesis, the primary event that leads to chromosome aberrations is not breakage but the formation of so-called **primary lesions** (Fig. 11.1A). Such lesions are regions of instability or labile sites

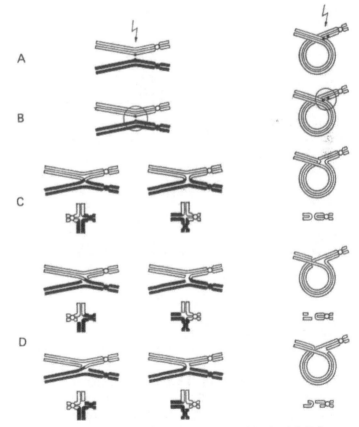

Fig. 11.1A–D. Diagrammatic representation of the exchange hypothesis. (*A*) Primary event: primary lesions. (*B*) Secondary event: exchange initiation. (*C, D*) Tertiary event: actual mechanical exchange process. (Modified after Rieger, 1966).

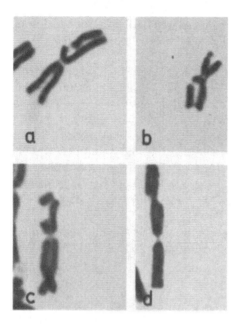

Fig. 11.2a-d. Apparent and real chromatid discontinuities in metaphase chromosomes of *Allium*. (*a, b*) Gaps as visible evidence of the primary event: lesions. (*c, d*) Definite displacement of chromosome fragment in real break. (From Kihlman, 1970. Reprinted by permission of Plenum Publishing Press, New York).

in the chromosomes or chromatids. The visible evidence of such lesions are the **gaps** (Figs. 11.2a and b) that are unstained Feulgen-negative regions in the chromosomes or chromatids. Evans (1968) explained that one can observe them at anaphase and see that a chromosome that has such a gap does not lose its fragment. Gaps do not represent discontinuities in the chromosome. Under phase contrast or with the interference microscrope, one can actually see a continuity between the two parts of the chromosome on either side of the gap. With a real break (Figs. 11.2c and d), a definite displacement of the fragment is visible, so that according to Evans there is no operational difficulty in distinguishing between gaps and breaks.

The secondary event, according to the exchange hypothesis, is the **exchange initiation** (Fig. 11.1B), an interaction between two lesion sites that is caused by the primary event. Such secondary sites are predisposed to, but have not yet reached actual exchange (Revell, 1959). If the two lesion sites are not close enough together, or are not receptive to each other at the same time, then the two primary lesions fail to interact and may be subject to repair. Consequently, the exchange initiation may or may not be followed by an actual **mechanical exchange process,** which is analogous to crossing over (tertiary event; Figs. 11.1C and D). If this tertiary event takes place, it either leads to a complete (both re-joins occur) or to an incomplete exchange (only one re-join occurs).

The exchange hypothesis postulates that all chromatid rearrangements produced by irradiation are the result of an exchange in which two strands cross one another. It has been developed and supported by a strong group of modern radiation biologists and is based on experimental evidence. Since the discovery of primary lesions or gaps, the actual incidence of chromosome and chromatid breaks is known to be only one tenth of earlier reports (Neary and Evans, 1958; Evans et al., 1959; Revell, 1959). Earlier scoring of chromatid breaks was high because of the inclusion of gaps (Thoday, 1951). The low frequency of actual breaks and the observation of chromatid exchanges (Fig. 11.3), in humans, is good evidence for the exchange hypothesis (Cohen and Shaw, 1964; Brinkley and Hittelman, 1975). Primary lesions are injured sites and possible subchromatid breaks that do not cause discontinuities and may allow the delay of chromatid exchange into the T_1 and T_2 or even later generations (T_1 = one generation after treatment, etc.). This is in harmony with observations that have been made particularly after treatment with chemical mutagens. Chromosome aberrations were observed several cell generations after treatment by Fahmy and Fahmy (1955), Slizynska (1963), Evans and Scott (1964), Moutschen (1965) and Müller (1965). If breaks were the immediate result of treatment, such long delay in chromosome rearrangement could not be explained.

11.2   Spontaneous and Induced Chromosome and Chromatid Aberrations

As early as 1937, Mather showed that the timing of irradiation determines whether the aberration involves a chromatid or a chromosome. If a single radiation event occurs after the S period, only one chromatid will be involved in the lesion. If such a radiation even occurs before the S period, both sister chromatids are affected because the lesion becomes replicated with the chromatid.

Swanson (1957) believed that all spontaneous aberrations are the result of naturally occurring radiations, but evidence that chromosomal anomalies can be produced by viral infections has accumulated. Such breaks and rearrangements in human chromosomes have been reported to be caused by measles (Nichols et al.,

Fig. 11.3. Chromatid exchange involving two human chromosomes. (From Cohen and Shaw, 1964. Reprinted by permission of the Rockefeller University Press, New York).

1962), chicken pox (Aula, 1963), meningitis (Makino et al., 1965), and Simian tumor virus SV_{40} (Moorhead and Saksela, 1963). Sometimes, as in the case of SV_{40} infection, the sites of the chromosome breakage appear to be nonrandom.

The induction of chromosomal aberrations by experimental procedures makes it possible to further inquire into the nature of all those changes that happen spontaneously. Methods of inducing chromosome aberrations include the application of various agents such as radiation (Wolff, 1961), chemicals (Shaw, 1970), viruses (Nichols, 1970), temperature changes (Hampel and Levan, 1964; Dewey et al., 1971), and mycoplasmas (Paton et al., 1965). A definite correlation between the use of drugs and the increase of chromosome aberrations in the human population has been established by Cohen et al. (1967a, 1967b). Novitski (1977) reports that LSD and marijuana have been implicated in chromosome breakage of humans. The effects of thiotepa, caffeine, and 8-ethoxycaffeine on the exchange frequency of sister chromatids in *Vicia faba* has been studied by Kihlman (1975).

A number of rare inherited diseases in humans are associated with an increase of chromosome aberrations in cultured fibroblasts and peripheral blood lymphocytes. These are the **human chromosome instability syndromes:** Fanconi's anemia (Schroeder et al., 1964; German and Crippa, 1966), Bloom's syndrome (German et al., 1965; German, 1969), and the Louis-Bar syndrome (Hecht et al., 1966; Gropp and Flatz, 1967). All three of these syndromes are inherited as autosomal recessives.

A method to demonstrate **sister chromatid exchange** (SCE) developed by Latt (1973) has made it possible to quantify the incidence of chromatid breakage. Figure 11.4 shows a human lymphocyte cell pretreated according to the **SCE method.** Several researchers demonstrated a close linkage between chromosome aberrations and sister chromatid exchanges in chromosome instability syndromes (Chaganti, 1974; Kato and Stich, 1976; Shiraishi et al., 1976; de Weerd-Kastelein, 1977).

Powerful mutagens such as ionizing radiation (alpha, beta, gamma rays from radioactive sources, x-rays, protons, neutrons) cause only slight increases in SCE frequency (Perry and Evans, 1975). Paradoxically, the effect of some weak carcinogens such as sodium saccharin can be easily measured by the increase in SCE's (Wolff and Rodin, 1978). A small but significant rise in the number of SCE's was observed after the exposure of fresh human lymphocytes to 30 minute treatments with diagnostic ultrasound (Liebeskind et al., 1979).

Chromatid-type damage, like **triradial chromosomes**[1] and chromosome-type damage, like dicentrics, were observed in Louis-Bar syndrome lymphocytes by Taylor et al. (1976) after radiation with x-rays in the G_0 phase. The hypothesis is that there is a defect of DNA repair in these patients that leads to radio-sensitivity. According to this hypothesis, Louis-Bar syndrome lymphocytes lack the full complement of functional polynucleotide ligases that are able to join breaks in one strand of a DNA double helix.

---

[1]Triradial chromosome: a three-armed chromosome configuration involving two non-homologous chromosomes arising from an interaction between an **isochromatid break** and a chromatid break.

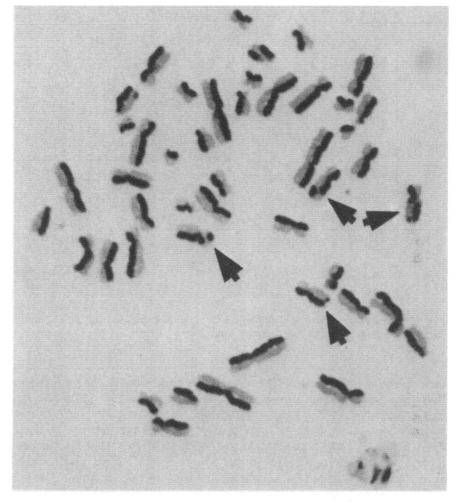

Fig. 11.4. Mitotic metaphase cell showing sister chromatid exchange (SCE) in human chromosomes after 2 rounds of replication in the presence of 5-bromodeoxyuridine (BrdU) followed by staining with Hoechst 33258 plus Giemsa. Arrows point to sister chromatid exchanges. (Courtesy of Dr. Cheng Wou Yu, Louisiana State University, Shreveport).

It is likely that DNA ligases play an important part in genetic recombination, repair of radiation-induced damage, and DNA synthesis, but exactly how is yet unknown. Gellert (1967) and Olivera and Lehman (1967) first isolated polynucleotide ligases that are thought to complement the enzyme polymerase in the process of DNA catalysis. Okazaki et al. (1968) demonstrated that the early products of DNA synthesis are relatively short DNA strands that must be *stitched* together in ligase reactions. Enzymes repairing or sealing single strand breaks are also called **repairases** (Kozinski et al., 1967), **repair polymerases,** or **antimutator polymer-**

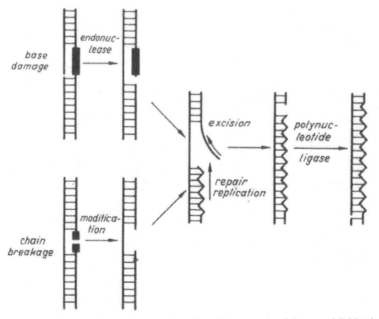

Fig. 11.5. Schematic representation of excision repair of damaged DNA bases and broken strands involving repair replication. (From Cleaver, 1974).

ases. Such polymereases are part of an **enzymatic proofreading mechanism** (Goodenough, 1978) that functions during chromosome replication to compensate for spontaneous and induced mutations. This mechanism provides for the insertion of nucleotides into the damaged DNA. It may occur after the excision of damaged DNA fragments (**excision repair;** Fig. 11.5). Such **repair replication** has been observed in eukaryotes after treatment with x-rays, UV-light, and chemical mutagens but it is not known in prokaryotes after ionizing radiations. Excision repair is generally considered to produce high fidelity. **Postreplication repair** (Rupp and Howard-Flanders, 1968), however, is considered to be error-prone. It does not act on primary lesions but on secondary lesions that originate as a consequence of unrepaired primary lesions. Streisinger et al. (1966) developed a model of such **misrepair** according to which, after DNA breakage, the DNA strand may **buckle,** and DNA strands that are either too long (addition) or too short (deletion) may be synthesized.

Drosophila became a classical cytogenetic object for the study of x-ray induced aberrations. The large size of salivary gland chromosomes of *Drosophila* allowed a very exact study of small and larger chromosome aberrations. Since the Morgan school had already made available a great amount of genetical data in *Drosophila* by this time (Chapter 1), the cytological changes could be easily checked against and correlated with this genetic information. Muller reported x-ray induced translocations and other chromosome aberrations in *Drosophila* as early as 1927. In 1929 they were demonstrated cytologically by Painter and Muller. During the following decades, the early results in *Drosophila* were confirmed in many plant and animal species, and by 1946 (Catcheside et al.) a classification system for the description of different chromosome aberrations in *Tradescantia* was completed. The meiotic chromosomes of *Tradescantia* and the mitotic chromosomes of *Vicia faba* were also ideal materials for the study of chromosome aberrations induced by radiation and chemicals. The large size of the chromosomes, the small number of them, and the ease of obtaining large numbers of cells for comparison made these species ideal objects for such investigations.

Induction of mutants by radiation has been used as a tool in plant breeding. According to a report of the International Atomic Energy Agency (IAEA), 68 useful mutant varieties of food and crop plants were released to farmers between the period from 1930 to 1971 (Nabors, 1976). Factors such as chemicals, infrared rays, moisture, temperature, and oxygen applied to the living plant tissue before, during, and after radiation change the effect of the radiation treatment (Nilan, 1956).

Berns et al. (1979) have demonstrated that the laser microbeam can be used to produce heritable deficiencies on preselected regions of individual chromosomes. They conducted extensive studies on the ribosomal genes of salamander *(Taricha)* and rat kangaroo *(Potorous)* cells in culture. These cells were chosen because they remain flat during mitosis, making all chromosomes easily identifiable during mitosis. Most human cells, for instance, round up during division (Berns, 1978).

11.3   Terminal Deficiencies

Bridges (1917) defined a **deficiency** as "a structural change of a chromosome resulting in the loss of a terminal acentric chromosome-, chromatid-, or subchromatid-segment and in the loss of the genetic information which this chromatin segment contains". Bridges' work (1923) on structural changes in *Drosophila* chromosomes was briefly mentioned in Chapter 1. In its classical sense, a deficiency is of a terminal nature and involves only a single chromosome break followed by a healing of its broken end (Fig. 11.6A). In contrast, a **deletion** involves an intercalary chromosome segment and requires two chromosome breaks (Fig. 11.6B). However, in practice, the term "deletion" frequently is used for both of these types of structural chromosome changes. Generally, in both instances a centric and an acentric chromosome segment are produced. The centric segment will persist during cell division, while the acentric fragment will be lost. Most chromosome aberrations that cause large deficiencies will lead to the death of the cell involved or will, at least, prohibit sexual reproduction. They are eliminated from the population and will not survive or become part of a permanent karyotype.

The terminal deficiency type of chromosome aberration is a category by itself since it does involve only one break. Therefore, a tendency for sister strand reunion (Section 11.1) of chromatids or reunion of broken chromosome ends exists which does not permit stability. Different tendencies for such sister strand reunion have been observed depending on biological material, treatment, or chromosome material involved. Healing of broken chromosome ends can occur in plants, but it seems to be very rare in animals. Simple chromosome breaks in *Drosophila* and other ani-

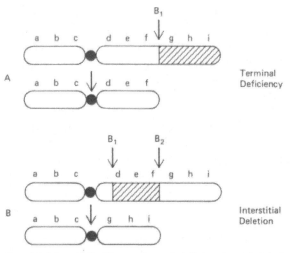

Fig. 11.6A and B. Illustration of deficiency and deletion. (*A*) A terminal deficiency caused by a single chromosome break. (*B*) An interstitial deletion caused by two chromosome breaks.

mals generally are not stable because of sister strand reunion. In plants, the treatment seems to make a difference. If maize was treated with ultraviolet radiation, terminal deficiencies mainly resulted. If x-rays were applied, only interstitial deletions were observed (Stadler, 1941; Stadler and Roman, 1948).

If the break occurs in the heterochromatic portion of the chromosome, it is more likely to heal than if it happens in the euchromatin. White (1956) and Southern (1969) reported, after studying the centromere in grasshoppers, that simple breaks through the centromere were stable throughout the spermatogonial mitosis. Centromere regions are generally heterochromatic. Khush and Rick (1968a) observed that the frequency of recovered x-ray-induced breaks is highest in heterochromatin and lowest in euchromatin.

The extent to which a terminal deficiency can be tolerated in animals has been tested in *Drosophila*. Demerec and Hoover (1936) have shown several deficiencies that demonstrate the loss of the left tip of the X chromosome (Fig. 11.7). Up to 11 bands are involved in this terminal deficiency including such loci as *y, ac,* and *sc*. If only 8 bands in this region are lost, this deficiency is lethal to the whole organism in the homozygous condition but not to individual cells. A loss of 4 bands is viable in the homozygous as well as in the hemizygous condition. If one considers that the X chromosome of *Drosophila* has more than 1000 bands and that the entire chromosome complement consists of approximately 5000 bands, then a loss of 8 bands seems to be minimal but, nonetheless, consequential. On the other hand, loss of heterochromatic segments can occur almost unnoticed. Large pieces of the Y chromosome of *Drosophila* may be deficient without any lethal effect. The effects of hemizygous deletions and that of duplications that cover the entire autosomal complement of *Drosophila* were reported by Lindsley et al., (1972). Aneuploidy (Section 16.2) in 57 **dosage-sensitive loci** leads to recognizable changes in the organism.

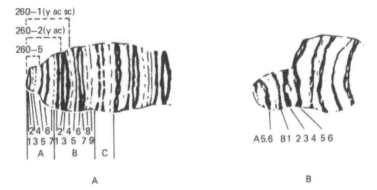

Fig. 11.7A and B. Deficiency at tip of X-chromosome of *Drosophila melanogaster.* (*A*) Normal tip. (*B*) Deficiency of 10 or 11 bands (260-1), which includes the genes *y, as,* and *sc.* (From Demerec and Hoover, 1936. Redrawn by permission of American Genetic Association, Washington, D.C.).

11.4   Interstitial Deletions

The loss of an intercalary or interstitial chromosome segment is referred to as a **deletion.** Painter and Muller in 1929 described the parallel cytology and genetics of such induced deletions in *Drosophila.* During the same year Serebrovsky suggested, on the basis of x-ray experiments, that perhaps all mutations are deletions or other chromosome aberrations. It is now known that deletions can vary from the absence of a single nucleotide to large chromosome segments. It is hard to determine where point mutations end and where deletions begin. The genetic proof for a deletion is its failure to back-mutate to the original form and to recombine in genetic crosses with two or more point mutations that do back-mutate and recombine.

McClintock (1938b) succeeded in the phenotypical demonstration of an interstitial deletion in maize that could be confirmed cytologically. This deletion is exceptional in that the deleted portion is centric rather than acentric. The chromosomal region involved included the locus of the gene Bm_l in the short arm of chromosome 5 close to the centromere. The recessive allel bm_l of this gene expresses brown midrib, producing a brown color in lignified cell walls. The absence of Bm_l as well as bm_l (interstitial deletion) also produces brown midrib. The heterozygous condition (Bm_lbm_l) causes variegated tissue. The deficiency was caused by x-raying pollen containing a normal haploid chromosome complement with the dominant gene Bm_l. By placing this pollen on silks of recessive plants ($bm_l\ bm_l$), two plants, variegated for Bm_l and bm_l, were found in a progeny of 466 plants. The cytological analysis of the variegated plants revealed an interstitial deletion in one homologue of chromosome 5 and a small ring-shaped chromosome (Fig. 11.8A). The deletion chromosome and the ring chromosome arose as a result of two breaks in the normal chromosome 5, one dividing the centromere, the other breaking the chromosome at a distance from the centromere of 1/20 (plant 1) or 1/7 (plant 2) of the total chromosome length. Proof for the assumption that the

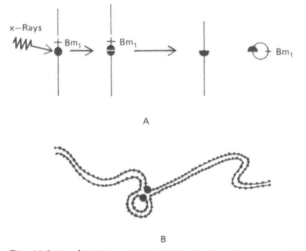

Fig. 11.8A and B. Interstitial deletion in the short arm of chromosome 5 of maize involving locus Bm_1 causing color change in the midrib of leaf sheath and blade, particularly in older leaves. (A) Demonstration of chromosome breakage by x-rays producing a deleted centric rod chromosome and a small ring-shaped chromosome. (B) Synaptic pachytene configuration of normal, deleted, and ring-shaped chromosomes.

ring-chromosome was indeed the region that was missing in the deletion chromosome was the discovery of the synaptic configurations in pachytene between normal, deleted, and ring-chromosomes (Fig. 11.8B). It was discovered that the ring chromosome could get lost from certain portions of the somatic plant tissue and that such portions would show brown streaks (variegated). Thus, it was assumed that the dominant Bm_1 locus was carried on the ring. Another phenomenon discovered in connection with these ring-chromosomes was the **breakage-fusion-bridge cycle** described in the next section.

11.5 Breakage-Fusion-Bridge Cycle

McClintock (1938a, 1938b, 1941a, 1941b, 1941c, 1942, 1944) found that the size of these ring-chromosomes changed through successive nuclear cycles. In order to change its size, the ring must obviously break. Figure 11.9 shows how ring-chromosomes change size in somatic tissue. It should be remembered that breakage was the original event that led to the formation of the ring (fusion). It is understood all along that the ring-chromosome has a centromere. If the ring reproduces itself in interphase and no sister strand crossing over occurs in prophase, then the two ring chromatids can separate from each other in anaphase without difficulty, reproducing two new equally sized ring-chromosomes that do not differ in size from the original one. However, if sister chromatid exchange (breakage + fusion) occurs in prophase, a ring of twice the size will be produced initiating the cycle. The ring will have two centromeres. Such a dicentric ring will behave like any other **dicentric chromosome** in that the two centromeres will move toward opposite poles in

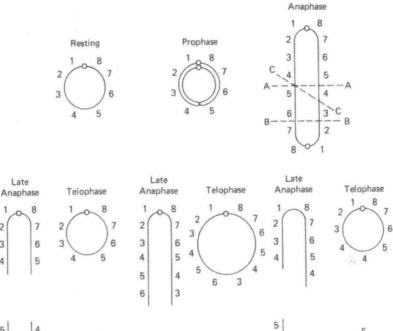

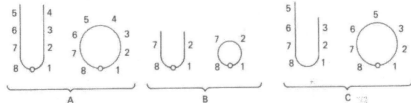

Fig. 11.9. Breakage-fusion-bridge cycle as observed by McClintock (1941) in maize. In the upper portion of the illustration, the drawing on the left is a ring-chromosome in an undivided resting state, the middle one is a replicated ring with a "crossover" between sister strands, and on the right is a drawing of a double-sized dicentric ring in an anaphase with A, B, and C representing three possible breakage situations. The bottom half represents the results of the three possibilities at late anaphase and telophase. (After McClintock, 1941b. Redrawn by permission from Genetics Society of America, Austin, Texas).

anaphase and will form an anaphase double bridge. Chromosome breakage in anaphase will occur subsequently and will complete one turn of the breakage-fusion-bridge cycle. The double-sized ring can break at different points along the ring-chromosome. Three possible different breakage situations (A, B, C) are shown in Fig. 11.9. The result are rings of different sizes in the next nuclear cycle, which stem from the fusion of the broken chromosome ends. If the different segments of the anaphase double-sized ring chromosome are numbered (Fig. 11.9), then it becomes obvious that duplications and deficiencies result from the uneven breakage of the ring. The cycle is initiated by **primary chromosome aberrations** (deletions) and results in **secondary chromosome aberrations** (deletions and duplications). The smaller rings that occur as a result of this cycle often are lost from the tissue. It is obvious in Fig. 11.10 that there is an alternation between breakage-fusion

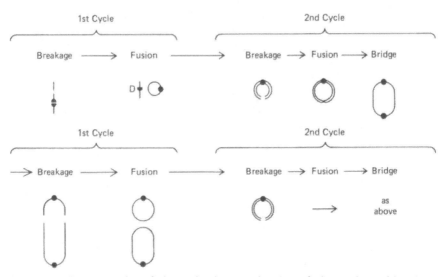

Fig. 11.10. Demonstration of alternation between breakage-fusion cycles and breakage-fusion-bridge cycles.

cycles and breakage-fusion-bridge cycles. Other examples of breakage-fusion-bridge cycles exist and do not necessarily involve ring-chromosomes.

11.6   Genetic and Cytological Tests of Deletions

Deletion mapping or **cytogenetic mapping** has already been mentioned in Chapter 4 as one of several possible ways of locating genes on chromsomes in *Drosophila* (Mackensen, 1935; Slizynska, 1938). **Deletion mapping** in its strict sense (Rieger et al., 1976) is the genetic localization of the positions of deletions in the linkage structures of eukaryotes and prokaryotes. This kind of mapping is based on three mutants (*a, b,* and *c*) that differ from the wild-type by a deletion being tested for recombination. If two mutants (*a* and *c*) mutually recombine and yield the wild-type but neither recombines with the third (*b*), then the three deletions are overlapping in the order *a, b, c* in the following fashion:

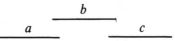

Deletion mapping also has been successfully applied in viruses by Benzer (1955) and in bacteria by Ames and Hartman (1963). It also may be useful for chromosomal localization of autosomal genes in humans (Nance and Engel, 1967).
Cytogenetic deletion mapping has recently been possible in bacteriophages that have much simpler genomes than the genomes of complex higher organisms. In bacteriophage, many mutants are readily available in which practically any region of the genome is deleted or replaced by nonhomologous DNA derived from the bacterial host or from other phages. These deletions can be used as genetic markers

and mapped genetically if they are not lethal. If by the DNA hybridization method (Harris and Watkins, 1965, Chapter 1) normal and deleted DNA genophores were combined to form double helices, a loop was formed at the site of the deficiency (Fig. 11.11). These **heteroduplex genophores** look remarkably similar to maize pachytene chromosomes (Westmoreland et al., 1969). An example of a heterozygous deletion in a *Drosophila* salivary gland cell is shown in Fig. 11.12.

Burnham (1962) suggested the possible use of deletion chromosomes in locating recessive genes. Smith et al. (1968) extended this principle for the possible localization of the gene for **cystic fibrosis** (*cf*) of the pancreas to the short arm of chromosome 5 in humans (Fig. 11.13). A patient with a heterozygous short arm deletion for chromosome 5 has only a single set of genes on the homologous segment of the missing piece (hemizygosity). Should this single set of genes contain a recessive

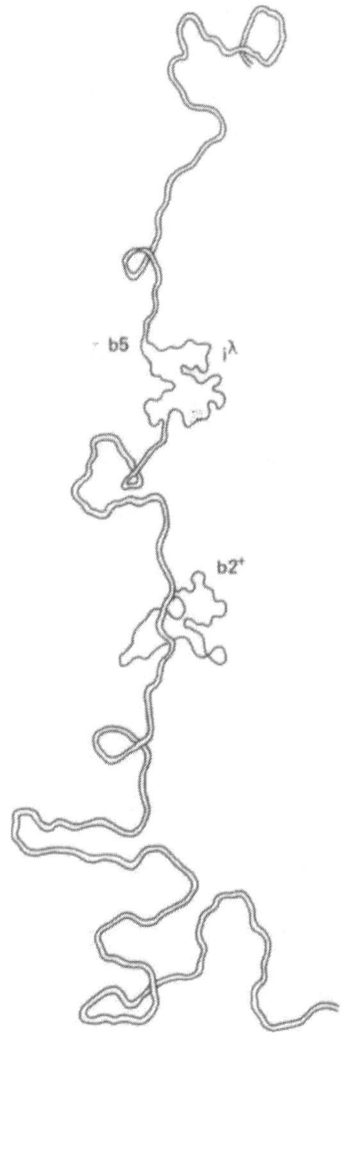

Fig. 11.11. A drawing of an electronmicrograph of viral genophore aberrations produced by heteroduplex formation. A loop was formed at the site of the deficiency (b2⁺) where normal and deleted DNA genophores were combined to form double helices. (From Westmorland et al., 1969. Redrawn by permission of the American Association for the Advancement of Science, Washington, D.C.).

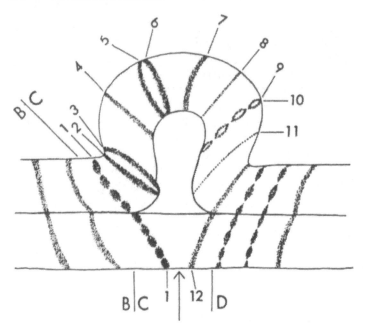

Fig. 11.12. Diagram of heterozygous deletion in a *Drosophila* salivary gland cell. (From Principles of Human Genetics, Third Edition, by Curt Stern. W. H. Freeman and Company. Copyright © 1973).

cf gene, the patient would express cystic fibrosis, since there would be no normal allelic gene to counteract the adverse effect of the mutant gene.

11.7   Human Deletion Syndromes

In 1963 Lejeune et al., for the first time, could link a clinical syndrome to a chromosome deletion in humans. They discovered that the loss of a short arm segment in chromosome 5 (5p-) of the B group (see Fig. 2.3) resulted in an abnormal cry of the affected baby resembling that of a suffering cat. This phenomenon results from a malformation of the larynx. It is also called the **cri du chat syndrome.** Other symptoms associated with this syndrome are severe facial malformations and microcephaly and, above all, mental retardation. The severity of this syn-

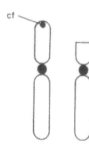

Fig. 11.13. Possible detection of gene for **cystic fibrosis** (*cf*) in short arm of human chromosome 5 by way of deletion chromosome.

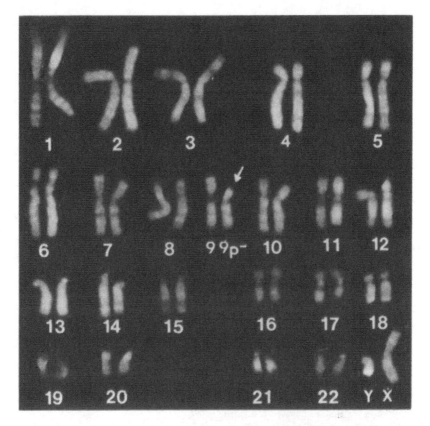

Fig. 11.14. Quinacrine fluorescent karyotype of a C group deletion in human chromosome 9 [46, X/Y, del(9) (pter—p22)]. (Courtesy of Dr. Penelope W. Allderdice, Faculty of Medicine, Memorial University, St. John's, Newfoundland, Canada).

drome varies from patient to patient and is thought to depend on the extent of the deletion. Many instances of this syndrome have now been demonstrated.

Another deletion syndrome in the human B chromosome group involves the short arm of chromosome 4 (4p-) (Wolf et al., 1965a, 1965b; Wolf and Reinwein, 1967; Hirschhorn et al., 1965, and others). This syndrome seems to occur much less frequently than the 5p- syndrome. None of the children with 4p- seem to have the characteristic cat cry. They are much more grossly malformed than the 5p- subjects, and facial anomalies are similar to the chromosome 5 deletion.

The possibility of a distinct 13 q- deletion syndrome that involves the D group of human chromosomes was postulated by Allderdice et al. in 1969. This conclusion was based on their own observations as well as on earlier similar findings (Gey, 1967; Laurent et al., 1967; Mikelsaar, 1967). The deletion involved about 20% of the long arm of chromosome 13. The syndrome resembles anomalies that were earlier described for D ring chromosome cases. It is characterized by microcephaly; eye, ear, and nose abnormalities; marked facial asymetry; and the absence of thumbs. By 1977 (Nielsen et al.), ten cases of deletion syndrome involving the long arm of chromosome 13 were known in humans. Three of them were terminal deficiencies, four were interstitial deletions, and three were unspecified cases of 13 q-. Only one case (described by Nielsen, et al., 1977) involved a person older than two years, a 65-year-old mentally retarded woman with a karyotype 46, XX, del (13) (q21-q31).

Alfi et al. (1973, 1974) and Allderdice et al. (1976) described four cases of C group deletion in chromosome 9 [46, del (9)(pter-p 22)]. A karyotype of this syndrome is presented in Fig. 11.14. The most striking facial feature of this syndrome was trigonocephaly (flat and triangular head) (Fig. 11.15). A deletion in the long arm of one of the 3 E group chromosomes (17q-, 18q-; Fig. 2.3) was first detected by De Grouchy et al. in 1964. Lejeune et al. (1966) found two cytologically and clinically similar cases and thus establsihed this Eq- syndrome. The clinical observations associated with this syndrome included mental retardation, growth and development failure, microcephaly, anomalies of ears and eyes, and genital abnormalities in males. Curran et al. (1970) showed a patient with several of these but lacking genital abnormalities. Deletions in the short arm of chromosome 18 (E group) have been found repeatedly but such an 18p- deletion could be associated with a syndrome only in about 50% of the observed cases (Ferguson-Smith, 1967).

A G group deletion syndrome was discovered in 1960 by Nowell and Hungerford. It is often related to the *Philadelphia* or Ph[1] chromosome. Originally, chromo-

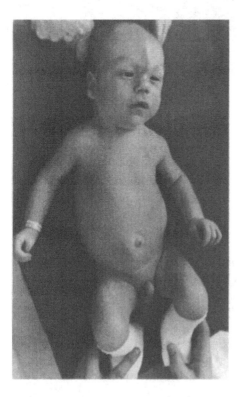

Fig. 11.15. Frontal view of proband with chromosome 9 deletion syndrome exhibiting trigonocephaly. (From Allderdice, P. W. et al.: 9 pter → 22 deletion syndrome: A case report. In: Bergsma, D., Schimke, R. N. (eds.): "Cytogenetics Environment and Malformation Syndromes." New York: Alan R. Liss for March of Dimes Birth Defects Foundation, BD: OAS XII (5): 151–155, 1976).

some 21 was thought to be involved but it is not known which of the two G group chromosomes are deleted. The deletion involves about 61% of the DNA of a normal G group chromosome (Gq-), which is most of the long arm. The abnormal chromosome is usually found in heterozygous condition in blood and marrow cells of chronic myeloid leukemia (CLM) patients. In tissues other than the hemopoietic system (responsible for blood cell production), the chromosomes are generally normal (Tough et al., 1961). Several cases with two Gq- chromosomes (homozygous deletion) have been found and seemed not to affect the characteristics of the disease (Dougan and Woodcliff, 1965). The discovery of the Gq- syndrome was rapidly confirmed by another research team (Baikie et al., 1960; Tough et al., 1961). The positive evidence of a direct linkage between the Gq-chromosome and CML was not immediately obvious, but by 1971, Hamerton could state that there is little doubt that most, if not all, adequately diagnosed cases of CML carry the Ph^1 chromosome.

Deletions that involve an X chromosome have been reported by several authors (Jacobs et al., 1960; Fraccaro et al., 1960, Hamerton, 1971b). These deletions can involve the short arm (XXp-) or the long arm of the X chromosome (XXq-). Since the Turner syndrome (Chapter 16) in humans seems to be determined by the hemizygous condition of the short arm of the X chromosome, individuals with XXp-express this sex anomaly. This syndrome is also referred to as **ovarian disgenesis** (or female infertility). The XO Turner syndrome is usually chromatin-negative, while the XXp- condition is chromatin-positive in that the deleted X usually forms a small Barr body. According to Hamerton (1971b) only about 10% of all chromatin-negative individuals are mixoploid (e.g., 45, X/46, XX), while over 80% of all chromatin-positive individuals are mixoploid (e.g., 46 XXp-/46XX).

Several deletions in the satellites of human chromosomes (groups D and G) have been observed that were not associated with syndromes. The satellites are considered to be heterochromatic. A loss of chromatin in these regions obviously is not of any genetic consequence. Ferguson-Smith and Handmaker (1961) showed that the manifestation of satellites varies from cell to cell.

Chapter 12
Chromosome Duplications

12.1　Types of Chromosome Duplications

A duplication is a structural change in chromosomes that causes doubling of a chromosome segment. The size of the doubled segment can vary considerably. Chromosome duplications are generally more tolerated by an organism than chromosome deletions.

Duplications can occur within a chromosome or among nonhomologous chromosomes and, consequently, are called **intrachromosomal** or **interchromosomal** duplications. According to Swanson (1957), there are three types of duplications:

1. tandem duplications
2. reverse tandem duplications
3. displaced duplications

The first two duplication types are intrachromosomal, the third is interchromosomal. Figure 12.1 illustrates these three types. The chromosome segment "def" is the duplicated segment in all three cases. Modifications of the first two examples (A, B) can occur if the duplicated segment is shifted to a different position in the same arm or in the other arm. Duplications may occur in different ways. McClintock (1941a, 1944) demonstrated how meiotic pairing and crossing over in a **reverse tandem duplication** heterozygote can initiate a breakage-fusion-bridge cycle (Section 11.5). The duplication involved the short arm of chromosome 9 of maize and included genes for *colorless aleuron* (*C*:26), *shrunken endosperm* (*sh*:29), and *waxy pollen* and endosperm (*wx*:59) (Fig. 12.2). Pachytene association of chromosomes in duplication heterozygotes led to two possible ways of pairing. In one instance the reverse tandem chromosome segment folded up and resulted in pairing within the same chromosome. Another way of pairing involved both homologous chromosomes, as shown in the figure. The result of this pairing was a bridge and a fragment in anaphase I. Depending on where the break in anaphase I would take place, the resulting gametes would carry smaller or larger deficiencies or duplications. The tendency for **sister chromatid reunion** at broken chromosome ends would lead to a new breakage-fusion-bridge cycle during the next mitotic division (Fig. 12.2).

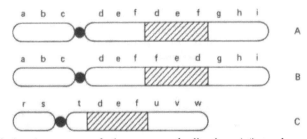

Fig. 12.1A–C. Drawing of the three types of chromosome duplication: (*A*) tandem duplication, (*B*) reverse tandem duplication, (*C*) displaced duplication.

12.2   Origin of Chromosome Duplications

Chromosome duplications can generally occur in three different ways (Rieger et al., 1976):

1. by primary structural changes of chromosomes,
2. by unequal crossing over of chromatids,
3. by crossing over in inversion or translocation heterozygotes (Chapters 13 and 14).

Primary structural changes resulting in duplications involve three breaks (B_1, B_2, and B_3; Fig. 12.3). Two breaks (B_1 and B_2) in a normal chromosome (Fig. 12.3A) result in a deleted centric chromosome (abc·ghi) and an acentric fragment (def) (Fig. 12.3B). The third break (B_3) could occur in the homologous partner chromosome (Fig. 12.3C). If the fragment is inserted into the partner chromosome, a **tandem duplication** results (Fig. 12.3D).

The result of such an **interchromosomal transposition** is a deficiency-duplication individual. If such an individual mates with a normal one, a duplication heterozygote could result that has a normal and a duplication chromosome (Fig. 12.3D).

The possible origin of such primary structural changes has been described in a unique way by McClintock in maize. She called it the **Activator-Dissociation system** (see Chapter 1). Since this system involves a type of **position effect,** it will be discussed in more detail under that heading in this chapter (section 12.3).

Chromosome duplication as a result of **unequal crossing over** is illustrated in Fig. 12.4. Usually crossing over occurs between homologous chromatids at loci that exactly correspond to each other in gene content (alleles). Such loci are responsible for the same biochemical and developmental processes. When the mechanism of pairing (Section 7.2.2) and crossing over (Section 4.2) is less specific, chromosome aberrations can occur as a deviation from the normal process. Such less specific pairing is observed particularly in areas in which heterochromatic chromosome segments are involved. Riley and Law (1965) called this **heterochromatic fusion** or **nonspecific pairing,** depending on the type of chromatin involved. The cytological evidence of nonspecific pairing was given by McClintock (1933) during her studies of maize pachytene.

Unequal crossing over was first observed in *Drosphila* by Sturtevant in 1925, involving the well-known *Bar* locus (*B,* chrom. 1:57.0). It produces one chromatid containing a chromatid segment twice (duplication) and the second lacking that chromatid segment (deletion). In Fig. 12.4A, two normal homologous chromo-

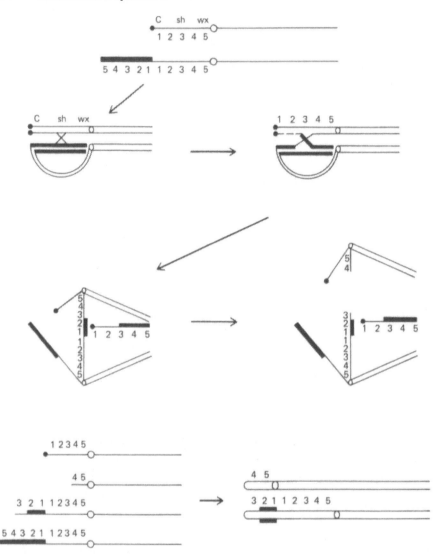

Fig. 12.2. Diagram of a **reverse tandem duplication** in the short arm of chromosome 9 of maize initiating a breakage-fusion-bridge cycle. Explanation in text. (Modified from McClintock, 1941a. Redrawn by permission of the Genetics Society of America, Austin, Texas).

somes are shown with breakpoints (B_1 and B_2) indicated in two non-sister chromatids (regions fg and cd). Figure 12.4B shows the four chromatids after crossing over is accomplished. Figure 12.4C shows the four resulting chromosomes that will be distributed to the four gametes. Two of them are normal (abc·defghi), one duplicated (abc·defdefghi), and one deleted (abc·ghi).

The formation of an abnormal hemoglobin (Hb-Lepore) in man was postulated by Baglioni (1962) as a result of the process of unequal crossing over and deletion. The *Bar* duplication in *Drosophila* will be discussed in more detail because it gave rise to the concept of the position effect.

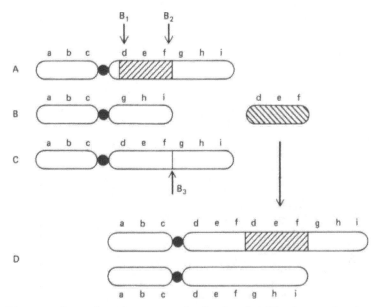

Fig. 12.3A–D. Diagram of a tandem chromosome duplication. (*A*) The first two break-points (Br₁ and Br₂) in the normal chromosome result in (*B*) a deleted centric chromosome (abc.ghi) (*C*) and an acentric fragment (def). If the third break occurs in a homologous chromosome (Br₃), the acentric fragment (def) could insert into the partner chromosome resulting in (*D*) a tandem duplication.

12.3   Position Effect

Geneticists talk about a **position effect** when genes or chromosome segments that are placed in new chromosomal neighborhoods cause a change in the phenotype of the individual affected due to their new position. Two types of position effects are recognized (Lewis, 1950):
1. **S-type of position effect**
2. **V-type of position effect**

The S-type (or stable type) of position effect was the first discovered and is the one associated with chromosome duplication. This type is confined to euchromatic regions of the chromosome. One of the oldest examples for the stable type is the *Bar* locus in *Drosophila*. The *Bar* effect is associated with a duplication of region 16A1 to 16A6 of the X chromosome and contains five bands, two of which are doublets. If this region is duplicated *(Bar)*, the facets in the fly's compound eye are reduced in number. If unequal crossing over (such as demonstrated in Fig. 12.4) occurs between two homologous chromosomes, both having a duplicated *Bar* region, chromosomes can result that carry the 16A1 to 16A6 region in triplicate (Fig. 12.5). This situation allows a comparison of four doses of 16A1 to 16A6 in two different combinations. These two different combinations are called *Bar* (homozygous *Bar*) and heterozygous *Bar double* in Fig. 12.6 (Morgan et al., 1935).

Bar produces an average number of 68 facets in the compound eye and heterozygous *Bar double* produces only 45. Obviously, the gene expression is stronger

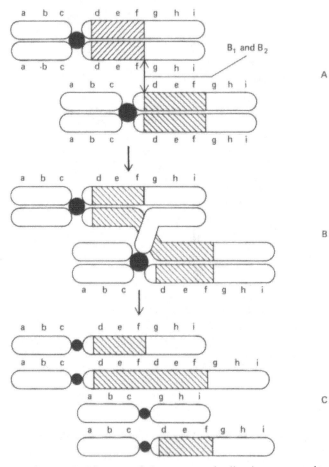

Fig. 12.4A–C. Diagram of chromosome duplication as a result of **unequal** crossing over. (*A*) Two normal homologous chromosomes with breakpoints (Br₁ and Br₂) in non-sister chromatids (regions fg and cd). (*B*) The four chromatids after crossing over occurred at the breakpoints. (*C*) The resulting four chromosomes. Two chromosomes are normal (abc.defgh), one is duplicated (abc.defdefghi), and one is deleted (abc.ghi).

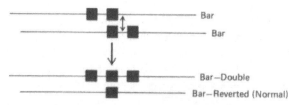

Fig. 12.5. Origin of *Bar-double* by unequal crossing over in the *Bar*-locus of the salivary gland X chromosome of *Drosophila melanogaster*. (From Morgan et al., 1935. Redrawn by permission of the Carnegie Institution of Washington, Cold Spring Harbor, New York).

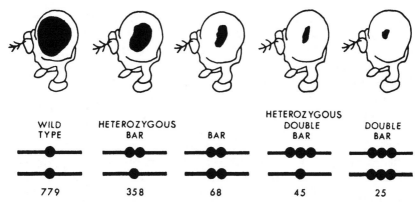

Fig. 12.6. Illustration of the different sizes of compound eyes of the female *Drosophila melanogaster* as caused by the varying numbers of facets. The size of the eye is influenced by the position effect. (From King, 1965. Redrawn by permission of Oxford University Press, New York).

when the genes are adjacent in the same chromosome than when they are on separate homologous chromosomes.

The V-type (or variegated type) of position effect was closely associated with the production of reverse tandem duplications observed in maize by McClintock in 1951 (Section 12.1). This type is limited to genes that are present in the heterozygous state and results in heterochromatinization and repression of a wild-type gene if this gene is transferred into the vicinity of heterochromatic chromosome segments. In contrast to the S-type, the V-type is subject to a large degree of fluctuation. The phenotype expresses a mixture of cell patches of both the wild-type and the recessive phenotype. The result is a **somatic mosaicism** that Schultz (1936) called **variegation.** This type of position effect is often associated with the translocation and inversion type of chromosome aberration and will be discussed further in the following chapters. The classical example for the V-type position effect is the **Activator-Dissociation system** (*Ac-Ds*).

12.3.1 The *Ac-Ds* System

This system was discovered by McClintock (1950a, 1950b, 1951, 1953,) in maize and depends on the action of two separate loci, the *Ac* locus (**Activator**) and the *Ds* locus (**Dissociation**). *Ds* only functions in the presence of *Ac*. If both loci are present, chromosome breakage is increased in the organism. Breakage has led to such chromosome aberrations as deficiencies, duplications, inversions, translocations, and ring-chromosomes.

Ac and *Ds* are visualized as blocks of heterochromatin that can move to different sites of the chromosome complement. This phenomenon was called **transposition** by McClintock. *Ds* was discovered first and was close to the locus *wx* on chromosome 9. Other locations of *Ds* were discovered later. No standard position was found for *Ac*. Since *Ac* does not have a mutating effect on neighboring genes as *Ds*, it is more difficult to map *Ac*. However, *Ac* is also capable of transposition.

Not only the location of *Ac* and *Ds* in the chromosome complement but possibly also the amount of heterochromatin involved can change. Such a dosage effect could be particularly well studied in triploid endosperm of maize where from 0 to 6 dosage factors of *Ac* could be observed. The larger the number of *Ac* dosage factors were, the later *Ds* was expressed during endosperm development. As a matter of fact, in kernels where four *Ac* dosage factors were present, the time of *Ds* chromosome breaks was so much delayed that none occurred before the endosperm growth had been completed. The earlier during development the *Ds* locus became active, the larger were the patches of variegated tissue in the endosperm.

Systems similar to the *Ac-Ds* system have been observed in maize, *Drosophila,* and bacteria. An example of the variegated type of position effect in *Drosophila* is the variegated eye color gene in the X chromosome (Glass, 1933; Baker, 1963).

The similarity between the maize and bacterial systems was emphasized by McClintock (1961, 1965) and Peterson (1970). McClintock adapted the terms used in bacteria to the maize system. The action of the **structural gene** (e.g., *wx* in maize) may come under the control of a foreign element (*Ds*) at the gene locus that would be comparable to an **operator gene.**

The control of time and frequency of occurrence of change in action of the structural gene is determined by the *Ac* **regulator gene** (Jacob and Monod, 1961a and b; Chapter 1).

12.4   Other Phenotypic Effects

The phenotypic expression of duplications generally is not as strong as that of deficiencies. Few duplications have unique phenotypic effects. As is the case with deficiencies in plants, the gametophyte is more easily affected by a duplication than the sporophyte. As described in the case of the *Bar* locus in *Drosophila,* some duplications not only increase the genetic effect of a gene, but they actually behave like dominant mutations. Several of the duplicated chromosome segments behave like dominant mutations in the heterozygous condition. Examples are the *Theta, Pale,* and *eyeless-Dominant* duplications in *Drosophila.* The *Theta* (*Th*) duplication was discovered by Muller and Painter (1929) and involves the duplication of the left end of the X chromosome including the loci for *y, sc,* and *bb.* The duplicated segment is attached to the centromere region at the right end of the X chromosome (Bridges and Brehme, 1944). This duplication causes the development of interalar (between the wings) bristles that are not ordinarily present in *Drosophila melanogaster.*

The *Pale* (*P*) character is a result of an interchromosomal or displaced duplication such as shown in Fig. 12.1C. It involves a transposition of a chromosome 2 segment into an interstitial position of chromosome 3. It is the result of an aneuploid segregant from a T(2;3)P translocation (Morgan et al., 1935). This was the first discovery of a translocation in *Drosophila melanogaster.* The phenotypic expression of this heterozygous duplication is a dilution of the eosin eye color.

The *eyeless-Dominant* character (*ey^D*) also discovered by Muller (Patterson and Muller, 1930) involves an unidentified segment transposed into an interstitial posi-

Fig. 12.7A and B. Drawing of
the eyeless dominant character
(*ey^D*) in *Drosophila* caused by
an unidentified segment transposed into an interstitial position in the middle of chromosome 4. (*A*) head; (*B*) first pair
of legs. (From Patterson and
Muller, 1930. Redrawn by permission of the Genetics Society
of America, Austin, Texas).

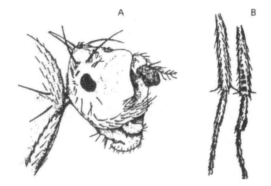

tion in the middle of chromosome 4 (Sturtevant, 1936). It is suspected that the
transposed segment is a reversed repeat since it forms a buckle bending back on
itself in synapsis. The phenotypic expression of this heterozygous duplication
includes small, irregularly outlined eyes that are displaced toward the top rear of
the head (Fig. 12.7). The homozygous condition produces complete lethality during
the larval period.

12.5   Human Chromosome Duplication Syndromes

Chromosome duplications per se in humans had not been discovered until recently.
They have been demonstrated in connection with pericentric inversion progenies.
They are designated as duplication-deletion syndromes (Stevens, 1974; Welch,
1974; Allderdice et al., 1975). The pericentric inversion, which is the progenitor of

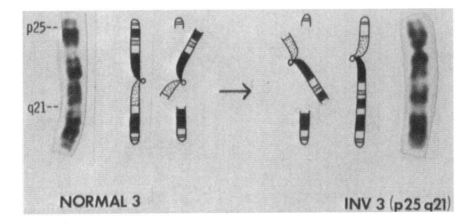

Fig. 12.8. Diagram and actual photographs of pericentric inversion in human chromo-
some 3 showing postulated two breaks (p25 and q21) in a normal chromosome 3. (From
Allderdice et al., 1975. Reprinted by permission of the University of Chicago Press,
Chicago).

this duplication-deletion syndrome, involves the major part of the short arm of chromosome 3 and about half of the long arm (p25 to q21[1]; Fig. 12.8). If crossing over occurs close to the centromere in such an inversion heterozygote (see Fig. 13.4), one of the recombinants would include a heterozygous deletion for the tip of the short arm of chromosome 3 (p25 to pter[2]) and a duplication for the distal half of the long arm (q21 to qter). This abnormality has been identified by banding patterns in the karyotypes of one fetus and of four children of known inv (3) (p25q21) carriers. The syndrome includes facial dismorphy and congenital anomalies.

Duplication in the human X chromosome is often found in the form of isochromosomes and has been discussed in Chapter 10 (Section 10.3). As mentioned in Chapter 11, variation in the amount of heterochromatin does not find expression in the phenotype. Consequently, duplication or increase in size of satellites, which has been observed in human chromosomes (D and G groups), does not manifest itself morphologically or physiologically. Giant satellites in humans were first reported by Tjio et al. (1960). Other reports confirmed this observation of satellite inertness (Ellis and Penrose, 1961; Cooper and Hirschhorn, 1962; Handmaker, 1963).

---

[1]p25 = break position on short arm
q21 = break position on long arm
[2]*ter* for "terminal" = chromosome end

Chapter 13
Chromosome Inversions

Inversions are probably the most common type of chromosome aberrations found in natural animal and plant populations (Darlington, 1937; Dobzhansky, 1941). During the discussion of chromosome duplications, we saw that chromosome segments can be separated from their mother chromosome by breaks and then can be reinserted into another homologous chromosome in the reverse order (reverse tandem duplications; Chapter 12). If no duplication is involved in such a process, the chromosome aberration is called an **inversion.** Just as in the case of deletions and duplications, organisms can be heterozygous for an inversion, homozygous for an inversion, or homozygous for the standard order of genes in the chromosome (Fig. 13.1).

Chromosome inversions have no effect on mitotic divisions, but they do affect meiosis. If an inversion is in the heterozygous condition, pairing of chromosomes cannot occur in a simple linear fashion. But if the inverted chromosome segment has the proper size, a loop can form that satisfies the pairing requirements. Depending on the occurrence of the inversion in relation to the centromere, two different kinds of chromosome inversions are known: (1) pericentric inversion and (2) paracentric inversion. In the pericentric type of inversion, two chromosome breaks occur, one on each side of the centromere, involving both chromosome arms. In the paracentric type, both breaks occur in the same arm (Fig. 13.2). The paracentric type of inversion is more common in submetacentric and subtelocentric chromosomes.

13.1   Pericentric Inversions

In a **pericentric inversion,** the centromere is included in the inverted region. This usually results in a morphological change of the chromosome due to change in centromere position and arm ratio. This kind of chromosome aberration can easily be detected in the karyotype (Fig. 13.2). Depending on the size of the inverted chromosome segment, different meiotic pairing configurations may be formed in the **inversion heterozygote.** If the inverted region is small, the homologous chromosomes may not pair in that particular region of the chromosome (Fig. 13.3A). If the inverted region is large enough for maneuvering, the two relatively inverted segments of two homologous chromosomes in an inversion heterozygote can pair

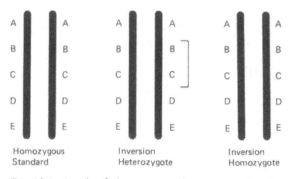

Fig. 13.1. A pair of chromosomes homozygous for the standard order of genes, hetero-zygous for an inversion, and homozygous for that same inversion. (Adapted from Srb et al., 1965).

gene by gene by forming an **inversion loop** (Fig. 13.3B). If the inverted region includes the major length of the chromosome, only the relatively inverted region may pair while the uninverted chromosome ends outside the inversion may remain unpaired (Fig. 13.3C).

In instances where the inverted regions in the inversion heterozygote are extremely small (Fig. 13.3A), no further changes would be encountered during meiosis. But if the inverted region is extremely large (Fig. 13.3C), crossing over in the inverted segment can result in unbalanced recombinant chromosomes. In the case of inversion loop formation, crossing over inside the loop will lead to further complications. One crossover will produce deficiency-duplication chromatids. An example of such development is shown in Fig. 13.4. This figure demonstrates the meiotic conse-quences of a pericentric inversion in chromosome 3 of humans as discussed in Chapter 12 (Allderdice et al., 1975). In this instance the crossover must have

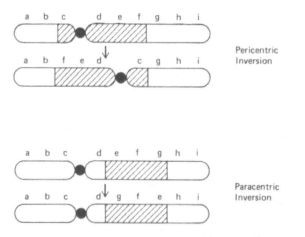

Fig. 13.2. Drawing of pericentric and paracentric inversions. In a pericentric inversion, the centromere is included in the inverted region. In a paracentric inversion the centro-mere is not included in the inverted region. The paracentric inversion is more common in submetacentric and subtelocentric chromosomes.

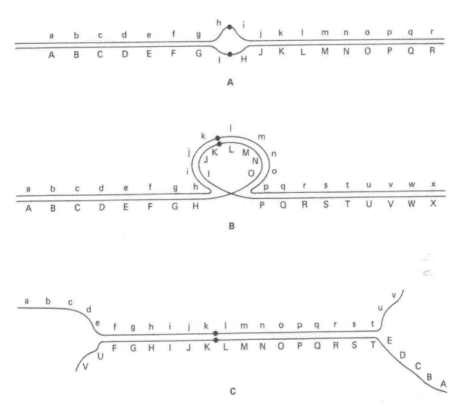

Fig. 13.3A–C. A drawing of inversion heterozygote chromosomes when paired in meiosis. (*A*) Example of chromosome pairing when the inverted material is very small in content. (*B*) Example of the **inversion loop.** This condition occurs when the inverted region of the chromosome is large enough for maneuvering. One chromosome buckles while its heterozygote partner loops, and then these chromosomes pair gene for gene. (*C*) Example of an inversion that includes the major length of the chromosome. The majority of the chromosome pairs in the inverted region leaving the uninverted ends of the heterozygote chromosomes to dangle.

occurred fairly close to the centromere (between D and E. Letters are used arbitrarily in Fig. 13.4B). The four recombinant chromosomes resulting from such a crossover are shown in Fig. 13.4C. Two of these involve duplication-deficiencies. One of them shows a larger duplicated segment (ghi), and it lacks a small segment (a). This chromosome (shown in the square) has the constitution ihgbcd.EFGHI. The other recombinant chromosome has a smaller duplicated segment (a) and lacks a larger segment (ghi). The chromosome with the smaller deficient segment has the better chance to survive since deficiencies are more likely to be detrimental than duplications. Indeed, the autosomal monosomic condition for the large deficiency has not been identified in leucocyte tissue culture. The recombinant with the q duplication and the p deletion (Fig. 13.4) has been identified by banding patterns in karyotypes from one fetus and four children of known inv(3) (p25q21) carriers (Allderdice et al., 1975). The deleted segment reaches from the end of

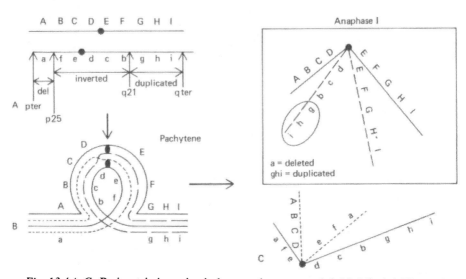

Fig. 13.4A–C. Pericentric inversion in human chromosome 3 (p25q21). (*A*) Illustration of the two chromosomes of an inversion heterozygote (capital letters: normal chromosome; small letters: inversion chromosome). (*B*) Meiotic pairing configuration in inversion heterozygote with crossing over in inversion loop (dE) close to the centromere. (*C*) The four recombinant chromosomes resulting from a crossover. One of the chromosomes in the square shows a larger duplicated segment (ghi) and a lacking small segment (a).

the short arm (pter) to band p25. The duplication reaches from the end of the long arm (qter) to band q21 (see Fig. 13.4A).

Another case of a family with presumptive pericentric inversion heterozygosity of chromosome 2 in humans was reported by De Grouchy et al. (1966). Figure 13.5 shows the unfortunate reproductive history of the maternal grandmother of this family; she showed normal intelligence but had two spontaneous abortions and a pair of stillborn twins. Here, as in the previous case described by Allderdice et al. (1975), the most likely cause of chromosome imbalance in this family was suspected to be crossing over within the inversion loop, which results in a chromosome carrying a duplication-deficiency.

Instances of extremely small pericentric inversions in humans were recorded by de la Chapelle et al. (1974). They reported such an inversion in chromosome 9 (p1q13) of 35 individuals that were related to each other. Two similar pericentric inversions in chromosome 10 (p11q21 and p11q11) were detected by them in two sibs of 11 and 8 individuals each. They concluded that minor pericentric inversions are readily propagated and do not lead to mitotic or meiotic disturbances. They estimated the incidence of these inversion types in the Finnish population as being above 1%, which exceeds previous reports.

Other presumptive pericentric inversions in men have been reported by Jacobs and Ross (1966) in the Y chromosome, by Gray et al. (1962) and Schmid (1967) in the G group, and by Court Brown (Court Brown et al., 1966; Court Brown, 1967), Ferguson-Smith (1967), and Jacobs et al. (1968) in the C group.

If two crossovers involving the same two chromatids (**two-strand double crossing**

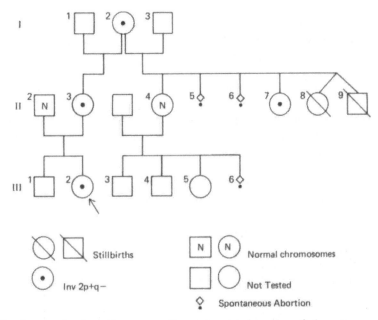

Fig. 13.5. A family tree drawing of a presumptive pericentric inversion of chromosome 2 in humans. Because of crossing over in an inversion loop, duplication-deficiency carrying chromosomes result. (From De Grouchy et al., 1966. Redrawn by permission of Academic Press, Inc., New York).

over) occur within the inversion loop, no duplication-deficiency chromatids will be formed. A second crossover cancels the effect of the first one (Fig. 13.6). Any odd number of crossovers within the inversion loop involving the same two chromatids (strands) will cause duplication-deficiency chromatids or gametes. Any even number of such crossovers will cancel this effect. If more than two chromatids are involved in such crossing over (e.g., **three-strand double crossing over, four-strand double crossing over**) within the pericentric inversion loop, then duplications and deficiencies do occur and the effect is not canceled.

As demonstrated, crossing over leads to duplications and deficiencies in the

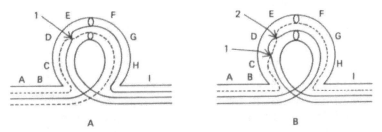

Fig. 13.6. (*A*) Drawing of two-strand double crossing over when only one crossover occurs in the inversion loop. Duplication-deficiency chromatids result. (*B*) If two crossovers occur in the inversion loop, the second crossover cancels the cytological effect of the first crossover.

gametes produced by pericentric inversion heterozygotes. Such chromosome aberrations have led to reduced gamete fertility (Alexander, 1952; Patterson and Stone, 1952). This may be the reason for the low frequency of observed pericentric inversions in the genus *Drosophila* as compared to the occurrence of paracentric inversions in this genus. However, small pericentric inversions are apparently much more frequent in some animal groups than previously suspected. Gene rearrangements induced by such inversions have been detected which cause polymorphic karyotypes within the species *Rattus rattus* and *R. norvegicus* (Yosida and Amano, 1965; Yosida et al., 1965).

Different karyotypes originated by pericentric inversions causing chromosome polymorphism have been observed in some natural grasshopper populations (White, 1958; Lewontin and White, 1960; White et al., 1963). Polymorphic populations can be in a karyotype equilibrium. In such an equilibrium the frequencies of the different karyotypes do not change conspicuously from generation to generation.

Many plant genera are also known to have inversions, but their frequency is much less than in animals (Snow, 1969). Only three pericentric inversions have so far been identified in fungi such as *Neurospora* (Newmeyer and Taylor, 1967; Turner et al., 1969).

13.2   Paracentric Inversions

Paracentric inversions occur more frequently than pericentric inversions in natural populations. As mentioned before in this case the centromere is not included in the inverted segment (Fig. 13.2). A photograph of a paracentric inversion heterozygote in the X chromosome of *Drosophila* is shown in Fig. 13.7. In contrast to pericentric inversion heterozygotes, paracentric ones produce **anaphase bridges** and **acentric fragments** in meiosis if crossovers occur within the inversion loop. Consequently, in natural or irradiated organisms the occurrence of anaphase bridges and acentric fragments is an indication that paracentric inversions may be present. But anaphase bridges and acentric fragments do not only occur as a consequence of this type of inversion. Other possible reasons for their occurrence could be spontaneous breakage and fusion of chromosomes during meiosis (Haga, 1953) as well as chromosome breakage and sister chromatid reunion as observed in the meiosis of rye, inbred for 27 generations (Rees and Thompson, 1955).

As discussed for pericentric inversions, the occurrence of crossing over in the inversion loop will also have an effect on the chromosomes of paracentric inversion heterozygotes. Depending on the number of crossovers within and outside the inversion loop and on the number of chromatids involved in the crossing over, anaphase bridges and acentric fragments will be single or double and can occur in anaphase I or anaphase II (Table 13.1). Only some of the possible combinations will be illustrated here for demonstration of the meiotic and genetic consequences. A master diagram illustrates the crossover points involved in these examples (Fig. 13.8). Crossover point I involves chromatids 2 and 4, crossover point II involves chromatids 1 and 4, crossover point III involves chromatids 2 and 3, and crossover point IV involves chromatids 1 and 3. Crossover points I to III are inside the inversion

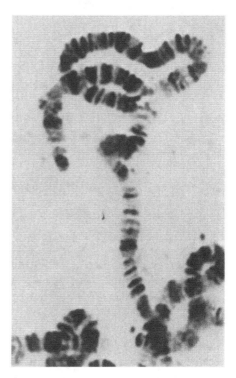

Fig. 13.7. Photomicrograph of a paracentric inversion heterozygote [In (1)d1-49] in the X chromosome of *Drosophila*. (Courtesy of Dr. Joseph B. Morton, Cargill Inc., Lubbock, Texas).

loop while crossover point IV is located between the centromeres and the inversion loop. If a crossover occurs only at point I (two-strand single crossing over), a bridge and an acentric fragment result in anaphase I (Fig. 13.9). Since the acentric fragment does not inherit a centromere it will not be included in any daughter nucleus, and its genetic information will be lost. It may still be visible in the second meiotic division but since it is eventually abandoned it is not shown in the diagram. The size of the acentric fragment gives an indication of the size of the chromosome segment that is inverted. The acentric fragment represents the length of the

Table 13.1. Inversion heterozygotes with different types of crossing over and consequences in anaphase I and II.

Strands	Crossing over	C-I*	-I-†	AI	AII
2	single	X	—	normal	normal
2	single	—	X	1B + 1F	normal
4	double	—	XX	2B + 2F	normal
3	double	—	XX	1B + 1F	normal
3	double	X	X	1F	1B
4	triple	X	XX	2F	2B

*C-I = crossover between centromere and inversion region
†-I- = crossover within inversion region
 B = bridge
 F = fragment

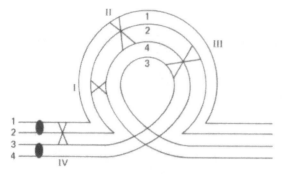

Fig. 13.8. Master diagram for inversion loop crossovers. Crossover point I involves chromatids 2 and 4; point II involves chromatids 1 and 4; point III involves chromatids 2 and 3; and crossover point IV involves chromatids 1 and 3.

inverted region plus twice the length of the uninverted region from the distal breakpoint to the end of the chromosome. In Fig. 13.9 the four differently indicated lines show the four chromatids in pachytene after crossing over has occurred. Each of the four crossover products is indicated with a uniform but different type of line. These were originally supposed to be drawn in color but the expense of reproduction prohibited this procedure. The student is encouraged to draw in his own colors. This procedure is in contrast to other approaches that use different colors for the same chromosome to indicate the genetic changes. One could call our procedure **cytological coloring**. The genetic changes are indicated by the use of letters (**genetic lettering**). The two sister chromatids of the upper chromosome have been designated with small letters (a to i) and the two sister chromatids of the lower chromosome have been marked with capital letters (A to I). The crossover point in the AI (anaphase I) drawing of Fig. 13.9 is indicated where the lettering changes from small letters to capital letters and vice versa (point cD in the dicentric chromosome indicated by the solid line and point Cd in the acentric chromosome indicated by the dot-dash line). The dicentric chromosome is deficient for segment HI and duplicated for segment AB. Bridges in AI can break at any point between the two centromeres of the dicentric chromosome (solid line). In Fig. 13.9 breakage (BP) is assumed between E and F. The break products of AI are two highly deficient monocentric chromosomes that are shown in the AII (anaphase II) drawing and in the pollen (solid line chromosomes). Gamete abortion can result from two such deficiencies. Two gametes are fertile in this case. As mentioned for inversion heterozygosity of the pericentric type, two reciprocal chiasmata within the inversion loop, which involve the same two chromatids (two-strand double crossing over), cancel each other's effect (Fig. 13.6). Complementary chiasmata in the inversion loop such as in four-strand double crossing over result in a double chromatid bridge and two acentric fragments in AI (Fig. 13.10; Table 13.1). As a consequence, all four pollen will have deficient chromosomes resulting in possible pollen sterility. If this condition (four-strand double crossing over within the inversion loop) is combined with two-strand single crossing over in the region between the centromere and the inversion loop, then two acentric

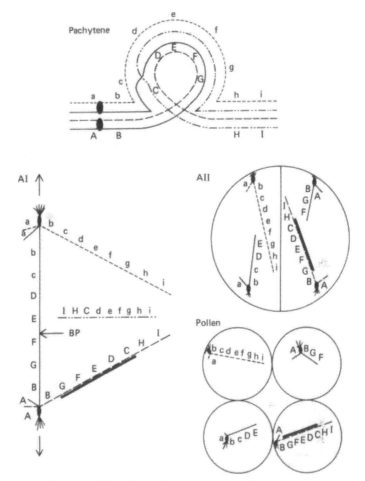

Fig. 13.9. Demonstration of cytological and genetic consequences of a crossover occurring at point I of Fig. 13.8. (two-strand single crossing over). The first illustration shows the 4 chromatids in pachytene. Each crossover product is shown with a different line. The following illustrations show the events in AI, AII, and at the pollen stage. Genetic changes are shown by the use of letters (**genetic lettering**). A bridge (solid line) and an acentric fragment (dot-dash line) result in AI. Two pollen can abort because of deficient chromosomes (solid lines). Heavy solid line = inverted region.

fragments and two looped chromosomes will result in AI (Fig. 13.11). Bridges will occur in each of the two AII cells. Each of the four resulting pollen will have one deficient chromosome; this will result in pollen sterility. As demonstrated in Figs. 13.9 to 13.11, chromatids involved in crossing over are generally eliminated by pollen or embryo sac abortion in plants or by zygote and embryo abortion in animals.

In maize as well as in *Drosophila,* a phenomenon has been observed that prevents the inclusion of dicentric bridges in the megaspore and the egg nucleus. Carson

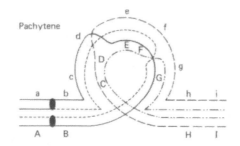

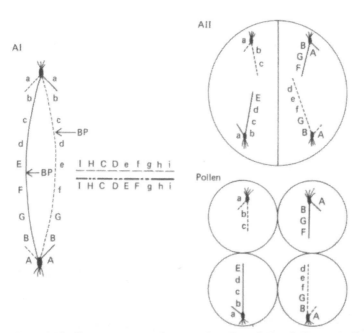

Fig. 13.10. Crossovers occurring at points II and III of Fig. 13.8 (four-strand double crossing over). A double chromatid bridge (solid line and short-dash line) and two acentric fragments (long-dash line and dot-dash line) result in AI. All four pollen receive deficient chromosomes resulting in possible pollen abortion.

(1946) studied more than 2500 eggs of *Sciara*. He found that the dicentric chromatids formed after two-strand single crossing over in the inversion loop usually do not break but remain as a link between the two inner nuclei (Fig. 13.12). As a result, the dicentric bridge always passes into the polar bodies (**bridge elimination mechanism**) and the noncrossover chromatids are always included in the functional egg nucleus. Moderate amounts of naturally occurring inversion polymorphism have been shown in mosquitoes (Kitzmiller, 1976). Twenty-seven different autosomal inversions have been reported from three different localities in Bulgaria in the mosquito species *Anophiles messeae* (Belcheva and Mihailova, 1972).

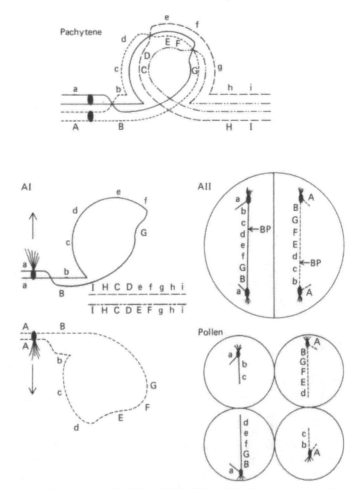

Fig. 13.11. Crossovers occurring at points II, III, and IV of Fig. 13.8 (four-strand double crossing over within the inversion region and two-strand single crossing over between the centromere and the inversion region). Two looped chromosomes (solid line and short-dash line) and two acentric fragments (long dash line and dot-dash line) result in AI. Bridges result in each of the two AII cells. All pollen will have one deficient chromosome.

13.3 Complex Inversions

If more than a single inversion is found in a chromosome, they are known as **complex** or **multiple inversions.** The types of inversions involved in a complex inversion (Rieger et al., 1976) may be:

1. independent inversion (Fig. 13.13A)
2. direct tandem inversion (Fig. 13.13B)
3. reversed tandem inversion (Fig. 13.13C)
4. included inversion (Fig. 13.13D)
5. overlapping inversion (Fig. 13.13E)

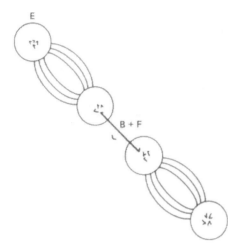

Fig. 13.12. Illustration of **bridge elimination mechanism.** Decentric bridges resulting from inversions remain as a link between two inner nuclei. Bridges always pass into polar bodies and noncrossover chromatids always are included in functional egg nucleus.

In **independent inversions,** a chromosome inversion occurs in independent sections of the chromosome, and the two resultant inverted segments are separated from each other by an uninverted chromosome segment. **Direct tandem inversions** are the result of two successive inversions involving chromosome segments that are directly adjacent to each other. In **reversed tandem inversions,** the two inverted segments are adjacent to each other but mutually interchanged. In an **included inversion,** a segment that is part of an inverted segment is inverted once again. An **overlapping inversion** is the result of part of an inverted chromosome segment being inverted a second time together with an adjacent segment that was not included in the first inversion segment.

Many complex inversions have been discovered in wild populations of *Drosophila.* Over 40 different inversions were found in various chromosomes of *D. willistoni,* which demonstrates a great degree of **chromosome polymorphism.** Several

Fig. 13.13. (A) Diagram of an independent inversion. The top chromosome shows the ▶ first inversion region (bc. Br_1 to Br_2), the middle chromosome shows the second inversion region (fgh. Br_3 to Br_4), and the bottom chromosome shows the final condition after both inversions have occurred. (B) Diagram of a direct tandem inversion. The top chromosome shows the first inversion region (bc. Br_1 to Br_2), the middle chromosome shows the second adjacent inversion region (def. Br_3 to Br_4), and the bottom chromosome shows the resultant chromosome after both inversions have occurred. (C) Diagram of a reversed tandem inversion. The top chromosome shows the first inversion region (de. Br_2 to Br_3) and an additional breakpoint (Br_1), the middle chromosome shows the second adjacent inversion region (bc. Br_4 to Br_5), and the bottom chromosome shows the final condition. (D) Diagram of an included inversion. The top chromosome shows the first inversion region (cdefgh. Br_1 to Br_2), the middle chromosome shows the included inversion region (fe. Br_3 to Br_4), and the bottom chromosome shows the condition after both inversions have occurred. (E) Diagram of an overlapping inversion. The top chromosome shows the region of the first inversion (bcde. Br_1 to Br_2), the middle chromosome shows the overlapping second inversion region (dcbfg. Br_3 to Br_4), and the bottom chromosome depicts the final condition.

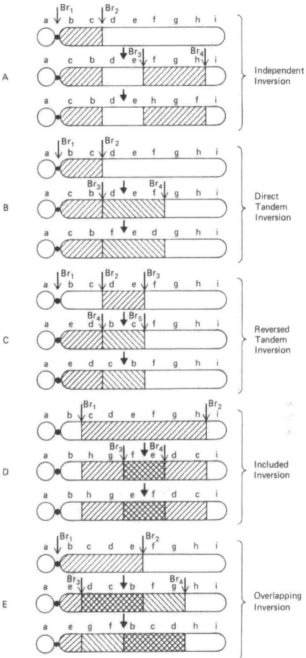

inverted chromosome types can occur within one and the same fly population. In *D. pseudoobscura* and *D. persimilis* chromosome inversions are also very frequent. They occur mainly in the third but also in the X chromosome. A female in *D. willistoni* was discovered that was heterozygous for 16 different inversions. Studies in *D. pseudoobscura* were carried out by Dobzhansky (1941), Koller (1936), Strickberger and Wills (1966), Pavlovsky and Dobzhansky (1966), and Crumpacker and Salceda (1968).

There are two reasons why inversion heterozygosity does not result in a high degree of sterility in *Drosophila*. In the males there is an **achiasmatic mechanism** and in the females a **bridge elimination mechanism.** In the achiasmatic mechanism the chromosomes are meiotically paired but no synapsis and crossing over exist. Consequently, no inversion loops are formed in pachytene, and bridges and fragments do not occur in anaphase I or II, thus there is no gamete abortion. The bridge elimination mechanism has already been descirbed (Fig. 13.12).

13.4   Inversions as Crossover Suppressors

A crossover suppressor can be a structural chromosome change that suppresses or reduces the frequency of meiotic crossing over. The best known examples for such effects are inversion heterozygotes. There are two different factors that reduce the occurrence of crossing over in inversion heterozygotes:

1. Crossing over inside and around the inverted segment is reduced as a result of incomplete pairing. The most drastic case is the example of a very short inversion segment that eliminates pairing altogether (Fig. 13.3A). But also in cases in which the segment is large enough so that a loop is formed, crossing over can be reduced inside and around the loop as a consequence of incomplete pairing.
2. The products of crossing over in an inversion loop are mostly inviable and are not recovered (Figs. 13.9 to 13.11). This makes it appear as though an inversion segment in an inversion heterozygote is completely free of crossovers.

In this second instance (2), the crossover frequency is inversely proportional to the length of the inverted chromosome segments. Almost complete crossover suppression has been observed in short inversions of *Drosophila* species where two-strand single crossing over provides total elimination of the crossover products. Incomplete crossover suppression is accompanied with longer inversions where two-strand double crossing over can occur and the second crossover cancels out the abortion effect of the first. Some balanced crossover products will thus have a chance to be included in gametes.

Muller (1928) was the first to make use of the crossover suppressor phenomenon in *Drosophila*. He developed the *ClB* method. In this method he used a special female stock that had: (1) an X chromosome with a large inversion as a crossover suppressor (*C*) preventing exchange between the *ClB* chromosome and the male chromosome to be examined (black in Fig. 13.14), (2) a recessive lethal gene, preventing homozygosity for the *ClB* chromosome (*l*), and (3) a *Bar* duplication, which permitted identification of individuals that carried the *ClB* chromosome in a heterozygous condition.

Fig. 13.14. Schematic drawing of the *ClB* method for the demonstration of recessive sex-linked lethal factors in *Drosophila*. (From Rieger and Michaelis, 1958).

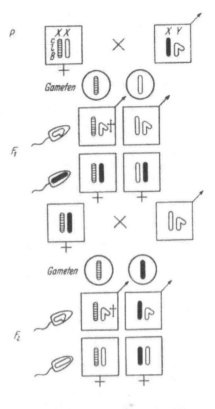

In this method the *ClB* females are mated with irradiated male flies (Fig. 13.14). Half of the males in the F₁ generation die because they possess the *ClB* chromosome in the hemizygous condition. There is no compensating *L* allele in the Y chromosome. Half of the F₁ females are phenotypically marked (*Bar*) by the *ClB* chromosome and also possess the X chromosome of the x-irradiated males. Recombination between the two X chromosomes is restricted because of the inversion. The *Bar* females are individually crossed with normal males, and each progeny is tested separately. Because all *ClB* sons of these F₂ crosses die, all surviving F₂ males possess the x-rayed X chromosome from the grandfather that is to be examined (black X). If the x-irradiation resulted in a mutation, all F₂ males will show the phenotypic mutant effect. If the result was a lethal mutation, no F₂ males will be observed. Because of the isolated treatment (separate test bottles) of the fertilized *ClB* F₁ females, the frequency of lethal and nonlethal mutations in the X chromosomes of the treated grandfather can be easily calculated. This method showed Muller and many other investigators the effect of specific mutagens and the resulting mutation rates.

A refinement of this tool is the M-5 method (Demerec, 1948; Spencer and Stern, 1948). This method makes use of the Muller-5 stock, which has greater crossover suppression than the *ClB* stock and does not operate with a lethal factor. The crossover suppressing inversion is of the included inversion type (Fig. 13.13D), which is a small inversion included in a larger inversion in the X chromosome. This stock has three marker factors, the dominant *Bar* (*B*); *scute,* which is a hair factor (*sc*^{s}); and *white apricot,* which is an eye color factor (*w*^{a}). The recessive markers help to identify any crossovers that may occur between the Muller-5 chromosome and the irradiated chromosome. Crossover suppressors are now also developed for mutagenicity tests in mice (Evans and Phillips, 1975).

13.5   The Schultz-Redfield Effect

The presence of an inversion in the chromosome complement does not always produce only crossover suppression within the inverted segment of the heterozygote. It also can have an effect on other chromosome pairs of the same complement.

In Chapter 4 (Section 4.3) the phenomenon of crossover interference was discussed. This is based solely on a mechanical principle which states that the occurrence of a crossover in one area of a particular chromosome suppresses a crossover in an adjacent region (Whitehouse, 1965). The possibility of an interchromosomal effect of the crossover in one chromosome pair on a crossover in another such pair is automatically excluded in such mechanical consideration. However, in 1919 Sturtevant found that crossing over increased between the eye color gene *purple* (*pr*:54.5) and the wing shape gene *curved* (*c*:75.5) on chromosome 2 of *Drosophila melanogaster* due to a dominant third chromosome gene. In 1933 Darlington observed reciprocal effects on chiasma frequency between a B chromosome bivalent and the rest of the chromosome set in rye. When the B-bivalent had high chiasma frequency, the chiasma frequency in the other bivalents was reduced and vice versa. More exact evidence for such interchromosomal relationship was later given by Morgan et al. (1932, 1933), Glass (1933), Komai and Takahu (1942), Steinberg and Fraser (1944), Schultz and Redfield (1951), Redfield (1955, 1957), Levine and Levine (1955) for *D. melanogaster* and by MacKnight (1937) for *D. pseudoobscura.* In all these later cases, heterozygous inversions in one chromosome pair caused increased crossing over in other nonhomologous chromosome pairs of the same complement. Schultz and Redfield in 1951 placed this effect on a quantitative basis and were credited with its discovery. White and Morley (1955) speculated that the Schultz-Redfield effect could be the result of a genetic **homeostatic effect,** which keeps the crossover frequency close to an optimal value within the population, the effect being not only interchromosomal but also intrachromosomal. For example, Carson (1953) demonstrated this in *D. robusta* where crossing over was increased in the noninverted region of the inversion chromosome pair. Rieger and Michaelis (1958) speculated that the Schultz-Redfield effect probably is not limited to the genus *Drosophila* but may be a more widely occurring phenomenon associated with heterozygous paracentric inversions.

Chapter 14
Chromosome Translocations

The most common type of translocation is the reciprocal translocation. Brown (1972) goes so far as to say that all translocations observed are, with no known exception, reciprocal. In order to survive, all cells ought to have a balanced set of genes. Cells with reciprocal translocations provide such conditions. However, since other types of translocations have been discussed in the literature they will be briefly mentioned here also.

14.1 Types of Translocations

The nomenclature on chromosome translocations has not always been consistent. For the sake of simplicity the terms used here are arranged according to the number of breaks occurring. According to this system, four different classes of chromosome translocations can be distinguished:

1. simple translocations (one break involved)
2. reciprocal translocations (two breaks involved)
3. shift type translocations (three breaks involved)
4. complex translocations (more than three breaks involved).

The **simple translocation** (Fig. 14.1) is probably of more historical than practical value. However, the discussion will soon show why it is still included here. As mentioned, a simple translocation would be the result of a single break in a chromosome arm and the transfer of the acentric chromosome fragment to the end of another nonhomologous chromosome. Evidence of such an event came first from Muller and Painter (1929) when they found that a group of third linkage group genes in *Drosophila, roughoid eye* (*ru*:0.0) to *scarlet eye* (*st*:44.0), were linked to the genes of the second linkage group. The remaining genes in the third linkage group, *pink* (*p*:48.0) to *Minute-g* (*Mg*:106.2), remained independent of the genes in the second linkage group. The cytological proof of this genetic observation was also given by Painter and Muller (1929).

Such an event seems to be in contrast to the earlier stated premise that telomeres seem to seal off the ends of chromosomes so that they cannot join with other broken chromosome ends (Section 2.2.4). However, what seems to be a simple translocation may in fact be a reciprocal translocation in that the very end of chromosome 2 in *Drosophila,* including the telomere, may have broken off and may have exchanged position with the acentric fragment of chromosome 3. In this way one

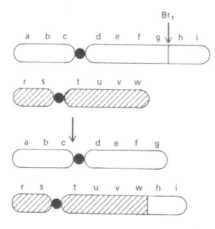

Fig. 14.1. Illustration of a simple transloca-tion. The top two chromosomes are nonhom-ologous chromosomes with Br_1 indicating the break point on top chromosome. The bottom two chromosomes are the same nonhomolo-gous chromosomes after translocation of the acentric fragment (hi).

really deals with a two-break situation that meets the requirements of a true exchange. The small fragment of chromosome 2 may not have carried any detect-able genes, thus simulating a condition of a simple translocation. Today, it is gen-erally believed that true simple translocations cannot occur and that all earlier reported cases are really reciprocal translocations (Burnham, 1956).

A **reciprocal translocation** involves the mutual exchange of broken chromosome fragments between nonhomologous chromosomes. As mentioned, this is the main type observed, and it will be discussed in great detail. It is dependent on two break events (Fig. 14.2).

The **shift type translocation** (Fig. 14.3) or **transposition** (Section 12.3.1) depends on three breaks. It can happen in three different ways:

Intrachromosomal shifts

1. The broken chromosome segment can be inserted into the same chromosome arm but at a different location (Fig. 14.3A).
2. The broken chromosome segment can be shifted to an intercalary position in the other arm of the same chromosome (Fig. 14.3B).

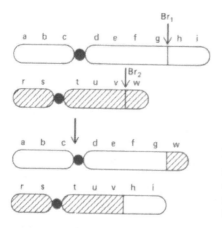

Fig. 14.2. Diagram of a reciprocal transloca-tion. The top two chromosomes are nonhom-ologous chromosomes with break points Br_1 and Br_2. The bottom two chromosomes are the same nonhomologous chromosomes after the acentric fragments (hi, w) have translocated.

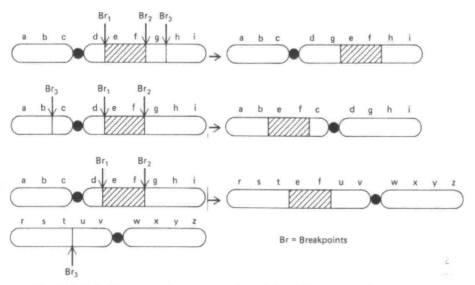

Fig. 14.3A–C. Diagrammatic representation of the shift type translocation (transposition). (*A*) Broken chromosome segment (ef) is inserted in same chromosome arm but at different location (g↓h). (*B*) Broken chromosome segment (ef) is shifted to an intercalary position (b↓c) in other arm of same chromosome (*A and B:* intrachromosomal shifts). (*C*) Broken chromosome segment (ef) is shifted to an intercalary position (t↓u) of a nonhomologous chromosome (interchromosomal shift).

Interchromosomal shifts

3. The broken chromosome segment can be shifted to an intercalary position in one of the two arms of a nonhomologous chromosome (Fig. 14.3C). The first translocation found by Bridges (1923) was such an interchromosomal shift.

Complex translocations are those in which three or more breaks are involved. In the progeny of an irradiated *Drosophila* male, Kaufmann (1943) found an individual with a complex arrangement involving at least 32 breaks. The treatment involved x-rays of 4000 r followed by infrared radiation for a period of 144 hours.

14.2   Origin of Translocations

Translocations can occur naturally as well as by induction. As mentioned at the beginning of this discussion on variation in chromosome structure (Part VI), structural chromosome changes are generally considered to depend on breakage of chromosomes and on reunion of chromosome segments.

Translocations along with other chromosome aberrations were reported by Beadle (1937) in the progeny of maize that had the gene *sticky* [*st*:(55)] on chromosome 4. In such mutants the chromosomes adhere to each other in anaphase I and rupture during anaphase movement producing structural chromosome changes such as translocations. A similar stickiness effect has been reported by McClintock (1950a, 1950b, 1951, 1953) in which this property is also genetically controlled by the

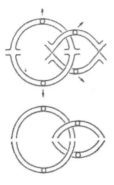
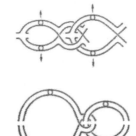

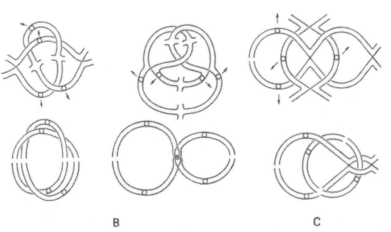

Fig. 14.4A–C. Illustration of different types of chromosome interlocking. Upper rows of *A, B,* and *C:* diplotene. Lower rows of *A, B,* and *C:* metaphase I. (Modified after Darlington, 1965).

Activator-Dissociation system described in Chapter 12 (Section 12.3.1). Translocations also have been found in plants that were grown from aged wheat and barley seed (Gunthardt et al., 1953). Nilsson (in Kostoff, 1938) in *Oenothera* and Navashin and Gerassimova (1935) in *Crepis* and other plants found that with aging seed, structural chromosome changes significantly increased.

Other possible causes for chromosome translocation have been reported from interlocking of bivalents that subsequently can break in anaphase I (Sax, 1931; Sax and Anderson, 1933; Burnham, 1962). Interlocking takes place during zygotene or pachytene of first meiotic prophase when a nonhomologous chromosome passes through a loop of two homologous chromosomes that are in the process of pairing. Different examples of such interlocking are shown in Fig. 14.4. Interlocked bivalents are occasionally found in *Tradescantia* and in other genera in which translocation rings are found.

Kostoff (1938) speculated that translocations originated from spontaneous segmental association in heterochromatic chromosome regions. If Kostoff's suggestion were true, then naturally occurring translocations should show a high frequency of breakpoints in heterochromatic chromosome segments.

14.3   Reciprocal Translocations

Figure 14.2 shows the mutual exchange of two chromosome segments between two chromosome pairs of nonhomologous origin. If the ends of these chromosomes are numbered (1.2, 3.4) then 4 chromosomes that all have different end combinations (1.2, 3.4, 1.4, 2.3) will result. No two chromosomes of this group can pair along their entire length, but all four can come together in a pairing configuration (**quadruple**) that allows partial pairing of homologous chromosome segments. Quadruples resemble **quadrivalents,** which are formed when all four chromosomes are homologous. Sybenga (1972) supported the use of separate terms for these two types of configurations that resemble each other in shape but originate differently. In quadrivalents all four chromosomes are homologous or equivalent to each other, while in quadruples only certain segments are homologous. An organism in which quadruple pairing occurs is called a **translocation heterozygote.** In pachytene such a configuration can appear as a cross (Fig. 14.5). Sometimes such a cross still can be prevalent in the following stage of diplotene (Fig. 14.6). As can be seen, the pairing partners in this figure change at the translocation breakpoints (Fig. 14.5). Consequently, such figures can reveal the location of the breakpoints. This is particularly true in instances where exact pairing occurs as in the polytene chromosomes of Diptera. However, pachytene configurations in maize and tomato do not always show complete synapsis of the pairing partners involved. For instance, in the heterozygous translocation strain T2-6a of maize (Burnham, 1932), which has translocation breaks in the long arm of chromosome 2 and in the short arm of chromosome 6 (satellite chromosome), pairing can be gene by gene (Fig. 14.7A), asynaptic or nonhomologous near the center of the cross (Fig. 14.7B). Translocation breaks can occur at any point along the chromosome, possibly even in the centromere region. The position of the breakpoints will determine the future fate of the translocation quadruple. If the breakpoints are located close to the chromosome ends (distal area), the chance of crossover formation between the breakpoint and the chromosome end is reduced. Possible interstitial crossovers (between the centromere and the breakpoint) can lead to duplication and deficiency gametes and consequently are not recovered (Burnham, 1962). If no crossovers form in the distal area, the quadruple will break up into two open bivalents by the end of prophase I (diakinesis) and meiosis will continue mechanically normal. But if crossovers are formed between the translocation breakpoints and the chromosome ends, the quadruple configurations will persist through diakinesis into metaphase I.

Different kinds of quadruple configurations can arise depending on the formation of crossovers in interstitial (between breakpoint and centromere) or distal (between breakpoint and chromosome end) chromosome segments. It should be remembered here that chiasmata always move away from the centromeres and not toward them (Section 7.2.4; Fig. 14.5, arrows). Possible diakinesis configurations originating

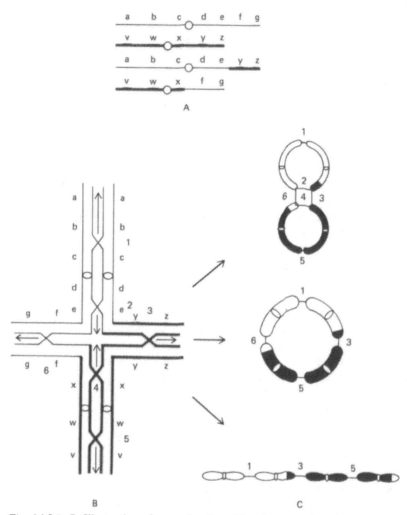

Fig. 14.5A–C. Illustration of a quadruple pairing in a translocation heterozygote. (*A*) The two homologous chromosome pairs involved in the reciprocal translocation. (*B*) Pachytene configuration appearing as a cross. Numbers 1 to 6 designate chiasmata. (*C*) Depending on the location of the chiasmata 8-, ring-, or rod-shaped quadruples can form in diakinesis.

from different pachytene situations are shown in Fig. 14.5. If, for instance, crossovers occur in locations 1, 2, 3, 4, 5, and 6, diakinesis configurations resemble a number 8. If chiasmata occur only in locations 1, 3, 5, and 6, a ring of 4 chromosomes is formed in diakinesis. If chiasmata occur in locations 1, 3, and 5, a chain of 4 chromosomes can form. The crossover positions in locations 1 to 6 are minimum requirements for the diakinesis configurations indicated above. Any additional crossovers in the same chromosome segments will lead to identical configurations.

Chromosome orientation of quadruples in metaphase I is critical. Bivalents in nor-

Fig. 14.6. Diakinesis of barley showing translocation cross involving chromosomes 4 and 5. (Courtesy of Mrs. Christine E. Fastnaught-McGriff, Department of Plant and Soil Science, Montana State University, Bozeman, MT).

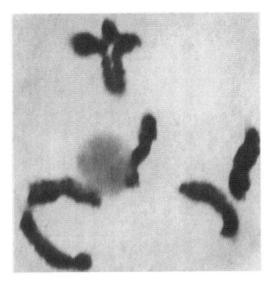

mal meiosis have only two centromeres that are arranged in coorientation and that distribute the two chromosomes involved to opposite poles. However, several types of orientation in the metaphase plate are possible when a quadruple is formed since not two but four centromeres are involved.

Theoretically, the following types of chromosome orientation are distinguished:

1. Coorientation
   a. alternate-1 orientation
   b. alternate-2 orientation
   c. adjacent-1 orientation
   d. adjacent-2 orientation
2. Noncoorientation

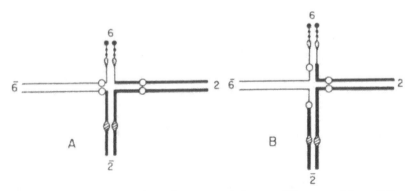

Fig. 14.7A and B. Line interpretations show synapsis in two different positions of the center of the cross of the T2-6a translocation heterozygote in maize. The thin lines represent chromosomes 6. The thick lines chromosomes 2. (*A*) Association of homologous parts in the center of the cross at the original exchange breakpoints. (*B*) Association of nonhomologous segments in the center of the cross not at exchange points. (From Burnham, 1962. Redrawn by permission of Charles R. Burnham, St. Paul, Minnesota).

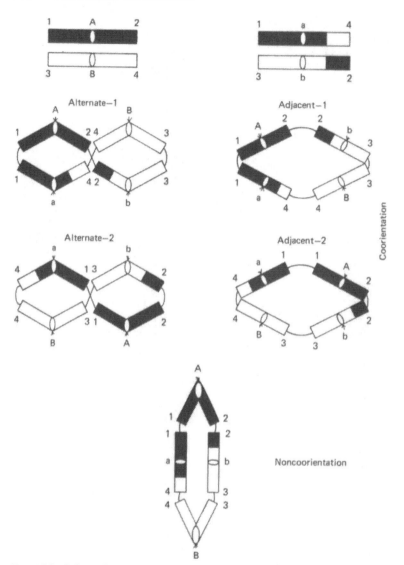

Fig. 14.8. Orientation types of quadruples in metaphase I. Numbers designate chromosome ends, and letters designate centromeres. Explanation in text.

Chromosomes of quadruples in **coorientation** generally are distributed in even numbers to opposite poles. In alternate-1 and adjacent-1 orientation, homologous centromeres (A,a, Fig. 14.8) coorient and wander to opposite poles just like during normal bivalent separation. In alternate-2 and adjacent-2 orientation, nonhomologous centromeres (A,b) coorient and pass to opposite poles (Fig. 14.8). Apparently there is no guarantee in quadruples as there is in bivalents, that homologous centromeres coorient and go to opposite poles. Consequently, alternate-2 and adjacent-2 orientations actually occur. Endrizzi (1974) could morphologically differentiate all four coorientation types in three different translocations of cotton.

In all three the ratio of alternate-1 to adjacent-1 was 1:1. In one translocation, the ratio of alternate 2 to adjacent 2 was 1:1 but in another 2:1.

In order that balanced combinations result in quadruples, alternate orientation has to occur. Only if the two translocated chromosomes (centromeres a and b, Fig. 14.8) pass to one pole and the two nontranslocated chromosomes pass to the other (centromeres A and B) will there be balanced chromosome complements in the gametes. This condition is met in the alternate-1 and alternate-2 orientations indicated in Fig. 14.8. All gametes that do not meet this condition will have duplicated and deleted chromosome complements that are caused by adjacent-1 and adjacent-2 orientation. In the case of adjacent-1 in Fig. 14.8, one gamete will be duplicated for segment 2 and deficient for segment 4 and the other gamete will be duplicated for segment 4 and deficient for segment 2. In the case of adjacent-2, one gamete will be duplicated for segment 1 and deficient for segment 3, while the other gamete will be duplicated for segment 3 and deficient for segment 1. Such gametes are also referred to as **Dp-Df gametes** (duplication-deficiency). In general, **semisterility** results in translocation heterozygotes because of the formation of balanced and unbalanced gametes. The proportion of fertile and aborted gametes is close to 1:1 in several species. This is expressed, for instance, in half of the seeds missing in an inflorescence of a plant. Examples of such plants are maize, petunia, peas, and sorghum. In mammals, reciprocal translocations have been investigated in mice. Varying degrees of semisterility were demonstrated by Carter et al. (1955), Ford et al. (1956) and Slizinsky (1957). In humans the frequency of chromosome imbalance among reciprocal translocation progeny is likely to be less than the theoretical 50% (Hamerton, 1971a).

However, it has been observed that there is **preferential** or **directed segregation** of quadruple chromosomes in some species. In these instances alternate orientation and disjunction of chromosomes range from 70% to 95%. Examples of directed segregation are *Hordeum, Secale, Datura, Triticum, Oenothera,* and several insects.

Species that have directed chromosome segregation seem to meet some specific requirements for movability as far as the chromosomes are concerned. Factors that seem to influence quadruple orientation are:
1. length of the chromosomes
2. position of the breakpoints
3. number and position of chiasmata
4. degree of chiasma terminalization
5. position of centromere

Alternate segregation seems to be increased if the chromosomes involved in the quadruple are uniform in length. Also, short chromosomes are easier to maneuver on the metaphase plate. If chromosomes are too short, they are often too rigid for alternate orientation. In *Oenothera,* for instance, the chromosomes are all of the same length.

The position of the breakpoints seems to influence chiasma formation in the quadruple. It is generally believed that chiasma formation is reduced in the interstitial segments (between centromere and breakpoint) (Sybenga, 1972). This partially may be the case because heterochromatin near the centromeres does not allow active pairing activity, which is a prerequisite for crossing over and chiasma for-

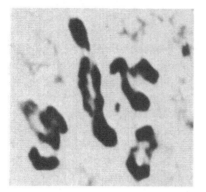

Fig. 14.9. Noncooriented quadruple in a T5-7g translocation heterozygote of barley ($\times$ 2244). (Schulz-Schaeffer, unpublished).

mation. If the breakpoints are distal, the interstitial segments are large, and crossing over is reduced to a minimum. This reduces the number of chiasmata, and terminalization can be accomplished by metaphase I. If no interstitial chiasmata are present by metaphase I, the quadruple will be quite maneuverable on the metaphase plate. It is obvious that a median position of the centromere also would enhance flexibility of the quadruple.

In **noncoorientation** (Fig. 14.8) the two centromeres on opposite sides of the quadruple (e.g., A, B) are cooriented and are positioned equidistant from the equatorial plate. The two intermediate centromeres (e.g., a,b) are noncooriented and are stretched out between the other two, seemingly not attached to the poles by centromeres. Figure 14.9 shows a noncooriented quadruple in barley. In anaphase I the two cooriented chromosomes pass to opposite poles while the noncooriented ones either pass to the same pole (3:1 segregation) or pass to opposite poles (2:2 segregation). In 3:1 segregation of the quadruple the gametes become aneuploid (Fig. 14.10). After fertilization, this leads to trisomy or monosomy (Chapter 16). Noncoorientation always will lead to unbalanced gametes. In the case of 2:2 segregation, normal and translocated chromosomes will pass to the same poles (e.g., A, a) and duplication-deficiency gametes will result, as in adjacent orientation.

Another centromere orientation phenomenon is **reorientation.** Often the initial orientation at the metaphase plate is not appropriate, and, therefore, reorientation is necessary for controlled chromosome segregation. This may mean the loss of a chromosomal spindle fiber connection to one pole followed by the formation of a new connection to the opposite pole (Rieger et al., 1976). Such a phenomenon has been observed in *Tipula oleracea* in living and fixed material (Bauer et al., 1961; Rohloff, 1970).

14.4   Translocations in Humans

A prominent structural change in human populations is the reciprocal translocation (Hamerton, 1971b). Since meiosis is not readily accessible for study in humans, the spotting of reciprocal translocations as translocation quadruples is

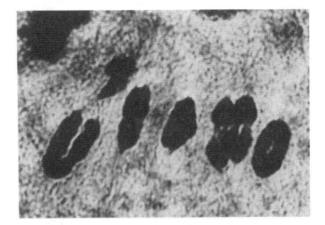

Fig. 14.10. 3:1 segregation of a quadruple in a T5-7g translocation heterozygote of barley (× 2231). (Schulz-Schaeffer, unpublished).

not a good scoring method. Consequently, translocations in humans have been mainly detected by karyotyping. The modern methods of chromosome banding have been extremely helpful in this respect (Section 2.3). A frequent structural rearrangement in man is the so-called **Robertsonian translocation** or centric fusion type (Robertson, 1916). This was confirmed through a cytogenetic survey of 11,680 newborn infants by Jacobs et al. (1974). These findings verified the earlier observation by Court Brown et al. (1966) that Robertsonian translocations ranked highest in a high risk population including state hospital patients and patients attending a subfertility clinic. They had studied 1,870 individuals, and the frequency of Robertsonian translocations was 0.43%, while that of other reciprocal translocations was 0.16%. Other reports of Robertsonian translocation were by Hamerton et al. (1961), Kjéssler (1964), Hamerton (1966), and Hultén and Lindsten (1970).

A Robertsonian translocation is the centric fusion between two acrocentric chromosomes, which results in the reduction of the chromosome number (2n = 45). The least complicated way would be an interchange between the long arm acentric fragment of one chromosome and the short arm acentric fragment of the other chromosome (Fig. 14.11). The two acrocentric chromosomes break close to the centromere. The two long arms fuse and result in a metacentric chromosome. The two short arms form a very small chromosome that may be lost without any genetic damage to the organism. The reason that such breakage occurs more frequently here than at other parts of the chromosome lies in the inherent nature of heterochromatin. Since heterochromatin is located close to the centromeres, breakage happens more often in that region (Section 2.2.1). This phenomenon is remindful of the findings of McClintock (1950a, 1950b, 1951, 1953) who demonstrated a close association of heterochromatin with chromosome breakage (Section 12.3.1).

The human chromosomes most often involved in Robertsonian translocations are the acrocentrics of the D group (13 to 15) and of the G group (21 and 22). They

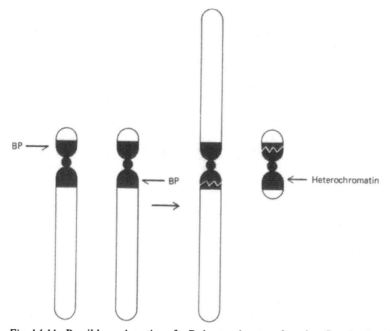

BP

BP

Heterochromatin

Fig. 14.11. Possible explanation of a Robertsonian translocation. Breakpoints (BP) occur in the heterochromatic regions close to the centromere in the short arm and long arm, respectively, of two acrocentric chromosomes. Translocation products are a long armed metacentric chromosome and a very small metacentric chromosome that may be lost without genetic damage to the organism.

are usually closely associated with each other in the cell because they are the organizers of the nucleoli that fuse during the prophase of meiosis (Section 7.2). The close proximity of these chromosomes favors interaction when lesions occur. Since the D and G chromosomes are involved in Robertsonian translocations of humans, Patau (1961) designated them as D/D, D/G, or G/G translocations, and this nomenclature has since been restricted to such types of **centric fusion.** Apparently the most frequent reports of centric fusion type translocations are the D/G type. The reason for this is the association of this translocation with an increase in frequency of children with Down's syndrome who can be easily identified morphologically (Court Brown, 1967). The most common D/G centric fusion seems to be the one between chromosomes 14 and 21 (Hamerton, 1971b). If a Robertsonian translocation occurs between chromosomes 14 (D group) and 21 (G group) in humans, the result will be three chromosomes that can associate in meiosis, the translocated chromosome 14^{21} and the two nontranslocated chromosomes 14 and 21. If regular coorientation occurs, a trivalent will separate in such a fashion that the central 14^{21} chromosome will pass to one pole while the other two chromosomes pass to the opposite one (Fig. 14.12). Such disjunction will result in balanced gametes since each will receive the essential parts of both chromosomes 14 and 21. However, if nondisjunction will occur, two adjacent

Fig. 14.12. Metaphase I configuration of a trivalent formed by human translocated chromosome 14^{21} and two nontranslocated chromosomes 14 and 21.

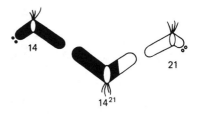

chromosomes ($14 + 14^{21}$ or $14^{21} + 21$) can pass to the same pole. If a gamete with such a combination gets fertilized, it will result in a zygote that carries the equivalent of an extra chromosome, though the chromosome number will be normal ($2n = 46$). Such cases have often been recorded as trisomics, but they are not termed right. Down's syndrome results from the $14^{21} + 21$ combination. Since the duplicated element is the long arm of the smaller chromosome 21, the chromosome abnormality can survive more readily. Another type of Robertsonian translocation has been reported between the short arm of the Y chromosome and the long arm of chromosome 15 (Šubrt and Belhová, 1974) in four generations of male progeny. Balanced chromosome polymorphism for Robertsonian translocations also has been reported in animals such as the goat (Soller et al., 1966), European wild pig (McFee et al., 1966), house mouse (Evans et al., 1967; Léonard and Deknudt, 1967; White and Tjio, 1968), and cattle (Gustavsson, 1966; Gustavsson et al., 1968).

Heterozygotes must produce balanced gametes in most of these instances of Robertsonian translocation heterozygosity in humans and animals since semisterility is rarely reported. A balanced translocation of part of the long arm of chromosome 13 [13q (q21-qter)] attached to the long arm of chromosome 4 (4 q-) is shown in Figs. 14.13 and 14.14 (Vigfusson, unpublished). This mother was normal because of the balanced condition. But her child had a displaced duplication (Section 12.1) resulting in two normal 13 chromosomes and a duplicated 13q21-13qter segment attached to chromosome 4, inherited from the mother.

Other types of reciprocal translocations in humans have been recorded such as t(A;A) between two of the 3 group A chromosomes by Lee et al. (1964), Summitt (1966) and Lejeune et al. (1968b), others between chromosomes of the A group and those of the B group, or t(A;B), by Court Brown et al. (1964), De Grouchy (1965), De Grouchy et al. (1966) and by Walzer et al. (1966). Many other combinations (A;C, A;G, B;B, etc.) are possible. Many of these cases were found because the patient was mentally retarded or had congenital malformation.

Two recent discoveries of 46 chromosome reciprocal translocations are a t(B;D) involving a translocation of the distal half of the long arm of chromosome 14 onto the short arm of chromosome 5 (Fig. 14.15) and a t(A;C) involving a translocation of most of the long arm of chromosome 2 onto the long arm of chromosome 8 (Fig. 14.16). Both photographs indicate the usefulness of modern banding techniques in identifying translocations. The fibroblast cell cultures in these two cases were established from skin biopsies of a 17-year-old normal male who presented emotional and mental problems and from a 24-year-old normal male who was a bal-

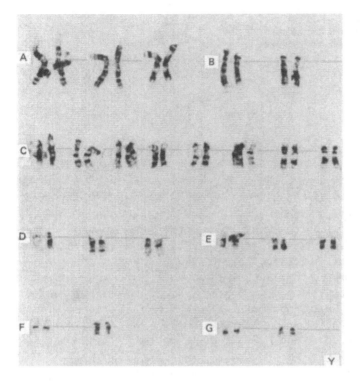

Fig. 14.13. Karyotype of a human with a balanced translocation of part of the long arm of chromosome 13 attached to the long arm of chromosome 4. (Courtesy of Dr. Norman V. Vigfusson, Department of Biology, Eastern Washington University, Cheney).

anced translocation carrier, respectively. Stocks of both are stored in the **Human Genetic Mutant Cell Repository** at the Institute for Medical Research, Camden, New Jersey.

Every arm of the human chromosome complement has been reported as being involved in chromosome translocations (Absate and Borgaonkar, 1977). By 1977 at least 490 so-called "simple translocations," 224 reciprocal translocations, 84

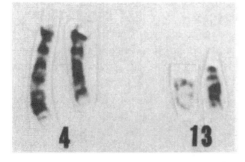

Fig. 14.14. Photomicrographs of human chromosome pairs 4 and 13 showing reciprocal translocation involving one chromosome of each pair. (Courtesy of Dr. Norman V. Vigfusson, Department of Biology, Eastern Washington University, Cheney).

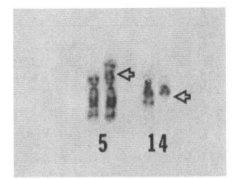

Fig. 14.15. Photomicrographs of human chromosome pairs 5 and 14 showing a reciprocal translocation of the distal half of the long arm of chromosome 14 onto the short arm of chromosome 5. (From Borgaonkar et al., 1977. Reprinted by permission of S. Karger AG, Basel).

Robertsonian translocations, 24 tandem translocations, and 15 complex translocations have been reported in human chromosomes (Borgaonkar, 1977).

14.5   Complex Heterozygosity

The possibility of a reciprocal translocation changing the meiotic behavior of an organism was discussed in Section 14.3. This will be even more drastic if more than one translocation has occurred. Figure 14.17 shows two rings of 4 chromosomes in barley where two reciprocal translocations are involved. The mechanics of events leading to **complex heterozygosity** are best explained in the case of two reciprocal translocations leading to the formation of a **hexaple,** the equivalent of a quadruple, but where six chromosomes are united in a pairing configuration instead of only four (Fig. 14.18). Two reciprocal translocations in the same chromosome complement form such a configuration of six chromosomes if they involve the same chromosome pair. This happens if a reciprocal translocation occurs between a member of a ring of four chromosomes and another chromosome pair. In this case it is not important if the two translocations involve the same chromosome (Fig. 14.18A) or both homologous chromosomes of the same pair (Fig. 14.18B). In Fig. 14.18A, chromosomes A(1.2), B(3.4) and C(5.6) are the nontranslocated chromosomes. The capital letters designate the chromosomes, while the numbers designate the chromosome arms. Chromosomes A'(3.6), B'(1.4) and C'(5.2) are the translocated chromosomes where A' is involved in two translocations. With alternate-1 disjunc-

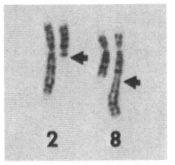

Fig. 14.16. Photomicrographs of human chromosome pairs 2 and 8 showing a reciprocal translocation of most of the long arm of chromosome 2 onto the long arm of chromosome 8. (From Worton et al., 1977. Reprinted by permission of S. Karger AG, Basel).

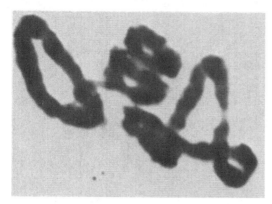

Fig. 14.17. A metaphase I cell of a translocation heterozygote of barley. The two rings of 4 chromosomes each indicate two reciprocal translocations (3^{II} and 2^{IV}. n = 7). (Courtesy of Mrs. Christine E. Fastnaught-McGriff, Department of Plant and Soil Science, Montana State University, Bozeman).

tion (homologous centromeres to opposite poles A,A'; etc.), the translocated chromosomes (A', B', C') all pass the the same pole, and complete compensation for displaced chromosome ends guarantees fertile gametes. In Fig. 14.18B, the two translocations are shared by both partners of a homologous pair (A, A'). There are four translocated chromosomes, A(1.6), A'(2.3), B'(1.4), C'(2.5), and two nontranslocated ones, B(3.4) and C(5.6). Even though translocated chromosomes pass to both poles in alternate-1 disjunction, complete compensation for displaced chromosome ends is guaranteed also in this instance. Tuleen (1972) found rings of six chromosomes in barley intercrosses and could demonstrate by locating the breakpoints that the two translocations involved the same chromosome.

If a third translocation involves a chromosome of a hexaple, a ring of eight chromosomes can occur. This process can continue until all chromosomes of the complement are involved in what is known as a translocation complex.

The best known case of **complex heterozygosity** is the genus *Oenothera* (Cleland, 1962, 1972). Here, not only forms with rings of 6 chromosomes are present but also with rings or chains of 8, 10, 12, and even all 14 chromosomes.

14.6   *Oenothera* Cytogenetics

Oenothera (2n = 14) is one of several plant genera that has developed mechanisms favoring the formation and frequency of translocation heterozygotes in the population. Renner (1914, 1917) first discovered that in this genus "there are several different **genetic factor complexes** which are combined in pairs in the various species and these complexes segregate as wholes in meiosis, each gamete carrying one or the other." They were named after him—**Renner complexes.** Prior to that in 1908, Gates had first observed **multiples,** chromosome associations of more than two, in *O. rubrinervis.* Belling (1925, 1927) in his **interchange hypothesis** concluded that the multiples had to be the result of chromosome translocations. This explained the cytological nature of the Renner complexes. Within the pairing configuration of *O. lamarckiana,* for instance, the paternal and maternal chromosomes are arranged alternately joined together by reciprocal translocations. In

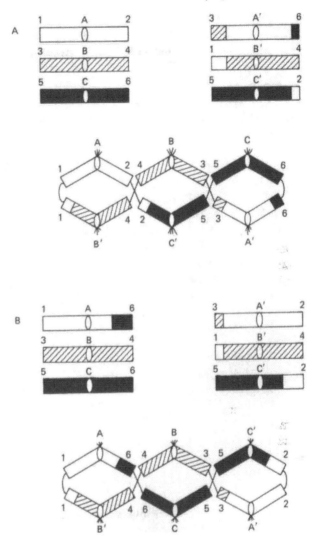

Fig. 14.18A and B. Diagrammatic representation of 2 reciprocal translocations involving 3 chromosome pairs leading to a ring of 6 chromosomes in metaphase I. (*A*) Two translocations involve the same chromosome (3.6). (*B*) Two translocations involve both chromosomes of a homologous pair (1.6, 3.2).

metaphase I, they are arranged in alternate orientation so that eventually one pole receives all the paternal chromosomes while the other one receives only the maternal ones. Consequently, only two kinds of gametes result that are identical to those from which the plant was formed. The complex is comprised of those chromosomes that are distributed in meiosis as a unit (Fig. 14.19). In *O. lamarckiana* the two complexes that segregate to opposite poles in meiosis are called *gaudens* and *velans*. On outcrossing to other forms, it became evident that the *gaudens* complex carries genes for red spots on leaves that are broad and express nonpunctate stems

Gametes
Produced

Fig. 14.19. Complex heterozygosity in *Oenothera*. Two Renner complexes (white: from father; black: from mother) are distributed in anaphase I to opposite poles. (From Cleland, 1962).

and green buds. The *velans* complex possesses genes that prohibit the expression of red spots on leaves that are narrow and express punctate stems and red-striped buds. But self-fertility in these species guarantees maximal heterozygosity and that the complexes do not break up but stay together. **Balanced lethals** (Muller, 1917) insure that only heterozygotes are formed. Such a system consists of two or more linked recessive lethal genes that are permanently maintained in the heterozygous condition such as l_1L_2/L_1l_2. All homozygotes abort because of the double recessive lethal effect (e.g., l_1L_2/l_1L_2 or L_1l_2/L_1l_2). In *O. lamarckiana,* for instance, both megasporocytes and microsporocytes develop *gaudens* and *velans* gametes (Fig. 14.20). During the formation of zygotes, only the *gaudens-velans* (G-V) zygotes survive, while the *velans-velans* (V-V) and *gaudens-gaudens* (G-G) zygotes abort. This phenomenon is known as **zygotic lethality**. In *O. muricata* the two complexes are called *rigens* (R) and *curvans* (C). Here the inactivation occurs earlier, during gametogenesis already. In male gametogenesis the game-

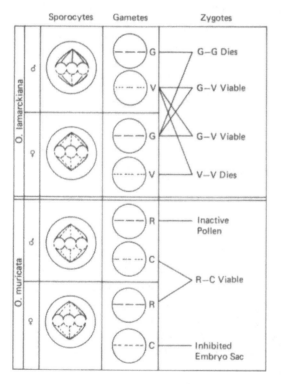

Fig. 14.20. Diagrammatic representation of zygotic lethality in *Oenothera lamarckiana* and of the Renner effect in *O. muricata*. Explanation in text. (From Swanson, 1957. Redrawn by permission of Prentice-Hall, Inc., Englewood Cliffs, N.J.).

tophytes carrying the *regens* complex do not survive so that only *curvans* (C) pollen are effective. In female gametogenesis the gametophytes carrying the *curvans* complex do not survive, and only *rigens* embryo sacs develop. Consequently, only *rigens-curvans* (R-C) zygotes form. This phenomenon, a case of megaspore competition, was also discovered by Renner (1921) and was called the **Renner effect** by Darlington (1932). In summary, then, the devices that guarantee complex heterozygosity in *Oenothera* are:

1. reciprocal translocations
2. balanced lethals
3. self pollination

14.7   Other Systems with Complex Heterozygosity

Another well-investigated system of complex heterozygosity is the genus *Rhoeo* (2n = 12). Studies in this genus have been performed by Darlington (1929a, 1929b), Kato (1930), Sax (1931), Anderson and Sax (1936), Simmonds (1945), Tschermak-Woess (1947), Walters and Gerstel (1948), Stearn (1957), Flagg (1958), Carniel (1960), and Wimber (1968). In contrast to *Oenothera,* the chromosomes in this system are not all of equal length. Sax presented an idiogram that was prepared from mitotic chromosomes (Fig. 14.21). The idiogram also demonstrates that the positions of the centromeres in *Rhoeo discolor* are not median in every instance. This may have resulted from unequal reciprocal translocations. Sax concluded that there were translocated chromosomes in both complexes. He designated the arms of the nontranslocated chromosomes with the same letter, but the left arm with a capital letter and the right arm with a small letter (e.g., Aa). DE is an example of a translocated chromosome that possesses the left arm (or any portion thereof) of an originally nontranslocated chromosome Dd and the right arm (or any portion thereof) of an Ee chromosome. In the metaphase I ring formation, the 12 chromosomes will be attached to each other in the following way: Aa-aB-Bb-bC-Cc-cD-DE-Ee-ed-dF-Ff-fA (where fA is also attached to Aa). If these chromosomes are arranged in alternate or zig-zag formation then the two complexes pass to opposite poles, each gametic complex con-

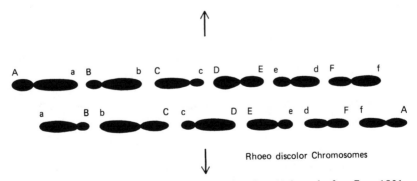

Fig. 14.21. The two Renner complexes of *Rhoeo discolor.* (Adapted after Sax, 1931. Redrawn by permission of Cytologia, Tokyo).

taining six chromosomes in the following fashion: first complex Aa Bb Cc DE ed
Ff; second complex aB bC cD Ee dF fA (see Fig. 14.21). As can be seen, both
complexes have a completely compensated set of genetic material (Aa to Ff).
Even though the alternate orientation of *Rhoeo* is under fairly rigid control, complete configurations of 12 chromosomes (rings or chains) occur only in about 70%,
two separate chains in 15%, three chains in 14%, and four chains in 1% of all male
meiocytes (Walters and Gerstel, 1948). Lin (1979) in a study of *Rhoeo spathacea*
var. *concolor* observed complete rings in 30.3%, complete chains in 41.2%, two separate chains in 17.6%, three chains in 10.1%, and four chains in 0.8% of all meiocytes. The complex heterozygotes of *Oenothera* in contrast as a rule have rings
involving all chromosomes. Balanced lethals in *Rhoeo* presumably guarantee that
all progeny results in complex heterozygotes.
Another case of permanent complex heterozygosity was discovered in the genus
Isotoma (2n = 14) by James (1965, 1970). He observed that localization and
effective terminalization of chiasmata was pronounced. Translocation heterozygotes occur in almost all natural populations of central and western Australia.
They are common to the three species of *I. petraea, I. axillaris,* and *I. anaethi-
folia.* Meiosis is highly irregular and large rings of multiples occur in 80% of all
PMC's and EMC's. In eastern Australia the general condition is seven bivalents,
but rings of 4 and 4 or 4 and 6 were also discovered.
An interesting case in which permanent translocation heterozygosity was linked
with sex determination was reported in East African mistletoes (Wiens and Barlow, 1973, 1975; Barlow and Wiens, 1975, 1976; Wiens, 1975). They reported that
in the 21 dioecious African species of the genus *Visca,* 11 (57%) have translocation
heterozygosity. All species so far studied, except *V. fischeri,* have translocation
polymorphism with closed ring formation involving rings of four, six, or eight chromosomes. *V. fischeri,* however, has a sex-determining chain of nine chromosomes.
Regular alternate orientation and disjunction in this chain results in a 4/5 segregation of these chromosomes that are transmitted along with the segregation
products of seven bivalents forming 11- and 12-chromosome genomes. Male
plants have 2n = 23 and female plants 2n = 22 chromosomes. Such sex determining apparatus may be classified as a multiple X-multiple Y system (4X/5Y)
where the male is the heterogametic sex.
Multiples with chromosome associations of more than four also occur naturally
in such plant genera as *Hypericum* (Hoar, 1931; Hoar and Haertl, 1932), *Clarkia*
(Anderson and Sax, 1936), *Chelidonium* (Nagao and Saki, 1939), *Paeonia* (Stebbins and Ellerton, 1939; Walters, 1942), and *Gaura* (Bhaduri, 1942). In animals
they were observed in the scorpions (Piza, 1950) and in the cockroaches (John
and Lewis, 1958). Sybenga (personal communication) believes that centromere
orientation is sequential including possible orientation. One centromere orients
after the other, which insures predominant alternate orientation. Occasional adjacent orientation within the chain can be corrected. When in such a case the pull
from one centromere lapses, the centromere of the adjacent chromosome is able
to reorientate and for a new orientation may select the opposite pole. The third
centromere may then reorient until all chromosomes in the chain are arranged in
zig zag orientation (Sybenga, 1975).

Complex heterozygosity also can be produced experimentally. For instance, Yamashita (1947, 1950, 1951) produced various translocations in diploid *Triticum* species by x-ray treatment. After crossing different homozygous translocation lines, he was successful in establishing a line in which all 14 chromosomes were united in one translocation complex. Similar results were accomplished in *Campanula, Hordeum, Tradescantia,* and *Zea* (see review by Burnham, 1956). Darlington and LaCour (1950) produced a system of translocation heterozygosity in *Campanula persicifolia* in which all chromosomes of this species were involved. In most of these systems total sterility results because alternate orientation usually does not occur regularly.

14.8 Chromosome Mapping via Translocations

The translocation itself behaves like any other genetic factor. Its genetic expression, **semisterility,** shows linkage with genes in two different linkage groups and behaves like a dominant character in test crosses. In 1930 Burnham found linkage between the gene *waxy* (*wx*:59, chromosome 9) and partial sterility caused by translocation in maize.

The genetic expression of semisterility will be demonstrated here with the data of Brink and Cooper (1931) obtained from a linkage test involving semisterile-1, *brachytic* (*br*:75) and *fine striped* (*f*:80), both located on chromosome 1 of maize. Semisterile-1 plants $\left(\dfrac{+\ +\ T}{br\,f\ +} \right)$ were backcrossed to nontranslocated plants $\left(\dfrac{br\,f\ +}{br\,f\ +} \right)$ as shown in Table 14.1 (see p. 240) in typical testcross fashion. A 1:1 ratio of semisteriles (SS) to fertiles (F) would have been expected in each of the four segregation classes if semisterility had been independent from *br* and *f*. Consequently, the **three-point cross** data show linkage with semisterility. Semisterility has all the characteristics of a gene located at the translocation breakpoint. Its location in relation to the other two genes can therefore be determined and mapped (Table 14.1). The location of *br* and *f* is known from other studies. Since *br* is located at position 75.0 and *f* at 79.7, the translocation point must be located distal from these at 87.2 since its position is 12.2 units from *br* and 7.5 units from *f*. The method of determining linkage between translocation breakpoints and new genes has been used profitably for assigning these genes to specific chromosomes or even chromosome arms or segments.

In Chapter 4 (Section 4.4) it was indicated that translocations can serve for the determination of gene position. Burnham (1957, 1962, 1966) described two methods, the **all-arms marker method** (Burnham and Cartledge, 1939; Burnham, 1954) and the **linked marker method** (Anderson, 1943, 1956). In the all-arms marker method, a tester set was developed in maize that included at least two translocation breakpoints in every arm of the complement (Burnham, 1966). The "all-arms tester set" included 22 translocations by 1962 (Burnham). Plants of a strain with an unmapped gene or a new mutant are being crossed with the trans-

Table 14.1. Three-point cross. Linkage test involving semisterility in chromosome 1 of maize (Brink and Cooper, 1931).

	+ + br f	+ f br f	br + br f	br f br f	Totals
SS[1]	333	1	17	25	376
F[2]	19	8	6	273	306
					682

			Percent
Noncrossovers:			
+ + T	333 ⎫	606	88.8
br f +	273 ⎭		
Single-crossovers in region 1:			
br + T	17 ⎫	25	3.7
+ f +	8 ⎭		
Single-crossovers in region 2:			
+ + +	19 ⎫	44	6.5
br f T	25 ⎭		
Double-crossovers:			
+ f T	1 ⎫	7	1.0
br + +	6 ⎭		
Total		682	100.0

[1]SS = semisterile plants
[2]F = fertile plants

1,2 = crossover regions 1 and 2.

location lines of the tester set. The semisterile F_1 is backcrossed to the parental stock with the recessive mutant character. The progeny is classified for semisterility of the pollen and for the mutant. If there is no linkage the percentage of mutants should be similar in the fertile and semisterile classes. A higher percentage of mutants in the fertile class is an indication of linkage between the mutant gene and the translocation point.

In barley, too, translocations served to determine linkage groups to which new genes belong (Swomley, 1957; Ramage and Suneson, 1958; Ramage, 1964). But first of all, in barley, translocations helped to assign linkage groups permanently to chromosomes. For instance, two linkage groups formerly thought to be independent (III and VII, Robertson et al., 1941) were both demonstrated to be on the newly assigned chromosome 7 (Kramer et al., 1954). They used linkage tests (F_2 data) from crosses between translocation stocks and genetic markers.

A tester series of translocations linked with the endosperm marker is an efficient method of locating genes. In one series in maize, translocations are used where one breakpoint is close to the *waxy* locus (*wx*:59.0) in the short arm of chromosome 9. The *wx wx* translocation stocks in this tester set are T1-9, T2-9, T3-9, T4-9, T5-9, T6-9, T7-9, T8-9, and T9-10 (Burnham, 1966).

The use of translocations for mapping the eight chromosomes of *Aspergillus* has been recently described by Käfer (1977). Here, as in maize, the various steps for mapping of translocation breaks are basically the same as those for mapping genes and centromeres. Translocations have been used for mapping genes to linkage groups in *Neurospora* (Perkins and Barry, 1977). The efficiency of assigning unmapped point mutants in this organism has been greatly increased by a tester strain with three independent translocations each with a closely linked genetic marker. The majority of new point mutants could be readily linked to one of the three markers by Perkins (Perkins et al., 1969; Perkins, 1972).

Probably the most efficient method of assigning genes to chromosome arms is the use of B-A translocations as described by Beckett (1978) in maize. This method is based on earlier work by Roman, Robertson, and Lin (Roman and Ullstrup, 1951; Rakha and Robertson, 1970; Lin, 1974). The method makes use of the phenomenon that B chromosome centromeres in maize do undergo nondisjunction at the second pollen grain division (Section 10.4). If a reciprocal translocation occurs between an A chromosome and a B chromosome, two new chromosomes result, an A^B with an A-centromere and a B^A with a B-centromere. Due to the nondisjunction in the pollen, hypoploid sperm will result that carry a deficient A^B chromosome. If the translocation occurred very close to the A chromosome centromere, the plant resulting from pollination with the A^B carrier will be hemizygous for almost an entire chromosome arm. Any recessive gene in the mother plant will immediately show up in the F_1 generation. In comparison with earlier mapping methods, the **B-A translocation method** offers the advantage of supplying information in a single generation that can be used to assign a gene to a specific chromosome segment. The method was not in widespread usage up to this point because a full tester set of B-A translocations involving all 20 chromosome arms of maize was not available. But substantial progress has been made in recent years. Beckett listed 71 B-A translocation stocks.

Part VII
Variation in Chromosome Number

Chapter 15
Haploidy, Diploidy, and Polyploidy

15.1 Haploidy

Haploidy is a general term for designating individuals or tissues (in mosaics) that have somatic cells with a gametic chromosome number (n). However, since particularly in plants, polyploid series occur that carry multiples of the basic chromosome set (x), the term *haploidy* should be subdivided into two major categories:
1. **monohaploids** (x)–individuals that can arise from diploid species
2. **polyhaploids** (2x, 3x, 4x, etc.)–individuals that can arise from any given polyploid species (4x → 2x; 6x → 3x, etc.)

If this relationship is understood, the term haploid is still a very useful, general term that can be used in discussions of this topic.

15.1.1 Origin of Haploids

Haploids can arise spontaneously or can be induced. The origin of spontaneous haploids is often obscure. They have occurred from time to time and have been reported in the literature. They usually arise by asexual development as a haploid of an individual that should be diploid (Chapter 19).

Among the animals, haploids have frequently been discovered in *Drosophila* (Castle, 1934; Bridges, 1925). Other references on spontaneously and induced occurring haploidy in animals are salamander (Fankhauser, 1937), newt (Fankhauser and Griffiths, 1939), frog (Briggs, 1952), mouse (Edwards, 1954), axolotl (El'Darov, 1965), Anura (Hamilton, 1966), chicken (Bloom, 1970), and onionfly (Heemert, 1973). Usually in animals, haploidy produces physiologically abnormal individuals that die during embryogenesis.

Spontaneous plant haploids have been found in tomatoes (Morrison, 1932), and cotton (Harland, 1936) and more recently in coffee (Visheveshwara, 1960), beets (Fisher, 1962), barley (Tsuchiya, 1962), flax (Plessers, 1963), coconut (Ninan and Raveendrananatth, 1965), pearl millet (Powell, 1969), rape (Thompson, 1969; Stringham and Downey, 1973), *Theobroma* (Dublin, 1973), asparagus (Marks, 1973), and wheat (Lacadena and Ramos, 1968; Sendino and Lacadena, 1974).

Kimber and Riley (1963) reported 36 species of 26 genera and 10 families in which haploidy occurred spontaneously.

Several methods of obtaining spontaneous and induced haploids have been described in the literature. They are:

1. interspecific and intergeneric hybridization
2. irradiation and chemical treatment
3. selection of twins
4. alien cytoplasm
5. isolation following pollination involving a pollen or seed parent carrying a marker
6. anther and pollen culture
7. chromosome elimination

One of the first accounts of experimental results on haploid production following hybridization was given by Jørgensen in 1928. He crossed *Solanum nigrum* with *S. luteum* and 7 of the 35 resulting plants were *S. nigrum* haploids. This was an example of **matroclinal pseudogamy** for which a male gamete was required for endosperm formation to stimulate embryo development. However, the embryo was developed directly from the egg without fertilization. Chase (1947, 1949a, 1949b, 1949c, 1952a, 1952b) and Coe (1959) isolated male stocks in wheat and maize which after intraspecific hybridization produced high frequencies (2% to 3%) of haploids in these species. The method of interspecific hybridization found wide application in potato breeding. The potato is thought by some to be an auto-tetraploid $(4x = 48)$. Diploidy in this species would offer less complicated inheritance $(2x = 24)$. Hougas et al. (1958, 1964) reported a method by which they could produce high frequencies (tenfold increase) of haploids and also could spot them easily by using a pigmented pollinator. They crossed tetraploid *Solanum tuberosum* with diploid *S. phureja*. All the nonpigmented plants were suspected haploids. Other reports on the success of this method were by Frandsen (1967), Budrin (1969), and Cipar and Lawrence (1972). Reviews on the subject of interspecific and intergeneric crossing for haploid production were carried out by Magoon and Khanna (1963), Kimber and Riley (1963), and Chase (1969). A total of 39 species were reported by Rowe (1974) to produce haploids after wide hybridization.

The method of x-irradiation for haploid production has been tried by various researchers in several crops. A few should be mentioned such as the work on tobacco (Goodspeed and Avery, 1929; Webber, 1933; Ivanov, 1938; Badenhuizen, 1941), wheat (Katayama, 1934; Yefeiken and Vasilev, 1936), *Crepis* (Gerassimova, 1936a, 1936b), snapdragon (Ehrensberger, 1948), and *Oenothera* (Linnert, 1962). In the experiments of Yefeiken, Ehrensberger, and Linnert, normal plants were pollinated with irradiated pollen. Some irradiated pollen must lose its ability to fertilize, thus stimulating the unfertilized egg to parthenogenetic development. Gamma radiation treatment of pollen for the production of haploidy in poplar trees has also been reported (Winton and Einspahr, 1968; Stettler, 1968).

Induction of haploidy by treatment of pollen with a vital dye, toluidine blue (TB) has been reported for *Vinca rosea* (Rogers and Ellis, 1966), tomato, maize (Al-Yasari, 1967; Al-Yasari and Rogers, 1971), and poplar (Winton and Stettler, 1974). But even greater success was reported when the TB treatment was applied to the pistils after pollination, at a time when the pollen tubes had developed but had not engaged in fertilization. With such postpollination spray treatment, 282 maternal haploid seedlings were scored from a total of 1,192 seedlings raised (23.6%) (Illies, 1974).

The method of screening twin seedlings for the selection of haploids has been successful in quite a number of species. Morgan and co-workers (Morgan and Rappleye, 1950, 1954; Campos and Morgan, 1960) showed that the frequency of twin seedlings in *Capsicum* is controlled by the genotype of the female parent. Through selection for favorable genotypes they were successful in raising the percentage of twins to a level of several percent in this species. The frequency of haploids depends on that of the twin seedlings. Such seedlings arise from **polyembryonic** seed. Polyembryonic seeds can produce haploid-haploid, diploid-diploid, or haploid-diploid twins. It is believed that in haploid-diploid twins a normal diploid zygote has developed together with a haploid synergid into two embryos. Morgan and Rappleye found 30% haploid-diploid twins among polyembryonic pepper seeds. Results in other species are much lower. Wilson and Ross (1961) found only 5% in bread wheat. Lacadena (1974) reviewed the subject and reported the occurrence of twin seedlings for 42 plant species.

The method of using alien cytoplasm as a means for the production of haploids was suggested by Kihara and Tsunewaki (1962). They backcrossed a hybrid *Aegilops caudata* x *Triticum aestivum* var. *erythrospermum* with wheat and obtained a frequency of 53% haploids, while no haploids were found in lines without cytoplasmic substitution. The wheat with the *Aegilops* cytoplasm was obtained by this backcross method. The method of haploid production that involves isolation of haploid material following pollination with a pollen or seed parent carrying a marker was already briefly mentioned in connection with the interspecific hybridization approach, specifically in the potato program (Hougas et al., 1958, 1964). But this approach also has been applied in cases of intraspecific hybridization.

Earlier studies with this method were carried out in maize. Randolph and Fischer, in 1939, used a seed parent with the genetic constitution A_1 b pl r^g y_1 and pollen with different genotypes to screen for parthenogenesis in tetraploid maize (2n = 40). The anthocyanin *purple* plant color gene (A_1:111, chrom.3) requires the dominant *Booster* gene (B:49, chrom.2) and another dominant *anthocyanin* gene (Pl:48, chrom.6) to express purple plant color. Since the seed parent carried both b and pl in the recessive condition, the plants inherited green plant color. The recessive genes for *colorless aleurone* (r^g:57, chrom.10) and for *white endosperm* (y_1:13, chrom.6) were also carried by the seed parent. One of the pollen parents had the constitution A_1 B Pl R^g Y_1. Any purple F_1 plants with colored aleurone and yellow endosperm were likely to be tetraploids. Any green F_1 plants with colorless aleurone and white endosperm were parthenogenetic suspects or polyhaploids (2n = 20). With a similar approach Randolph (1939) discovered 23 polyhaploid parthenogenotes among 17,165 tetraploid maize plants (1:750). Chase (1949a) found wide variation of haploid frequency in maize from 1:4,500 to 1:145 with an average of 1:900. The highest haploid frequency detected was 1:100, which is supposedly 20 times the average frequency in maize of 1:2000 (Stadler, 1942). Seedling markers for haploid screening have been used for several other crop species. In tobacco, Burk (1962) used a recessive *yellow green* (*yg*) seedling marker in the female parent. In tomato, Ecochard et al. (1969) used three recessive seedling markers. In cotton, Turcotte and Feaster (1969) used several multiple gene markers either for the seed or pollen parent. Bingham (1971) used hypocotyl pigmentation as a seedling marker in alfalfa to screen haploids in crosses between tetraploids and diploids.

Anther and pollen culture methods for the production of haploids have recently been discussed by Sunderland (1974) and Nitsch (1974). Guha and Maheshwari (1964, 1966) are credited with the discovery of a method for the production of haploid plants directly from pollen by culturing anthers of *Datura innoxia*. This work was followed up by the extensive work of J. P. Nitsch and his colleagues in France on tobacco (Bourgin and Nitsch, 1967; Nitsch et al., 1968; Nitsch and Nitsch, 1969). Anthers in culture can yield haploid plants either by the direct formation of embryo-like products from pollen grains or by the formation of callus and subsequent plant regeneration. Sunderland (1974) stated that simplicity of operation, ease of induction, and high induction frequencies are some of the merits of these methods. He claims that given optimal culture conditions, induction frequencies of up to 100% can be obtained in *Datura* and *Nicotiana*. In *Datura* haploid induction took place within 24 hours of culture (Sunderland et al., 1974) and in *Nicotiana* within several days (Sunderland and Wicks, 1971). The incidence of induction in a single anther in *Datura* was more than a thousand haploids and in *Nicotiana* even higher. In *Nicotiana* and *Datura* the growth rate was comparatively high. Production time from anther inoculations to mature plant stage was 3 to 4 months.

Many other cultivated species have now been used for this kind of haploid production. Some of these are mentioned in Table 15.1. It is important to notice that the bulk of the plants recovered from these experiments is not haploid. The best results in this respect have been obtained with tobacco (Sunderland, 1970; Collins and Sunderland, 1974) for which most of the resulting plants were haploids. Diploids, triploids, tetraploids, and hexaploids were reported by several authors. In the cereals, a higher percentage of the resulting plants were albinos or green-albino chimeras. Another difficulty in cereals is the low percentage of callus formation from anther culture that varied from 0.04% in maize (Murakami et al., 1972) to 32% in one barley genotype (Grunewaldt and Malepszy, 1975).

Knowledge of the anther stage at which haploid induction can take place is important. Sunderland (1974) reported that this stage can be precisely defined and lies between the quartet stage and a stage just past the first pollen mitosis in those plants that have been investigated.

The last method of haploid production mentioned here is chromosome elimination. This method was first reported by Kasha and Kao (1970). It resulted from crossing cultivated barley, *Hordeum vulgare* ($2x = 14$), with its wild relative *H. bulbosum* ($2x = 14$). Fertilization in this hybrid and subsequent mitotic elimination of the *H. bulbosum* chromosomes in the developing embryo was observed by Subrahmanyam and Kasha (1973). Since Kasha and Kao's original report, several other successful attempts with this approach have been made. The yield of haploids from this interspecific cross has steadily increased. Kasha and Kao obtained 23 haploid seedlings from 209 cultured embryos (11.0%), while Jensen (Kasha,

Table 15.1. Cultivated species in which anther culture for haploid production has been applied.

Species	Year	Authors
Tobacco	1967	Bourgin and Nitsch
	1968	Nitsch et al.
	1969	Nitsch and Nitsch
Rice	1968	Nakata and Tanaka
	1969	Tanaka and Nakata
	1968,1971	Niizeki and Oono
Barley	1971,1973	Clapham .
	1975	Grunewaldt and Malepszy
Tomato	1971,1972	Sharp et al.
	1972	Gresshoff and Doy
	1973	Debergh and Nitsch
Asparagus	1972	Pelletier et al.
Potato	1972	Irikura and Sakaguchi
	1973	Dunwell and Sunderland
Wheat	1973	Ouyang et al.
	1973	C. C. Wang et al.
	1973	Chu et al.
	1973	Picard and Buyser
Triticale	1973	Y. Y. Wang et al.
	1974	Sun et al.
Eggplant	1973	Raina and Iyer
Pepper	1973	George and Narayanaswamy
	1973	Y. Y. Wang et al.
	1973	Kuo et al.
Pelargonium	1973	Abo El-Nil and Hildebrandt

1974) reported 215 from 314 cultured embryos (68.5%). Apparently, the *H. bulbosum* chromosome loss from the embryos in this procedure is gradual. Subrahmanyam and Kasha (1973) found that 3 to 5 days after pollination, 40% of the dividing embryo cells were haploid; but 11 days after pollination, 94% were haploid. This chromosome elimination after wide hybridization in plants is similar to that reported from somatic cell hybridization in mammals (Weiss and Green, 1967; Allderdice et al., 1973). One of the suspected reasons for chromosome elimination is the difference in duration of the somatic cell cycles in the two parents involved (Gupta, 1969; Lange, 1971; Subrahmanyam and Kasha, 1973). Apparently, the somatic cell cycle in *H. bulbosum* is longer than in *H. vulgare* (Barclay, Finch and Bennet, personal communication. Cited in Riley, 1974).

15.1.2   Meiotic Behavior of Monohaploids

As mentioned in Section 15.1, one has to distinguish between two major classes of haploids, monohaploids, and polyhaploids. **Monohaploids** have only one basic genome (x) and, consequently, are meiotically very irregular. In order to have meiosis functioning properly there must be two homologous chromosomes present for each chromosome type of the complement. Meiosis in a number of monohaploids has been studied in sorghum (Schertz, 1963; Reddi, 1968), rye (Heneen, 1965), maize (Ting, 1966, 1969; Ford, 1970; Weber and Alexander, 1972), rice (Chu, 1967), tomato (Ecochard et al., 1969), pearl millet (Manga and Pantalu, 1971), barley (Sadasivaiah and Kasha, 1971, 1973), and tobacco (Collins and Sadasivaiah, 1972). Pairing of chromosomes in monohaploids (**intragenomic pairing**) has been thought to be the consequence of chromosome duplication and genetic redundancy. Such pairing has been observed in rice, tomato, maize, and barley. It is interesting that even synaptonemal complexes have been observed under the electron microscope in studies of pachytene in monohaploids of tomato (Menzel and Price, 1966), maize (Ting, 1969, 1971), and petunias and snapdragons (Sen, 1970). Such complexes were similar in nature to those observed in the corresponding diploid forms. In spite of these synaptonemal complexes often being formed in pachytene, the chromosomes appear mostly as univalents in the diakinesis of these monohaploids. However, bivalents and multivalents were occasionally observed. Ting (1966) found cells with one or more bivalents in 50.9% of all cells observed in maize. Sadasivaiah and Kasha (1971, 1973) even observed quadrivalent structures. Metaphase I is striking in monohaploids in that the spindle is highly disorganized. The chromosomes are mostly bivalents, and trivalents also occur. In anaphase I the distribution of the chromosomes to the opposite poles is usually at random. Spindles seem to function weakly in some cases, but the distribution mechanism is not yet very thoroughly studied.

The pairing mechanism of nonhomologous chromosomes in the meiosis of monohaploids is not yet understood. However, Rieger (1957) had an interesting theory which states that all chromosomes have a certain tendency for pairing in meiotic prophase. If homologous chromosomes are present in the cell, then those are preferentially paired. If such homologues are not present in the cell then the tendency to pair is satisfied by forces that can unite nonhomologous chromosome segments. This reminds one of the precocity theory of Darlington (1932), mentioned in Chapter 1, which states that single chromosomes are in an unsatisfied, or unsaturated, state electrostatically, and in order to become saturated they must pair.

15.1.3   Meiotic Behavior of Polyhaploids

Since the meiosis of polyhaploids should be considered, the nature of different kinds of polyploidy must be briefly mentioned here, though this topic is going to be more fully discussed in Section 15.3. Polyploids can be either **autoploid** or **alloploid,** depending on their origin. In true autoploids all basic genomes (x) have the same origin (e.g., AAAA). In alloploids, basic genomes are of different origin (e.g.,

AABB). Consequently, polyhaploids can be classified into **autopolyhaploids** (e.g., AA) or **allopolyhaploids** (e.g., AB) (Kimber and Riley, 1963). Meiotic studies of polyhaploids may or may not help to determine which of these two types of polyhaploidy are present in a given polyploid species. If no pairing of chromosomes or chromosome segments takes place (e.g., AB) the species under consideration is generally thought more likely to be an allopolyploid. Pairing of chromosomes in a polyhaploid is indicative of some chromosome relationship either of **homoeologous** or of **homologous** nature. Homoeology (Huskins, 1932) is partial homology during which some chromosome segments may pair and others do not pair. In Section 15.1.2 it was mentioned that even in monohaploids, some chromosome pairing may occur (**intragenomic pairing**). However, since homologous chromosome pairing is thought to happen preferentially, the majority of pairing observed in polyhaploids probably is caused by partial or complete homology between genomes (**intergenomic pairing**). Most naturally occurring polyploids are of alloploid (e.g., AABB) or **segmental alloploid** ($A_1A_1A_2A_2$) nature (Stebbins, 1950). Consequently, the majority of autopolyhaploids are derived from artificially induced autopolyploids. Since this hypothesis is not generally accepted, the literature is full of proposed cases of natural autopolyploids and, consequently, autopolyhaploids that are derived from them (e.g., Kimber and Riley, 1963). The author believes that most naturally derived polyploids are products of hybridization (e.g., A_1A_1 x $A_2A_2 \rightarrow A_1A_2$) and subsequent chromosome doubling ($A_1A_2 \rightarrow A_1A_1A_2A_2$). Since the races involved in the original hybridization (e.g., A_1A_1 and A_2A_2) have generally grown in separate environments, they very likely have undergone different genetic and cytological changes ($A \rightarrow A_1$, $A \rightarrow A_2$). A high incidence of chromosome pairing associated with high seed fertility in polyhaploids of alfalfa (Stanford and Clement, 1955; Bingham and Gilles, 1971; Stanford et. al., 1972) and potato (Ivanovskaja, 1939), for instance, has led to the widespread conclusion that these cultivated species are autopolyploids. However, the pairing of chromosomes is not entirely a measure of genome relationships, since this process is independently under a rigid genetic control (Riley and Law, 1965). In alfalfa Bingham and Gilles observed incomplete pairing (average:$6^{II} + 4^{I}$) in one polyhaploid strain. Ten of the 35 pollen mother cells had 6 or more univalents and one had 16. One other strain showed occasional bridges at anaphase I (indication for inversion during $A \rightarrow A_1$ changes). These strains also showed varying degrees of female and male sterility indicating cryptic changes in the A_1 and A_2 genomes, which are not identifiable on the basis of abnormal meiosis. In potato, Yeh et al. (1964) found from 16% to 26% cells with univalents in polyhaploids. Most of those present findings lead one to the conclusion that alfalfa and potato are also segmental allopolyploids ($A_1A_1A_2A_2$) rather than autopolyploids (AAAA).

The most important groups of allopolyhaploids, consequently, are segmental allopolyhaploids and allopolyhaploids. Allopolyhaploids are obviously the least questionable group. Absence of pairing in polyhaploids may be a valid means of concluding allopolyploidy in the parents. But under certain circumstances pairing can be induced between partially homologous chromosomes of allopolyhaploids. This was detected by Riley (Riley and Chapman, 1958; Riley, 1960) in wheat, which is considered to be an alloploid (AABBDD). Five different nullisomic 5B allo-polyhaploids showed greatly increased chromosome pairing that apparently was caused by the absence of one or more genes that are responsible for the prevention of homoeologous pairing located in chromosome 5B. A similar system was reported for oats (Gauthier and McGinnis, 1968). Limited pairing in allopolyhaploids was found in tobacco (Lammerts, 1934; Collins and Sadasivaiah, 1972), *Brassica* (Ramanujam and Srinivasachar, 1943) cotton (Barrow, 1971), oats (Nishiyama and Tabata, 1964), the grass *Festuca arundinacea* (Malick and Tripathi, 1970), and other species.

15.1.4 Possible Use of Haploids

The main reason for plant breeders to obtain haploids has been to develop a new and rapid method of breeding homozygous diploids or polyploids. Since in a haploid every gene is presented only once, doubling of the chromosome should theoretically result in **complete homozygosity.** Repeated inbreeding for homozygosity in plants takes many generations, but doubling of haploids results in immediate homozygosity. This doubling may occur naturally or may be induced by tissue wounding, heat treatment, colchicine or other chemical application, or decapitation. A recent development is the possibility of obtaining homozygous diploids directly from anther cultures. Niizeki and Oono (1971) reported that they had obtained diploid rice plants directly from pollen and not from somatic cells. Narayanaswamy and Chandy (1971) obtained 70% diploids, 23% triploids, and 7% haploids from anther cultures of haploid *Datura metel.* Engvild (1974) received 20% diploids in tobacco when the anthers were cultured at the nuclear pollen stage.

In maize breeding, haploids obtained by natural parthenogenesis have been used with success in the production of homozygous diploid strains of commercial value (Yudin and Khvatova, 1966; Gyulavari, 1970; Petrov and Yudin, 1973).

In maize (Thompson, 1954) and in barley (Park et al., 1974), it has been established that the genetic variability for factors such as yield, heading date, plant height, etc., is the same for the **monoploid method** as for other conventional breeding techniques.

In the Solanaceae, haploid plant breeding is well in progress. Morrison (1932) reported the development of a diploid, commercial tomato strain from a haploid source. Similar results in tomato were reported by Kirillova (1965). Other work in progress is reported for pepper (Y. Y. Wang et al., 1973) and some tobacco species (Nöth and Abel, 1971; Nitsch, 1972).

15.2 Diploidy

Just as the term *haploidy* is used in a general and a specific way, the term *diploidy* has also two different meanings. In its general sense it designates organisms with two homologous chromosome sets. Each type of chromosomes is always represented twice. Sex chromosomes are exceptions. In this wider sense even polyploids could be designated as **functionally diploid** (2n). On the other hand, the term diploidy is used more specifically as a distinction between monohaploids (x), diploids (2x), and polyploids (3x, 4x, etc.).

15.2.1   Diploidization

The term "*diploidization*" describes a selection process that causes polyploids that are originally meiotically irregular, having multivalents and bridges, to become meiotically regular like "good diploids." The end product of such a process if often called an **amphidiploid,** which is a diploid-like polyploid or a functional diploid, as described in Section 15.2.

15.3   Polyploidy

Polyploid individuals or populations have more than two basic genomes or chromosome sets (3x, 4x, 6x, etc.). They are particularly prominent in the plant kingdom. Among the angiosperms, 30% to 35% of the species are polyploid (Darlington and Janaki-Ammal, 1945; Stebbins, 1950). Almost 75% of the Gramineae are polyploid. Polyploidy is thought to be a product of interspecific hybridization. Polyploids occupy different niches than their related diploids. It seems that polyploids possess a wider ecological range of tolerances. They often can occupy habitats that cannot be occupied by diploids (Swanson, 1957).

15.3.1   Classification of Polyploidy

For the classification of polyploids, the use of **genome formulas** is a handy tool. In the genome formula a capital letter represents a group of chromosomes that is generally referred to as the basic genome or chromosome set. Such a chromosome set corresponds to the basic chromosome number that, for instance, in the Gramineae is 7 or 10. Many species of Gramineae have multiples of 7 chromosomes. The genome formula for wheat, for instance, is AABBDD. The letter "A" represents a basic genome of 7 chromsomes.

For the characterization of different types of polyploids, the best representation is still the one by Stebbins (1950) (Fig. 15.1). According to this scheme there are four major types of polyploids recognizable:

1. Autopolyploids   = AAAA
2. Segmental Allopolyploids = $A_1A_1A_2A_2$
3. Genome Allopolyploids   = AABB
4. Autoallopolyploids   = AABBBB

Autopolyploids have chromosome sets (A) or basic genomes in which the chromosomes are entirely homologous to each other, which results in complete pairing in meiosis. Terms like **autotriploidy, autotetraploidy, autopentaploidy,** and **autohexaploidy** indicated that there exist three, four, five, and six homologous basic genomes per cell.

Segmental allopolyploids (Stebbins, 1947b) are characterized by **homoeology** (Huskins, 1932) or **partial homology,** also called **residual homology** by Stephens

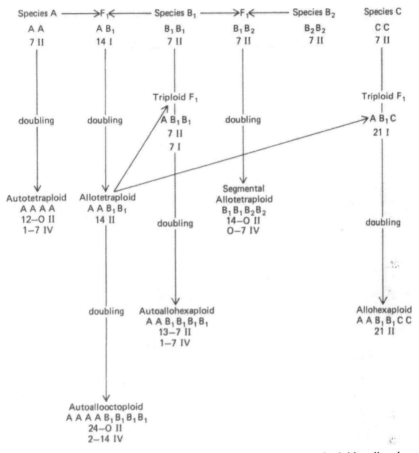

Fig. 15.1. Diagram illustrating genome relationships between autopolyploids, allopolyploids, segmental allopolyploids, and autoallopolyploids. (From Elliott, 1958. Modified from Stebbins, 1950. Redrawn by permission of Columbia University Press, New York).

(1942). This type of homology may indicate that only some of the members of the chromosome set are homologous with those of the other set or sets, while the others are nonhomologous or only partially homologous. This kind of polyploidy includes a wide array of types that ranges all the way from nearly autoploid to the other extreme, almost alloploid.

Genome allopolyploids are believed to be derived by hybridization of parents that had striking structural dissimilarity between their basic genomes. Chromosomes in meiosis are limited to bivalent pairing. Genome allopolyploids received their name because the entire genome, for instance, of parental species A is different from that of parental species B, forming after hybridization and doubling a new functional diploid with two new AB sets (AABB). Genome allopolyploids also can occur at different ploidy levels and can produce allotriploids (AAB), allotetraploids (AABB), allopentaploids (AABBC), allohexaploids (AABBCC), etc.

Crucial for further classification of these alloploids is the basic chromosome number of the contributing diploid parents. Darlington and Janaki-Ammal (1945) distinguished between **di-, tri-,** or **polybasic polyploids.** Typical examples for dibasic polyploids are brown mustard *(Brassica juncea)*, rape *(B. napus)*, and *B. carinata* (Fig. 15.2). These three *Brassica* species are believed to have derived from the three basic genomes of black mustard *(B. nigra* = N), *(B. campestris* = C), and cabbage *(B. oleracea* = O). The basic chromosome numbers are:

B. nigra: x = 8 (N)
B. oleracea: x = 9 (O)
B. campestris: x = 10 (C)

The genome formulas for the dibasic polyploids are:
B. juncea = NNCC (2n = 36)
B. napus = CCOO (2n = 38)
B. carinata = OONN (2n = 34)

However, many alloploids are **monobasics** in which the contributing diploid parents have identical basic chromosome numbers.

15.3.2 Autopolyploidy

There have been different opinions about the frequency of autopolyploids in nature. This stems in part from the different interpretation of what actually is an autoploid. For instance Müntzing (1936) included segmental allopolyploids in the autoploid category and stated that autopolyploidy is quite common in plants. However, Stebbins (1950) believed that autoploids are relatively rare in nature. As a matter of fact, the only clear-cut case of autoploidy in nature, according to Stebbins, is *Galex aphylla,* which was reported by Baldwin (1941). According to Stebbins, several cultivated crops can be classified as being either autoploid (segmental alloploid inclusive) or alloploid (Table 15.2).

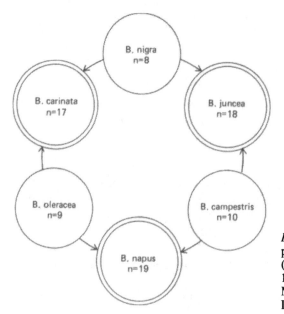

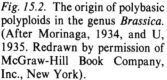

Fig. 15.2. The origin of polybasic polyploids in the genus *Brassica.* (After Morinaga, 1934, and U, 1935. Redrawn by permission of McGraw-Hill Book Company, Inc., New York).

Table 15.2. Crop plants listed according to their classification as autoploids (segmental alloploids respectively) or alloploids (modified from Elliott, 1958)

Common name	Scientific name	x-number	2n-number
A. *Autoploids*			
(incl. segmental alloploids)			
Potato	*Solanum tuberosum*	12	48
Coffee	*Coffea arabica*	11	22,44,66,88
Banana	*Musa sapientum*	11	22,33
Alfalfa	*Medicago sativa*	8	32
Peanut	*Arachis hypogea*	10	40
Sweet potato	*Ipomea batatas*	15	90
B. *Alloploids*			
Tobacco	*Nicotiana tabacum*	12	48
Cotton	*Gossypium hirsutum*	13	52
Wheat	*Triticum aestivum*	7	42
Oats	*Avena sativa*	7	42
Sugar cane	*Saccharum officinarum*	10	80
Plum	*Prunus* spp.	8	16,24,32,48
Loganberry	*Rubus loganobaccus*	7	42
Strawberry	*Fragaria grandiflora*	7	56
Apple	*Malus sylvestris*	17	34,51
Pear	*Pirus communis*	17	34,51

Phenotypically, autoploids are generally larger in size than their diploid counterparts, but there can be exceptions to this. The cytoplasm and the nucleus of autoploids are larger than those of diploids; this will lead to giant growth characteristics, provided the cell number also increases proportionally. Very often, however, the cell number does not match with those of the diploids, particularly in artificially produced autoploids.

15.3.2.1 Autoploidy in Plant Breeding. Several methods have been applied to produce polyploids in cultivated crops. A few should be mentioned here (Briggs and Knowles, 1967):

1. decapitation
2. indoleacetic acid
3. twin seedlings
4. heat treatments
5. colchicine
6. other chemicals

Decapitation and generation of callus tissue with or without the use of indoleacetic acid to stimulate regrowth has been successfully used in *Solanum* and *Nicotiana*. The production of twin seedlings was one of the earliest methods of obtaining polyploids. Twin embryos are occasionally found in a low frequency among germinating seedlings and often yield **heteroploid** plants. Heteroploidy (Winkler, 1916) is the phenomenon that shows deviation from the normal chromosome number. The twin seedling method was already mentioned during the discussion of haploid production (Section 15.1.1). Müntzing (1937) was one of the first to recognize that twin seedlings could also be autoploids. High temperature treatments for short periods have been employed for the production of polyploids in plants such as maize (Randolph, 1932).

The previously mentioned chromosome doubling procedures did not yield a high incidence of success. As mentioned in Chapter 1, Blakeslee and Avery in 1937, and Nebel during the same year, discovered the value of the alkaloid colchicine in producing polyploids. Colchicine is a spindle fiber poison or suppressant. It inhibits the spindle mechanism at mitosis, resulting in multiples of the normal chromosome number. A high number of artificial polyploids has been developed in the last 50 years. Table 15.3 shows some of the crops that were developed from artificial and spontaneous autoploids. Other chemicals for the induction of polyploids have not been as effective as colchicine. Among them are chloral hydrate, ether, chloroform, acenapthene, phenylurethane, and nitrous oxide (Briggs and Knowles, 1967).

The benefits of polyploid plant breeding have increased the size of plant organs such as roots, leaves, flowers (Fig. 15.3), fruits, and seeds (Fig. 15.4). Also the chemical characteristics of some plants have been substantially changed by polyploid breeding. Tetraploid maize has about 40% more vitamin A content than its diploid counterpart (Randolph and Hand, 1940). In sugar beets large roots are desirable for the total sugar harvest per hectare. Unfortunately, the sugar content generally decreases with increase in root size. But Peto and Boyes (1940) found that the sugar content in triploids decreases less with increasing root size than in diploids. The tetraploid perennial ryegrass *(Lolium perenne),* which is used in forage production, has more sugar content and dry matter than the diploid (Sullivan and Myers, 1939).

One common disadvantage of polyploidy breeding is the reduced pollen production and the increased seed sterility in the polyploids. Genic imbalances in the polyploids are believed to be the reason. According to Gottschalk (1978) only 32% of 135 autotetraploid species that have been developed from diploids have a pollen fertility of 90% to 100% of the control values of the diploids. Pollen fertility of 42% was only 10% to 11% of the control values or lower.

15.3.2.2 Autotriploids. When autotetraploids are crossed with diploids, autotriploids can be derived. Such individuals have each chromosome in triplicate. Meiotically they behave as do multiple primary trisomics (Section 16.2.4.1). During meiosis the three homologous chromosomes potentially can form **trivalents.** Such trivalents can assume different configurations. The prerequisite for trivalent formation is interpairing of all three chromosomes in synapsis. As it was explained previously, the pairing of homologous regions during synapsis can generally bring only two homologous segments together in a two-by-two fashion in one particular chromosome region. However, Moens (1968) in triploid lily and Comings and Okada (1971) in triploid chicken observed that the three homologous chromosomes were, over short segments, connected by a double synaptonemal complex. In meiosis three homologous chromosome regions are generally not synapsed side by side as it happens in salivary gland cells of triploids (somatic synapsis, Fig. 9.7). If two of the three homologues pair along their entire length, trivalents cannot form, but one chromosome will be left completely unpaired. The result will be a bivalent and a univalent (Fig. 15.5A). If interpairing involves all three chromosomes (Fig. 15.5B and C), one basic requirement for trivalent formation is

Table 15.3. Cultivated crops that were developed from artificial and spontaneous autoploids

Common name	Scientific name	Basic chrom. no. (x)	Normal chrom. no. (2n)	Autoploid chrom. no. (2n)	References
Maize	Zea mays	10	20	40	Randolph, 1932
Rye	Secale cereale	7	14	28	Müntzing 1951, 1954
Alsike clover	Trifolium hybridum	8	16	32	Levan, 1942a
Red clover	Trifolium pratense	7	14	28	Levan, 1942a
Sugar beet	Beta vulgaris	9	18	27,36	Schlösser, 1936; Frandsen, 1945; Hagberg and Åkerberg, 1962
Turnip	Brassica campestris	10	20	40	Elliott, 1958
Banana	Musa paradisiaca	11	22	33,44	Dodds and Simmonds, 1938
Apple	Malus sylvestris	17	34	51	Einset and Lamb, 1951
Narcissus	Narcissus bulbocodium	7	14	21,28,35,42	Fernandes, 1934
Garden tulip	Tulipa gesneriana	12	24	36	Upcott and LaCour, 1936
Garden hyacinth	Hyacinthus orientale	8	16	24,32	Darlington et al., 1951
Gladiolus	Gladiolus spp.	15	30	45,60,75,90, 120	Bamford, 1935
Garden dahlia	Dahlia variabilis	8	32	64	Lawrence, 1931a
Florists' cyclamen	Cyclamen persicum	12	48	96	De Haan and Doorenbos, 1951
Snapdragon	Antirrhinum majus	8	16	32	Stebbins, 1957
Easter lily	Lilium longiflorum	12	24	48	Emsweller, 1951
Watermelon	Citrullus vulgaris	11	22	33	Kihara, 1951b
Common day lily	Hemerocallis fulva	11	22	33	Mookerjea, 1956
Grape	Vitis vinifera	19	38	76	Olmo, 1952
Orchid	Paphiopedilum insigne	13	26	39	Mehlquist, 1947
Forsythia	Forsythia intermedia spectabilis	14	28	42,56	Hyde, 1951
Iris	Iris mesopotamica	12	24	48	Sturtevant and Randolph, 1945
Perennial ryegrass	Lolium perenne	7	14	28	Myers, 1939; Sullivan and Myers, 1939

Fig. 15.3A and B. Comparison between (*A*) diploid (2n = 16) and (*B*) tetraploid (2n = 32) snapdragons. (Courtesy of W. Atlee Burpee Co., Doylestown, PA).

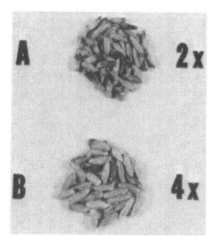

Fig. 15.4A and B. Comparison of seeds and spikes of (*A*) diploid (2n = 2x = 14) and (*B*) tetraploid (2n = 4x = 28) rye. (Courtesy of Dr. Herb Luke, Department of Plant Pathology, University of Florida, Gainsville).

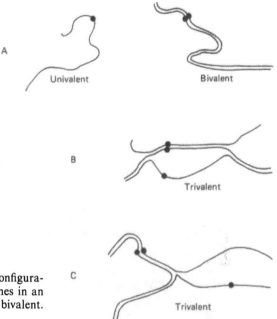

Fig. 15.5A–C. Possible pairing configurations of 3 homologous chromosomes in an autotriplod. (*A*) Univalent and bivalent. (*B, C*) Trivalents.

present. The other requirement is chiasma formation in the paired segments. If no chiasma is formed in a paired region, the homologous chromosomes separate in that region during diplotene. Four different possible trivalent configurations in autotriploids are shown in Fig. 15.6A.

The region where one chromosome (e.g., chromosome 1) changes its pairing association from one pairing partner (e.g., chromosome 2) to another (e.g., chromosome 3) is called the point of **partner exchange** (Darlington, 1929b). It has been observed that there is a reduction of chiasma frequency around the point of partner exchange (Sybenga, 1975).

Chromosome separation in meiosis I from trivalents is irregular. Daughter nuclei will receive either one or two chromosomes from any given trivalent. Therefore, each sister nucleus will have a haploid set of chromosomes plus additional ones. Consequently, most of the gametes resulting from autotriploid individuals do not have balanced chromosome complements and are not viable. If progeny survives from triploids it is mostly **aneuploid** (Chapter 16).

The fact that high sterility results from triploids has been explored in polyploid breeding. Triploid bananas (2n = 33) are vigorous but seedless and are therefore preferred for food consumption. Likewise, triploid watermelons, which were developed in Japan, have undeveloped seeds that naturally are of great advantage. Such seeds are no more objectionable than those in cucumbers (Kihara, 1951b) (Fig. 15.7). Triploid offspring (3x) has been produced from 2x x 2x interspecific *Citrus* crosses caused by fertilization of unreduced egg cells. This method could be utilized for breeding seedless *Citrus* cultivars (Geraci et al., 1975).

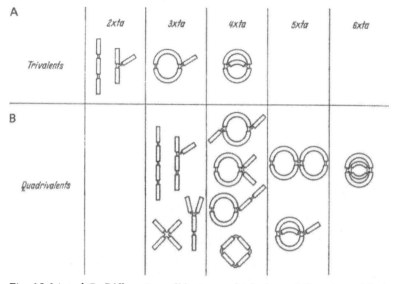

Fig. 15.6A and B. Different possible types of trivalents (*A*) and quadrivalents (*B*) in meiosis I (xta = chiasmata). (From Rieger et al., 1976).

According to Kuliev et al., (1975), triploidy in humans was the most frequent karyotype anomaly in a cytogenetic study of more than 4000 abortuses. A total of 323, about 8.1%, fetuses were found to be triploid. This is in close agreement with data from Jacobs et al. (1978), who found 28, or 8.2%, triploids among 340 spontaneous abortions. In a review of triploidy in humans, Niebuhr (1974) reported clinical data on about 230 triploid abortuses and 33 live-born triploid infants. Triploids surviving for more than a few days were all diploid-triploid mosaics.

One was a 3-year-old boy with micrognathia (small skull), syndactyly (grown together fingers or toes) and mental retardation. Blood cultures were almost completely diploid, but in primary skin cultures, about 92% of the cells were triploid and 8% diploid (Böök and Santesson, 1960; Böök et al., 1962). Other malformations observed in human triploids are hydrocephalus or relatively large heads, malformed ears and eyes, and cleft palate. Jacobs and co-workers calculated that 66.4% of all human triploids were the result of **dispermy** (see Chapter 1, Boveri), 23.6% the result of fertilization of a haploid ovary by a diploid sperm, and 10% the result of a diploid egg fertilized by a haploid sperm.

15.3.2.3 Autotetraploids. In autotetraploids there exist four homologous sets of chromosomes. Theoretically, the affinity between the four homologous chromosomes is equal. If one would assume only one single point of chromosome pairing initiation, there would be only a chance for bivalent formation since pairing is in two-by-two fashion. If there would be two such initiation points, then partner exchange could occur, and a chance for **quadrivalent** formation would exist. Just as in triploids, the kind of pairing configurations will be determined by the kind of pairing and the number and location of the chiasmata formed. Examples of the

Fig. 15.7A and B. (*A*) Diploid watermelon (2n = 2x = 22) with numerous seeds. (*B*) Triploid (2n = 3x = 33) variety with few and imperfect seeds. (Courtesy of W. Atlee Burpee Co., Doylestown, PA).

possible shapes of quadrivalents in diakinesis are shown in Fig. 15.6B. That figure shows that there are 11 different shapes possible for quadrivalents. Any other combinations of univalents, bivalents, and trivalents add to the possible expression of pairing in meiosis I. According to quadrivalent analysis, the 11 configurations shown in Figure 15.6B do not occur at random. For instance, in tomato and perennial ryegrass, more than 50% of the configurations are rings of four homologous chromosomes, and about 30% are chains (Dawson, 1962). Most autotetraploids have an abundance of these two kinds of quadrivalents. The arrangement of these multivalents in metaphase is similar to that of those discussed under translocations (Chapter 14). If all centromeres are oriented toward the poles and attached by the spindle fibers, one speaks of **coorientation.** If only two of the four centromeres are oriented toward the poles (**noncoorientation**), **false univalents** (Upcott,

Table 15.4. Numbers of chromosomes (n) in pollen from tetraploid tomato (modified from Dawson, 1962)

Item	Number									Total no. of cells
n number	20	21	22	23	24	25	26	27	28	
No. of cells observed	2	1	14	36	285	31	8	2	1	380
Percentage of cells with certain chromosome number	0.5	0.3	3.7	9.4	75.0	8.2	2.1	0.5	0.3	

1938) can occur, which will upset the chromosome distribution pattern. False univalents arise from multivalents as contrasted to true univalents that are the consequence of asynapsis or desynapsis. Because of this uneven chromosome distribution the progeny of autotetraploids seldom contain a full balanced chromosome complement. Unbalanced gametes are subject to inviability, which causes sterility. The percentage of such sterility is lower than in triploids. Dawson (1962) shows some typical data of the number of chromosomes in pollen from artificially produced autotetraploid tomato as compiled by three researchers (Table 15.4). The normal n-number of tetraploid tomato is 24 (75%). Artificially produced tetraploids, therefore, will always produce a high number of aneuploids. In Russian cultivars of tetraploid rye (2n = 28), the percentage of aneuploids was 10.5% while in a comparatively old Swedish tetraploid red clover strain, it was 35% (Sakharov and Kuvarin, 1970; Ellerström and Sjödin, 1974).

Autotetraploidy in humans has the same fate as reported for autotriploidy, only it does not have the same frequency. Hamerton (1971b) reported four cases that died as fetuses. Three of these were 92, XXXX and one 92, XXYY.

15.3.2.4 Mendelian Inheritance in Autotetraploids. Mendelian inheritance in autopolyploids becomes increasingly more complicated at higher ploidy levels. **Tetrasomic inheritance** shall here serve as an example of such inheritance patterns. In autotetraploids as contrasted to diploids each gene is represented four times rather than only twice. Tetrasomic inheritance is a better indicator of autoploidy than the formation of multivalents, which can either occur or not occur or can even be present in diploids. Multivalents (or multiples) can occur also as a consequence of reciprocal translocations and are thus indistinguishable from those that have arisen as a result of complete homology in autoploids. As a case in point, autotetraploids of different *Lotus* species and of *Raphanus sativus* have very few quadrivalents (Somaroo and Grant, 1971; Savosjkin and Cheredeyeva, 1969).

Blakeslee et al. (1932) coined a nomenclature for the five possible genotypes in autotetraploids:

AAAA–**Quadruplex** for dominant gene *A*
AAA a–**Triplex** for dominant gene *A*
AA a a–**Duplex** for dominant gene *A*

Aaaa–**Simplex** for dominant gene *A*
aaaa–**Nulliplex** for dominant gene *A*

In a quadruplex individual, for instance, the dominant gene *"A"* is represented four times, in a triplex three times and so on down the line. In a true autotetraploid, each gene should theoretically be subject to tetrasomic inheritance. One of the prerequisites for tetrasomic inheritance is that all four chromosomes involved segregate at random in meiosis I (**random chromosome assortment**). As with gene expression in diploids, the expression of dominance can also vary in tetraploids. In **complete dominance** the heterozygous genotypes (*Aaaa, AAaa,* and *AAAa*) are phenotypically identical with the homozygous dominant genotype (*AAAA*). In **incomplete dominance** the phenotypes of the different heterozygous genotypes (*Aaaa, AAaa,* etc.) have different degrees of intermediate expression between those of the quadriplex (*AAAA*) and nulliplex (aaaa) genotypes. Two different genes in *Primula sinensis* can serve as a demonstration for these two phenomena (Dawson, 1962). They are the gene for *green stigmas* (*G*) and the gene for *suppression of anthocyanin formation (d)* (Table 15.5). In the case of complete dominance, tetrasomic inheritance ratios can be expected if random segregation occurs between all four chromosomes that carry the gene in question. Random chromosome assortment only holds true for genes close to the centromere. If genes are distant from the centromere, crossing over will interfere and will change segregation to a chromatid basis.

In order to understand Mendelian Inheritance in autotetraploids, one first should be aware of the possible gamete combinations. In a duplex for gene *A* (*AAaa*), for instance, the following gametes are produced:

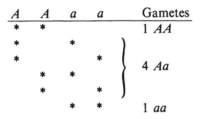

Table 15.5 Complete and incomplete dominance expression in tetraploid Chinese primula (*Primula sinensis*) (Dawson, 1962).

Complete dominance		Incomplete dominance	
Genotype	Phenotype	Genotype	Phenotype
GGGG		*DDDD*	Full color
GGGg	Green	*DDDd*	Recognizably different
GGgg	Stigmas	*DDdd*	intermediate colors
Gggg		*Dddd*	
gggg	Red stigmas	*dddd*	White

The gamete ratio in this instance is $1AA:4Aa:1aa$, assuming that the two sister chromatids of a chromosome always move into different gametes (chromosome segregation). Relative genotype frequencies can be calculated from such gametic ratios. If a duplex individual ($AAaa$) is selfed, one can expect the following offspring:

Gametes	$1AA$	$4Aa$	$1aa$
$1AA$	$1AAAA$	$4AAAa$	$1AAaa$
$4Aa$	$4AAAa$	$16AAaa$	$4Aaaa$
$1aa$	$1AAaa$	$4Aaaa$	$1aaaa$

According to this **Punnett square**, the phenotype frequencies in a strictly complete dominant fashion are 35 dominants to one recessive (35:1). However, in actual experimental data this ratio is usually not that high but approaches more a ratio of $21A:1a$. As mentioned, crossing over between the gene and the centromere disturbs the typical tetrasomic ratio and causes a phenomenon called **double reduction** (Fig. 15.8). This phenomenon causes an increase in recessive gametes. It occurs when at the end of meiosis the two sister chromatids of a chromosome end up in the same gamete. Figure 15.8 demonstrates double reduction in a duplex ($AAaa$). The chromatids carry the genes $a_1a_1/A_3A_3/a_2a_2/A_4A_4$ on four homologous chromosomes that are separated by dashes. This is the situation before crossing over. After crossing over, the four homologous chromosomes carry the genes $a_1A_3/a_1A_3/a_2A_4/a_2A_4$. Without double reduction (Fig. 15.8A) all gametes are dominant ($4Aa:0aa$). With double reduction in two of the four gametes (Fig. 15.8B), the gamete ratio is three dominants to one recessive ($1AA:2Aa:1aa$). With double reduction in all four gametes (Fig. 15.8C), the gamete ratio is one dominant to one recessive ($2AA:2aa$). As indicated with increasing double reduction the number of recessives rises.

If random chromosome assortment does not occur, as in tetrasomic inheritance, the gamete ratio decreases from 5:1 ($1AA:4Aa:1aa$) to 3:1 ($1AA:2Aa:1aa$). If the genes on the two chromosome pairs (1, 2) are designed A_1a_1/A_2a_2 and only A_1 can pair with (and segregate from) a_1 and not with A_2 and a_2, then the following gametes are produced:

A_1	A_2	a_1	a_2	Gametes
*	*			$1\ AA$
*			*	$2\ Aa$
	*	*		$2\ Aa$
		*	*	$1\ aa$

Since only A_1 carrying chromosomes segregate from A_1's, and A_2's from A_2's, only A_1 and A_2 carrying chromosomes can end up in the same gametes and not A_1's and A_1's.

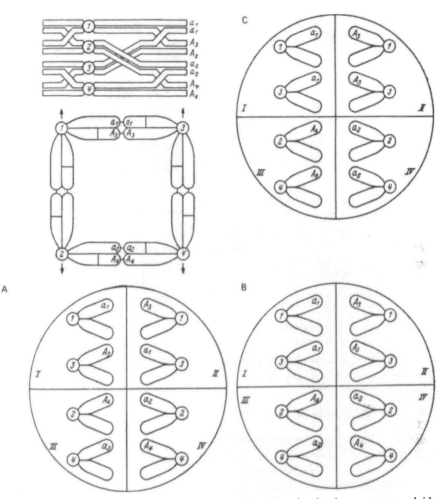

Fig. 15.8A, B and C. Schematic representation of double reduction in an autotetraploid. (*A*) No double reduction. (*B*) Double reduction in two of four gametes. (*C*) Double reduction in all four gametes. (From Rieger and Michaelis, 1958).

If an individual with such gametes is selfed, the following progeny results:

Gametes	1 *AA*	2 *Aa*	1 *aa*
1 *AA*	1 *AAAA*	2 *AAAa*	1 *AAaa*
2 *Aa*	2 *AAAa*	4 *AAaa*	2 *Aaaa*
1 *aa*	1 *AAaa*	2 *Aaaa*	1 *aaaa*

A typical **dihybrid ratio** of 15 *A*:1 *a* is the result. This expression is also called **duplicate gene expression.** This phenomenon is caused by two identical allele pairs ($A_1A_1A_2A_2$) that have the same phenotypic expression but are located on two different chromosome pairs.

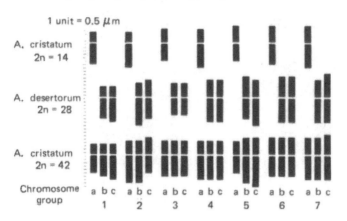

Fig. 15.9. Idiograms of *Agropyron cristatum* (2n = 14), *A. desertorum* (2n = 28) and *A. cristatum* (2n = 42) arranged in chromosome groups (1 to 7) in order to facilitate morphological comparison of basic genomes. (From Schulz-Schaeffer et al., 1963. Reprinted by permission of the American Society of Agronomy, Madison, Wisconsin).

Thus, tetrasomic inheritance (35:1 to 21:1) in a duplex (*AAaa*) is easily distinguishable from duplicate gene expression (15:1). Duplicate gene expression may indicate some homoeology (segmental allopolyploidy), but only tetrasomic inheritance demonstrates autopolyploidy.

15.3.3   Segmental Allopolyploidy

Segmental as well as **genome allopolyploids** are both thought to be derived by hybridization of diploids and subsequent chromosome doubling, as indicated in Fig. 15.1. Species $B_1(B_1B_1)$ may hybridize with species $B_2(B_2B_2)$. Both species may have originated from the same population (BB) but may have become subject to geographical isolation. Different environments may have favored different chromosomal and genetic changes ($B_1 \leftarrow B \rightarrow B_2$). Such development was suggested in the crested wheatgrass complex (Schulz-Schaeffer et al., 1963). The possible chromosomal changes in this complex are diagrammed in Fig. 15.9. Chromosomes of *Agropyron cristatum* (A_1A_1) are designated with the letter *a*. Chromosomes of *A. desertorum* ($A_1A_1A_2A_2$) are designated with the letters *b* and *c*. Chromosomes *a* and *b* are similar in morphology within their respective chromosome groups (1 to 7). All *a* and *b* chromosomes belong to the A_1 genomes, and all *c* chromosomes belong to the A_2 genomes. *A. desertorum* could, consequently, be an amphiploid hybrid between diploid *A. cristatum* (A_1A_1) and an unknown diploid species with the genome formula A_2A_2. Reciprocal translocations are one type of a segmental chromosome change that could have contributed to the genome change ($A_1 \rightarrow A_2$). Translocations of this type are evident in the *c* chromosomes of chromosome groups 1, 2, 5, and 7 in *A. desertorum* and *A cristatum* (2n = 42). They should be interpreted as three reciprocal translocations with two sets of homologues in common.

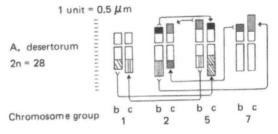

1 unit = 0.5 μm

A. desertorum

2n = 28

Chromosome group   1   2   5   7

Fig. 15.10. It is assumed that b and c chromosomes were originally alike and that b's changed into c's. The changes that occurred are thought to be due mainly to reciprocal translocations. Possible interchanges of this kind are shown in this figure. A segment of the short arm of chromosome 2b could have been exchanged with a segment of chromosome 5b while a segment of the long arm of chromosome 2b may have been exchanged with a segment of the short arm of chromosome 7b. A similar interchange could have occurred between the long arms of 1b and 5b. A segment of the long arm of chromosome 1b could subsequently have been lost by a deletion. The resulting chromosomes are 1c, 2c, 5c, and 7c. These chromosomes are also shown in the idiogram in Fig. 15.9. (From Schulz-Schaeffer et al., 1963. Reprinted by permission of the American Society of Agronomy, Madison, Wisconsin).

The chromosomes that are involved in two translocations are of the chromosome groups 2 and 5, because translocations seem to be reflected by different arm lengths in both arms of the c chromosomes. The possible nature of these translocations is demonstrated in Fig. 15.10.

15.3.4   Genome Allopolyploidy

A typical allopolyploid or alloploid species being derived by hybridization from two or more diploid species has no pairing between chromosomes of its parent species. But only pairing between chromosomes of homologous genomes occurs. Bivalent formation is typical for such species and, thus, the meiotic behavior is just like in diploids (functional diploidy, Section 15.2). Alloploidy occurs frequently in nature. As mentioned earlier, many of the most valuable crop plants are segmental alloploids or alloploids, such as wheat, oats, cotton, and sugar cane. According to Gottschalk (1977) the number of experimentally produced alloploid hybrids has increased sharply during recent years to more than 1000. Gottschalk further reports that the common opinion, that **polyploidization** of interspecific and intergeneric hybrids restores fertility, does not generally hold true. Many alloploids are either sterile or have very low fertility. The classical intergeneric alloploid was made by Karpechenko in 1928 between radish (*Raphanus sativa,* 2n = 18) and cabbage (*Brassica oleracea,* 2n = 18). There is absolutely no affinity between the chromosomes of these two parents in the hybrid, and a high degree of sterility occurs in the F_1 hybrid. Karpechenko (1928) obtained a number of allotetraploid hybrids (RRBB) in the F_2 generation that arose by the formation and union of unreduced gametes.

If the genomes of the two parents in an artificial alloploid are very distantly related, one speaks of wide hybridization. In wide hybridization one often encounters the phenomenon of **chromosome elimination** or **chromosome diminution.** This phenomenon has also been called **Rückregulierung** (Rieger and Michaelis, 1958) or downward adjustment, which is the tendency of polyploid or mixoploid tissues to return to the original chromosome number of one of the diploid parents. The mechanism of this adjustment is not very well known yet. It is suspected that one of the ways to accomplish this is the formation of multipolar spindles with subsequent elimination of occurring unbalanced cells (Rieger and Michaelis, 1958). In autoploids this cytological instability is not very frequent. For instance, in the progeny of artificially produced tetraploid barley the frequency of diploid individuals is only 1 in 5000 to 6000 (Müntzing, 1957). If, however, genomes of very remote species are combined by **somatic cell hybridization** or **cell fusion** (see Chapter 1; Harris and Watkins, 1965) then chromosomes of one of the parents are usually successively eliminated in the alloploid cell cultures (Ephrussi and Weiss, 1969; Zepp et al., 1971). In human-mouse somatic cell hybrid cultures Weiss and Green (1967) observed that after 100 to 150 generations only two or three human chromosomes remained.

Most alloploid species are tetraploid. Hexaploid alloploids are less common. It is still very difficult to obtain ploidy levels higher than octoploid experimentally. Examples of very high ploidy levels in nature are given in Table 15.6.

Much attention has been directed to an intergeneric alloploid combination between wheat and rye, called *Triticale*. The first fertile *Triticale* hybrid was achieved by the German private plant breeder Wilhelm Rimpau in 1888 (Jenkins, 1969). Since then many attempts were made to develop a new crop from such a combination. The hope is to combine into one plant type the nutritional and baking qualities of wheat with the drought tolerance, adaptation to poor soils, and disease resistance of rye. Early *Triticale* breeding programs were based on octoploid combinations such as AABBDDRR in which three wheat genomes (ABD) were combined with the rye genome (R). However, it later was discovered that the hexaploid combinations such as AABBRR were more successful. The most

Table 15.6 Natural and artificial alloploids with very high chromosome numbers

Species	Basic no. (x)	Somatic no. (2n)	Ploidy level	Reference
Hisbiscus radiatus x *H. diversifolius*	9	216	24 x	Menzel and Wilson, 1963
Galium grande	11	± 220	20 x	Dempster and Stebbins, 1968
Poa litorosa	7	265	38 x	Hair and Benzenberg, 1961
Kalanchoe spp.	17,18	500	30 x	Baldwin, 1938
Aulacantha	—	1000–2700	—	Grell and Ruthmann, 1964
Schizaea dichotoma	—	1080	—	Brown, 1972
Ophioglossum reticulatum	—	1260	—	Ninan, 1958

promising hexaploid Triticales are the so-called secondary types that have been obtained from intercrosses between hexaploid and octoploid Triticales (Pissarev, 1963; Kiss, 1966) and between hexaploid Triticales and hexaploid wheats (Nikajima and Zennyozi, 1966; Larter et al., 1968; Jenkins, 1969; Zillinsky and Borlaug, 1971). Some of the agronomically interesting hexaploid Triticales are not Triticales in the real sense of this term but are more properly considered to be **substitutional hexaploid Triticales** since some of their R-genome chromosomes are substituted by D-genome chromosomes (Gustafson and Qualset, 1974). However, the highest yielding fertile lines all appear to have seven pairs of rye chromosomes (Gustafson and Qualset, 1975).

Hexaploid combinations of wheat with *Agropyron* are being developed that could combine all the useful agronomic characteristics of durum wheat (AABB) with the vegetative vigor, disease resistance and winter-hardiness of *Agropyron* (Schulz-Schaeffer and McNeal, 1977). A very extensive program of wide hybridization among the grasses of the tribe Triticeae has been carried out by Dewey (1965, 1970, 1975). One of the goals of this program is to synthesize new **amphiploid** species and to evaluate their breeding potential (Asay, 1977). Dewey's work is also a practical application of Kihara's (1930) principle of genome analysis, which was discussed in the first chapter. The genome relationships in the genera *Elymus, Agropyron,* and *Hordeum,* as worked out by this analysis, are shown in Fig. 15.11.

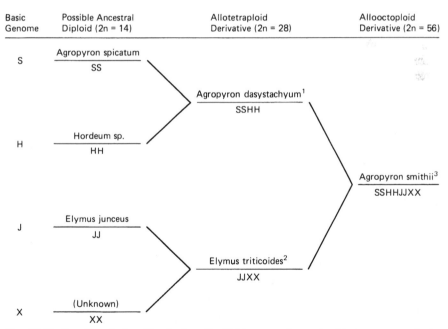

Fig. 15.11. Species relationships in the tribe Triticeae as determined by genome analysis. ([1]Dewey, 1965. [2]Dewey, 1970. [3]Dewey, 1975. Courtesy of Dr. Douglas Dewey, Crops Research Laboratory, USDA, Utah State University, Logan).

15.3.5   Complications with Polyploidy in Man and Animals

It was mentioned in the preceding sections that polyploidy in man is extremely detrimental. Three major reasons for the lack of polyploidy in animals have been mentioned (White, 1973):

1. disturbance of sex determining mechanism
2. cross fertilization barrier
3. histological barrier

Muller (1925) proposed that the reason for the paucity of polyploidy in animals compared with plants is that in bisexual forms, polyploidy could upset the sex-chromosome mechanism. If the diploid mechanism is XY:XX the tetraploid one consequently should be XXYY:XXXX. It could be argued that during meiosis of the male the two Y chromosomes would pair and also the two X chromosomes with the result that only XY gametes would form. If such a male gamete would fertilize a female egg (XX), only XXXY progeny would result and consequently only one sex. This idea was voiced during a period when Bridges' Theory (Section 5.1.1) of balance between X chromosomes and autosomes based on *Drosophila* was accepted for animals in general. But Muller's argument did not hold ground when it was established that in bisexual plants *(Melandrium album)* and in some animal groups (Urodeles and mammals) the control mechanism of sex differentiation is the presence or absence of the Y chromosome. XXXY male individuals can backcross to XXXX females in the following fashion:

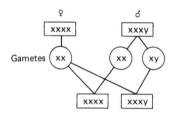

Bogart and Tandy (1976) state that the prior assumption that bisexual polyploid animals are not able to overcome sexual imbalances in gametogenesis is not true anymore because of the recent discovery of several bisexual natural polyploid animal populations. They found diploid and tetraploid populations of African anuran frogs included in the same bisexual species and thought they had established the fact that polyploidy is a general phenomenon in frogs and may appear in any genus.

The second barrier, cross fertilization, is closely linked to the necessity of bisexuality in animals. With the low incidence of a tetraploid in a diploid population, the simultaneous occurrence of another tetraploid individual as mating partner is unlikely. If such a tetraploid individual mates with a diploid it will produce sterile triploid progeny. The high proportion of triploids among chromosomally abnormal aborted human fetuses verifies this assumption (Hamerton, 1971b).

The third barrier, caused by histological complications, is explained by the more complex nature of animals. It is thought that polyploidy interferes with the developmental pattern of animals during tissue differentiation. Fankhauser (1945) gave evidence for such complications in amphibians. The size and the number of cells in such vital tissues as the brain, the spinal cord, and the nervous system are smaller than those of diploids. A strong argument against the possible effectiveness of such a histological barrier is the high frequency of possible polyploidy in parthenogenetic animals (White, 1973).

A different phenomenon is the quite regular occurrence of specialized polyploid tissues in otherwise diploid animals. Rat liver, for instance, has 5% octoploid, 40% tetraploid, and 55% diploid cells (Dawson, 1962). This phenomenon is also referred to as **mixoploidy** (Nemec, 1910) or **chromosome mosaicism** (Section 17.1.3). The rectal glands of *Drosophila* are predominantly octoploid.

Chapter 16
Aneuploidy

The previous chapter dealt with the effect of genomic changes in the number of chromosomes in living organisms. In this context we can speak of a phenomenon of **genome balance** that preserves certain requirements of function. It has been observed that loss or gain of one or more chromosomes may influence the meiotic pairing ability of chromosomes within balanced genome sets (Person, 1956; Tsuchiya, 1959, 1960; Schulz-Schaeffer et al., 1973). Genome imbalance is not only expressed as meiotic disturbance but also morphologically. A change in the relative proportion of different genes has a phenotypic effect. If a chromosome is added to or missing from a normal genomic multiple, this change is often visible in the organism. If a whole genome is added, the polyploid is often indistinguishable from the diploid form. For instance, there is little difference between the diploid and tetraploid forms of *Galex aphylla,* both of which exist in nature (Baldwin, 1941). In this chapter the effect of chromosome number changes that are not genomic in nature is being studied.

16.1   Euploidy

Euploidy is the term for cells, tissues, and individuals that have either the basic chromosome number of a genus (x) or complete multiples thereof (2x, 3x, 4x, etc.). Some genera have more than one basic chromosome number and are recognized for this fact. Examples are the genera *Crepis* (x = 4, 5, 6, 8), *Carex* (x = 6, 7, 8, 9, 10, 13, etc). *Brassica* (x = 8, 9, 10, 11), and *Viola* (x = 6, 10, 11, 13). In these genera it is more difficult to distinguish euploidy from **aneuploidy,** which deviates from euploidy in that it has incomplete multiples of the x-number. Euploidy is often necessary for survival. For instance, some euploid diploids can not afford to lose single chromosomes. Monosomics of barley do not survive. Every chromosome is necessary for the genome balance. However, monosomics in maize have been isolated. Haploid barleys have been reported and are viable (Clavier and Cauderon, 1951; Suzuki, 1959; Tsuchiya, 1962; Fedak, 1972; Kasha, 1974). In polyploids, aneuploidy is a common phenomenon and often goes unnoticed phenotypically. The extra genomes function as genetic buffers in such a polyploid system.

16.2   Aneuploidy

As already indicated, aneuploidy is any deviation from a euploid condition. This condition can be expressed either as an addition of one or more entire chromosomes or chromosome segments to a genomic number (1x, 2x, 4x, etc) or as a loss of such chromosome material.

Aneuploidy can be caused by any of the four following disturbances (Rieger et al., 1976):

1. Loss of chromosomes in mitotic or meiotic cells, often caused by lagging chromosomes or **laggards,** which are characterized by retarded movement during anaphase: This results in **hypoploid** chromosome numbers (e.g., $4x-1$, $4x-2$, $2x-1$, $2x-2$, etc.).
2. Non-disjunction of chromosomes or chromatids during mitosis or meiosis: This is a failure of such genetic units to separate properly and results in their not being distributed to opposite cell poles (Fig. 16.1). It can cause **hypo-** or **hyperploid** chromosome numbers (e.g., $4x-1$, $4x+1$, etc).
3. Irregularities of chromosome distribution during the meiosis of polyploids with uneven numbers of basic genomes such as in triploids, pentaploids, etc. (e.g., 3x, 5x): In such polyploids, some chromosomes are often present as univalents. They are distributed randomly to either pole or may be lost in anaphase I or anaphase II.
4. The occurrence of multipolar mitosis, resulting in irregular chromosome distribution in anaphase: Such **multiform aneuploidy** (Böök, 1945) can result in cells with different aneuploid chromosome numbers, causing the formation of tissue with chromosome mosaicism.

16.2.1   Nullisomy

Hypoploids are individuals, tissues, or cells that are deficient for one or more chromosomes. One class of hypoploids are the nullisomics, which have one or more pairs of homologous chromosomes missing. Nullisomics usually are not found in natural populations but have to be obtained by intercrossing or selfing of monosomics (e.g., $6x-1$). This can occur by the fusion of two gametes that are lacking the same chromosome. Selfed monosomics produce disomic, monosomic, and nullisomic progeny. Since male gametes lacking a chromosome usually have a low survival rate during fertilization or are less competitive, the percentage of nullisomics from monosomic selfing is quite low. In wheat only from 0% to 10% of the 20-chromosome male gametes can compete with 21-chromosome male gametes during pollen-tube growth. Therefore, only a small percentage of the progeny of selfed monosomics are nullisomic. Sears (1953) reported that monosomic 3B yields up

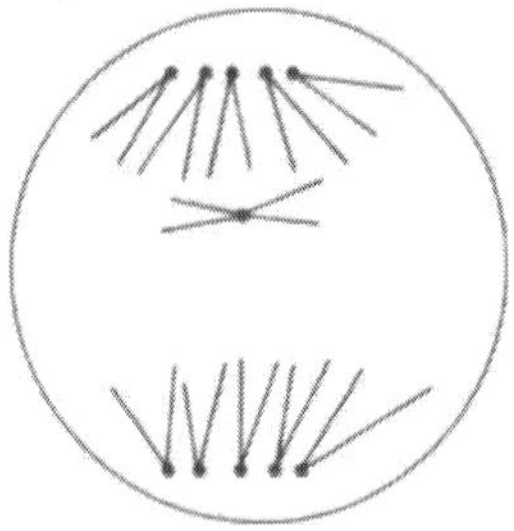

Fig. 16.1. Nondisjunction of chromatids during mitosis.

to 10% nullisomics, and several other monosomics yield as little as 1% after selfing. Nullisomics are generally weak individuals that are difficult to maintain. They are reduced in fertility, size, and vigor. In wheat all 21 possible nullisomics (the **nullisomic series**) have been obtained by Sears. Only nullisomic strains 7B and 7D can be maintained easily as nullisomic stocks. Nullisomic series are not of great agronomic importance, but they can be used for genetic studies. The wheat nullisomics differ from each other morphologically and thus demonstrate the genetic effect of the missing chromosome pair (Fig. 16.2). **Nullisomic analysis** can be used to assign dominant genes to specific chromosomes (Dawson, 1962). The **disomic**[1] individual that shows a certain homozygous dominant character (e.g., A) can be crossed with a nullisomic series the members of which all show the recessive character. The offspring of these crosses will all be heterozygous (Aa) or **hemizygous**[2] dominants ($A0$). If the heterozygotes (Aa) are selfed, they will segregate 3A to 1a. The hemizygous monosomics ($A0$) will, upon selfing, result in a majority of dominants ($AA + A0$) and a small proportion with the recessive character being nullisomic. The nullisomic that upon crossing produces hemizygous F_1's and a ratio deviating from the normal 3:1 in the F_2 is the one that designates the carrier of the dominant gene (A) in question (Fig. 16.3).

16.2.2 Monosomy

As mentioned, monosomics are organisms with one missing chromosome ($6x - 1$, $4x - 1$, etc.) (Fig. 16.4). Monosomics have been discovered in humans, animals, and plants. Three types of monosomy can be recognized (symbols show wheat situation):

1. *Primary Monosomy.* One chromosome is missing. The remaining homologue to the missing chromosome is a structurally normal chromosome. Rieger et al. (1976) also recognize this term (symbol: $20'' + 1'$. 2n = 41).
2. *Secondary Monosomy.* One homologous chromosome pair is missing and is replaced by a **secondary chromosome** or isochromosome for one arm of the missing pair. Kimber and Riley (1968) and Khush (1973) call this monoisosomy (symbol: $20'' + i'$. 2n = 41).
3. *Tertiary Monosomy.* As a result of pollen irradiation, two nonhomologous chromosomes are broken in the centromere region. Two arms of these non-homologues unite and form a **tertiary chromosome** with a functional centromere. The other two arms are lost. A plant fertilized with such pollen becomes a tertiary monosomic (Khush and Rick, 1966). Such a plant really is a **double monosomic** with a tertiary chromosome (TC) addition (symbol: $19'' + 1' + 1' + TC$. 2n = 41) (Fig. 16.5).

Monosomics have been used extensively in wheat breeding for the purpose of **chromosome substitution.** The first **monosomic series** in wheat was established by Sears (1954) in the cultivar "Chinese Spring." The sources for such monosomics are:

1. *Asynapsis* as caused by nullisomy: This was the major source in wheat. Of 212 monosomics recovered, 114 (53.8%) were obtained from progeny of asynaptic nullisomic 3B (Sears, 1954).

---

[1]Disomic–Individuals (e.g., 6x) with complete sets of homologous chromosomes as opposed to monosomics ($6x - 1$), nullisomics ($6x - 2$), etc.
[2]Hemizygous–Genes not present as pairs of alleles but only once as a result of aneuploidy or loss of chromosome segments.

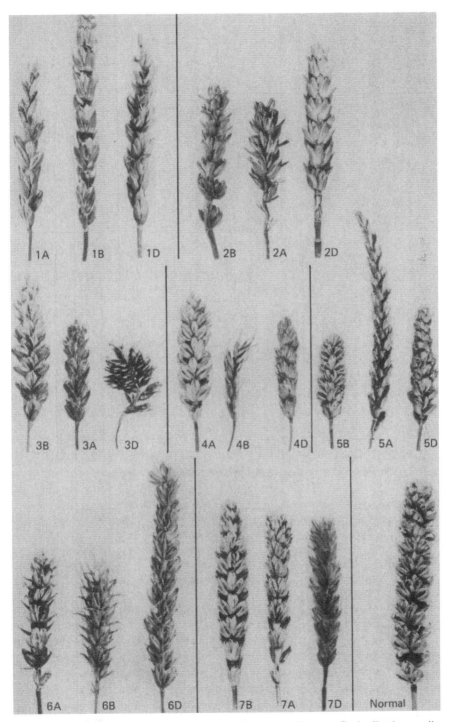

Fig. 16.2. Seven homoeologous groups of 3 nullisomic "Chinese Spring" wheat spikes each (A,B,D), compared with a normal spike. (Courtesy of Dr. Ernest R. Sears, USDA, SEA, Cereal Genetics Research Unit, Columbia, Missouri).

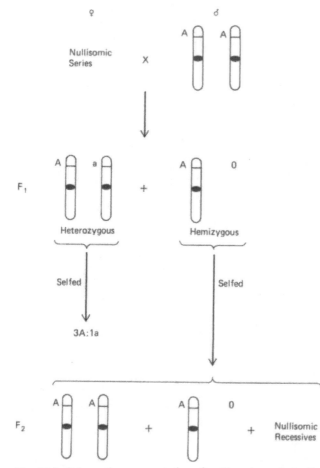

Fig. 16.3. Schematic representation of nullisomic analysis. Explanation in text.

2. *Polyhaploid progeny:* Of the 212 monosomics obtained by Sears (1954), 66 (31.1%) were derived from two different polyhaploid individuals.
3. *Chromosome loss* as a result of **non-disjunction** during meiosis or during the early mitotic divisions of a diploid zygote.
4. *Unequal chromosome distribution* (non-coorientation) during meiosis of translocation heterozygotes (see Section 14.3).

Chromosome substitution lines in wheat have been produced since the first monosomic series became available. Rieger et al. (1976) defined chromosome substitution as the exchange of a single chromosome or a chromosome pair by chromosomes of the same complement (from a different variety, for instance) or by chromosomes of the complements of other species or genera (**alien chromosome substitution**). One of the purposes for chromosome substitution could be that, after demonstrating that disease resistance or some other desirable agronomic characteristic is conditioned by a gene or genes carried by a certain chromosome, this desirable chromosome could be substituted into an otherwise acceptable cultivar.

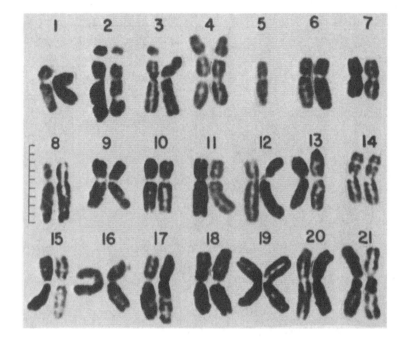

Fig. 16.4. Karyotype of monosomic common wheat, *Triticum aestivum L.* Chromosome pairs 1 to 4 are arranged according to the length of their satellites. Chromosome 5 (4D) and pairs 6 to 21 are arranged according to the length of their short arms. The photographs of the chromosomes were taken from a cell of a plant that was monosomic for chromosome 4D. Scale units at left represent 1 μm. (From Schulz-Schaeffer and Haun, 1961. Reprinted by permission of Verlag Paul Parey, Hamburg).

Assuming that "Chinese Spring" is that improved acceptable cultivar, it can be used as a monosomic female recipient for repeated backcrossing until the desirable chromosome of the male donor variety is transferred into the Chinese Spring background. Such a technique takes advantage of the fact that monosomes have much greater transmission through male than through female gametes (Fig. 16.6). In order to improve varieties in such a way, the monosomic series of Chinese Spring had first to be transferred to other such desirable cultivars. This has been accomplished in more than a score of cultivars.

In recent years chromosome substitution, with the help of aneuploids, has been used less because of the enormous effort in work and time required. Usually, the same objective can be achieved through the backcross method (Hurd, 1976).

Monosomics were first found in tobacco (Clausen and Goodspeed, 1926). They also were detected in oats, tomato, maize, and cotton. Since there is some homoeologous pairing between genomes in tobacco, trivalents occur in 25% of the monosomics. In wheat, monosomics do not form trivalents.

In humans the most common single abnormality in chromosomally abnormal fetuses is the **Turner syndrome** (45,X) (Hamerton, 1971b). It occurs in 0.03% of all female births. This is the monosomy for the X chromosome. Autosomal mono-

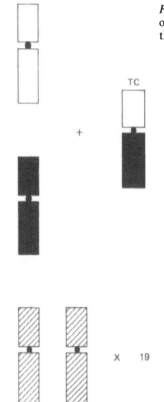

Fig. 16.5. Schematic representation of tertiary monosomy. The double monosomic condition is caused by the formation of a tertiary chromosome (TC).

somics in humans are very rare. Al Aish et al. (1967) reported one single complete G-group monosomy but most others are mosaics.

In *Drosophila* only one type of autosomal monosomic is known, the haplo-IV's or individuals with only one chromosome 4. This is a loss of 54 bands of a total of about 5000 in the entire *Drosophila melanogaster* complement. Haplo-IV's are not as robust and healthy as normal flies. There is no monosomy for chromosomes 2 and 3. Monosomics for the sex chromosomes, so called XO flies, also exist in *Drosophila*. They are males but are sterile. XO types have also been reported in mice, and are female and fertile.

As mentioned before, monosomics in diploids are rare but have been isolated in maize as early as 1929 (McClintock, 1929c; Einset, 1943). Only more recent efforts have made it possible to use monosomics in genetic studies. The establishment of the complete monosomic series in maize has been particularly helpful (Weber, 1974). Plewa and Weber (1975) used monosomic analysis for the study of fatty acid composition in embryo lipids of maize.

	♂ 21 chromosomes 96% (90–100%)	20 chromosomes 4% (0–10%)
♀ 21 chromosomes 25%	disomic 21 ^{II} 24%	monosomic 20 ^{II}$_{+1}$ ^{I} 1%
20 chromosomes 75%	monosomic 20 ^{II}$_{+1}$ ^{I} 72%	nullisomic 20 ^{II} 3%

Fig. 16.6. Breeding behavior of a monosome. (Courtesy of Dr. Ernest R. Sears, USDA, SEA, Cereal Genetics Research Unit, Columbia, Missouri. From Elliott, 1958. Redrawn by permission of McGraw-Hill Book Company, New York).

16.2.3   Telosomy

A telosomic is an individual that has as part of its chromosome complement one or more telocentric chromosomes (Chapter 2), which are otherwise known as **telosomes** (Endrizzi and Kohel, 1966). Since aneuploids are used more in wheat breeding than in work with any other species, a thorough nomenclature has been developed for this species that could serve here as an example (Kimber and Sears, 1968). **Monotelosomics** are monosomics that have as the unpaired univalent chromosome a telosome. The symbol for monotelosomy in wheat nomenclature takes into account the 20 paired chromosomes (20″) plus the unpaired telosome (t′).

Monotelosomics have greatly facilitated the mapping of genes in wheat (Sears, 1969). Telosomics for most of the 42 different wheat chromosome arms were available by 1969. The distance of a gene from the centromere can be determined by the use of a telosomic because the centromere becomes a marker. The telosome is usually transmitted poorly or not at all through the pollen in competition with the corresponding complete chromosome. If a dominant marker gene (A) is located on the only arm of the telosome in a **monotelodisomic** (Fig. 16.7), the only way to recover it in a testcross is after crossing over. The frequency of recovery is a measure of crossing over between the marker gene (A) and the centromere. Telosomes have been used for determining the chromosome arm location in other polyploid species such as cotton (White and Endrizzi, 1965; Endrizzi and Kohel, 1966) and oats (McGinnis et al., 1963).

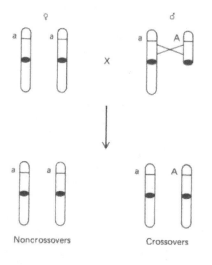

Fig. 16.7. Monotelosomic facilitated mapping of genes in wheat.

Noncrossovers   Crossovers

Most of the presently available telosomics of wheat are maintained as ditelosomics for which the chromosome concerned is represented by a homologous pair of telosomes (20″+t″) (Sears, 1966).

16.2.4 Trisomy

Trisomics are individuals with one or more (**doubletrisomics,** etc.) extra chromosomes in an otherwise disomic chromosome complement. Trisomy is very common in plants. It was first discovered in Jimson weed by Blakeslee in 1921 (Chapter 1). Organisms of this kind are hyperploids. There are five major kinds of trisomy recognized:

1. *Primary Trisomy.* The additional chromosome is completely homologous to one of the chromosome pairs of the complement.
2. *Secondary Trisomy.* The additional chromosome is a **secondary chromosome** or an isochromosome (see Section 10.3).
3. *Tertiary Trisomy.* The additional chromosome is a translocated or **tertiary chromosome** consisting of two nonhomologous chromosome segments.
4. *Compensating Trisomy.* A chromosome is missing and is genetically compensated by two other modified chromosomes.
5. *Telosomic Trisomy.* The additional chromosome is a telocentric chromosome.

These five types of trisomics can be distinguished cytologically in meiosis. Given the proper zygotene pairing and conditions for chiasma formation, primary trisomics can form chains of three chromosomes in meiosis (Fig. 16.8A) but never rings of three. (Other possible trivalent configurations are shown in Fig. 15.6A.) Secondary trisomics can form rings of three (Fig. 16.8B) and tertiary trisomics can form chains of five but never rings of five chromosomes (Fig. 16.8C). Telosomic trisomics can be of several different constitutions. If one deals with a monotelotrisomic (20″ +t2‴ in wheat), a chain of three chromosomes can be formed but never a ring of three (Fig. 16.8D).

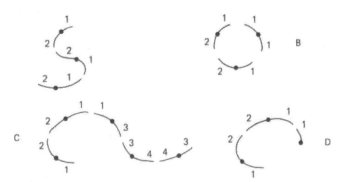

Fig. 16.8A–D. Cytological identification of trisomics in meiosis, (*A*) Primary trisomic. (*B*) Secondary trisomic. (*C*) Tertiary trisomic. (*D*) Telosomic trisomic.

16.2.4.1 Primary Trisomy. The first detailed morphological description of a complete series of 12 primary trisomics (2n = 25) was presented by Blakeslee (1934) in *Datura*. Each type in the series had its distinct fruit capsule morphology (Fig. 16.9). Primary trisomics can also be distinguished morphologically in *Avena sativa* (Azael, 1973), *Avena strigosa* (Rajhathy, 1975), *Potentilla argentea* (Asker, 1976), *Pennisetum* (Manga, 1976), and in many other species. Morphological differences between trisomics are not large enough to be distinguished in *Clarkia* (Vasek, 1956, 1963), *Collinsia* (Dhillon and Garber, 1960; Garber, 1964), *Triticum* (Sears, 1954), and *Nicotiana* (Clausen and Goodspeed, 1924). In maize only two trisomics, triplo-3 and triplo-5, could be identified morphologically. The rest could not be distinguished from each other nor from the disomics (McClintock, 1929a; McClintock and Hill, 1931; Rhoades and McClintock, 1935). Additional series of primary trisomics have been established in maize (McClintock, 1929a), barley (Tsuchiya, 1958b, 1961), tomato (Lesley, 1932), rye (Kamanoi and Jenkins, 1962), and in other species. The main source for primary trisomics is 3x types and subsequent hybridization between 3x and 2x types. Other sources for obtaining primary trisomics are non-disjunction, progenies of triploids and tetrasomics, and translocation heterozygotes. Primary trisomics have been also obtained by the use of **ionizing radiation** and after colchicine and other chemical treatments. Gottschalk and Milutinovic (1973) in peas and Palmer (1976) in soybeans found certain desynaptic mutants that have a genetically conditioned occurrence of meiotic univalents, which is a potential source of trisomics. Primary trisomics are excellent tools for assigning linkage groups to specific chromosomes. As a matter of fact, the most extensive genetic studies with the aid of aneuploids have been conducted with trisomics. Geneticists made use of the fact that gene segregation in primary trisomes is different from that of any other chromosome in the complement.

The crossing scheme for this kind of gene mapping is typically the following (Fig. 16.10):

1. A plant homozygous recessive for a given gene (*a*), the linkage group of which is to be determined, is crossed to all plants of the primary trisomic series.

Fig. 16.9. Fruit capsules and chromosomes (1.2, 3.4, etc.) of 12 primary trisomics of the Jimson weed, *Datura stramonium,* compared with a normal fruit capsule. (From Blakeslee, 1934. Reprinted by permission of American Genetic Association, Washington, D.C.)

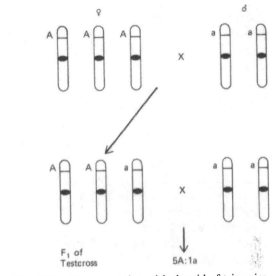

Fig. 16.10. Gene mapping with the aid of trisomics.

2. The trisomic F_1 plants are identified by cytological analysis or selected by morphological characteristics if that is possible. They are then backcrossed to the homozygous recessive.
3. The F_1 testcross plants are analyzed genetically for their segregation. If there is a striking deviation from the normal phenotypic $1A{:}1a$ testcross ratio, then the linkage group in question (marked by gene a) can be assigned to the primary trisome that produced this ratio.

As seen in Fig. 16.9, the trisomic carrying AAA produced AAa trisomics in the F_1. Genotypes of such trisomics are called duplex (Section 15.3.2.4). A plant with a duplex genotype produces the following gametes: $1AA{:}1a{:}2A{:}2Aa$. The testcross ratio consequently results in $5A{:}1a$. This ratio may be modified depending on the distance of the a locus from the centromere permitting double reduction (Section 15.3.2.4). It also depends on the transmission of hyperploid gametes through the female. In maize where this transmission was expected to be about 33% for chromosome 10, a testcross ratio of 3.8:1 was obtained for a trisomic of the $R_1R_1r_1$ genotype (R_1:57. Colored aleurone and plant) (McClintock and Hill, 1931).

This **trisomic method** for assigning linkage groups to chromosomes has been applied in *Datura* (Avery et al., 1959), *Antirrhinum* (Rudorf-Lauritzen, 1958), maize (McClintock and Hill, 1931), spinach (Janick et al., 1959), barley (Tsuchiya, 1958b, 1959a, 1959b, 1960, 1961; Tsuchiya and Takahashi, 1959, 1960), tomato (Rick et al., 1964), and *Petunia* (Smith et al., 1975).

The first discovered human trisomic syndrome was the one involving the G-group of chromosomes called mongolism or Down's Syndrome, which has a frequency of about 1 in 600 to 700 births. It is also the most frequent autosomal aberration in humans. This trisomy involves one of the smallest human chromosomes which is significant in that most larger duplications of gene complexes cannot very often survive. The frequency of this syndrome at conception is estimated at 1 in 140. Spontaneous abortion is the explanation for the reduction in frequency of G-trisomy at birth. As a matter of fact, from 60% to 100% of all chromosomally abnormal fetuses are believed to be spontaneously aborted (Hamerton, 1971b). The incidence of mongolism is known to increase with the age of the mother. The reason for this relationship is not yet entirely clarified, but one could conceive that the oocyte decreases in meiotic efficiency as maternal age increases. Since the increase in G-group chromosomes to five probably is caused by non-disjunction, one can speculate that the chromosomes with increasing age of the oocyte increase in stickiness, which could prohibit their separation at anaphase and facilitate their combined transport to a single cell pole. The smallness of the G chromosomes probably also contributes to their greater difficulty in separating. Recent cytogenetic evidence seems to indicate that trisomy 21 can also originate from paternal chromosome non-disjunction (Erickson, 1978). Many researchers have described the clinical features of Down's Syndrome (Øster, 1953; Penrose, 1961; Hanhart, 1960; Benda, 1960; Beckman et al., 1962; Gustavson, 1964; Penrose and Smith, 1966). Some of these features are severe mental retardation, saddle nose, and slanting eyes.

Other examples of human primary trisomy are the Edward's Syndrome (E-trisomy), Patau's Syndrome (D-trisomy), and C-trisomy or C-trisomy mosaicism. The Edward's Syndrome (47,XX or XY, 18+) is the second most common autosomal trisomy found in live birth. Based on three surveys of hospital new borns, the overall minimum frequency is about 1 in 3,500 (Hecht et al., 1963; Marden et al., 1964; Taylor and Moores, 1967). Eighty percent of the cases die within the first two months after birth and all usually die before one year of age. The first case was reported in 1960 (Edwards et al.). By 1971 about 150 individuals with this syndrome had been reported (Hamerton, 1971b). These cases show a high degree of mental retardation, short sternum (breastbone), and laterally flattened head. The children are small and weak.

The Patau Syndrome (47,XX or XY,13+) is characterized by deafness, myoclonic seizures (irregular, involuntary contraction of muscles), eye defects, cleft palate (split roof of mouth), and mental deficiency. Usually the largest of the D-group chromosomes is involved in the duplication and consequently chromosome 13 is suspect (Fig. 16.11). Autoradiograph and measurement studies have confirmed this observation (Giannelli, 1965a and b; Büchner et al., 1965; Giannelli and Howlett, 1966; Yunis and Hook, 1966). Patau first described the clinical features of this chromosome syndrome (Patau et al., 1960). About 75 children with this syndrome were described by 1971. The incidence is about 1 in 5000 live births. According to Taylor (1968), the mean survival time is about 90 days. There is no evidence that there is trisomy for any of the other D chromosomes.

C-trisomy and C-trisomy mosaicism have been reported by several researchers (El Alfi et al., 1963; Laurent et al., 1971; Malpuech et al., 1972; De Grouchy et al., 1971). Presently, this is the only trisomy of a large human chromosome that seems to be viable. Most of the cases found up to now are mosaics. Bijlsma et al. (1972) and Kakati et al. (1973) have attempted to establish this trisomy as a syndrome. Since the wide application of chromosome banding has become a common practice, C-trisomy has been associated with individual C-group chromosomes. Trisomies 8

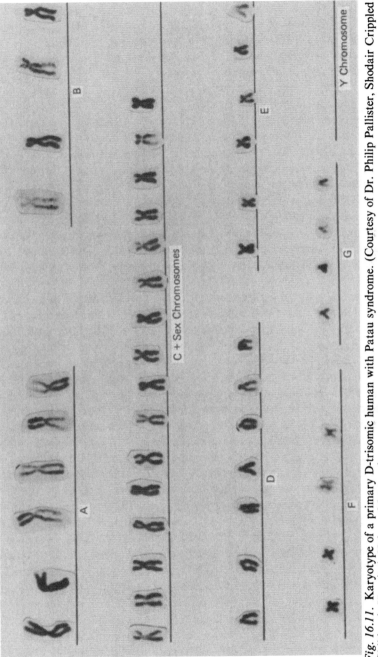

Fig. 16.11. Karyotype of a primary D-trisomic human with Patau syndrome. (Courtesy of Dr. Philip Pallister, Shodair Crippled Children's Hospital, Helena, Montana)

and 9 are well established. Only a few cases of complete trisomy 8 have been described (Caspersson et al., 1972; Kakati et al., 1973; De Grouchy et al., 1974; Jacobsen et al., 1974; Sperber, 1975; Gagliardi et al. 1978). A wide variety of congenital malformations was reported in these individuals. Mace et al. (1978) summarized the present situation for trisomy 9. About 25 cases of trisomy for the short arm of chromosome 9, two cases of complete trisomy 9, and one case of a mosaic condition have been reported.

Sex-chromosome trisomics have a relatively high frequency in man. One of the earliest clinical syndromes linked to chromosome aneuploidy after the establishment of the right chromosome number in humans was *Klinefelter's Syndrome*. At least two X chromosomes and one Y (XXY) are a common feature of all these syndromes. But other combinations such as XXXY, XXXXY, XXYY, and XXXXYY have also been observed. Mixoploids such as XXY/XX, XXY/XY, XXY/XXXY, and XXXY/XXXXY also exist. The incidence of live births in the population is about 1 in 500. This is a higher frequency than mongolism. XXY males show incomplete sexual expression. Some Klinefelter males have been reported to have mild mental or psychotic disorders (Mosier et al., 1960; Anders et al., 1968).

Trisomics for the X chromosome (47,XXX) are females with double sex chromatin. Seventy cases had been observed by 1971. They do not have any sexual abnormalities. According to Lubs and Ruddle (1970), they occur one in 727 female new borns. The **double Y syndrome** (47,XYY) has often been associated with supermaleness. Non-disjunction of Y chromosomes in anaphase of meiosis II is the most likely explanation for its origin. According to Lubs and Ruddle (1970), the incidence of this trisomy in male infants was one in 570. They pooled the results of three studies surveying 6,746 male infants. This trisomy is similar to the 48,XXYY tetrasomy that was found in high proportion among males in institutions for the criminally insane (Casey et al., 1966, 1968). Several workers found evidence that XYY trisomy was often also associated with aggressive, tall males who were in prison (Close et al., 1968; Telfer et al., 1968). However, Witkin et al. (1976) maintained that according to their studies there is no evidence that XYY men are especially aggressive. Hamerton (1971b) suggested that males with the double Y syndrome suffer from considerable inherent psychosocial disorders that make it difficult if not impossible for them to adjust to a normal social environment. Sex chromosome syndromes are generally less upsetting to the **genome balance** than autosomal ones because Y chromosomes are almost entirely heterochromatic and genetically inert (Section 5.2) and the X chromosomes are, with the exception of one, all heterochromatinized (Section 5.3.1).

In animals a number of trisomic cases have been reported. Trisomy of a small acrocentric autosome in chimpanzee resembled Down's Syndrome (McClure et al., 1969), and XXY sheep showed testicular hypoplasia (arrested development of testes) typical for Klinefelter's Syndrome (Kilgour and Bruere, 1970). Mouse trisomics were found to be phenotypically normal but sterile or semisterile (Cattanach, 1964; Griffen and Bunker, 1964, 1967). In *Drosophila melanogaster*, chromosome 4 trisomy (or triplo-IV) is viable and fertile in the female sex (Sturtevant, 1936). This is the only chromosome in *Drosophila* that can survive in the triplicate state. It constitutes only about 2% of the total chromosome complement. Grass-hoppers were found to be trisomic for various autosomes (Callan, 1941; Lewis and John, 1959; Hewitt and John, 1965; Sharma et al., 1967).

16.2.4.2 Secondary Trisomy. Secondary trisomy sometimes occurs in the progeny of normal plants but mostly among offspring of plants with univalent chromosomes (Burnham, 1962). Misdivision of the centromere is the origin of the isochromosome, which distinguishes secondary trisomics (Section 10.3).

In *Datura* (2n = 24) 24 different secondary trisomics are possible. If each chromosome arm is numbered, the types are:

1.2, 1.2, 1.1 or 1.2, 1.2, 2.2
3.4, 3.4, 3.3 or 3.4, 3.4, 4.4
5.6, 5.6, 5.5 or 5.6, 5.6, 6.6
etc.

Fourteen had been identified by Blakeslee and Avery by 1938. Secondary trisomics are the least investigated among the four major kinds mentioned above (Section 16.2.4). Khush (1973) stated that the production of isochromosomes is predominantly a chance event, although experimental methods can be used to produce them. Sen (1952) obtained two monoisodisomics[1] in tomato in progenies of pollen treated with formaldehyde and ammonia vapor. Khush and Rick (1967a) received five monoisodisomics from pollen irradiation. Such plants can serve as a source of secondary trisomics. Other secondary trisomics are known for maize (Rhoades, 1933), tomato (Khush and Rick, 1968b, 1969), wheat (Sears, 1954), and oats (Rajhathy and Fedak, 1970; Rajhathy, 1975).

Advanced studies with secondary trisomics have been carried out by Khush and Rick in tomato (2n = 24). They isolated 9 of the possible 24 secondary trisomics, which brings the total number for tomato up to 10 (Moens, 1965). They found that most of the morphological characteristics of the primary trisomics are exaggerated in the secondaries since one arm is represented four times in the chromosome complement rather than only three times as in the primaries. The segregation ratios of the secondary trisomics are different from those of the primary trisomics. In the primary trisomics, the three homologous chromosomes can entirely substitute for each other accounting for the unique trisomic segregation ratios. But in secondary trisomics, the segregation is primarily disomic but complicated by the existence of an extra isochromosome. The transmission of the isochromosome varies depending on the chromosome arm involved. In *Datura* that transmission to the progeny after selfing ranged from 2% for the 1.1 secondary to 31% for the 5.5 secondary (Blakeslee and Avery, 1938). Any spore that receives an isochromosome instead of a normal chromosome aborts because it is deficient of one chromosome arm. The segregation testcross ratio of secondary trisomics of *AAAa* constitution depends on the location of the recessive marker. If the recessive marker is located on the isochromosome, none of the trisomic progeny will be

---

[1] Monoisodisomic–one chromosome is missing but is replaced by an isochromosome for one of the arms of its homologue. Tomato has 2n = 24. The monoisodisomic symbol is 23″ + i1″.

recessive. If the recessive marker is located on one of the normal homologues, the expected trisomic testcross ratio will be 1:1. In Khush and Rick's (1968b, 1969) data, the percentage of recessive secondary trisomics in the testcross F_1 was lower than expected because of lower viability of these trisomics.

Khush and Rick could clarify the relationship between four tomato chromosomes and their corresponding genetic linkage maps by the trisomic method using secondary trisomics in tomato.

According to Khush (1973), secondary trisomics can be used as efficient tools in linkage mapping. The segregating progenies can give data on the chromosomal and arm location of a genetic marker, the centromere position, and the proximity of the marker to the centromere.

Feldman (1966) obtained six doses of the pairing suppressor *Ph* of wheat by producing triisosomic 5BL ($20'' + i'''$) and was, thereby, able to deduce the method of action of this important gene.

16.2.4.3   Tertiary Trisomy.   As mentioned, the additional chromosome in a tertiary trisomic is a tertiary or translocation chromosome (Section 14.3). They regularly occur in the progeny of translocation heterozygotes. In spite of their cytogenetic value, they have been studied in only a few species. Avery et al. (1959) established 30 different tertiary trisomics in *Datura*. Other tertiary trisomics were identified in *Oenothera* (Catcheside, 1954), barley (Ramage, 1960; Prasad and Das, 1975; Prasad, 1976), maize (Burnham, 1930), rye (Sybenga, 1966), tomato (Khush and Rick, 1976b), and in peas (Müller, 1975). As mentioned in Section 16.2.4.1, primary trisomics are ideal tools for assigning genetic markers and entire linkage groups to specific chromosomes. Tertiary trisomics and telotrisomics are tools to determine arm location and approximate distance from the centromere. In a tertiary trisomic, the genetic ratios are modified only for genes located in one chromosome arm, since it has only one extra arm or part of such an arm for a particular chromosome.

In a tertiary trisomic test, the recessive gene to be located, (e.g., *a*) has been previously identified with a specific chromosome by the primary trisomic test (Fig. 16.10). The recessive gene is then incorporated into the corresponding tertiary trisomic, which is duplicated for one arm or part of one arm of the identified chromosome. One of the two normal homologues of the resulting trisomic will carry the recessive gene (*a*) while the other homologue and the tertiary chromosome will carry the normal alleles (Fig. 16.12). If crossing over is ignored, the disomic fraction in the testcross progeny of such a trisomic may segregate $1A:1a$ and all the trisomics would be normal (*A*). Such ratios would indicate that the gene under investigation is located in the duplicated arm. If the gene is not located in the duplicated arm, both the disomic and the trisomic fraction of the progeny will segregate $1A:1a$.

In the tomato, seven tertiary trisomics were studied, five of which were used for genetic tests. Marker genes were assigned to specific arms of chromosomes 1, 4, 5, 7, 9, and 10.

Ramage (1964, 1965) described a method by which one could use tertiary trisomics with genetic recessive male sterile genes (*ms*) in the production of hybrid

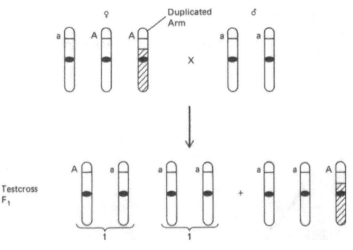

Fig. 16.12. The tertiary trisomic test.

barley. He called this method the **balanced tertiary trisomic system.** According to Ramage, balanced tertiary trisomics are "tertiary trisomics set up in such a way that the dominant allele of a marker gene, closely linked with the interchange breakpoint, is carried on the extra chromosome; and the recessive allele is carried on the two normal chromosomes that constitute the diploid complement". If the dominant marker allele is responsible for male fertility (Ms), the tertiary trisomic has the genetic constitution Ms ms ms (Fig. 16.13). All functional pollen with a normal haploid chromosome complement from such a plant carries only the recessive ms gene, since Ms ms pollen is not able to compete. Pollen with only the translocated chromosome (Ms) carries a duplication and a deficiency (Dp-Df) and is not viable. Since there is also lowered transmission of the extra translocated Ms marked chromosome through the egg, the progeny consists of only 30% balanced tertiary trisomics but 70% disomics. In Fig. 16.13, the extra translocated chromosome has a second marker, in this case a dominant gene for a red plant color. All balanced tertiary trisomics have a red plant phenotype and are male fertile. All diploids would have a green plant phenotype and would be male sterile. The red marker gene or any other similar marker can be used for separating the male sterile from the male fertile plants.

However, all of the presently available commercial hybrid barleys do not have to rely on any extra color marker gene (Ramage, 1975). The male parent rows are a normal barley cultivar, which is a good pollen producer. The female parent rows are the selfed progeny of trisomic plants (Fig. 16.13) containing about 30% male fertile balanced tertiary trisomics (Ms ms ms) and 70% male sterile diploids. The male fertile trisomics are shorter, weaker, and later flowering. Therefore, this female parent produces almost pure stands (95% to 100%) of male sterile diploid plants in commercial hybrid seed production. In order to completely assure crowding of the male fertile trisomic individuals in the female parent rows, the seed from the balanced tertiary plants is sown at a specially heavy rate (25 to 30 kg/hectare).

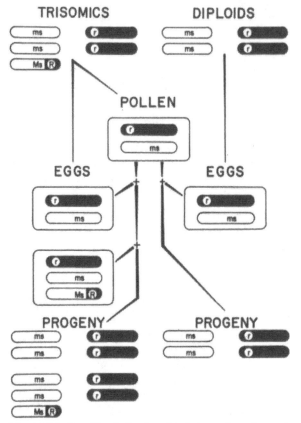

Fig. 16.13. Breeding behavior of a balanced tertiary trisomic marked with a dominant mature plant character (*Ms*). (From Ramage, 1965. Reprinted by permission of the American Society of Agronomy, Madison, Wisconsin)

16.2.4.4 Compensating Trisomy. A compensating trisomic was found by Blakeslee (1927) in the progeny of a translocation heterozygote of *Datura* that involved two reciprocal translocations and a ring of 6 chromosomes. The translocation involved the chromosomes 1.2, 5.6, and 9.10 of *Datura* (2n = 24). The translocated chromosomes were of the constitution 10.2, 1.9, 1.6, and 5.2 (Fig. 16.14). If chromosomes 9.10, 1.9, 5.6, and 2.5 of the translocation ring combine with a normal gamete, a plant results with only one 1.2 chromosome, the missing 1.2 chromosome being compensated for by the 1.9 and 2.5 chromosomes. Such a

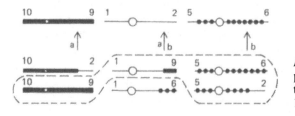

Fig. 16.14. Illustration of the possible origin of compensating trisomy. (From Burnham, 1962)

plant is trisomic for the .5 and .9 chromosome segments. A chain involving 7 chromosomes may occur in the first meiotic division of such a plant: 9.10-10.9-9.1-1.2-2.5-5.6-6.5.

16.2.4.5 Telosomic Trisomy.   In telosomic trisomy the additional chromosome (6x+t, 2x+t, etc.) is a *telosome,* which is homologous to one chromosome arm in the otherwise normal disomic complement. In wheat (2n = 42) nomenclature this is called a monotelotrisomic or is symbolized as 20″ + t2‴. If the telosome is identified the chromosome number can follow the symbol. For instance, if the extra telocentric wheat chromosome is 5A, the symbol would be 20″ + t2‴5A.
Telotrisomics have been reported in maize (Rhoades, 1936, 1940). *Datura* (Blakeslee and Avery, 1938), tobacco (Goodspeed and Avery, 1939), wheat (Moseman and Smith, 1954), barley (Tsuchiya, 1960; Singh and Tsuchiya, 1977), rye (Kamanoi and Jenkins, 1962), and tomato (Khush and Rick, 1968c). In maize and wheat the telotrisomics could be used to determine the arm location of various genes. In most species it is difficult to identify the telosomes as to their arm homology. However, in tomato and maize identification is possible because accurate pachytene analysis can be carried out (Section 2.2). Genetic segregation ratios for telotrisomics are very similar to those in tertiary trisomics. They are only modified for the markers located on the duplicated telocentric arm.
By 1968, six monotelotrisomics were discovered and identified in tomato. Inheritance studies with three of them facilitated arm assignment of marker genes. In barley 7 monotelotrisomics were used to analyze over 50 genes on 7 chromosomes. Since telocentric chromosomes are shorter than complete chromosomes, the transmission rate was higher through female gametes than in primary trisomics.

16.2.5   Tetrasomy

Tetrasomics are organisms in which one chromosome is present four times in an otherwise disomic chromosome complement (e.g.. 6x+2). In wheat the symbol is 20″ + 1⁗. Sears (1952c) used tetrasomic wheats in order to establish the so-called **homoeologous groups** in that species. Particular tetrasomics after combination with nullisomics can cancel the morphological expression of certain nullisomics. From the study of nullisomic-tetrasomics, Sears concluded that there are seven such chromosome groups of three homoeologous chromosomes each. Each tetrasomic compensated to some degree for either of the other two nullisomics. Sears synthesized all 42 possible nullisomic-tetrasomic combinations within each of the 7 homoeologous groups, and each showed some superiority over the nullisomics. In many cases the compensation was complete. After this study the chromosomes of wheat were reclassified from a Roman numeral numbering system (I−XXI) to an Arabic numbering system with capital letters following numbers to designate genomic relationships (1A-7A, 1B-7B, 1D-7D). Similar **chromosome compensation** between nullisomics and tetrasomics has also been demonstrated in oats.

Part VIII
Variation in Chromosome Function and Movement

In Parts V, VI, and VII the possible deviations from the normal chromosome types, structure, and number were described and discussed. In the forthcoming three chapters, the variation in the function and movement of the chromosomes is considered.

Chapter 17
Variation in Function of Autosomes

Both chromosome function and movement are highly coordinated and precisely efficient processes. In this context Swanson (1957) commented: "The fact that cell division is not a unitary process means that the steps that normally occur in orderly succession are subject to disturbance and open to attack. The natural causes of upsets in cell and chromosome behavior can be examined, as well as their consequences to the particular individual and to the population at large."

17.1 Somatic Segregation

Each cell division normally leads to the formation of two cytologically and genetically identical daughter cells. But, due to cytological and genetical disturbances, cell division can lead to unlike daughter cells and, consequently, to unlike tissues. The results are phenomena like mosaicism, chimeras, variegation, and mixoploidy. Genetic mosaics caused by intrachromosomal changes, for instance, are the result of **somatic crossing over.**

17.1.1 Somatic Crossing Over

Somatic crossing over occurs during mitosis of somatic cells and leads to the segregation of heterozygous alleles. Its prerequisite is somatic chromosome pairing (i.e., pairing of homologous chromosomes) as discussed in Section 9.2. Somatic crossing over, like meiotic crossing over, occurs in the four-strand stage of chromosomes and is common in many dipterans. If, for instance, a certain tissue of the fly *Drosophila* has a gene for yellow body color represented in a heterozygous condition (*Bb*) and crossing over occurs between the centromere and this gene, then a daughter cell can originate that carries the gene *b* in a homozygous recessive condition (Fig. 17.1a). The tissue that develops from this cell would show a yellow (*bb*) **single spot.** In a more critical test, two recessive genes for body color, *a* and *b*, are located on the same chromosome and are involved in somatic crossing over. The crossover location is between the centromere and the first gene (*a*). The result can be two adjacent cells of which one is homozygous for gene *a* (*aa*) and the other for gene *b* (*bb*). The tissues that develop from these two adjacent cells will form a so-called **twin spot, twin patch,** or **double spot** (Fig. 17.1B). Figure 17.2

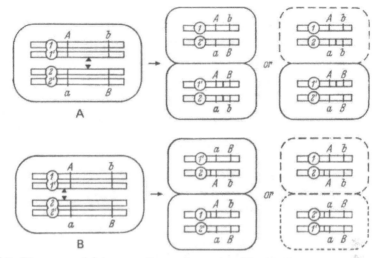

Fig. 17.1A and B. Diagram explaining somatic crossing over and its genetic consequences in a double heterozygote (*Ab/aB*). (*A*) Crossing over is located between the two loci *A* and *B*, which results in a *bb*-spot. (*B*) Crossing over is located between the centromere and the first locus, *A*, which results in a twin spot, *aa, bb*. (From Rieger et al., 1976)

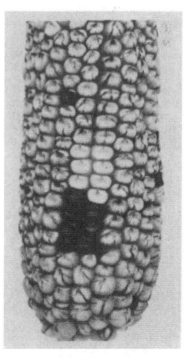

Fig. 17.2. Twin spot in maize, consisting of light variegated and self colored kernels, on a variegated ear. (From Brink and Nilan, 1952. Reprinted by permission of the Genetics Society of America, Austin, Texas).

shows a twin spot in the ear of maize. Somatic crossing over has been demonstrated in *Drosophila* (Stern, 1936), maize (Jones, 1937), and asexual fungi (Pontecorvo, 1958). (Haendle, 1971a, 1971b, 1974) showed that somatic crossing over in *Drosophila* can be induced by x-rays and that it is dependent on the dosage.

Vig (1973a, 1973b) studied the effect of inhibitors of DNA synthesis on the induction of somatic crossing over in soybeans. Somatic crossing over was induced only by those chemicals (caffeine and actinomycine D) that are known to allow rejoining of chromosomes. He saw this as evidence that somatic crossing over is caused by a specific event in DNA repair rather than by mere inhibition of DNA synthesis. Soybean seems to be an ideal object for the study of somatic crossing over. Its frequency of somatic crossing over is almost 10 times higher than in tobacco (Evans and Paddock, 1976). Twin spots composed of a dark green ($Y_{11} Y_{11}$) and a yellow ($y_{11} y_{11}$) component can be observed adjacent to each other on the light green ($Y_{11} y_{11}$) leaves in the areas of complementary exchange for these genes. Vig suggested that this genetic system in soybeans should be given a wider try at least for preliminary testing of the effect of mutagens.

Zimmerman et al. (1967) suggested that somatic crossing over may be responsible for some form of cancer in humans. Somatic crossing over leads to homozygosis of recessive genes that when phenotypically expressed may be detrimental or lethal and might lead to malignant growth.

17.1.2   Chromosomal Chimeras

These are cytogenetically heterogeneous tissues that lie side by side in an organism and lead to the formation of mosaics. They are caused by changes in chromosome structure or number and can therefore be called **chromosomal chimeras**. A chimera can be defined as "an organism, usually a plant, that is not genetically uniform throughout" (Cramer, 1954). In chromosomal chimeras, distinct adjacent tissue layers have different chromosome structures or numbers. They have been reported in *Nicotiana, Solanum, Datura,* and *Crepis,* etc. They may be classified according to their different structural origin:

1. sectorial chimeras (Fig. 17.3A)
2. mericlinal chimeras (Fig. 17.3C)
3. periclinal chimeras (Fig. 17.3B)

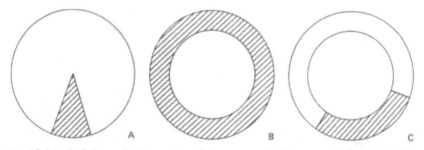

Fig. 17.3A–C. Schematic illustration of chromosomal chimeras. (*A*) Sectorial chimeras. (*B*) Periclinal chimeras. (*C*) Mericlinal chimeras. (From Swanson, 1957. Redrawn by permission of Prentice-Hall, Inc., Englewood Cliffs, N.J.).

In **sectorial chimeras,** different tissues occupy distinct sectors of the plant and are not limited to tissue layers. Instead, the heteroploid tissue extends from the center of the affected plant part (root, stem, or leaf) to the epidermis. This type was discovered and described in *Datura* (Blakeslee et al., 1939, 1940). In this type one branch of the plant may become tetraploid and another diploid, depending on the origin of specific branches. In an investigation by Brumfield (1943) on the faba bean *(Vicia faba)* and *Crepis,* involving chromosome rearrangements induced by x-rays, most of the chimeras obtained were of the sectorial type. The prevalence of sectorial chimeras and the almost complete absence of periclinal chimeras in this study seemed to be caused by the method of treatment that involved x-rays. Only single apical cells were affected by the treatment that supposedly gave rise to a chimeral sector behind the apical meristem. The sector usually involved about one-third of the root's cross section including root cap, epidermis, cortex, and central cylinder. Figure 17.4, for instance, shows how, from a stem with a 2x-4x sectorial chimera (A), pure 4x (B), sectorial (C), pure 2x (D), periclinal (E), and mericlinal (F) branches can arise. If the plant can be propagated asexually, one could produce plants that are composed entirely of tissues with different chromosome numbers. Much of the information about the behavior of sectorial chimeras stems from **gene differential chimeras** (Rieger et al., 1976). These are chimeras that could arise, for instance, from somatic mutation of a gene to its recessive allele. A periclinal chimera can arise from a sectorial one, such as if a superficial strip of the 2x component overlaps the 4x component (Fig. 17.4). If such an overlap is extensive and the budding branch originates within its periphery, such a newly originated branch will be periclinal (Neilson-Jones, 1969).

Mericlinal chimeras (Fig. 17.3C) are interrupted periclinal chimeras in which, for instance, only part of the covering layer or epidermis is involved in the tissue differentiation. Swanson (1957) believed that this type is probably the most commonly found although the most unstable insofar as perpetuation is concerned.

Periclinal chimeras (Fig. 17.3B) are probably the most stable ones. They may have

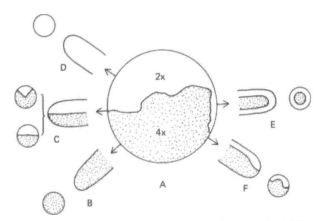

Fig. 17.4. Possible origin of uniform branches (B and D), sectorial (C), periclinal (E), and mericlinal branches (F) from the stem of a sectorial chimera. (From Neilson-Jones, 1969. Reprinted by permission of Methuen and Co., LTD., London).

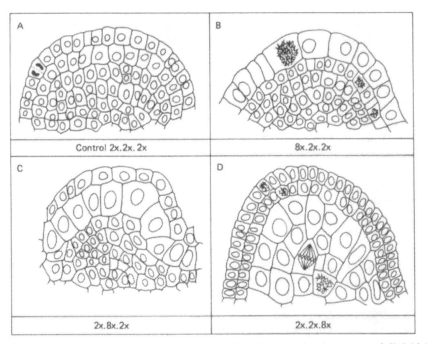

Fig. 17.5A–D. Drawings of longitudinal sections through the shoot apex of diploid Jimson weed, *Datura stramonium* L., showing three layers of periclinal chromosome chimeras. (A) Diploid layers (2x) of first tunica, second tunica, and corpus. (B) Octoploid first tunica (8x), diploid second tunica (2x) and diploid corpus (2x). (C) Diploid first tunica (2x), octoploid second tunica (8x) and diploid corpus (2x). (D) Diploid first and second tunica (2x) and octoploid corpus (8x). (After Satina et al., 1940. Redrawn by permission from: Colchicine - in Agriculture, Medicine, Biology, and Chemistry by O. J. Eigsti and P. Dustin, Jr. © 1955 by the Iowa State University Press, Ames, Iowa 50010).

entire chimeral layers of tissue that can be one, two, or more cells in depth. The differing tissue can occupy either the core of the plant structure, it can be sandwiched between two layers, or it may involve the covering layer such as the epidermis. Most chimeras produced by colchicine are of the periclinal type. Such chimeras were described in *Datura* (Satina et al., 1940; Satina and Blakeslee, 1941). The cells that are affected by the spindle fiber poison, colchicine, are in the actively dividing meristem. The stage most susceptible to the action of colchicine is late mitotic metaphase. The cells of one particular germ layer are all in metaphase, while the neighboring layers are in earlier or later division stages. Consequently, only one cell layer will be affected by the colchicine, while the others will remain unchanged (Neilson-Jones, 1969). In *Datura*, periclinal chimeras of 2x + 4x and 2x + 8x constitution were produced by treating germinating seeds with colchicine. The diagrams of longitudinal sections through the shoot apex of *Datura* in Fig. 17.5 show the various combinations of different ploidy levels in the three germinal layers of the shoot apex (first tunica, second tunica, corpus; the tunica is the outermost of the growth regions of the apical meristem). Similar chimeras were studied by Dermen (1941, 1945, 1953, 1960) in peach, apple, and cranberry. Blakeslee (1941) explained the use of polyploidy in periclinal chimeras to label the different germ layers as to their contribution to the development of a given plant organ. Chromosomal chimeras also have been observed as part of normal tissues and are often referred to as polysomaty (Section 17.1.4).

17.1.3   Chromosomal Mosaics

In men and animals, the term **mosaic** is generally used for a phenomenon that is similar to a chimera in plants. **Chromosome mosaics,** like chromosomal chimeras, may have cells differing in chromosome structure or number. Many human chromosomal mosaics have been mentioned in previous chapters. They are often referred to as mixoploids. An example of aneuploidy as a normal phenomenon in a human tissue is the *endometrium* (mucous membrane lining the uterus) in which chromosome numbers range from $2n = 17$ to $2n = 103$ (Hughes and Csermely, 1966). Tetraploid cells have been found along with diploid ones in rat liver (Alfert and Geschwindt, 1958) and in certain mammalian brain cells, including the Purkinje cells of the cerebellum (Cohn, 1969).

17.1.4   Polysomaty

This phenomenon is actually identical to endopolyploidy, which was discussed earlier (Section 9.1). **Polysomaty** is a term that was coined by Langlet (1927) to designate normal tissues that contain diploid and polyploid cells adjacent to each other. The terms "chimeras" and "mosaics" are generally used for anomalies, but the literature is not consistent. The terms "chromosome chimeras" and "chromosome mosaics" do not necessarily always imply changes in chromosome number only, but also in chromosome structure. The term "polysomaty," however, is restricted to tissues in which euploid chromosome numbers at various ploidy levels occur together. Such change in ploidy can be explained by the origin of polysomaty through the process of endomitosis (Section 9.1; Fig. 9.1). Endomitosis always implies that polyploidization occurs in differentiating tissue. This is another form of somatic segregation that occurs both in animal and plant tissues. Polysomaty apparently was first discovered in plants by Stomps in 1910. He observed that many cells in the periblem (cortical region) of the spinach root regularly had twice the typical somatic number. This finding was confirmed by Litardière (1925) in hemp. More extensive studies demonstrating polysomaty in spinach ($2n = 12$) have been carried out by Lorz (1937), Gentcheff and Gustafsson (1939), and Berger (1941), who showed that the degree of polysomaty extended from 4x to 8x to even 16x in some cases. Other plant species in which polysomaty has been demonstrated are maple (Meurman, 1933), melon (Ervin, 1941), and 39 species and varieties of Liliales (Sen, 1973).

Polysomaty in animals was first discovered by Holt (1917) in the alimentary tracts of the mosquito, *Culex pipiens.* Hertwig (1935) showed that the phenomenon occured in the nurse cells of the *Drosophila* ovary where nuclear volumes corresponded to cells having 2x, 4x, 8x, 16x, 32x, 64x, and 128x constitutions. Similar findings by Geitler (1937, 1939, 1941) in the salivary glands of the water insect *Gerris lateralis* were reported previously (Section 9.1).

17.1.5   Somatic Reduction

Rieger et al. (1976) defined **somatic reduction** as the spontaneous or induced reductional segregation of chromosomes in tissues other than those that are involved in meiosis. This phenomenon was first described in the cottony-cushion scale insect *Icerya purchasi* (Hughes-Schrader, 1925, 1927). It also occurs during the normal reproductive development of some insects. In the ileum of the mosquito, *Culex pipens,* some highly polyploid cells (32x), which have arisen by endomitosis, are reduced to lower degrees of polysomaty in a series of reductional divisions during the beginning of pupal metamorphosis. During prophase the homologous chromosomes form large bundles of chromonemata that result from somatic pairing and multiple replication. During metaphase the chromonemata bundles become dispersed, and the chromosomes assume the role that daughter chromatids usually have. As a consequence of a series of such reductional mitoses, the number of epithelial cells is greatly increased, but their size is decreased, and their chromosome number reduced from 32x to 2x (2n = 6) (Berger, 1937, 1938; Grell, 1946a, 1946b). Huskins (1948) and Huskins and Steinitz (1948a, 1948b) induced somatic reduction in *Allium* and *Rhoeo* root tips by applying indole acetic acid. Somatic reduction parallels the meiotic process in that synapsis occurs between homologous chromosomes.

17.2   Variations in Mitosis

Swanson (1957) stated that it is logical to assume that the entire process of cell division is under very accurate control by certain genes or groups of genes. It would be impossible to demonstrate such genetic control of cell division if genes responsible for chromosome behavior would not mutate and produce variations that could be studied cytologically and tested genetically.

Meiosis is a much more complicated and delicate process than mitosis. Consequently, genetically controlled variations in cell division are much less frequent for mitosis than for meiosis. One of the earliest accounts of mutations affecting the integrity of chromosomes in mitosis (chromosome breakage) was a recessive gene *sticky* [*st,* chrom. 4:(55)[1]] of maize. This gene also caused a kind of chromosome agglutination resulting in a sticky appearance of the chromosomes during meiosis in a certain strain of maize (Beadle, 1932a, 1937). Another gene in maize that affected stickiness in mitotic tissue was discovered by Schwartz (1958). This is st^e in which stickiness is observed mainly in the endosperm.

Another gene of maize, *polymitotic* (*po_1,* chrom. 6:4), causes **polymitotic divisions** in mitosis and was also discovered by Beadle (1931, 1933a). This gene affects postmeiotic mitoses. During the first microspore division the chromosomes do not reduplicate and split, but cytokinesis occurs in rapid succession separating the chromosomes into smaller and smaller cells. This is a reversal of the phenomenon of endopolyploidy (Section 9.1) during which the chromosomes reduplicate and split but cytokinesis does not occur. The result of these polymitotic divisions

---

[1]( ) – indicates probable gene position, based on insufficient data.

is many small cells with one or no chromosomes. As many as five divisions have been observed that can cause the formation of 32 small cells from 1 microspore. Since there are only 10 chromosomes in maize microspores (n = 10), many resulting daughter cells are left without a chromosome. The consequence of such polymitosis is complete male sterility, but the gene is not as effective during the female gametophyte formation (see Fig. 8.7). About 10% of all female gametophytes contain 10 chromosomes and are viable. Similar polymitotic behavior has been reported for *Rhumohra* (Bhavanandan, 1971), *Alopecurus* (Johnson, 1944) and the spider plant *Chlorophytum elatum* var. *variegatum* (Koul, 1970).

Genes in *Drosophila* causing mitotic variation were detected by their genetical consequences. These are the so-called *Minute* factors that increase the normally low incidence of **somatic crossing over** of *Drosophila* (Section 17.1.1) and the concomitant mosaicism (Stern, 1936). Generally, a *Minute* gene increases the frequency of somatic crossing over of some other gene located on the same chromosome. For instance, *Minute-n* [*M(1)n,* chrom. 1:62.7] strongly affects the crossover frequency on the chromosome arm section to the right of *singed* (*sn,* chrom. 1:21) but not so much to the left of it. Somatic crossing over induced by *Minutes* had a higher frequency in the centromeric regions than in regions distant from the centromeres. Extra Y chromosomes also increased the frequency of somatic crossing over. These observations suggest a direct relationship between somatic crossing over and heterochromatin since Y chromosomes and centromeric regions are largely heterochromatic (see Section 2.2.1). Swanson (1957) suggested that, because of the genetic effect of the *Minutes,* the heterochromatic regions may become stickier, so that there could be an increased chance for somatic crossing over during somatic pairing.

Genetic control of the production of chromosome aberrations as discovered by McClintock has been discussed previously (see Chapter 1 and Section 12.3.1). The cytological and genetic effects of the *Ds* locus were manifested only in the presence of the activator *Ac* in maize.

17.3   Variations in Meiosis

The number of reported genetically controlled meiotic abnormalities is large. A recent comprehensive review of the subject is proof of this fact (Baker et al., 1976). The abnormalities naturally influence all the major phenomena of meiotic behavior such as:

1. synapsis
2. crossing over
3. chromosome contraction
4. spindle formation

By 1976 Baker et al. could report about 32 meiotic mutants in *Drosophila* involving 29 loci. Among plants, *Pisum* is one of the most thoroughly analyzed objects. Gottschalk and Klein (1976) alone have reported 58 mutants showing genetically conditioned meiotic anomalies. Among them 34 are *ds* mutants, 7 *as* mutants, 1 shows asynaptic and desynaptic effects, 13 are *ms* mutants, and 3 mutants causing less specific meiotic disturbances.

17.3.1 Asynapsis and Desynapsis

Both these processes represent asynaptic mutations and lead to the reduction or loss of synapsis. However, while in **asynapsis,** the homologous chromosomes either fail to pair completely during zygotene or pair very incompletely; in **desynapsis** the homologues pair initially but fall apart during early diplotene or just after it, but usually before metaphase I. The asynaptic condition caused by major genes can be of varied origin:

1. asynapsis owing to gene mutations
2. asynapsis in progenies of varietal or species hybrids
3. asynapsis owing to loss or addition of a chromosome or chromosome pair

Futhermore, asynapsis can be induced or influenced by external environmental conditions, especially by mutagens. The same genetic and environmental influences are valid for desynapsis.

Beadle (1930, 1933b) discovered a gene in maize that upsets synaptic pairing. He called this gene *asynaptic* (*as,* chrom. 1:53). The effect of this gene was that most of the chromosomes did not pair during zygotene and as a result occurred as univalents rather than bivalents at metaphase I. Normal synapsis and crossing over are guaranteed only if two doses of *As* are present (*As As*) (Baker and Morgan, 1969; Nel, 1973). Evidence for genetic control of asynapsis was also discovered in *Matthiola* (Philip and Huskins, 1931), *Drosophila* (Gowen, 1933), *Datura* (Bergner et al., 1934), peas (Koller, 1938), *Oenothera* (Catcheside, 1939), wheat (Smith, 1939), rye (Prakken, 1943), tomato (Soost, 1951), rice (Katayama, 1961), maize (Miller, 1963; Sreenath and Sinha, 1968) sorghum (Stephens and Schertz, 1965), broad bean (Sjödin, 1970), rape (Stringham, 1970), cotton (Weaver, 1971), and *Lolium* (Omara and Hayward, 1978).

The structure that spans the region between two synapsed chromosomes is the **synaptonemal complex** (see Section 7.2.2.1). This complex ultrastructure is completely lacking in a *Drosophila* mutant in which crossing over in homozygous females is practically eliminated or almost completely reduced in the entire chromosome complement. The name of that mutant is *c(3)G*[17] (chrom. 3:57.4) (Meyer, 1964; Smith and King, 1968). Such a lack of synaptonemal complexes as well as the lack of pairing of chromosomes in pachytene may be a critical criterion for distinguishing between asynapsis and desynapsis. Desynapsis is controlled by *ds* genes that lower the chiasma frequency or prevent chiasma formation entirely. Both *as* and *ds* genes are similar in their action. They both cause disturbances in micro- and megasporogenesis. A third group of genes that also affect fertility in higher plants, the *ms* genes, are only effective in microsporogenesis (Section 17.4). Desynapsis is a widespread phenomenon in the mutant collections of countless plant species.

Desynaptic mutants have been reported in *Crepis* (Richardson, 1935), wheat (Li et al., 1945; Bozzini and Martini, 1971), peas (Gottschalk and Jahn, 1964; Gottschalk and Baquar, 1971), sorghum (Magoon et al., 1961; Sadasivaiah and Magoon, 1965), oats (Thomas and Rajhathy, 1966), rice (Misra and Shastry, 1969), *Lolium* (Ahloawalia, 1969), cabbage (Konvička and Gottschalk, 1971; Gottschalk and Konvička, 1972), soybeans (Palmer, 1974), *Allium* (Gohil and Koul, 1971; Kaul,

1975), barley (Scheuring et al., 1976), and *Pennisetum* (Singh et al., 1977; Koduru and Rao, 1978; Rao and Koruru, 1978).

Evidence of probably genetically determined asynapsis comes from the study of an azoospermic but otherwise healthy and normally developed man (Chaganti and German, 1974). Pachytene pairing and chiasma formation at diakinesis were disturbed. Univalents were observed in diakinesis. No chiasmata were seen. Almost all spermatocytes were in pachytene. The patient's mother's brother and mother's sister's son were also infertile.

17.3.2   Variation in Crossing Over

Beadle (1933b) assumed that crossing over in the asynaptic mutant of maize would be greatly reduced because there was very little chromosome pairing in zygotene as manifested in pachytene. But Rhoades (1947) could demonstrate that the frequency of crossing over, and of double crossing over in particular, was much higher than normal in this mutant. For instance, in the ws_3-lg_1-gl_2 region of chromosome 2 (*white sheath, ws_3*:0; *liguleless, lg_1*:11; *glossy, gl_2*:30), double crossing over was increased 25 times. The same observation was made for the C_1-sh_1-wx region of chromosome 9 (*aleurone color, C_1*:26; *shrunken* endosperm, sh_1:29; *waxy* endosperm, *wx*:59). If chromosomes do not pair in zygotene, one could speculate that pairing and crossing over could happen during premeiotic cell divisions.

Premeiotic crossing over was assumed for male *Drosophila* for which crossing over does not seem to occur during meiosis in the primary spermatocytes (see Section 8.2.1), but somatic crossing over does occur in the spermatogonia (Whittinghill, 1937, 1947). Other meiotic mutants that reduce and/or change the distribution of crossing over in *Drosophila* can be classified into three categories:

1. Reduction of crossing over without changes in the distribution pattern. [An example is mutant *mei*-9 (Baker and Carpenter, 1972; Carpenter and Sandler, 1974).]
2. Reduction of crossing over with changes in the distribution pattern [Mutants involved are *mei*-41, *mei*-218, *mei*-251, *mei*-S282, *mei*-B, *a b o*, and *mei*-68$^{L/1}$ (Bridges, 1929; Lindsley et al., 1968; Baker and Carpenter, 1972; Lindsley and Peacock, 1976; Valentin, 1973; Carpenter and Sandler, 1974; Parry, 1973).]
3. No reduction of crossing over but changes in the distribution pattern. [An example is mutant *mei*-352 (Baker and Carpenter, 1972).]

All of these genes reduce crossing over when they are homozygous recessive in females. If genes $c(3)G^{17}$ or $c(3)G^{68}$ are homozygous recessive, crossing over is almost completely absent (Carlson, 1972; Hall, 1972). But, when they are heterozygous, they show a nonuniform increase in crossing over (Hinton, 1966; Hall, 1972).

In a homozygous recessive mutant for gene *rec-1* of *Neurospora crassa*, crossing over in the *his-1* locus some distance away is increased ten-fold above normal (Catcheside, 1977). As pointed out earlier, chiasmata are the visible evidence for meiotic crossing over (see Section 4.2.1). Consequently, the frequency and distribution of chiasmata in meiotic mutants have been used as a possible indicator for crossover disturbances. In a recessive rye mutant investigated by Prakken (1943), pachytene pairing was almost normal but the total number of chiasmata was reduced from an average of 12.6 to a range from 2.6 to 6.4. The distribution of chiasmata shifted toward the distal ends of the chromosomes. Other mutants in which the total number of chiasmata was reduced were reported in *Crepis* and broad bean (Richardson, 1935) and in tomato (Soost, 1951). Decreasing chiasma frequency was observed in inbred lines of rye as homozygosity increased (Lamm, 1936; Müntzing and Adkik, 1948). Similar results from inbreeding were obtained in maize (Blanco, 1948) and *Drosophila* (Blanco and Mariano, 1953). However, an inbred line of mice showed a higher frequency of chiasmata (Slizinsky, 1955). In a broad bean mutant, Sjödin (1970) found a change in distribution of the chiasmata. Similar observations were made in maize (Beadle, 1930, 1933b; Miller, 1963), peas (Koller, 1938; Klein, 1969), *Oenothera* (Catcheside, 1939), wheat (Li et al., 1945) tomatoes (Moens, 1969a), and Scotch pine, *Pinus sylvestris* (Runquist, 1968). Failure of chiasma formation in one particular chromosome (chromosome IV) was observed in *Hypocheris radicata* (Parker, 1975).

Meiotic mutants in humans are difficult to discover since pedigree data are harder to obtain. But patients with Down's syndrome (see Section 16.2.4.1) have been reported to have an increased number of chiasmata per cell (Hultén and Lindsten, 1973). This obviously is not necessarily a result of aneuploidy, since human XYY trisomics and aneuploids with extra unidentified small centric chromosomes have normal chiasmata counts (Evans et al., 1970; Hultén, 1970; Hultén and Lindsten, 1973).

17.3.3   Variation in Chromosome Size

Evidence that the size of chromosomes must be under genetic control was given by Thomas (1936) in perennial ryegrass and by Lamm (1936) in rye lines derived from inbreeding. In the garden stock, *Matthiola incana,* a mutant exists in which the chromosomes are long while in the normal forms they are short (Lesley and Frost, 1927). A reverse situation was discovered in sweet pea where the mutant form showed short chromosomes while the normal situation was characterized by long chromosomes (Upcott, 1937). Moh and Nilan (1954) discovered a meiotic mutant in barley *(sc)* that was characterized by extremely condensed diakinesis chromosomes. In addition, the mutant had relatively well-spread pachytene chromosomes that proved to be favorable for **pachytene analysis** (see Section 2.2) (Blickenstaff et al., 1958). A gene for long chromosomes was found in barley (Burnham, 1946; McLennan, 1947; McLennan and Burnham, 1948).

17.3.4   Variation in Spindle Formation

Clark (1940) discovered a meiotic mutant in maize that is called *divergent spindle* (*dv*, location unknown). When this gene is homozygous recessive (*dv dv*), the spindle fiber apparatus in meiosis I cells forms parallel or divergent fibers instead of those converging to the two poles. The result is that the chromosomes fail to gather at the poles, but individual chromosomes or smaller chromosome groups come together and form separate nuclei. The spindles at the second meiotic division may again be divergent. Consequently, there may be more than four spores

Fig. 17.6. Multiple microsporo-cytes (syncytes) in barley caused by a recessive gene. Multiple sporocyte with 113 bivalents and 4 quadrivalents. (From Smith, 1942. Reprinted by permission of Burlington Free Press, Burlington, Vermont).

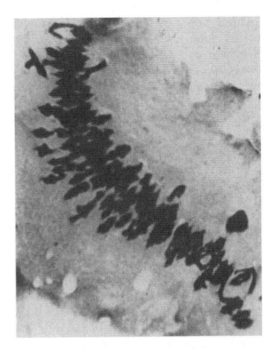

at the quartet stage (Fig. 7.27), and 42% to 95% of the microspores are multi-nucleate. A divergent spindle mutant was also discovered in crested wheatgrass (Tai, 1970).

Similar divergent spindles were found in a mutant that formed multiple sporocytes in barley (2n = 14) (Smith, 1942). Meiocytes were found with 14, 21, 28, 56, 112, and higher numbers of chromosome pairs. Such meiocytes are also referred to as *syncytes*[1] (Levan, 1942b), which are formed by *cytomixis*[2] (Gates, 1911). Fusion of some of the chromosomes was thought to have taken place prior to meiosis, since in many cases multivalents had formed during synapsis. Cell walls seemed to be absent in these so-called cells and sometimes all, or at least part, of the contents of an entire anther locule were included in one syncyte. Some of the chromosomes seemed to have fused as late as metaphase I. Very long metaphase plates were the result of this phenomenon (Fig. 17.6).

Multiple spindle formation was reported in *Clarkia* (Vasek, 1962). About half of all meiocytes possessed two complete spindles at metaphase I.

17.3.5   Other Variations in Meiosis

In a mutant of *Datura* (2n = 24) called *dyad* (*dy*), which resulted from pollen treatment with radium, no second meiotic division occurred (Satina and Blakeslee, 1935). At the end of telophase I, the normally short period of interkinesis (see

---

[1]*Syncyte*–a polyploid or multinucleate cell formed usually by inhibition of cytokinesis; it leads to the formation of a *syncytium* (Haeckel, 1894), which is a mass of protoplasm lodging many nuclei not separated by cell membrane.

[2]*Cytomixis*–the fusion of the chromatin of two or more cells.

Section 7.5) was replaced by a prolonged regular interphase during which a postmeiotic chromosome replication took place. Mitosis in the first male gametophyte division produced diploid nuclei that became diploid gametes.

Precocious chromosome division during meiosis I occurred in three mutants in tomato (Lamm, 1944; Clayberg, 1959) and *Alopecurus* (Johnson, 1944). Normally chromatids do not separate until anaphase II. In the mutants the chromatids separated during anaphase I, telophase I, or interkinesis. If a metaphase II plate was formed, the chromatids moved to the opposite poles at random.

Another group of meiotic mutants are those that result in frequent chromosome breakage. In some of these, bridges and fragments have often been observed at anaphases I and II. These mutants were demonstrated to occur in peas (Klein, 1969; Klein and Baquar, 1972). Plants with high frequencies of anaphase bridges and fragments without clear evidence of mutant origin have been observed in *Scilla* (Rees, 1952, 1958), *Solanum* (Lamm, 1945), *Picea* (Anderson, 1947), *Matricaria, Hyoscyamus* (Vaarama, 1950), *Sorghum* (Magoon et al., 1961), *Allium* (Koul, 1962), and *Podophyllum* (Newman, 1967). Other mutants that are often associated with chromosome breakage during meiosis are the *sticky* mutants. The first one discovered was mentioned previously in connection with chromosome breakage during mitosis (Section 17.2). This was the *sticky* gene in maize discovered by Beadle. Adherence of the chromosomes at anaphase I, and clumping, and sometimes breakage were characteristics that were associated with this stickiness. With this study Beadle could demonstrate that genes affecting chromosome behavior segregate and recombine in the same fashion as those that influence the morphology of plants or animals. Other sticky meiotic mutants sometimes associated with chromosome breakage were reported in *Alopecurus* (Johnson, 1944), durum wheat (Martini and Bozzini, 1965), broad bean, tomato (Moens, 1969a), *Brassica* (Stringham, 1970), *Collinsia* (Mehra and Rai, 1970, 1972), and peas (Klein, 1971).

17.4   Male Sterility

Aberrant meiotic behavior caused by nuclear genes, as discussed in Sections 17.3, can be the reason for genetically determined male sterility. Other phenomena associated with male sterility can be, for instance, pollen abortion, failure of anther dehiscence, anther abortion and distortion, and **pistillody** of the anthers (the metamorphosis of anthers into pistils) (Gottschalk and Kaul, 1974). Microsporogenesis is really the victim of disturbances and not megasporogenesis. The relatively unprotected haploid male gametophyte is probably more vulnerable than the protected embryo sac (Swanson, 1957). More than 400 *ms* mutants in over 50 species have been reported so far (Gottschalk and Kaul, 1974). In most of the male sterility mutants in maize, abortion occurred after meiosis, but, in a few, male sterility was caused by aberrant meiotic behavior (Beadle, 1932b). For instance, male gametophyte development in genetic male steriles ms_2, ms_7, and ms_{11} breaks down between the fifth and tenth days after meiosis (Madjolelo et al., 1966). Most male sterility genes are recessive (*ms ms*). One of them in maize is a dominant gene

(Ms_{21}). Twenty different male sterility genes in maize had been described by 1935 (Emerson et al., 1935). The numbering system indicates 43 by 1979 (Golubov-skaya, 1979). Some of the maize male sterility genes are ms_1 (Chrom. 6:17), ms_2 (Chrom. 9:67), ms_8 (Chrom. 8:14), ms_{10} (Chrom. 10), ms_{17} (Chrom. 1:23), and ms-si [Chrom. 6:(19)].

Genetic male sterility in plants has not occurred widely in nature, but it has been screened for in almost all crops. Besides in maize, genetic male sterility has been found, for instance, in barley (Suneson, 1940; Hockett and Eslick, 1971), tomato (Rick, 1948), lima bean (Allard, 1953), potato (Okuno, 1952), and cotton (Rich-mond and Kohel, 1961).

Genetic male sterility is of great value to plant breeders. It is increasingly being used for the production of hybrid varieties. Marker genes closely linked with the male sterility genes make roguing of the male-fertile plants easier since the progeny of a cross between a male-sterile plant (ms ms) and a male-fertile plant (Ms ms) always produces 50% undesired male-fertiles (Ms ms). In some instances, male-fertile plants can be identified before pollen shedding and can be removed. In watermelon, a sterile, homozygous recessive, male-sterile mutant has as a marker *glabrous leaves,* which allows the nonmarked male-fertiles to be removed (Watts, 1962). In lettuce, male steriles (ms_1 ms_1, ms_2 ms_2, ms_3 ms_3) have narrow sharply cut leaves that can be recognized prior to flowering (Lindqvist, 1960).

17.5   Preferential Segregation of Chromosomes

Meiotic segregation of chromosomes toward the poles at anaphase I is generally at random (see Section 7.3). But such behavior is difficult to prove unless the two chromosomes of a bivalent are morphologically marked or unless there is genetic proof of nonrandom segregation of genes that are on different chromosomes. The randomness of chromosome segregation was proved in several grasshopper species in which heteromorphic, homologous chromosome pairs could be studied (Caroth-ers, 1917, 1921).

Nonrandomness of chromosome segregation in meiosis has been observed in several cases, has been called **preferential segregation,** and leads to **segregation distortion** (Sandler and Hiraizumi, 1961). An instance in which preferential segregation was proved both by a marker chromosome as well as by distortion of genetic ratios was observed in maize. An abnormal chromosome 10 was discovered by Longley (1938) in certain maize strains grown by North American Indi-ans. This chromosome has a largely heterochromatic knobbed segment added to the long arm of chromosome 10. Burnham (1962) noticed that this terminal seg-ment superficially resembles the well-known B chromosome of maize. In plants that are heterozygous for abnormal 10, 70% of the megaspores carry this distinc-tive marker chromosome (Rhoades, 1942, 1952). This preferential movement has been explained as a purely chromosome-related phenomenon. The abnormal 10 chromosome has a tendency to form a neocentromere (see Section 2.1.2). The spindle fiber, instead of attaching to the normal position at the primary constric-tion of chromosome 10, pulls this chromosome at a new location close to the abnormally knobbed end. This movement is precocious in that abnormal 10 passes quickly to the nearest pole before the other chromosomes normally separate at anaphase. This movement is thought to cause the abnormal 10 to be preferentially distributed to the basal megaspore that, in the linear quartet (see Fig. 8.7), becomes the functional embryo sac. Not only the abnormal 10 segregates preferentially, but other chromosomes that possess knobs also can segregate nonrandomly in the heterozygous presence of one abnormal 10 (Longley, 1945). The knobs of these chromosomes are activated by the presence of abnormal 10 and become neocentromeres that move precociously in anaphase. Tests were conducted involving genes closest to the knob on the short arm of chromosome 9. In the heterozygote, the knobbed chromosome carried the alleles for *aleurone color* (C_i:26), *shrunken endosperm* (sh_i:29), and *waxy endosperm* (*wx*:59). The knob had a near *0* position on the chromosome map. The homologous chromosome without the knob carried the alleles c_1, SH_1, and *Wx*. From these map data, it can be seen that C_1 (26) was closest to the distal knob and *wx* (59) closest to the centromere (see genetic maize map, Fig. 4.8). C_1 showed the greatest percentage of preferential segregation, sh_1 was similar in this respect, but *wx* showed very little. Preferential segregation is increased for loci that are involved in frequent crossing over between the centromere and the knob and are in close proximity to this distal knob. This relationship is best illustrated in Fig. 17.7. Preferential segregation in translocation heterozygotes has already been discussed (see Section 14.3).

Another mechanism involving preferential segregation is often referred to as **meiotic drive.** This was defined by Sandler and Novitski (1957) as a meiotic phenomenon that modifies the breeding structure from the expected ratios by changing the frequencies of alleles in a population. An example for this phenomenon was

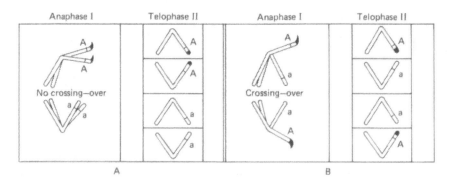

Fig. 17.7A and B. Preferential segregation caused by neocentric activity of the knobbed abnormal 10 chromosome in maize. (*A*) During meiosis in the embryo mother cell the neocentromere affects the knob orientation (gene A) by pulling the chromatid on which it is located to the extremity of the AI pole. When no crossing over occurs, one chromosome carries the neocentromere and the other does not. Preferential segregation is negligible. (*B*) When crossing over occurs, both chromosomes have one neocentric and one normal chromatid. The neocentric chromatids move to the extremes of the AI pole and are preferentially included in the terminal (functional) quartet cell during A II. (From Sybenga, 1972. Redrawn by permission of Elsevier/North-Holland Biomedical Press, Amsterdam).

orginally thought to be the gene for *Segregation Distorter* in *Drosophila (SD,* chrom. 2:55.0. Centromere location). This gene is located close to the heterochromatic-euchromatic junction and according to one interpretation may involve chromosome breakage in that region (Sandler and Novitski, 1957). Similar alleles of this gene have been found by other workers (Mange, 1961; Greenberg, 1962). The females *(SD SD* or *SD sd)* of this mutant behave normally. The homozygous dominant males *(SD SD)* are sterile. The heterozygous male *(SD sd)* produces a majority of *SD*-bearing functional sperm, often more than 95%. Generally, one would expect a 1:1 ratio of *SD-* to *sd*-bearing sperms.

The earliest explanation of this phenomenon was the **dysfunctional sperm hypothesis** of Sandler et al. (1959). According to this hypothesis, the *sd*-bearing chromosome (second *Drosophila* chromosome), through the action of *SD,* breaks at a specific location, that after chromosome replication is subject to reversed sister-strand reunion (see Section 11.1), causing chromatid bridges and death or non-function of the cells they tie together. The cytological evidence for this hypothesis could not be verified subsequently.

An alternative to this hypothesis was the **functional pole hypothesis** of Peacock and Erickson (1965) that states:

a. that *Drosophila melanogaster* normally produces two functional and two nonfunctional sperms from each spermatocyte.
b. that the *sd*-bearing chromosome is directed by *SD* to the nonfunctional cell pole during spermatocyte division.

The functional pole hypothesis was discredited by Hartl et al., (1967) and independently by Nicoletti et al., (1967). They established that the number of offspring produced by *SD sd* males was about half as great as that produced by normal *sd sd* males. A condition for the functional pole hypothesis would be that the *SD sd* males will produce no less functional sperm than normal males. More recent thought again favors the dysfunctional sperm hypothesis. *SD* could induce some kind of physical alteration in *sd,* too small to be detected cytologically, and by this action render *sd* nonfunctional (Hartl and Hiraizumi, 1976).

Another gene was discovered that seems to have a regulating function in regard to *SD.* It occurs together with *SD* and is called *Stabilizer of Segregation Distorter (St-SD,* Chrom. 2: close to *bw,* 104.5). In the absence of *St* the activity of *SD* is more variable (Sandler and Hiraizumi, 1960). Hartl and Hiraizumi (1976) believe that the action of *St* appears to be really a cumulative effect of minor modifier genes.

Zimmering et al. (1970) have suggested five possible mechanisms of unequal transmission of homologues:

1. a supplementary replication of the meiotically driven chromosome and the associated loss or degeneration of its homologue
2. impairment of sperm function by abnormal chromosome behavior (e.g., chromosome breakage) with the sperm that carries the favored chromosome remaining unimpaired
3. sperm competition or malfunction, depending on the genetic constitution of an organism
4. preferential segregation of a favored homologue to the functional pole at meiosis
5. differential acceptance of two different kinds of sperm by the ovum

Genetic proof for nonrandom segregation sometimes leads to the conclusion that preferential chromosome segregation is involved. Such genetic segregation was observed in the progeny of crosses between laboratory stocks of mice and was referred to as **genetic affinity** (Michie, 1953, 1955; Parsons, 1959; Wallace, 1953, 1957, 1958, 1959, 1961). It was found that genes located on different chromosomes (V and XIII) tended to segregate together at meiosis and pass to the same pole. This results in genetic linkage between genes of nonhomologous chromosomes. Similar interdependence between chromosomes in their movement toward the pole was reported for crane fly spermatocytes (Forer and Koch, 1973).

Chapter 18
Variation in Function of Sex Chromosomes

The so-called "normal function" of the sex chromosomes was discussed in Chapter 5. As there are genetic factors that upset the normal function of the autosomes, there are naturally also those that upset the normal function of sex chromsomes. Since the sex chromosomes constitute a distinct group with a much specialized task of specific genetic expression their special treatment is warranted and is presented in this chapter.

18.1   Variation in Sex Ratio

The normal sex ratio is determined by the number of males vs. females at birth, which generally is 1:1 or close to it. Deviations from this ratio can be caused by several factors. Sturtevant and Dobzhansky (1936) discovered a **sex ratio gene** in *Drosophila pseudoobscura* and *D. persimilis.* This gene caused abnormal frequency of daughters (more than 90%) if it was present in the male parent of a cross. During the first meiotic division of the spermatocyte the X and Y chromosomes did not pair but the X chromosome split twice and separated mitotically (Fig. 18.1) during meiosis I so that each daughter cell received one X chromosome. The Y chromosome did not divide during meiosis I but passed to one of the first division poles, became enclosed in a vesicle, and degenerated. Since the X chromosome had split twice during meiosis I (probably two replications during premeiotic interphase), it could again separate in meiosis II, distributing one X chromosome to each sperm. Novitski et al. (1965) and Polansky and Ellison (1970) reinvestigated the *sex ratio* gene in *D. pseudoobscura* and found that the mechanism was different. During anaphase I the X and Y chromosomes regularly passed to opposite cell poles. Following meiosis I the Y chromosome degenerated leading to nonfunctional sperm formation. Consequently, each primary spermatocyte produced only two instead of four functional sperms. Thus, the X chromosome did not split twice as was suggested by Sturtevant and Dobzhansky.

Another case of the distortion of the sex ratio in favor of females was reported by Novitski and Hanks (1961) and Erickson and Hanks (1961). Males of *Drosophila melanogaster* containing the gene *Recovery Disrupter* [*RD(1),* chrom. 1:62.9] may produce approximately 67% female progeny, due to a reduction in the recovery of the Y chromosome. The mechanism involved causes a fragmentation of the

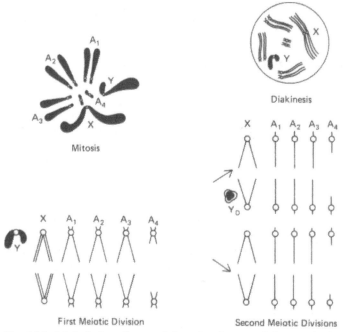

Fig. 18.1. Meiosis in a "sex ratio" male of *Drosophila pseudoobscura*. (A_{1-4} = autosomes). According to an older concept, the X splits in both meiotic divisions. The Y is heteropycnotic and eventually disintegates (Y_D). (From White, 1954. After Sturtevant and Dobzhansky, 1936. Redrawn by permission of the University Press, Cambridge).

Y chromosome during meiosis (Erickson, 1965). A second *RD* chromosome [*RD(2)*] has been discovered on chromosome 2, but the map location has not been determined (Wallace in Lindsley and Grell, 1968). This factor is thought to be another example of *meiotic drive* (see Section 17.5).

18.2   Different Sex Chromosome Systems

In Chapter 5 the basic form of sex determination in animals and in some plants was discussed. In the basic system, one sex has a pair of chromosomes that, microscopically, are similar (XX) and the other sex has visibly different chromosomes (XY). The closest deviation from this XY:XX system is the XO:XX system. The O in this formula merely indicates the absence of the Y chromosome. As mentioned, Henking in 1891 found the first one of such systems in the insect *Pyrrhocoris apterus* (Chapter 1). In meiosis I of such XO organisms, the X chromosome oriented itself in the metaphase plate without forming a bivalent. Usually, if autosomal univalents occur in meiosis, they do not line up on the metaphase I plate. Here, the X chromosome moves randomly to one of the two cell poles and later is recovered in 50% of the gametes. Autosomal univalents often lag behind in ana-

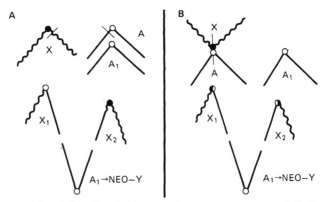

Fig. 18.2A and B. Two possible origins of multiple sex chromosome systems. (*A*) A reciprocal translocation occurs between the X chromosome of an XO:XX system and one of a pair of autosomes (A of AA_1). The two new translocated chromosomes become X chromosomes (X_1X_2). The second autosome (A_1) becomes the neo-Y chromosome. (*B*) The X and the A chromosome fuse at the centromere (**centric fusion**). The centromere then divides transversely instead of longitudinally resulting in two new X chromosomes each possessing one arm of the old X and one arm of A (X_1, X_2). (From Hughes-Schrader, 1950. Redrawn by permission of Prentice-Hall, Inc., Englewood Cliffs, New Jersey).

phase and do not always get included in the daughter nuclei. A great majority of species in the insects of the Orthoptera and Odonata have the XO:XX system (White, 1973).

Bivalent pairing and chiasma formation in meiosis I apparently are not always necessary requirements for chromosome distribution, as is the case for autosomes. For instance, in some Tipuloidea species, the X and Y chromosomes do not synapse but distribute regularly to the poles in meiosis I (Wolfe, 1941). Sometimes the sex chromosomes form a very brief end-to-end association in diakinesis as in the hemipter *Rhytidolomia senilis* (Schrader, 1940b). This transitory chromosome association was called **touch-and-go pairing** by Wilson in 1925.

Translocations between one of the sex chromosomes and an autosome can lead to multiple sex chromosome systems. If in the XO:XX system, the X in the XO sex translocates with one of the autosomes, an X_1X_2Y condition can arise (Fig. 18.2A, B). The two translocated chromosomes will become X_1 and X_2 while the non-translocated homologue of the autosome pair involved (A_1) becomes the neo-Y chromosome. Such sex chromosome trivalents (X_1X_2Y) have been observed in the mantids (Hughes-Schrader, 1950). The same phenomenon has been observed in 14 genera of grasshoppers (Helwig, 1941, 1942). In such a system the autosomal neo-Y chromosome becomes confined to the male sex. The female has one more chromosome than the male ($X_1X_2Y:X_1X_1X_2X_2$ system). Other examples for this system are *Drosophila miranda* (MacKnight and Cooper, 1944) and the Rhodesian pygmy mouse (Matthey, 1965).

Another possbility is a translocation between the Y chromosome and an autosome, which leads to the XY_1Y_2 condition. In such a system the male has one more chromosome than the female (XY_1Y_2:XX system). Examples for this system are *Drosophila americana* (Spencer, 1940), the gerbil (Wahrman and Zahavi, 1955), and some bats (Baker and Hsu, 1970).

Complex systems of higher magnitude can be based on the XO:XX or XY:XX systems. $X_1X_2X_3X_4O$ males occur in the aphid *Euceraphis betulae* (Shinji, 1931). All X chromosomes pass to the same pole during meiosis I, and the secondary spermatocytes that do not have any X chromosomes degenerate. Only one type of sperm is formed. The scale insect, *Matsucoccus gallicola,* even has six X chromosomes in the male ($X_1X_2X_3X_4X_5X_6O$, Hughes-Schrader, 1948).

Complex systems based on the XY:XX scheme have been found in the beetles of the genus *Blaps*. *B. walti, B. mortisage,* and *B. mucronata* had $X_1X_2X_3Y$ (Nonidez, 1915; Guénin, 1949; Lewis and John, 1957), *B. gigas* had $X_1X_2X_3X_4Y$ (Guénin, 1949), and the most complex sex chromosome system known is recorded for *B. polychresta*—$X_1X_2X_3X_4X_5X_6X_7X_8X_9X_{10}X_{11}X_{12}Y_1Y_2Y_3Y_4Y_5Y_6$ (Guénin, 1953).

18.3 Cytogenetics of *Sciara*

Striking meiotic anomalies with interesting departure in sex determination have been observed in the fungus gnats, small two-winged flies of the family Sciardidae feeding on fungi. The thorough studies of the species *Sciara coprophila* are fairly representative of the chromosome mechanism in the genus *Sciara* in general (Metz, 1931, 1933, 1934, 1936, 1938a, 1938b; Metz and Schmuck, 1929, 1931; Schmuck and Metz, 1932).

Sciara has a basic complement of four homologous chromosome pairs, three autosome pairs, and one X pair ($6A + 2X$, ♀ soma). But in the germ line there are, in addition, one to several **limited chromosomes** (l) that are limited to the germ line. During the growth of the nurse cells and of the primary spermatocytes, these chromosomes are **positively heteropycnotic** (see Section 2.2.1) in that they are compact and tightly coiled and darker appearing than the basic set. They are also larger in length and diameter and apparently without specific genetic activity (Fig. 18.3). The meiosis of the females is normal. Bivalents are formed and genetic evidence for crossing over exists (Schmuck and Metz, 1932). Spermatogenesis, however, is very irregular (Metz et al., 1926; Metz, 1933). The homologous chromosomes do not pair during meiotic prophase I, and the chromosomes remain univalents.

The chromosome complement in the male germline initially consists of $3A^{II}$* 3X $+3l+$ (Fig. 18.3). Before the first spermatocyte division, an X chromosome and an occasional superfluous l-chromosome become eliminated. The first meiotic division spindle is a **monaster** that performs a **monocentric mitosis.** The maternal chromosomes (white in Fig. 18.3) become attached to the spindle and move to a single pole. This fact is based on genetic as well as on cytological evidence (Metz, 1938a). The paternal chromosomes, in spite of also being attached to the mon-

---

*$3A^{II}$ = 3 pairs of autosomes

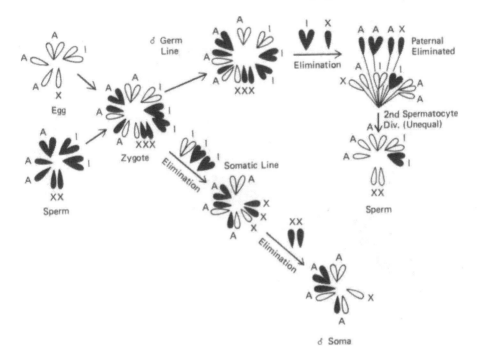

CYTOGENETICS OF SCIARA

Fig. 18.3. Meiotic and mitotic anomalies with departure in sex determination in *Sciara coprophila:* A–autosomes; 1–limited chromosomes; X–sex chromosomes; ◠–maternal chromosomes; ◖–paternal chromosomes. (Modified from Crouse, 1943. Redrawn by permission of the University of Missouri Agricultural Experiment Station, Columbia).

aster, back away from that pole and become collected into a tiny bud that later pinches off and becomes discarded. The maternal and paternal (black and white in Fig. 18.3) l-chromosome pass with the maternal A and X chromosomes into the secondary spermatocytes. Since a paternal l-chromosome does not possess any specific genetic function, no paternal genes become included in the secondary spermatocyte.

During the second spermatocyte division, the only remaining X chromosome divides. Both halves move precociously to the same pole (**non-disjunction**) so that each sperm always possesses two X chromosomes, since the cells without X chromosomes degenerate.

As mentioned, the somatic tissue (soma) from which the germ line branches off consists of three pairs of autosomes ($3A^{II}$), 3 X chromosomes (3X) and one or more l-chromosomes (3 are shown in Fig. 18.3). The l-chromosomes become discarded at the fifth or sixth cleavage division after zygote formation. One or both paternal X chromosomes are eliminated shortly thereafter, during the seventh or eighth cleavage division. This elimination will determine if the individual becomes a male or female. If both paternal X chromosomes are eliminated, the soma becomes male (XO, Fig. 18.3). If only one of the two paternal X chromosomes is elimi-

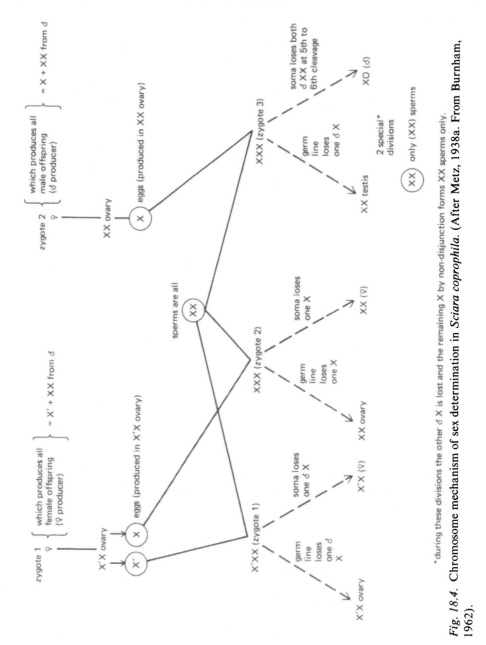

Fig. 18.4. Chromosome mechanism of sex determination in *Sciara coprophila.* (After Metz, 1938a. From Burnham, 1962).

nated, the soma becomes female (XX). Consequently, males and females in *Sciara* differ in their somatic chromosome number (*Sciara coprophyla:* ♂ 2n = 7, ♀ 2n = 8).

The **chromosome elimination** of the l-chromosomes and the X chromosomes in the soma is similar in nature. The prophase is apparently normal. Along with the normal chromosomes, the l-chromosomes become attached to the mitotic spindle in metaphase and open out at the centromere but fail to separate at their distal ends. As a result the l-chromosomes remain in the equatorial plate and eventually degenerate. They are not included in the daughter nuclei. This elimination process may be closely associated with the heteropycnotic or heterochromatic nature of the l-chromosomes. Heterochromatin has a tendency to become sticky under certain conditions, which may lead to the inability of these chromosomes to divide normally.

In *Sciara* the genotype of the father has no influence on the sex of the progeny (Metz and Moses, 1928). In a normal XO:XX system, there are two kinds of sperm produced, one with an X and one without. In *Sciara* all sperm have 2X. In *S. coprophila,* for instance, the offspring of any pair mating are all of one sex (**unisexual progenies**), either male or female. The sex of the offspring consequently must depend on the genetic constitution of the mother. Some mothers produce only male and some only female offspring. There apparerently are no visible cytological differences between these two kinds of mothers. They must differ in an invisible genetic factor. The X chromosome carrying this factor could be designated as X'. Germ lines of female producing females would be heterozygous for this factor (X'X) and females producing males would be homozygous (XX). The heterozygous mothers (X'X) produce two kinds of eggs, X' and X. If X' eggs are fertilized by XX sperm, X'XX zygotes will result that after elimination of one paternal X will become females (X'X) that, in turn, produce all female offspring (X'XX). The homozygous mothers (XX) produce only one kind of egg (X). If they become fertilized by XX sperm, they produce XXX zygotes that after elimination of one X become XX females that produce only male progeny (Fig. 18.4).

Chapter 19
Apomixis and Parthenogenesis

Asexual reproduction occurs in plants as well as in animals. In plants, asexual reproduction is commonly known as apomixis.

19.1   Apomixis in Plants

Apomixis is the replacement of sexual reproduction (**amphimixis**) by various types of asexual reproduction that do not result in the normal fusion of haploid gametes (Rieger et al., 1976).

Apomicts can be obligate or facultative. In an obligate apomict every plant of the species is always apomictic. In facultative apomicts sexual and apomictic reproduction occurs in the same plant. It is very difficult to classify a species as an obligate apomict because sexual or facultative apomicts may occur in nature that may not have been detected yet. For instance, Young et al. (1978) studied embryo sacs of formerly known obligatory apomict buffelgrass *(Cenchrus ciliaris)* and detected that about 10% of 1,300 pistils investigated showed single, fully differentiated, 8-nucleate embryo sacs that were indistinguishable from those of known sexual plants observed by the same method. Aposporous and sexually appearing embryo sacs were observed within the same plant. They concluded that this possibly indicates the presence of facultative apomixis in buffelgrass.

Nevertheless, some species have been classified as obligate apomicts. For instance, almost all apomictic Compositae are obligate rather than facultative apomicts (Stebbins, personal communication). Other examples are *Cooperia* (Coe, 1953) and the American polyploids of *Crepis* (Stebbins and Jenkins, 1939). According to Stebbins many species of the Rosaceae are mainly facultative apomicts, while the Gramineae are equally divided into both groups. Other examples of facultative apomicts are found among species of *Poa, Potentilla, Rubus, Citrus* (Brown, 1972), and *Pilsella* (Rosenberg, 1917).

Apomixis can be implicated in a situation in which two phenotypically different parents when crossed result in an F_1 progeny that are all phenotypically like the homozygous recessive female parent. In normal sexual reproduction, the progeny of such an F_1 cross should be phenotypically like the homozygous dominant pollen parent. The following illustrates the two conditions:

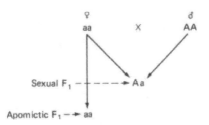

Additional evidence for apomixis is the maternal resemblance in the chromosome number after reciprocal crossing (Einset, 1947).

Apomixis can be subdivided into **agamospermy** and **vegetative reproduction.** In agamospermy there is seed formation but reproduction is asexual. In vegetative reproduction there is no seed formation and the new individual forms from a group of differentiated or undifferentiated cells (Fig. 19.1).

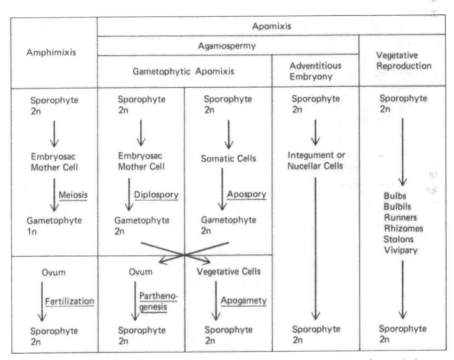

Fig. 19.1. Diagrammatic representation showing the interrelationships of apomictic processes as compared to normal amphimixis. (Modified after Stebbins, 1950).

19.1.1   Vegetative Reproduction

Vegetative reproduction can occur as the only means of plant propagation or in addition to the normal sexual reproductive process. Stebbins (1950) felt that vegetative reproduction should be considered apomixis only if it is the sole means of reproduction. This condition is met when the propagules (**bulbils**) occur within the inflorescence replacing the sexual flowers. This phenomenon is called **vivipary.** It is known in *Polygonum, Saxifraga, Allium, Agave, Poa,* and *Festuca.* In two cultivars of teosinte, a structure having many plant-like branches replaces the tassel under greenhouse conditions. Plants can be established by rooting these plant-like branches in sand. Some species do not reproduce sexually in northern latitudes because they are short-day sensitive. They then reproduce entirely asexually or vegetatively in northern latitudes. Gustafsson (1946) mentions *Elodea canadensis, Stratiotes aloides,* and *Hydrilla verticillata* as examples for such species.

Other means of vegetative reproduction, besides vivipary, are bulbs, runners, rhizomes, and stolons. Pangola digit grass (*Digitaria decumbens* Stent.) has been vegetatively propagated in Florida and Mexico on large acreages (more than 0.75 mill. hectares) by discing freshly cut hay into moist soil and also by sprigging stems and runners (Nestle and Creek, 1962; Hodges et al., 1967). Coastal Bermuda grass [*Cynodon dactylon* (L.) Pers.], a sterile hybrid between "Tift" Bermuda grass and an introduction from South Africa, has been established on some 5 million hectares in the southeastern United States by vegetative propagation (stolons and rhizomes), called sprigging, at a cost often less than that of seed establishment (Burton, 1956, 1973). Thousands of rhizomatous intermediate wheatgrass plants [*Agropyron intermedium* (Host) Beauv.], which are cytoplasmically fixed male steriles, have been vegetatively cloned by subdividing the sod so that they can be used in hybrid grass seed production (Schulz-Schaeffer, 1978).

19.1.2   Agamospermy

Agamospermy is a form of apomixis in which the ameiotic (without meiosis) and asexual (without fertilization) processes are carried out within the ovule. The simplest form of agamospermy is **adventitious embryony,** in which the embryo develops directly from the diploid sporophytic tissue of the nucellus or ovule integument without gametophyte (embryo sac) development (Fig. 19.2A). The gametophyte is usually formed but does not function. Other methods of agamospermy require the formation of an unreduced differentiated embryo sac in which diploid gametophytes are formed without meiosis. Such **gametophytic apomixis** can be either **aposporic,** when a diploid embryo sac is formed directly from a nucellar or integument cell (Fig. 19.2B), or **diplosporic,** when the embryo sac is formed from an unreduced megaspore cell (Fig. 19.2C).

19.1.2.1   Adventitious Embryony.   **Adventitious embryony** is similar to embryo formation in callus tissues on cut surfaces of stems as in tomatoes; of leaves, as in African violets; and of roots. Adventitious embryony is typical in species that are native to warm temperate or tropical climates. There are three types known:

1. those that depend on fertilization and endosperm development, such as *Citrus* (Frost, 1926, 1938a, 1938b)

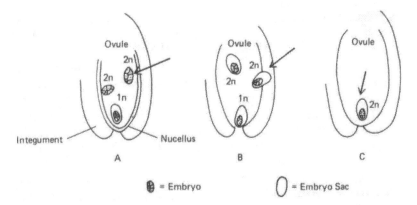

Fig. 19.2A–C. Comparison of different forms of agamospermy. (*A*) **Adventitious embryony:** 1n–normal haploid embryo sac; 2n–two adventitious embryos. (*B*) **Apospory:** 1n–normal haploid embryo sac with sexually produced embryo. 2n–two diploid aposporous embryo sacs containing apomictic embryos. (*C*) **Diplospory:** 2n–an embryo sac containing an apomictic embryo. (Modified from Brown, 1972. Redrawn by permission of the C. V. Mosby Company, Saint Louis, Missouri).

2. those that depend on endosperm development but not on fertilization, such as *Coelebogyne ilicifolia* (Schnarf, 1929)
3. those that depend neither on fertilization nor on endosperm development, such as the jointed cactus *Opuntia aurantiaca* (Archebald, 1939)

19.1.2.2 Somatic Apospory. This is one of the two forms of gametophytic apomixis. It means that the functional embryo sac does not develop from the megaspore but from a somatic cell. Two main types of somatic apospory were reported: the *Hieracium* type and the *Panicum* type.

The *Hieracium* type was found in some species of *Hieracium* (Rosenberg, 1906, 1907), *Artemisia* (Chiarugi, 1926), *Crepis* (Babcock and Stebbins, 1938), and in other genera. One or more somatic cells begin to enlarge, become vacuolate, and develop directly into the initial cell of the gametophyte. Three nuclear divisions result in the normal 8-nucleate embryo sac but the nuclei have the 2n somatic rather than the gametic chromosome number (n).

The *Panicum* type was reported for *Panicum maximum* (Warmke, 1954) and for other members of the Panicoideae (Emery, 1957; Emery and Brown, 1958; Simpson and Bashaw, 1969; Bashaw et al., 1970). One or more nucellar cells develop into an aposporous embryo sac. After vacuolation only two nuclear divisions result in a 4-nucleate embryo sac. All four nuclei remain in the micropylar region. Differentiation gives rise to two synergids, one egg and one polar nucleus, all having a somatic chromosome number (2n). Often more than one embryo sac forms in an ovule, but usually only one embryo sac functions in seed formation (Young et al., 1978).

19.1.2.3 Diplospory. In diplospory the embryo sac develops from an archespore cell, but meiosis is either missing or does not result in chromosome reduction. A meiosis that does not lead to chromosome reduction is called **apomeiosis** (Renner, 1916). The reasons for failure of reduction can be lack of chromosome pairing and of chromosome contraction, retardation of meiosis, and precocious meiosis. Apomeiotic divisions can range from almost meiotic to typical mitotic. In *Erigeron karwinskianus* 80% of the megasporocytes form restitution nuclei in anaphase I (Battaglia, 1950). The movement of the chromosomes in anaphase I is so erratic and scattered that by telophase I they are spread along the entire spindle (Fig. 19.3). The nuclear envelope forms adjacent to these scattered chromosomes and eventually will enclose all chromosomes within one single restitution nucleus rather than within the normal two. This leads to the unreduced chromosome number in some forms of diplospory.

Apomeiosis can also be caused by asynapsis. If the chromosomes do not pair, they remain as univalents and become enclosed in a diploid restitution nucleus. This system is typical for some species of *Taraxacum*. A mitotic division follows after restitution, and a dyad with the somatic chromosome number in each nucleus is formed. The embryo sac develops from one of the two dyad cells (Schnarf, 1929; Rosenberg, 1930; Gustafsson, 1932, 1935a, 1935b, 1937). Other examples of asynapsis with subsequent formation of two or four nuclei formed from divisions of the archesporial cell are *Artemisia, Eupatorium* and *Poa serotina*.

19.1.2.4 Pseudogamy.   Somatic apospory and diplospory are generally linked with **pseudogamy** (Focke, 1881) in that they require pollination. This kind of pollination does not lead to fertilization of the egg but is necessary to stimulate embryo formation. In pseudogamy a male gamete is necessary for embryo, endosperm and seed formation. Pseudogamy has been reported in species of *Poa, Rubus, Potentilla, Ranunculus, Hypericum, Parthenium, Citrus,* and *Allium*. Species that do not show pseudogamy are found in *Hieracium, Taraxacum, Antennaria, Crepis,* and *Calamagrostis* (Gustafsson, 1946, 1947a, 1947b). Some species can use pollen of related species for stimulation, as was discovered for *Pennisetum setaceum* by Simpson and Bashaw (1969).

19.1.2.5 Parthenogenesis.   **Parthenogenesis** is the development of an embyro from an ovum without the participation of a sperm. In parthenogenesis the embryo *always* develops from the female gamete or ovum. In **apogamety** a non-gametic 2n vegetative cell of the embryo sac (female gametophyte) produces an embryo (Fig. 19.1). The normal life cycle of a seed plant (Angiospermae) consists of an alternation of 2n sporophyte and 1n gametophyte generations (diplohap-

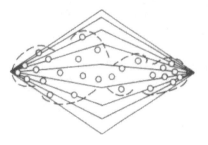

Fig. 19.3. Apomeiosis leading to the formation of a **restitution nucleus** at telophase I. The nuclear envelope (dashed line) forms around the scattered chromosomes and encloses all of them in a single restitution nucleus. (From Brown, 1972. Redrawn by permission of the C. V. Mosby Company, Saint Louis, Missouri).

lonts, Chapter 8, introduction). These generations are separed by the events of meiosis and fertilization that alternately reduce and restore the somatic chromosome number (amphimixis, Fig. 19.1). Agamospermy then is the by-passing of meiosis and fertilization in the process of embryo and sporophyte formation.

According to the embryological pathways chosen, there exist four different avenues leading to the formation of agamospermous seed (Fig. 19.1). These are:

1. diplosporic parthenogenesis
2. diplosporic apogamety
3. aposporic parthenogenesis
4. aposporic apogamety

Here we are concerned with the two forms of parthenogenesis (1 and 3). Diplosporic parthenogenesis occurs in quite a few species of the Compositae and in many other families. Grant (1971) gives a summary of the species in which it has been observed. Aposporic parthenogenesis occurs in eight different species of *Crepis*, in numerous species of *Hieracium*, in some Rosaceae and others.

19.1.2.6 Apogamety.   As already mentioned in apogamety, a vegetative cell in the embryo sac produces an embryo. Apogamous development of embryos has been reported for *Taraxacum, Hieracium, Alchemilla, Alnus,* and *Poa,* in which the vegetative embryo sac cell develops from one of the synergids (see Fig. 8.7). Embryo development from one of the antipodals has been observed in *Hieracium, Elatostema,* and *Allium* (Gustafsson, 1946, 1947a, 1947b).

19.1.2.7 Apomixis and Polyploidy.   Polyploidy is found in many apomicts. Many polyploids would not have survived without apomixis. Within groups of plants, the diploid species may have entirely sexual behavior while their polyploid relatives are mainly apomicts. But there are exceptions. In the genera *Allium, Agave,* and *Lilium,* diploid species with the vegetative form of apomixis, vivipary, do occur. But in a large number of genera, the viviparous species are nearly all polyploids. Examples are *Polygonum viviparum* (x = 10, 2n = c.88, c.100, c.110, c.132, Darlington and Wylie, 1955; Flovik, 1940; Löve and Löve, 1948; Skalinska, 1950), *Ranunculus ficara* (x = 7, 2n = 32, c.40. Böcher, 1938; Maude, 1939); *Cardamine bulbifera* (Stebbins, 1950), *Saxifraga* spp., various species of *Festuca, Poa, Deschampsia,* and other Gramineae.

Among the many groups of gametophytic apomicts, the polyploids far outnumber the diploids. Examples are *Potentilla, Hieracium,* and *Ranunculus* (Stebbins, 1950). Other groups are exclusively polyploid. Stebbins (1941) lists 24 gametophytic apomicts that are polyploids. Later, he mentions four additional ones, *Parthenium, Rudbeckia, Paspalum,* and *Crataegus* (Stebbins, 1950).

19.2   Parthenogenesis in Animals

In higher animals apomixis occurs almost exclusively as parthenogenesis. There are two kinds of parthenogenesis in animals: **haploid parthenogenesis** and **diploid parthenogenesis**.

Haploid parthenogenesis often occurs in the form of male haploid genetic systems in which the males arise by parthenogenesis from unfertilized eggs. **Male haploidy** is restricted to only a few higher taxa (Hartl and Brown, 1970). It occurs in some insect families of the orders Hymenoptera, Homoptera, Coleoptera, and Thysanoptera. It also occurs in the aquatic Rotifera and the Acarina mites. In the Hymenoptera, for instance, the haploid eggs parthenogenetically develop into males while the fertilized eggs develop into females. Spermatogenesis in the haploid males occurs without chromosome reduction. Usually only one equational meiotic division occurs. Haploid parthenogenesis is almost exclusively **facultative parthenogenesis,** since an egg may either be fertilized or develop parthenogenetically. In diploid parthenogenesis, only diploid females are produced from unfertilized eggs. They are genetically identical to their mothers. Two types of diploid parthenogenesis exist: **obligatory parthenogenesis** and **cyclical parthenogenesis.** In obligatory parthenogenesis, this is the only form of reproduction. The population is entirely made up of females. If an occasional male is found, its presence is not a prerequisite for species survival. Obligatory parthenogenesis in animals is often associated with polyploidy.

In cyclical parthenogenesis, diploid parthenogenesis alternates with sexual reproduction. Cyclical parthenogenesis has been demonstrated in the Trematoda, Rotifera, Cladocera, aphids, Diptera, Coleoptera, and Hymenoptera.

A classic example for cyclical parthenogenesis is the aphid *Tetraneura ulmi* (White, 1973). Fertilized eggs of this aphid overwinter in Europe, and in spring each egg develops by viviparous parthenogenesis into a small female nymph that forms a gall on the leaves of the primary winter host plant, the European elm. Inside the galls, the nymphs develop into adult wingless female aphids called **fundatrices** (Fig. 19.4). Each fundatrix parthenogenetically produces about 40 winged daughters called **emigrantes.** These make their way out of the galls and migrate to the roots of the secondary summer host food plants, which are various species of grasses. While on these plants, the female emigrantes parthenogenetically produce several generations of female **exules.** The last generation of exules includes winged sexual males and parthenogenetic females (**sexuparae**). These fly back to the primary winter host, the elm, where they parthenogenetically produce male and female **sexuales.** These then pair and produce the fertilized eggs that overwinter on the elm.

The female fundatrices, emigrantes, exules, sexuparae, and sexuales of *T. ulmi* all have 2n = 14. But the male sexuales have 2n = 13 (Schwartz, 1932). This constitutes an XO:XX sex-determining mechanism since the females have 2X and the males 1X. At the end of the warm season, the sexuparae parthenogenetically produce two kinds of eggs that will develop into female and male sexuales. From one kind of egg, a female with a normal 2n = 14 number of chromosomes develops by a single maturation division that will produce n = 6 + X eggs. The details of this ameiotic parthenogenetic egg maturation process had been unresolved for a long time. They have been elucidated by Cognetti (1961a, 1961b, 1961c, 1962) and Pagliai (1961, 1962) for *Brevicoryne brassicae, Macrosiphum rosae, Myzodes persicae,* and *Toxoptera aurantiae.* Synapsis and bivalent pairing occur normally. However, the bivalents then separate again into univalents without spindle for-

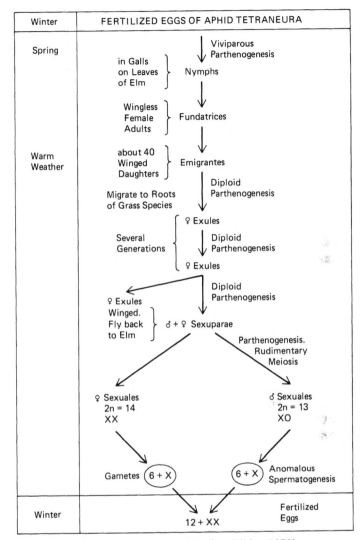

Fig. 19.4. Cyclical parthenogenesis in *Tetraneura ulmi*. (After White, 1973).

tion. Subsequently, a single mitotic maturation division occurs with the diploid chromosome set, and the formation of a diploid polar body aids in the elimination of the extra set of chromosomes. The polar kinesis has been described fairly early (Blochman, 1887; Stevens, 1905; Tannreuther, 1907). The other kind of egg develops by a **rudimentary meiosis.** Only the two X chromosomes pair and segregate meiotically. One of the two X chromosomes goes to one cell pole and remains single. At the other pole, all 12 autosomes and the other X chromosome come together and form the XO eggs that develop into male sexuales.

The spermatogenesis of the male sexuales is very anomalous. The autosomes pair and form a first meiotic metaphase plate with the X remaining univalent. During anaphase I, the X becomes stretched between the two poles, but eventually passes undivided into one of the two daughter nuclei. Two kinds of secondary spermatocytes become established, one possessing an X, the other without. The X-possessing secondary spermatocyte receives much more cytoplasm and undergoes meiosis II. The minus-X spermatocytes degenerate. Only one kind of sperm is produced, the plus-X ones; Ris (1942) studied this anomalous spermatogenesis in the aphid *Tamalia*.

The response of this parthenogenetic cycle to the weather is interesting. During the warm summer season, the parthenogenetic proliferation is observed. As the weather cools off, unfavorable conditions trigger the intervening sexual generation. With warmer spring temperatures, favorable conditions again trigger the parthenogenetic cycle.

19.2.1   Experimental Induction of Parthenogenesis in Animals

In the vertebrates, natural parthenogenesis is generally unknown, but some vertebrates have been artificially induced to reproduce parthenogenetically. Two basic requirements are necessary for the induction of parthenogenesis in animals. They are: egg activation and doubling of the chromosome number.

The egg can be activated by the induction of **gynogenesis** which is female parthenogenesis. After fertilization of the ovum, the male nucleus is eliminated and the resulting haploid individual has only a maternal chromosome set. Gynogenesis can be induced if a genetically inactivated spermatozoon is used to activate the egg. The spermatozoon can be inactivated by the application of x-rays, radium, ultraviolet rays, photodynamic chemicals in the presence of light, and nitrogen mustard. Gynogenesis is induced routinely in amphibians for the production of haploid cells (Selman, 1958).

Chromosome doubling can be achieved by the suppression of the polar body or of the first cleavage division. Abrupt heat or cold treatment gives best results. Such temperature treatments also may activate the egg so that parthenogenesis sometimes can be produced by a one step induction. Optimum temperature treatment lies within a very narrow range. The best time of chromosome doubling by heat or cold treatment is during the second polar body formation or at the first cleavage division.

Diploid parthenogenones (Beatty, 1957), which are products of diploid parthenogenesis, are subject to low viability. The number of deleterious recessive genes in heterozygous form is very high in animal populations. Beatty (1967) gives an example of this fact for the human population where one in 20,000 are homozygous albinos, yet about one in 70 individuals carries this gene in the heterozygous condition. In cattle (Falconer, 1960) one in 300 individuals manifests dropsy at birth but about one in 10 bulls is a heterozygous carrier of the gene. In haploid or diploid parthenogenesis, such deleterious genes affecting viability become uncovered and effective. Parthenogenesis causes the elimination of heterozygosity and the decrease of viability.

For instance, in haploid parthenogenones of amphibians, one can observe a typical **haploidy syndrome** (Hamilton, 1963). There is usually a delay of gastrulation. The embryos show microcephaly (abnormal smallness of head), lordosis (abnormally exaggerated forward curvature of the spine), and ventral odema (puffy swelling of the belly). After hatching, the animals are sluggish, inactive, and with abnormal muscle tissue. The animals have a high death rate, and almost all affected animals die at an early stage.

Adult parthenogenesis is known in fishes, amphibians, reptiles, and birds. Parthenogenetic embryos occur in mammals. Mice were inseminated with x-rayed and ultraviolet light treated spermatozoa, and embryos were examined in the blastocyst stage (Edwards 1954, 1957a, 1957b). Nitrogen mustard, toluidine blue, and trypaflavine also were applied (Edwards, 1954, 1958). At certain blastocyst mitoses at higher x-ray dosages, normal chromosomes were observed in haploid number, presumably the maternal ones, and in addition abnormal or fragmented ones, presumably the paternal ones. Development of the embryos was retarded. Similar experiments were carried out in rat, hamster, guinea pig, rabbit, sheep, and ferret (Beatty, 1967). Pincus (1939a, 1939b) claimed that he treated rabbit virgin eggs with hypotonic salts, transferred them into host females, and that two female young were born. Thirty years have passed since these first and only reports of the survival of such rabbit parthenogenones to birth, but no confirmation has been reported since.

Part IX
Extrachromosomal Inheritance

Parts II through VIII of this book dealt with the chromosomes as the hereditary determinants. However, it was recognized fairly early that the chromosomes are not the only carriers of genetic factors. Throughout the history of genetics, reports seemed to indicate that extranuclear elements could be possible agents of hereditary transmission. Wettstein in 1928 coined the expression **plasmon** with which he wanted to signify the cytoplasm as a hereditary agent. Part IX deals with such extrachromosomal genetic factors.

Chapter 20
Plastids, Mitochondria, Intracellular Symbionts, and Plasmids

A **plasmagene** (Darlington, 1939b) can be defined as an extranuclear hereditary determinant that shows non-Mendelian inheritance. Goldschmidt (1945) proposed that the term should be used only in such instances where a self-replicating unit in the cytoplasm produces definite genetic effects similar to those produced by genes in the chromosomes. The sum total of all plasmagenes constitutes the plasmon or the **plasmotype** (Imai, 1936). The plasmotype and the genotype are referred to as the **idiotype** (Siemens, 1921) or the entire genetic system of the cell. At the time the concept of the plasmon was established, the details of extra-chromosomal inheritance were not too well known. Now it seems to be clear that most of the cases of cytoplasmic inheritance could be included in one of the following three groups:
1. plastids and mitochondria
2. intracellular symbionts
3. plasmids

20.1   Plastids

The first evidence of cytoplasmic inheritance came from studies that involved plastid characteristics. In general, reciprocal crosses had shown the equality of genetic contributions from both parents. This was already shown by Kölreuter (1761–1766) (Chapter 1). But if cytoplasmic hereditary determinants are taken into account, the contribution from the egg is much greater than that from the sperm. Plastid characteristics are therefore inherited from the female parent. The discovery of such plastid inheritance was presented by Correns (1902, 1909). He studied the four-o'clock plant *Mirabilis jalapa*. The leaves of this plant have normal dark green chloroplasts but in the *albomaculatus* type, there are variegated leaf areas that have chlorophyll-deficient chloroplasts that cause pale green, pale yellowish, or white patches. Flowers on entirely dark green branches produce seeds that grow into normal dark green plants. Seeds that develop on variegated branches produce 3 kinds of progeny: green, variegated in variable proportions, and white. Seeds from entirely white branches produce progeny entirely deficient in chlorophyll. The pollen source is of no influence to the development of the progeny. Consequently, inheritance is entirely maternal and determined by cytoplasmic factors.

The evidence that hereditary material is included in the chloroplasts was finally shown in electron microscopic studies. In 1962 Ris and Plaut demonstrated by electron microscopic and cytochemical methods that chloroplasts in the alga *Chlamydomonas moewusii* actually contain DNA (Chapter 1). Chloroplasts are reported to have as much DNA as bacteria (Ellis, 1969). Reports of DNA in the alga *Acetabularia* are 10^{-16} grams per plastid (Green and Burton, 1970) and 10^{-14} grams per plastid in the algae *Chlorella euglena* and *Chlamydomonas* (Brown, 1972). The total DNA content of a chloroplast is usually about 10 times that of a mitochondrion (Granick and Gibor, 1967). The extranuclear hereditary determinants of chloroplasts can be likened to the genophores of bacteria. They very much resemble pure DNA and differ from nuclear chromosomes by carrying much less protein and by lacking histone in their structural organization. Plastid DNA as bacterial DNA is circular in structure (Manning et al., 1971; Sprey and Geitz, 1973; Ellis, 1974; Falk et al., 1974; Herrmann et al., 1974).

Replication of chloroplast and mitochondrial genophores was demonstrated by means of multiple **displacement loops** or **D-loops** (Kasamatsu et al., 1971; Kolodner and Tewari, 1973). These are short three-stranded closed circular DNA regions, approximately 10,000 base pairs apart. They expand undirectionally toward each other and upon meeting appear to initiate the formation of a double stranded replicative fork structure. Chloroplast DNA is unique in that it is specifically related to ribosomal and transfer RNA as revealed by DNA-RNA hybridization studies (Tewari and Wildman, 1970). Isolated chloroplasts can carry out protein synthesis (Heber, 1962). According to DuPraw (1970), intrachloroplast proteins sometimes account for 70% to 80% of the total leaf proteins. But protein synthesis in the chloroplasts is not entirely autonomous, as has been demonstrated in studies by Kirk (1966). Normally mutations in the plant nucleus chemically change the enzymes that synthesize chlorphyll. Consequently, not all chloroplast proteins are coded by chloroplast DNA. Apparently the synthesis of chloroplast structures is carried out by the combined effort of chloroplast genophores and nuclear genes.

20.2   Mitochondria

In 1940 Winge and Lautsen demonstrated mitochondrial inheritance of germination in yeast, *Saccharomyces*. Ephrussi et al. (1949) and Ephrussi (1953) induced the so-called *petite* (*p-*) type mutant in the yeast, *S. cerevisiae,* using the DNA-specific acriflavine dye, tetrazolium, and ultraviolet radiation. This caused respiration deficiencies affected by an inability of the mitochondria to synthesize certain respiratory enzymes. Slonimski (1949) demonstrated that mitochondrial respiration almost completely ceased; this was confirmed by Yotsuyangi (1962) who also noticed the lack of cytochromes of certain dehydrogenases and membranes. It is now well established that these deficiencies are caused by large deletions or even complete loss of mitochondrial DNA. Such loss has been genetically demonstrated by the absence of one, several, or all genetic markers in the mitochondrial genophore (Faye et al., 1973; Nagley and Linnane, 1972; Nagley et al.,

1973; Uchida and Suda, 1973). The deletions are compensated by repetition of nondeleted genophore segments of mitochondrial DNA (Hollenberg et al., 1972; Van Kreijl et al., 1972; Faye et al., 1973).

Additional evidence linking mitochondria to the inheritance of certain characteristics was the discovery of the so-called *poky* type mutants in the fungus *Neurospora crassa* (Mitchell and Mitchell, 1952). This slow growth characteristic, which cannot be supplemented by growth factors, is inherited only through the female and shows no transmission patterns through the male.

A significant finding similar to that in chloroplasts was the discovery of mitochondrial DNA under the electron microscope (Nass and Nass, 1962, 1963) (Chapter 1). Establishment of the fact that mitochondrial DNA like plastid DNA can be circular in shape followed soon (Kroon et al., 1966; Nass, 1966; Sinclair and Stevens, 1969). But not all mitochondrial DNA is circular as was reported for *Neurospora* (Shapiro et al., 1968) and *Phaseolus* (Kirschner et al., 1968). Restriction enzyme mapping of mitochondrial DNA has been reported for *Neurospora* (Bernard et al., 1975a, 1975b), yeast (Sanders et al., 1975a, 1975b), mouse, monkey, and humans (Brown and Vinograd, 1974; Robberson et al., 1974). It is now well established that the mitochondrial DNA molecules of fungi and plants are substantially larger ($50\text{-}70\text{x}10^6$ daltons) than those of vertebrates ($10\text{x}10^6$ daltons) (Dujon and Michaelis, 1974). That the gene sequence of DNA within the mitochondrion is completely different from that in the nuclear genome of the same species has been established by Tabak et al. (1973) and Flavell and Trampe (1973).

20.3   Intracellular Symbionts

Symbiotic organisms are those that have established such an intimate relationship with their host cells that they behave as if they were cellular inclusions. Those symbionts are subject to hereditary transmission and it is difficult to distinguish whether they are subject to heredity or infection.

20.3.1   The P-Particles of *Paramecium*

This is a group of some 10 different Gram-negative bacterial symbionts including *kappa, gamma, delta, pi, mu, lambda, alpha,* and *tau* that occur in the cytoplasm of *Paramecium aurelia* and are called collectively **P-particles** (Sonneborn, 1959). They were first reported by Sonneborn in 1938. They are comparatively large particles that consist of DNA, RNA, protein, and related substances. The first demonstration of their effect was presented by Sonneborn in 1943 when he showed that cytoplasmic *kappa* particles caused a killer trait in *Paramecium*. But the killer trait was also dependent on the genotype of the *Paramecium* cell. Reproduction of *kappa* particles occurs only in cells containing the genes K, s_1, and s_2. Cells with the genotype Kk contain only half the number of *kappa* particles that are contained in cells with genotype KK. About 400 particles per cell were postulated to be required for the killer effect. When Preer (1950) succeeded in stain-

ing the particles by the DNA specific Feulgen method, the expected number of *kappa* particles was observed microscopically. Some races of *Paramecium,* called **killers,** produced substances (paramecin) that had a lethal effect on members of other races, called **sensitives.** A medium that contains killers for a time and is then replaced by sensitives causes the sensitives to be killed. A certain killer strain, say *kappa,* is protected against the killer activity of its own homologous *kappa* particles, but it is sensitive to killer particles of other types like *pi* or *mu.*
The killer particles have different modes of killing sensitives. There are 6 different killer types (Siegel, 1953; Sonneborn, 1959):
1. *Vacuolizer.* The killer eliminates the sensitive by vacuolizing it.
2. *Humper.* The killer causes the sensitive to form humps.
3. *Spinner.* The killer makes the sensitive rotate.
4. *Paralyzer.* The killer paralyzes the sensitive.
5. *Rapid Lysis.* The killer eliminates the sensitives very quickly.
6. *Mate Killer.* Killing of sensitives occurs only during conjugation.
Work with symbionts in paramecium has been reported by Preer et al. (1972), Franklin (1973), Gibson (1973), Karakashian and Karakashian (1973), and Soldo and Godoy (1973). *Lambda* killer particles seem to contain multiple copies of each DNA sequence. This is in contrast to most free-living bacteria, which have only one or a few copies of a given DNA sequence. Soldo and Godoy speculated that this may be a consequence of adaptation resulting from prolonged intracellular existence. *Mu* particles could be removed from *Paramecium* with penicillin, which seems to prove that some killer particle cell walls are similar to those of certain bacteria (Stevenson, 1965; Franklin, 1973). *Kappa* particles, however, were unaffected by penicillin (Williamson et al., 1952). It has been discovered that a certain percentage of every *kappa* particle population contains so-called proteinaceous **R-bodies,** which are refractile inclusions consisting of thin ribbons of protein that are wound into a tight roll of 10 to 12 turns. R-bodies are responsible and essential for the toxic action of *kappa* particles on sensitive paramecia. After sensitives ingest *kappas,* the *kappas* begin to break down in their food vacuoles. The freed R-bodies then suddenly unroll or unwind into a long twisted ribbon $15\mu m$ long, 0.2–$0.5\mu m$ wide, and 12 nm thick. The food vacuole membrane breaks down and the killing process is initiated (Jurand et al., 1971). The remarkable structure and behavior of the R-bodies is unparalleled with any other bacterial structure known. According to Preer et al. (1966), no other bacterial structure is able to undergo such extensive and reversible changes in form. The bacterial nature of *kappa* particles has been established by their electron microscopic structure and chemical composition (Dippell, 1959; Smith-Sonneborn and Van Wagtendonk, 1964; Kung, 1971).

20.3.2   The Sigma Virus in *Drosophila*

This virus was discovered in *Drosophila melanogaster* in 1937 by L'Héritier and Teissier. It makes its host CO_2-sensitive. *Drosophila* can be easily anesthetized with CO_2 and usually recover fast and completely when the CO_2 is removed. But certain *Drosophila* strains were discovered that become permanently paralyzed by CO_2 exposure. Such strains occur naturally in different countries of the world

(Kalmus et al., 1954). Sensitive strains of *Drosophila* after brief CO_2 exposure become "drunk" or uncoordinated and some of the legs become paralyzed. Reciprocal crosses have been carried out that demonstrated that inheritance was mainly maternal. Repeated backcrossing of sensitive females to normal resistant males yielded sensitive offspring only. Backcrossing of normal females to sensitive males produced a few sensitive progeny, but mostly the trait was not passed on.

The virus can be injected in the form of extracts into normal resistant flies in order to induce CO_2 sensitivity. Such injection infected females do not regularly transmit their sensitivity to the offspring; injected males never do. Strains infected by injection are called **nonstabilized lines.** In the original **stabilized lines** the virus is believed to be located intracellularly in the germ line in a noninfectious form, maybe as naked nucleic acid, and is transmitted hereditarily (L'Héritier, 1962; Seecof, 1968). Infective virus could be produced by the maturation of the noninfectious form. *Drosophila* has nuclear genetic resistance to the *sigma* virus. Flies homozygous for a resistance factor *re* do not become infected after inoculation with *sigma* virus (Gay and Ozolins, 1968).

The *sigma* virus could not be isolated (Plus, 1962; Seecof, 1962), but electron microscopy has demonstrated it in *sigma*-bearing lines (Berkaloff et al. 1965). The virus particles are rod-shaped, 7 nm by 140 nm in size. Plus (1963) suspected that *sigma* is a DNA virus.

20.3.3   The Maternal Sex-Ratio Condition in *Drosophila*

The progeny of some *Drosophila* strains is entirely female. The original accounts of this trait were governed by nuclear genes as reported earlier (Section 18.1). But instances of maternally inherited sex ratios (*SR*) conditions are also known. Such conditions were reported for several species of *Drosophila*, specifically *D. bifasciata, D. prosaltans, D. willistonii, D. paulistorum, D. equinoxialis, D. nebulosa,* and *D. robusta* (Magni, 1954; Cavalcanti et al., 1958; Malogolowkin, 1958, 1959; Oishi and Poulson, 1970). Death of males was found mainly in early embryonic stages but usually as zygotes. Maternal sex ratio condition is either revealing total absence of males in the progeny, as in *D. willistonii,* or predominantly female with a few male offspring, as in *D. paulistorum.*

The sex ratio condition in *D. willistonii, D. equinoxialis, D. nebulosa,* and *D. paulistorum* has been linked to the occurrence of treponema-like *SR spirochaetes* that are $5\mu m$ to $6\mu m$ long and 0.1-$0.2\mu m$ wide in their filamentous stage and exhibit a typical spiral form (Malogolowkin, 1958; Malogolowkin et al., 1960; Poulson and Sakaguchi, 1960; Oishi and Poulson, 1970). If SR spirochaetes of any of these *Drosophila* species are mixed either *in vitro* or *in vivo,* spirochaetes of one or of both species die (Sakaguchi and Oishi, 1965; Sakaguchi et al., 1965; Oishi and Poulson, 1970). Oishi and Poulson have demonstrated that the cause of death of these spirochaetes is a DNA-containing spherical virus of 50nm to 60nm in diameter. Preer (1971) suggested that the same virus may be the agent that kills developing male *Drosophila* and causes the development of entirely female strains. In the species *D. bifasciata* no spirochaete bacteria are present, and a virus is suspected of killing the male zygotes instead (Ikeda, 1965; Leventhal, 1968). Another *Drosophila* species with sex ratio condition not caused by spirochaetes is *D. robusta* (Poulson, 1968). The cause of male death in *D. prosaltans* has not been completely resolved (Poulson, 1963). The SR spirochaetes are believed to be **mycoplasma**-related (Williamson, et al., 1977).

Mycoplasmas are the simplest known cellular organisms. Their size overlaps with the largest viruses and the smallest bacteria. They have a circular DNA molecule that is not separated from the remainder of the cell. Unlike viruses, they do not require host cells for duplication. Their plasma membrane is not surrounded by an elaborate cell wall as in bacteria (Novikoff and Holtzman, 1976).

20.3.4   The Milk Factor in the Mouse (MTV)

Bittner (1938, 1939) is credited with discovering an extranuclear factor responsible for the susceptibility of mice to mammary cancer. The factor was shown to be transmitted through the milk. It was believed to be related to the viruses, being a particulate nucleoprotein (Barnum et al., 1944). It is now known as the mouse mammary tumor virus (MTV). Moore (1967) and Hageman et al. (1968) established that the virion of MTV is the B particle, which was described by Bernhard (1958). The presence of MTV has been demonstrated in the gametes of mice by electron microscopy, immunofluorescence, and bioassay (Bentvelzen et al., 1970). This form of cancer was detected by outcrossing strains of mice that were inbred for many generations. About 90% of the mice of these inbred strains over 18 months of age had breast cancer. When females of these strains were crossed with inbred strains that had low incidence of cancer, 90% of the F_1 individuals had breast cancer. If the reciprocal cross was carried out, none of the F_1 had breast tumors. Mice of a cancerous line fed from birth by noncancerous foster mothers did not show evidence of tumors. But after injection with blood from cancerous mice, they did develop tumors (Woolley et al., 1943). Bentvelzen et al. concluded that host genes control the susceptibility to MTV.

20.3.5   Cytoplasmic Male Sterility (CMS)

CMS has been reported for 80 species, 25 genera, and six families (Edwardson, 1970). Viruses can be transmitted by the seed and induce pollen sterility. This has been demonstrated with the tobacco ringspot virus (TRSV) in *Petunia* (Henderson, 1931) and with the tomato ringspot virus (TMSV) in soybeans (Kahn, 1956). Atanasoff (1964a, 1964b) suggested that all cytoplasmically inherited traits such as cytoplasmic male sterility (CMS) could be due to viral infections even though the presence of a virus has not been demonstrated. The successful asexual transmission of CMS through plant grafts demonstrated in *Petunia* (Frankel, 1956, 1962, 1971; Edwardson and Corbett, 1961; Bianchi, 1962) and in sugar beets (Curtis, 1967) supports the possible assumption that CMS could be transmitted by virus. CMS has been demonstrated in many plant species. A plausible definition for cytoplasmic male sterility is a condition in which pollen sterility is at least partially caused by factors that are only passed on by the female and in which this pollen sterility is not abandoned during successive reproductive generations. The expression of this condition can also be influenced by chromosomal genes. The expression of cytoplasmic male sterility can often be changed by the conditions of the specific environment in which the plants are grown. The present usability or the potential use of CMS for the production of hybrid seed has been demonstrated in tobacco (Chaplin, 1964), maize (Rogers and Edwardson, 1952), sorghum (Ross, 1971), and wheat (Wilson and Ross, 1962). Other reports of economic possibilities with CMS are for flax (Chittenden, 1927), onion (Jones and Clarke, 1943), alfalfa (Davis and Greenblatt, 1967), petunia (Edwardson and Warmke, 1967), sugar beets (Theurer and Ryser, 1969), intermediate wheatgrass (Schulz-Schaeffer, 1970), cotton (Meyer, 1971), and other crops.

The phenomenon of cytoplasmic male sterility was first described by Correns in 1904 and was interpreted as a true case of CMS by Wettstein in 1924. There are 4 possible sources of CMS:

1. intergeneric hybridization and substitution backcrossing
2. interspecific hybridization and substitution backcrossing
3. intraspecific hybridization
4. spontaneous occurrence

A good example for the development of cytoplasmic male sterility through intergeneric hybridization is wheat. Kihara (1951a) crossed *Aegilops caudata* with *Triticum aestivum* and backcrossed the hybrid with *T. aestivum*. Since the female parent contributes the majority of the cytoplasm, the backcrossing accomplished a replacement of the *Aegilops* chromosomes by *Triticum* chromosomes, which were placed in the *Aegilops* cytoplasm. CMS was the result of this method.

Interspecific hybridization was the approach of obtaining CMS in tobacco when Burk (1960) placed the genome of *Nicotiana tabacum* into the cytoplasm of *N. bigeloni*. Cytoplasmic male sterility in onions was accomplished by intraspecific hybridization (Jones and Emsweller, 1937; Jones et al., 1939; Jones and Clarke, 1943; Jones and Davis, 1944; Jones, 1946). A recessive gene (*ms*) in the homozygous state caused plants to be male sterile when by appropriate crossing it was placed into a certain type of cytoplasm (*S*). The same gene had no effect in a different type of cytoplasm (*N*).

And finally, the most prominent example for spontaneous occurrence of CMS is maize. The Texas source (T-type) was isolated from the cultivar "Golden June", a variety of dent maize that was grown in the southwestern United States (Rogers and Edwardson, 1952). Another cytoplasm used in maize breeding is the S-type. Maize inbred lines that were sterile in the S-cytoplasm were not necessarily sterile in the T-cytoplasm. Because a higher percentage of corn belt inbred lines were completely sterile in the T-cytoplasm than in the S-source, most hybrid production was based on the T-cytoplasm (Duvick, 1966). It has been established that two plasmid-like DNA's are unequally associated with mitochondrial preparations from S-type maize (Pring et al., 1977).

20.4   Plasmids, Episomes, and Transposable Elements

Plasmids (Lederberg, 1952) in the strictest sense are extrachromosomal hereditary determinants that only occur in an autonomous condition and exist, replicate and are transferred independent of the chromosomes. This concept excludes the category of **episomes,** which, in contrast, are hereditary determinants that can alternate their autonomous existence with a condition in which they are attached to the chromosomes. Interpreted this way, an episome that is detached from its chromosome becomes a plasmid (Hayes, 1968; Novick, 1969; Richmond, 1970). The term episomes was coined by Jacob and Wollman and was adopted from *Drosophila* genetics (Thompson, 1931). **Insertion sequence elements** (*IS*) and **transposons** are much smaller than plasmids and episomes. Three classes of elements that are able to insert at different sites of DNA molecules can be distinguished:

1. *Conjugons* (Luria, 1963). These are specialized genetic promotor elements that are necessary during bacterial conjugation in order to establish contact between cells. Cells that possess conjugons (**donor cells**) can establish contact with related cells that lack such episomal conjugons (**recipient cells**).
2. *Temperate bacteriophages* (Jacob et al., 1953). These are bacterial viruses that can enter lysogenic bacteria and, in contrast to virulent or lytic bacteriophages that destroy their host by the process of lysis, they become integrated into the bacterial genophore or become attached to it without damage to the host. Phages when integrated by this process are called **prophages** (Lwoff and Gutman, 1950). Lysogenic bacteria become immune to further infection by an extrinsic homologous phage after integration of a prophage. The sites of integration of prophages into the bacterial genophore (**prophage sites**) are designated on the genetic map by a Greek letter.
3. *Transposable genetic elements.* These are the IS elements and transposons. They are small, nonreplicating transposable DNA that can be inserted into the DNA of temperate bacteriophages or plasmids. They have not been found to exist autonomously.

According to Rieger et al. (1976), conjugons have three functions:

1. the determination of the surface properties and synthetic abilities associated with the establishment of effective contacts between conjugation partners
2. mobilization of the genophore material that is to be transferred from the donor cell to the recipient cell
3. provision of the energy source required for genophore transfer

20.4.1 The F-Episomes

Also called the F-agent, this conjugon is an episome or a transmissible plasmid. At this point it is important that the student understands all these different terms because they are used interchangeably in the literature. The F-episome is part of a group of sex factors that are capable of inducing conjugation. The presence or absence of these factors in a bacterial cell determines its sex.

Depending on their organization, there are 4 kinds of bacterial cells (Demerec et al., 1966):

1. F^- cell
2. F^+ cell
3. Hfr cell
4. F' cell

An F^- cell is a bacterium that does not posses an F-episome. Such a bacterium is a genetic recipient cell or conjugation "female" that does not possess the ability of a genetic donor cell or "male." Such a cell cannot transmit an F-episome but it can be infected with such an episome (Fig. 20.1).

An F^+ cell is a bacterium that carries an F-episome extrachromosomally. Such

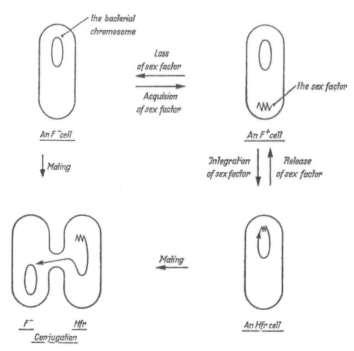

Fig. 20.1. Different manifestations of the F-episome in *Escherichia coli.* (From Scaife, 1963).

episome can be transferred to an F⁻ cell with high frequency. Such an F⁻ cell then becomes an F⁺ cell.

An Hfr cell is a bacterium that has an F-episome integrated in its bacterial genophore (Fig. 20.1). An F⁺ cell can convert into an Hfr cell by a process of integration (Demerec et al., 1966). The integrated F-episome confers on the Hfr cell the ability of **high frequency recombination** (Hfr). Due to transient coupling between F and the genophore, the frequency of genetic recombination in matings between F⁺ and F⁻ cells is about 10^{-6}, but in matings between Hfr and F⁻ cells, it is as high as 0.01% to 0.5%. The process of bacterial conjugation involves direct contact between the donor and the recipient, which is followed by the establishment of a cellular bridge that enables transfer of the entire male genophore, or only a segment, into the recipient cell. Transfer usually results only in a **merozygote** (Wollman et al., 1956), which is diploid for only part of the genophore and haploid for the rest of it. This process of mating is illustrated in Fig. 20.1. The transfer occurs only after breakage of the circular genophore. Breakage happens at one of the two insertion points of the F-episome. This point is referred to as "*O*" (origin) or head. This portion of the genophore is always the first to enter the recipient cell. The position of the Hfr-episome is at the end of the transferring linear genophore. Since the genophore usually transfers only in part, the integrated F-episome is included only rarely in the transfer. *O* always marks the head and Hfr the tail of the transferring genophore. But the sequence of the genetic markers of each Hfr-strain is strain-specific since Hfr can integrate at different sites of the genophore (Hayes, 1964). If the markers of the genophore are designated by the letters of the alphabet then the following sequences would be possible:

O-A-B-C-D-E-F-
O-B-C-D-E-F-A-
O-D-E-F-A-B-C-
O-F-A-B-C-D-E-

or other sequences.

The integration of the F-episome into the *Escherichia coli* genophore is not entirely random. Certain sites on the *E. coli* map are preferred.

An F′ cell is a bacterium that carries an extrachromosomal F-episome attached to a genophore fragment. F′ episomes are also called **F-merogenotes** (Clark and Adelberg, 1962) or **F-genotes** (Ramarkrishnan and Adelberg, 1965). An F′ episome carrying a genophore fragment can interact at a specific site on the bacterial genophore. An F′ cell is haploid except for a short genophore segment in which it is partially diploid and heterozygotic. Such a cell is also called a **heterogenote** (Morse et al., 1956a). F′ episomes arise when excision during release of the F-episome from an Hfr is not exact.

20.4.2   Colicinogenic Factors

Fredericq (1953, 1954) discovered extrachromosomal genetic elements in coliform bacteria that produce proteinaceous substances called **colicins** capable of killing sensitive members of other bacterial strains of the same or closely related species. He called these extrachromosomal elements **Colicinogenic factors** (Cf). He proved that Cf can be transferred from colicinogenic (col⁺) to noncolicinogenic (col⁻) strains by cell contact. The col-factors are a heterogeneous group of plasmids, and they are able to carry out a wide variety of activities such as colicin production and release, production of colicin immunity, and quiescent and vegetative reproduction. Herschman and Helinski (1967) described two quite distinct classes of col-factors: colicinogenic sex factors and nontransmissible col-factors. A well studied example of a colicinogenic sex factor is Col V-k 94. It can mobilize the genophore but it is not integrated. In addition to a sex factor, it carries a structural gene for a colicin and a gene for resistance or immunity to the colicin. Col V-k 94 was studied more extensively than any other colicinogenic sex factor. According to electron microscopic studies by Bradley (1967), nontransmissible col-factors such as colE1 are defective phages that have the ability to produce incomplete, noninfective phages.

20.4.3   Resistance Factors

The resistance or **R-factors** (Iseki and Sakai, 1953) of the coliform bacteria are capable of conferring resistance to antibiotics or to metal ions (Summers and Silver, 1972). Such factors confer selective advantage to organisms growing in the presence of antibiotics or having metal complexes such as mercurials and have been isolated from hospital environments (Joly and Cluzel, 1975). There are two classes of resistance factors: nontransmissible R-factors and transmissible RTF sex factors.

Nontransmissible R-factors consist of DNA that provides resistance to one or several antibacterial drugs or antibiotics such as chloramphenicol, sulphonamide, tetracycline or streptomycin and/or several metal ions such as Hg, As and Cd. For instance, in studies of *Shigella* bacteria it was found that strains resistant to all four of these commonly used drugs were more common than strains resistant to only one, two, or three. **Multiple drug resistance** is now an established fact. Resistance to kanamycin, neomycin, and the penicillins is very often included (Lewin, 1977).

In order for an R-factor carrying bacterium to be able to conjugate with other bacterial strains and to transfer drug resistance to them, the R-factor has to be linked to a **resistance transfer factor** (RTF) (Watanabe and Fukasawa, 1961). RTF often carries resistance to ampicillin only. Such an R-RTF complex is very similar to the F-episome (Section 20.4.1) but its transmission frequency from donor to recipient is not as high as that of the F-episome. The functional relationship between R-RTF complexes and F-episomes is also demonstrated by Watanabe's (1963) discovery that one class of R-factors, fi$^+$ (= **fertility inhibition**), inhibits the genetic expression of the F-episome when it is acquired by the F$^+$ recipient cell. Inhibition of the F-episome expression caused by an fi$^+$ R-factor seems to be caused by a single gene in the R-factor (Hirota et al., 1964).

The presence of antibiotics in livestock feed and as therapeutic agents as well as the use of heavy metals such as mercury as disinfectants has contributed to selection of R-factors, and transmission to human or animal pathogens leads to the difficulty in treating subsequent infections (Meyers et al., 1975).

20.4.4   Bacteriophages

In Section 20.4, two classes of episomes were described; conjugons and temperate bacteriophages. One of the best known is the small bacteriophage *lambda*. It is composed of a protein coat containing a single chromosome that consists of a single double stranded DNA molecule. This chromosome can be either linear or circular. Genetically, this chromosome acts as if it were linear having a definite head and a definite tail. The very ends of lambda DNA are single stranded and complementary to each other in nucleotide sequence. They are referred to as **sticky ends** or **cohesive ends** (Ris and Chandler, 1963; Hershey et al., 1963) and are from 10 to 20 nucleotides in length. They can pair and form a closed circular structure (Fig. 20.2). Apparently, there is a small amount of protein associated with the lambda DNA. The phage lambda is about 17 μm x 2nm in size. When lambda is in its integrated state in the bacterial host cell, it usually occupies a unique position within the bacterial genophore (Fig. 20.3). It has been demonstrated that other temperate phages also have their own specific attachment sites on the bacterial genophore (Jacob and Wollman, 1957). As part of the bacterial genophore, prophage genes function in precisely the same manner as bacterial genes. When lysogenic bacteriophages become excised (released) from the bacterial genophore, they can enter a lytic cycle (Wollman, 1953). Such a cycle can

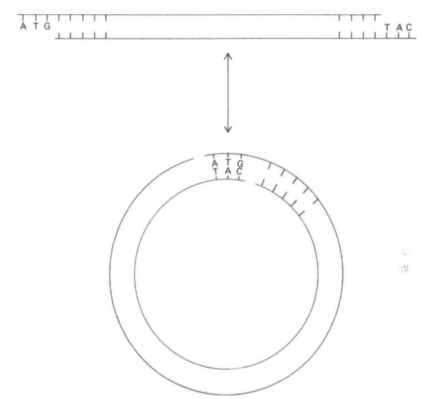

Fig. 20.2. Diagrammatic illustration of the lambda bacteriophage. (After Dr. C. A. Thomas, Jr. From Swanson et al., 1967. Redrawn by permission of Prentice-Hall, Inc., Englewood Cliffs, New Jersey).

be induced, for instance, by ultraviolet irradiation. Inexact excision of phages leads to a linkage of one or more bacterial genes to the phage and formation of specialized transducing phages. In **transduction** (Zinder and Lederberg, 1952), the phage transfers genetic material from a donor cell to a recipient cell. Since lambda is located close to the galactose genes (see *gal,* Fig. 10.1) of *E. coli,* these are some of the few genes that lambda usually transduces (Morse et al., 1956a, 1956b). The genetic material that can be transduced in such a manner is generally less than 1% of the total length of the bacterial genophore. If lambda picks up a *gal* segment from its previous host cell and transfers it to a new bacterium, the second host cell becomes diploid for the *gal* segment and becomes a **heterogenote.** Usually the phage leaves behind part of itself in the genophore of the first host cell and becomes defective. It cannot become released from the genophore and reproduce. But a defective lambda can be excised again if the new host cell contains another intact lambda helper phage. Transducing lambdas have been investigated physically and genetically. They have lower density, and large middle segments of the genetic map are missing (Arber, 1958; Weigle et al., 1959). Mapping of the lambda bacteriophage has been fairly extensive. According to Lewin (1977) it seems unlikely that any essential genes have not yet been identified. Campbell (1971) summarized genetic data obtained from several laboratories and presented a lambda gene map (Fig. 20.4). According to this map lambda has 26 essential and 9 nonessential genes. Other temperate bacteriophages that have been described are ϕ80, P1, P2, and Mu.

20.4.5   IS-Elements and Transposons

IS-elements are small, transposable DNA segments of 800 to 1400 base pairs. This compares with about 46,500 base pairs in the lambda bacteriophage. IS-elements can be inserted into bacterial genophores, temperate bacteriophages, or plasmids. IS-elements were first detected in the *E. coli galT* gene by Jordan et al. in 1968. They compared the density of mutated and wild-type λgal phages and could show that the mutated phages had increased density. After reversion to the wild-type, the density decreased to its usual value. Five of such IS-elements are now known in *E. coli*, others are known in *Salmonella* and *Citrobacter* (Starlinger, 1977). IS-elements can cause the effect of a special kind of mutation, called

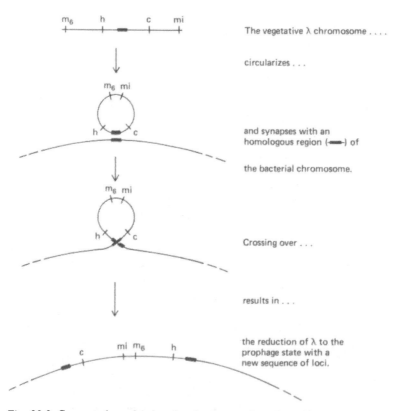

The vegetative λ chromosome . . . .

circularizes . . .

and synapses with an homologous region (——) of

the bacterial chromosome.

Crossing over . . .

results in . . .

the reduction of λ to the prophage state with a new sequence of loci.

Fig. 20.3. Suggested mechanism for the integration of lambda into the bacterial genophore. (From Franklin W. Stahl, The Mechanics of Inheritance, 2nd Ed., © 1969, p. 158. Reprinted by permission of Prentice-Hall, Inc., Englewood Cliffs, New Jersey).

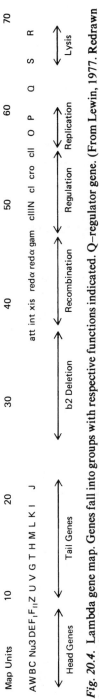

Fig. 20.4. Lambda gene map. Genes fall into groups with respective functions indicated. Q–regulator gene. (From Lewin, 1977. Redrawn by permission of John Wiley and Sons, New York).

polarity mutation (Franklin and Luria, 1961; Jacob and Monod, 1961b). A polarity mutation abolishes the function of the mutated gene and impairs the function of other genes within the same operon. But the polarity mutation caused by the insertion of an IS-element differs from a point polarity mutation in that it easily reverts to the wild-type condition. IS-elements have also been demonstrated under the electron microscope (Hu et al., 1975a, 1975b; Ptashne and Cohen, 1975; Saedler and Heiss, 1973; Saedler et al., 1975; Mosharrafa et al., 1976). They were shown by the **heteroduplex technique** originally developed by Davis and Davidson (1968) and Westmoreland et al. (1969). The strands of DNA molecules were separated and so-called **heteroduplex DNA molecules** were prepared by DNA-DNA hybridization (Hall and Spiegelman, 1961; Chapter 1). Heteroduplex DNA molecules are hybrid DNA double strands that consist of polynucleotide chains that originated from two different parental molecules. The presence of an IS-element in one of the two DNA strands can be seen under the electron microscope as a single strand loop.

IS-elements apparently do not exist in the form of autonomously replicating plasmids (Christiansen et al., 1973). Integration of IS-elements shows some site specificity that is intermediate between the strong specificity of lambda bacteriophage and the lack of it in Mu bacteriophage. Different IS-elements show different site specificity. IS *4* has been observed only at a single site in gene *galT* of *E. coli* (Shapiro and Adhya, 1969; Fiandt et al., 1972; Shimada et al., 1973; Pfeifer et al., 1977). IS *1* and IS *2* are less specific than IS *4*. Several other IS *4* sites have been reported. Integration sites of IS *2* and IS *3* in the F-plasmid have been mapped by Davidson et al. (1975) and by Hu et al. (1975a, 1975b, 1975c) (Fig. 20.5). IS-elements have been shown to be partially homologous to the inverted

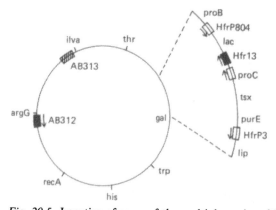

Fig. 20.5. Location of some of the multiple copies of IS-elements in the chromosome of *E. coli.* ■ represents IS2, ▤ represents IS3, ▨ is γ-δ. Arrows indicate the orientation of the IS-elements and the *Hfr* formed at these positions. The integration site of AB313 is taken from Ohtsubo et al. (1974) and the integration site of AB312 has been analyzed by Guyer (cited in Davidson et al., 1975). At the position of *Hfr P804*, seven other *Hfr* strains are known; at position of *Hfr 13* and the neighboring IS3, six other *Hfrs*, and at P3 three other *Hfrs* have been described. (From Davidson, Deonier, Hu and Ohtsubo, 1975. Redrawn by permission of the American Society of Microbiology, Washington, D.C.).

sequences that border the transposon for tetracycline resistance (Cohen and Kopecko, 1976). IS *1,* an element of 770 base pairs, has been analyzed by the nucleotide sequencing technique (Calos et al., 1978; Grindley, 1978).

Transposons (Hedges and Jacob, 1974) are very small DNA segments (100 to 1500 nucleotide pairs) consisting of one or several genes. They are capable of insertion and excision in DNA molecules without the requirement of a functional bacterial recombination system. They are responsible for resistance to antibiotics. They can be transposed from chromosome to chromosome in the same cell (Ptashne and Cohen, 1975; Berg et al., 1975; Berg, 1977a, 1977b). The nomenclature of transposons has been explained by Campbell et al. (1977). An example would be *Tn 9(Cm)* in which *Tn* is followed by the isolation number and the antibiotic (in parenthesis) to which the resistance is conferred, in this instance chloramphenicol. Gene mutation occurs through integration of transposons into a gene (Kleckner et al., 1975; Berg, 1977a). Reversion into wild-type generally occurs like that reported for IS-elements. Inexact excision of transposons can lead to deletions next to the transposon site (Foster, 1976; Campbell et al. 1977; Botstein and Kleckner, 1977; Brevet et al., 1977). Also duplications and inversions have been described occurring next to tetracycline transposon *Tn 10* (Botstein and Kleckner, 1977). Transposons also show different site specificities (Starlinger, 1977).

References

General References (Chapter 1)

Asimov, I. 1962. The genetic code. New Amer. Libr. World Lit., Inc., New York.

Baltzer, F. 1962. Theodor Boveri. Leben und Werk eines grossen Biologen. 1862–1915. Wissensch. Verl. Gesellsch. m. b. H., Stuttgart.

Brown, W. V. 1972. Textbook of cytogenetics. C. V. Mosby Co., Saint Louis, MO.

Cohn, N. S. 1969. Elements of cytology. 2nd ed. Harcourt, Brace and World, Inc., New York, N. Y.

Dampier, W. C. 1943. A history of science and its relations with philosophy and religion. Cambridge Univ. Press, Cambridge.

Darlington, C. D. 1964. Genetics and man. Schocken Books, New York.

Dawes, B. 1952. A hundred years of biology. Duckworth and Co., London.

De Beer, G. 1964. Charles Darwin. Doubleday, New York.

Dunn, L. C. (ed.) 1951. Genetics in the 20th century. Macmillan, New York.

Gabriel, M. L., and S. Fogel (eds.) 1955. Great experiments in biology. Prentice-Hall, Inc., Englewood Cliffs, N. J.

Gardner, E. J. 1960. Principles of genetics. John Wiley and Sons, Inc., New York.

Glass, B., O. Temkin and W. L. Strauss, Jr. 1959. Forerunners of Darwin, 1745–1859. Johns Hopkins Press, Baltimore.

Goldschmidt, R. B. 1956. Portraits from memory. Univ. Washington Press, Seattle.

Goldschmidt, R. B. 1960. In and out of the ivory tower; the autobiography of Richard B. Goldschmidt. Univ. Wash. Press, Seattle.

Hamerton, J. L. 1971. Human cytogenetics. Vol. I and II. Academic Press, New York, N. Y.

Hughes, A. 1959. A history of cytology. Abelard-Schuman, New York.

Jaffe, B. 1944. Men of science in America; the role of science in the growth of our country. Simon and Schuster, New York.

King, R. C. 1968. A dictionary of genetics. Oxford Univ. Press, New York, N. Y.

Nordenskiöld, E. 1935. The history of biology. A survey. Tudor Publishing Co., New York.

Rieger, R. A. Michaelis and M. M. Green. 1968. A glossary of genetics and cytogenetics, classical and molecular. 3rd ed. Springer-Verlag New York Inc.

Schrader, F. 1948. Three quarter-centuries of cytology. Science 107: 155–159.

Stern, C. 1968. Genetic mosaics, and other essays. Harvard Univ. Press, Cambridge.

Stubbe, H. 1972. History of genetics. Massachusetts Inst. Tech. Press, Cambridge, Mass.

Swanson, C. P. 1957. Cytology and cytogenetics. Prentice-Hall, Inc., Englewood Cliffs, N. J.

Taylor, G. R. 1963. The science of life. A picture history of biology. McGraw-Hill Book Co., Inc., New York, N. Y.

Wilson, E. B. 1925. The cell in development and heredity. 3rd ed. Macmillan Co., New York, N. Y.

Specific References

Abbé, E. 1886. Sitzber. Jen. Gesell. Med. Naturwiss. 1886:107.

Abo El-Nil, M. M., and A. C. Hildebrandt. 1973. Origin of androgenetic callus and haploid geranium plants. Can. J. Bot. 51:2107–2109.

Absate, M., and D. S. Borgaonkar. 1977. Chromosome arm involvement in interchanges. Lancet I:96.

Ahloawalia, B. S. 1969. Desynapsis in diploid and tetraploid clones of ryegrass. Genetica 40:379–392.

Al Aish, M. S., F. de la Cruz, A. Goldsmith, J. Volpe, G. Mella, and J. C. Robinson. 1967. Autosomal monosomy in man. Complete monosomy G(21–22) in a four-and-one-half-year-old mentally retarded girl. New Engl. J. Med. 277:777–784.

Albers, F. 1972. Cytotaxonomie und B-Chromosomen bei *Deschampsia caespitosa* (L.) P.B. und verwandten Arten. Beitr. Biol. Pflanz. 48:1–62.

Alexander, G. 1954. Biology. 7th ed. Barnes and Noble, New York.

Alexander, M. L. 1952. The effect of two pericentric inversions upon crossing over in *Drosophila melanogaster*. Univ. Texas Publ. 5204:219–226.

Alfert, M., and I. I. Geschwindt. 1958. The development of polysomaty in rat liver. Exptl. Cell Res. 15:230–235.

Alfi, O. S., G. N. Donnell, B. F. Crandall et al. 1973. Deletion of the short arm of chromosome 9(46,9p-): A new deletion syndrome. Ann. Genet. (Paris) 16:17.

Alfi, O. S., R. G. Sanger, A. E. Sweeny, and G. N. Donell. 1974. 46 del(9)(22:): A new deletion syndrome. *In:* Bergsma, D. (ed.). Clinical cytogenetics and genetics. Birth Defects: Orig. Art. Ser. 10:27–35.

Allard, R. W. 1953. A gene in lima beans pleiotropically affecting male-sterility and seedling abnormality. Proc. Amer. Soc. Hort. Sci. 61:467–471.

Allderdice, P. W., N. Browne, and D. P. Murphy. 1975. Chromosome 3 duplication q21→qter deletion p25→pter syndrome in children of carriers of a pericentric inversion inv (3)(p25q21). Amer. J. Human Genet. 27:699–718.

Allderdice, P. W., J. G. Davis, O. J. Miller, H. P. Klinger, D. Warburton, D. A. Miller, F. H. Allen, Jr., C. A. L. Abrams, and E. McGilvray. 1969. The 13q- deletion syndrome. Amer. J. Hum. Genet. 21:499–512.

Allderdice, P. W., O. J. Miller, P. L. Pearson, G. Klein, and H. Harris. 1973. Human chromosomes in 18 men-mouse somatic cell lines analyzed by quinacrine fluorescence. J. Cell. Sci. 12:809–830.

Allderdice, P. W. and T. A. Tedesco. 1975. Localization of the human gene for galactose-1-phosphate uridyltransferase to 3q21→qter by quantitative enzyme assay. Lancet 1975-II:39.

Allderdice, P. W., W. D. Heneghan and E. T. Felismino. 1976. 9pter→p22 deletion syndrome: A case report. Birth Defects: Orig. Art. Ser. 13:151–155.

Allen, N. S., G. B. Willson, and S. Powell. 1950. Comparative effects of colchicine and sodium nucleate. J. Hered. 41:159–163.

Al-Yasari, S. 1967. Studies of toluidine blue for inducing haploidy in *Lycopersicon esculentum* and *Zea mays*. Ph.D. Thesis, Univ. New Hampshire.

Al-Yasari, S. and O. Rogers. 1971. Attempting chemical induction of haploidy using toluidine blue. J. Am. Soc. Hort. Sci. 96:126–127.

Ames, B. N., and P. E. Hartman. 1963. The histidine operon. Cold Spring Harbor Symp. Quant. Biol. 28:349–356.

Anders, G., A. Prader, E. Hauschteck, K. Schärer, R. E. Siebenmann and R. Heller. 1960. Multiple Sex-Chromatin und komplexes chromosomales Mosaik bei einem Knaben mit Idiotie und multiplen Misbildungen. Helv. Paediat. Acta 15:515–532.

Anders, J. M., G. M. Jagiello, P. E. Polani, F. Gianelli, J. L. Hamerton, and D. M. Leiberman. 1968. Chromosome findings in chronic psychotic patients. Brit. J. Psychiat. 114:1167–1174.

Anderson, E. 1947. A case of asyndesis in *Picea abies*. Hereditas 33:301–347.

Anderson, E., and K. Sax. 1936. A cytological monograph of the American species of *Tradescantia*. Bot. Gaz. 97:433–476.

Anderson, E. 1968. Oocyte differentiation in the sea urchin *Arbacia punctulata*, with particular reference to the origin of cortica granules and their participation in the cortical reaction. J. Cell Biol. 37:514–539.

Anderson, E. G. 1943. Utilization of translocations with endosperm markers in the study of economic traits. Use of translocations in corn breeding. Maize Genet. Coop. Newsletter 17:4–5.

Anderson, E. G. 1956. The application of chromosomal techniques to maize improvement. Brookhaven Symp. Biol. 9:23–36.

Arber, W. 1958. Transduction charactères *Gal* par le bactériophage lambda. Arch. Sci. (Geneva) 11:259–338.

Archebald, E. E. A. 1939. The development of the ovule and seed of jointed cactus (*Opuntia aurantiaca Lindley*). South Afric. J. Sci. 36:195–211.

Asay, K. H. 1977. Forage and range program. Logan, Utah. 24th Grass Breed. Work Plann. Conf., Tifton, Georgia, p. 35.

Asker, S. 1976. Apomixis and sexuality in diploid and trisomic *Potentilla argentea* A. Hereditas 83:35–38.

Atanasoff, D. 1964a. Viruses and cytoplasmic heredity. Z. Pflanzenz. 51:197–214.

Atanasoff, D. 1964b. Phytopathol. Z. 50:336–358.

Aula, P. 1963. Chromosome breaks in leucocytes of chicken pox patients. Preliminary communication. Hereditas 49:451–453.

Aula, P., U. Gripenberg, L. Hjelt, E. Kivalo, J. Leisti, J. Palo, B. von Schoultz, and E. Suomalainen. 1967. Two cases with a ring chromosome in group E. Acta Neurol. Scandinav. 43 (suppl. 31):51–52.

Austin, C. R. 1969. Fertilization and development of the egg. *In:* Cole, H. H. and P. T. Cupps. (eds.) Reproduction in domestic animals. Academic Press, New York, pp. 355–384.

Avanzi, S., P. G. Cionini, and F. D'Amato. 1970. Cytochemical and autoradiographic analyses on the embryo suspensor cells of *Phaseolus coccineus*. Caryologia 23:605–638.

Avery, A. G., S. Satina, and J. Rietsema. 1959. Blakeslee: The genus Datura. Ronald Press, Co., New York.

Avery, O. T., C. M. MacLeod, and M. McCarty. 1944. Studies on the chemical nature of the substance inducing transformation of pneumococcal types. Induction of transformation by a desoxyribonucleic acid fraction isolated from *Pneumococcus* type III. J. Exp. Med. 79:137–158.

Azael, A. 1973. Beiträge zur Aufstellung eines Trisomen-Sortiments beim Hafer, *Avena sativa* L. Z. Pflanzenz. 70:289–305.

Babcock, E. B., and G. L. Stebbins. 1938. The American species of *Crepis*. Publ. Carnegie Inst. Wash. 504:1–199.

Badenhuizen, N. P. 1941. Experimentally produced haploids in *Nicotiana tabacum* by means of x-rays. Natourewetensch. Tijdf. Nederl.-Indie. 101:240–242.

Baer, K. E. von. 1827. De ovi mamalium et hominis genesi. Leipzig.

Baer, K. E. von. 1828. Über die Entwicklungsgeschichte der Tiere. Königsberg.

Baglioni, C. 1962. The fusion of two peptide chains in hemoglobin Lepore and its interpretation as a genetic deletion. Proc. Natl. Acad. Sci. U.S.A. 48:1880–1886.

Baikie, A. G., W. M. Court Brown, K. E. Buckton, D. G. Harnden, P. A. Jacobs, and I. M. Tough. 1960. A possible specific chromosome abnormality in human chronic myeloid leukemia. Nature 188:1165–1166.

Bajer, A. 1965. Subchromatid structure of chromosomes in the living state. Chromosoma 17:291–302.

Bajer, A., and J. Molè-Bajer. 1954. Acta Soc. Botan. Polon. 23:69. Cited in: Mazia, D. 1961.

Bajer, A., and J. Molè-Bajer. 1956. Cine-micrographic studies on mitosis in endosperm. II. Chromosome, cytoplasmic and Brownian movements. Chromosoma 7:558–607.

Baker, B. S., and A. T. C. Carpenter. 1972. Genetic analysis of sex chromosomal meiotic mutants in Drosophila melanogaster. Genetics 71:255–286.

Baker, B. S., A. T. C. Carpenter, M. S. Esposito, R. E. Esposito, and L. Sandler. 1976. The genetic control of meiosis. Ann. Rev. Genet. 10:53–134.

Baker, R. J., and T. C. Hsu. 1970. Further studies on the sex chromosome system of the American leaf-nosed bats (Chiroptera, Phyllostomatidae). Cytogenetics 9:138–141.

Baker, R. L., and D. T. Morgan, Jr. 1969. Control of pairing in maize and meiotic effects of deficiencies in chromosome 1. Genetics 61:91–106.

Baker, W. K. 1963. Genetic control of pigment differentiation in somatic cells. Am. Zool. 3:57–69.

Balbiani, E. G. 1881. Sur la structure du noyau des cellules salivaires chez les larves de Chironomus. Zool. Anz. 4:637.

Baldwin, J. T. 1938. Kalanchoe: The genus and its chromosomes. Amer. J. Bot. 25:572–579.

Baldwin, J. T. 1941. Galex: The genus and its chromosomes. J. Hered. 32:249–254.

Balinsky, B. I. 1970. An introduction to embryology. 3rd ed. W. B. Saunders Co., Philadelphia.

Balinsky, B. I., and R. J. Davis. 1963. Origin and differentiation of cytoplasmic structures in the oocytes of Xenopus laevis. Acta Embryol. Morph. Exp. 6:55–108.

Bamford, R. 1935. The chromosome number of Gladiolus. J. Agric. Res. 51:945–950.

Barlow, B. A., and D. Wiens. 1975. Permanent translocation heterozygosity in Viscum hildebrandtii Engl. and V. engleri Tiegh. in East Africa. Chromosoma 53:265–272.

Barlow, B. A., and D. Wiens. 1976. Translocation heterozygosity and sex ratio in Viscum fischeri. Heredity 37:27–40.

Barnum, C. P., Z. B. Ball, J. J. Bittner, and M. B. Visscher. 1944. The milk agent in spontaneous mammary carcinoma. Science 100:575–576.

Barr, M. L., and E. G. Bertram. 1949. A morphological distinction between the neurons of the male and female, and the behavior of the nucleolar satellite during accelerated nucleo-protein synthesis. Nature 163:676–677.

Barrow, J. R. 1971. Meiosis and pollen development in haploid cotton plants. J. Hered. 62:139–141.

Bashaw, E. C., A. W. Hain, and E. C. Holt. 1970. Apomixis, its evolutionary significance and utilization of plant breeding. Proc. 11th Internat. Grassl. Congr., pp. 245–248.

Bateson, W. 1902. Mendel's principles of heredity—a defense. Cambridge.

Bateson, W. 1907. Facts limiting the theory of heredity. Science 26:649–660.

Bateson, W., and R. C. Punnett. 1906. Report to the Evolution Committee of the Royal Society of London. London.

Bateson, W., and E. R. Saunders. 1902. Experimental studies in the physiology of heredity. Rep. Evolut. Comm. Roy. Soc., Rep. I:1–160.

Battaglia, E. 1950. L'alterazion della meiosi nella reproduzione apomittica di Erigeron kirwinskianus var. mucronatus. Caryologia 11:165–204.

Battaglia, E. R. 1952. Appaiamento cromosomico primario, appaiamento cromosomico secondario ed appaiamento cromatidio secondario nello meiosi. Atti della Soc. Toscana 59:166.

Battaglia, E. R. 1964. Cytogenetics of B-chromosomes. Caryologia 17:245–299.

Bauer, H. 1935. Der Aufbau der Chromosomen aus den Speicheldrüsen von *Chironomus thummi* Kiefer. (Untersuchungen an den Riesenchromosomen der Dipteren. I.) Z. Zellforsch. 23:280.

Bauer, H. 1939. Chromosomenforschung. (Karyologie und Cytogenetik). Fortschr. Zool., N. F. 4:584–597.

Bauer, H., R. Dietz, und C. Röbbelen. 1961. Die Spermatocytenteilungen der Tipuliden. III. Mitteilung. Das Berwegungsverhalten der Chromosomen in Translokationsheterozygoten von *Tipula oleracea*. Chromosoma 12:116–189.

Beadle, G. W. 1930. Genetical and cytological studies of Mendelian asynapsis in *Zea mays*. Cornell Univ. Agr. Exp. Sta. Mem. 129:1–23.

Beadle, G. W. 1931. A gene in maize for supernumerary cell divisions following meiosis. Cornell Univ. Exp. Sta. Mem. 135:1–12.

Beadle, G. W. 1932a. A gene for sticky chromosomes in *Zea mays*. Zeitschr. Ind. Abstamm. Verebungsl. 63:195–217.

Beadle, G. W. 1932b. Genes in maize for pollen sterility. Genetics 17:413–431.

Beadle, G. W. 1933a. Polymitotic maize and the precocity hypothesis of chromosome conjugation. Cytologia 5:118–121.

Beadle, G. W. 1933b. Further studies in asynaptic maize. Cytologia 4:269–287.

Beadle, G. W. 1937. Chromosome aberrations and gene mutation in sticky chromosome plants of *Zea mays*. Cytologia, Fujii Jub. Vol.:43–56.

Beadle, G. W. 1945. Biochemical genetics. Chem Rev. 37:15–96.

Beadle, G. W., and E. L. Tatum. 1941. Genetic control of biochemical reactions in *Neurospora*. Proc. Natl. Acad. Sci., U.S.A. 27:499–506.

Beasley, J. O. 1938. Nuclear size in relation to meiosis. Botan. Gaz. 99:865–871.

Beatty, R. A. 1957. Parthenogenesis and polyploidy in mammalian development. University Press, Cambridge.

Beatty, R. A. 1967. Parthenogenesis in vertebrates. *In:* Metz, C. B. and A. Monroy (eds.). Fertilization, comparative morphology, biochemistry, and immunology. Chapter 9, pp. 413–440. Academic Press, New York.

Beck, H., F. M. A. van Breugel and Ž. Srdić. 1979. A marked ring-Y-chromosome in *Drosophila hydei* with a new type of white variegation. Chromosoma 71:1–14.

Beckett, J. B. 1978. B-A translocations in maize: I. Use in locating genes by chromosome arms. J. Hered. 69:27–36.

Beckman, L., K.-H. Gustavson, and H.-O. Akesson. 1962. Studies of some morphological traits in mental defectives. Mongolism and unspecified mental deficiency. Hereditas 48:105–122.

Beermann, W. 1952. Chromosomenkonstanz und spezifische Modifikationen der Chromosomenstruktur in der Entwicklung und Organdifferenzierung von *Chironomus tentans*. Chromosoma 5:139–198.

Beermann, W. 1961. Ein Balbiani-Ring als Locus einer Speicheldrüsen-Mutation. Chromosoma 12:1–25.

Belcheva, R. G., and P. V. Mihailova. 1972. On chromosomal polymorphism in *Anophiles messeae* Fall. (Culicidae, Diptera) from various regions in Bulgaria. Yearb. Sofia Univ. (Fac. Biol.) 64:109–118. (in Russian).

Belling, J. 1925. A unique result in certain species crosses. Z. Ind. Abst. Vererb. 39:286–288.

Belling, J. 1927. The attachments of chromosomes at the reduction division in flowering plants. J. Genet. 18:177–205.

Belling, J. 1928. The ultimate chromomeres of *Lilium* and *Aloe* with regard to the number of genes. Univ. Calif. Publ. Bot. 14:307–318.

Belling, J. 1931a. Chromomeres of liliaceous plants. Univ. Calif. (Berkeley) Publ. Bot. 16:153–170.

Belling, J. 1931b. Chiasmas in flowering plants. Univ. Calif. Publ. Bot. 16:311–338.

Belling, J. 1933. Crossing over and gene rearrangement in flowering plants. Genetics 18:388–413.

Belling, J., and A. F. Blakeslee. 1924. The configurations and sizes of the chromosomes in the trivalents of 25-chromosome Daturas. Proc. Natl. Acad. Sci. U.S.A. 10:116–120.

Benda, C. E. 1960. The child with mongolism (congenital acromicria). Grune and Stratton, New York.

Beneden, E. van, 1875. La maturation de l'oeuf, la fécondation et les premières phases du développement embryonnaire des mammifères etc. Bull. Acad. Belg. Cl. XL. II. Serie, 686.

Beneden, E. van. 1883. Recherches sur la maturation de l'oeuf, la fécondation et la division cellulaire. Arch. Biol. 4:265–640.

Bennett, M. D. 1971. The duration of meiosis. Proc. Roy. Soc. Lond., Ser. B., Biol. Sci. 178:277–299.

Bennett, M. D. 1972. Nuclear DNA content and minimum generation time in herbaceous plants. Proc. Roy. Soc. Lond., Ser. B., Biol. Sci. 181:109–135.

Bennett, M. D., and J. B. Smith. 1972. The effects of polyploidy on meiotic duration and pollen development in cereal anthers. Proc. Roy. Soc. Lond., Ser. B., Biol. Sci. 181:81–107.

Bennett, M. D., H. Stern, and M. Woodward. 1974. Chromatin attachment to nuclear membrane of wheat pollen mother cells. Nature 252:395–396.

Bentvelzen, P., J. H. Daams, P. Hageman, and J. Calafat, 1970. Genetic transmission of viruses that incite mammary tumor in mice. Proc. Natl. Acad. Sci., U.S.A., 67:377–384.

Benzer, S. 1955. Fine structure of a genetic region in bacteriophage. Proc. Natl. Acad. Sci. U.S.A. 41:344–354.

Berg, D. E. 1977a. Insertion and excision of the transposable kanamycin resistance determinant Tn5. In: Bukhari et al., 1977. pp. 205–212.

Berg, D. E. 1977b. Detection of transposable antibiotic resistance determinants with phage lambda. In: Bukhari et al., 1977. pp. 555–558.

Berg, D. E., J. Davies, D. Allet, and J. D. Rochaix. 1975. Transposition of R factor genes to bacteriophage λ. Proc. Natl. Acad. Sci. U.S.A. 72:3628–3632.

Berger, C. A. 1937. Additional evidence of repeated chromosome division without mitotic activity. Amer. Nat. 71:187–190.

Berger, C. A. 1938. Multiplication and reduction of somatic chromosome groups as a regular developmental process in the mosquito, Culex pipiens. Publ. Carnegie Inst. 476:209–232.

Berger, C. A. 1940. The uniformity of the gene complex in the nuclei of different tissues. J. Hered. 31:3–4.

Berger, C. A. 1941. Reinvestigation of polysomaty in Spinacia. Bot. Gazette 102:759–769.

Berger, C. A., and E. R. Witkus. 1943. A cytological study of C-mitosis in the polysomatic plant Spinacia oleracea, with comparative observations on Allium cepa. Bull. Torrey Bot. Club 70:457–467.

Berger, R. 1971. Une nouvelle technique d'analyse du caryotype. C. R. Acad. Sci. 273(Series D):2620–2622.

Bergner, A. D., J. L. Cartledge, and A. F. Blakeslee. 1934. Chromosome behavior due to a gene which prevents pairing in Datura. Cytologia 6:19–37.

Berkaloff, A., J. C. Bregliano, and A. Ohanessian. 1965. Mise en évidence de virions dans des Drosophiles infectées par le virus héréditaire Sigma. C. R. Acad. Sci. (Paris) 260:5956–5959.

Bernard, U., A. Pühler, and H. Küntzel. 1975. Physical map of circular mitochondrial DNA from Neurospora crassa. IERS Lett. 60:119–121.

Bernard, U., E. Bade, and H. Küntzel. 1975. Specific fragmentation of mitochondrial DNA from Neurospora crassa by restriction endonuclease EcoRI. Biochem. Biophys. Res. Commun. 64:783–789.

Bernhard, W. 1958. Electron microscopy of tumor cells and tumor viruses: A review. Cancer Research 18:491–509.

Bernhard, W. 1959. Exptl. Cell. Res. Suppl. 16:17.

Bernhard, W., and E. DeHarven. 1960. L'ultrastructure du centriole et d'autres éléments de l'appareil achromatique. Proc. 4th Intl. Conf. Electron Microsc., Berlin, pp. 217–227.

Berns, M. W. 1978. The laser microbeam as a probe for chromatin structure and function. In: Stein, G., J. Stein and L. Klein-Smith (eds.). Methods in cell biology. Volume 18. pp. 277–294. Academic Press Inc., New York.

Berns, M. W., L. K. Chong, M. Hammer-Wilson, K. Miller and A. Siemens. 1979. Genetic microsurgery by laser: establishment of a clonal population of rat kangeroo cells (PTK$_2$) with a directed deficiency in a chromosomal nucleolar organizer. Chromosoma 73:1–8.

Beutler, E., M. Yeh, and V. F. Fairbanks. 1962. The normal human female as a mosaic of X-chromosome activity: studies using the gene for G-6-Pd-deficiency as a marker. Proc. Natl. Acad. Sci. U.S.A. 48:9–16.

Bhaduri, P. N. 1942. Further cytological investigations in the genus *Gaura*. Ann Bot. 6:229–244.

Bhavanandan, K. V. 1971. Supernumerary cell division during meiosis in *Rumohra aristata*. Cytologia 36:575–578.

Bianchi, K. 1963. Transmission of male sterility in *Petunia* by grafting. Genen en Phaenen 8:36–43.

Bijlsma, J. B., J. C. H. M. Wijffels, and W. H. H. Tegelaers. 1972. C8 trisomy mosaicism syndrome. Helv. Paediat. Acta 27:281–298.

Bingham, E. T. 1971. Isolation of haploids of tetraploid alfalfa. Crop Sci. 11:433–435.

Bingham, E. T., and C. B. Gilles. 1971. Chromosome pairing, fertility and crossing behavior of haploids of tetraploid alfalfa, *Medicago sativa* L. Can. J. Genet. Cytol. 13:195–202.

Birnstiel, M. L., and W. G. Flamm. 1964. Intranuclear site of histone synthesis. Science 145:1435–1437.

Birnstiel, M. L., H. Chipchase, and W. G. Flamm. 1964. On the chemistry and organization of nucleolar proteins. Biochim. Biophys. Acta 87:111–122.

Bittner, J. J. 1938. The genetics of cancer in mice. Quart. Rev. Biol. 13:51–64.

Bittner, J. J. 1939. Relation of nursing to the extrachromosomal theory of breast cancer in mice. Amer. J. Cancer 35:90–97.

Blackwood, M. 1956. The inheritance of B-chromosomes in *Zea mays*. Heredity 10:353–366.

Blakeslee, A. F. 1921. Types of mutations and their possible significance in evolution. Amer. Nat. 55:254–264.

Blakeslee, A. F. 1927. Nubbin, a compound chromosomal type in *Datura*. Ann. N.Y. Acad. Sci. 30:1–29.

Blakeslee, A. F. 1934. New Jimson weeds from old chromosomes. J. Hered. 25:81–108.

Blakeslee, A. F. 1941. Growth patterns in plants. Third Symp. Growth Develop., pp. 77–88.

Blakeslee, A. F., and A. G. Avery. 1937. Methods of inducing doubling of chromosomes in plants. J. Hered. 28:393–411.

Blakeslee, A. F., and A. G. Avery. 1938. Fifteen-year breeding records of 2n+1 types in *Datura stramonium*. Carnegie Inst. Wash. Publ. 501:315–351.

Blakeslee, A. F., J. Belling, and M. Fahrnham. 1932. Inheritance in tetraploid *Datura*. Bot. Gaz. 76:329–373.

Blakeslee, A. F., A. D. Bergner, S. Satina and E. W. Sinnott. 1939. Induction of periclinal chimeras in *Datura stramonium* by colchicine treatment. Science 89:402.

Blakeslee, A. F., S. Satina and A. G. Avery. 1940. Utilization of induced periclinal chimeras in determining the constitution of organs and their origin from the three germ layers in *Datura*. Science 91:423.

Blanco, J. L. 1948. Abnormalidades meióticas en relación con la consanguinidad en *Zea mays* L. Cons. Sup. Inv. Cient. Mision. Biol. Galicia, Madrid.

Blanco, J., and G. Mariano. 1953. Crossover reduction in the X chromosome of inbred lines of *Drosophila melanogaster*. Proc. IX Int. Genet. Cong. pp. 776–778.

Blickenstaff, J., P. Sarvella, and R. A. Nilan. 1958. Pachytene chromosome analysis in barley. Proc. X. Internat. Congr. Genet., Vol. 2:26–27.

Blochman, F. 1887. Morph. Jahrb. 12:544.

Bloom, S. E. 1970. Haploid chicken embryos: evidence for diploid and triploid cell populations. J. Hered. 61:147–150.

Böcher, T. W. 1938. Cytological studies in the genus *Ranunculus*. Dansk. Bot. Ark. 9:1–33.

Bogart, J. P., and M. Tandy. 1976. Polyploid amphibians: three more diploid-tetraploid cryptic species of frogs. Science 193:334–335.

Bogdanov, Y. F. 1977. Formation of cytoplasmic synaptonemal-like polycomplexes at leptotene and normal synaptonemal complexes at zygotene in *Ascaris suum* male meiosis. Chromosoma 61:1–21.

Bollum, F. J. 1963. Primer in DNA polymerase reactions. Progr. Nucleic Acid Res. 1:1–26.

Böök, J. A. 1945. Cytological studies in *Triton*. Hereditas 31:177–220.

Böök, J. A., and B. Santesson. 1960. Malformation syndrome in man associated with triploidy. Lancet I:858–859.

Böök, J. A., J. G. Masterson and B. Santesson. 1962. Malformation syndrome associated with triploidy—further chromosome studies of the patient and his family. Acta Genet. Statist. Med. 12:193–201.

Bopp-Hassenkamp, G. 1960. Electronmikroskopische Untersuchungen an Pollenschläuchen zweier Liliaceen. Z. Naturforsch. 15b:91–94.

Borgaonkar, D. S. 1975. Chromosomal variation in man. A catalogue of variants and anomalies. John Hopkins Univ. Press, Baltimore, Maryland.

Borgaonkar, D. S. 1977. Chromosome variation in man. A catalog of chromosomal variations and anomalies. 2nd ed. Alan R. Liss, Inc., New York.

Borgaonkar, D. S., M. M. Aronson, A. E. Greene, and L. L. Coriell. 1977. A(5;14) translocation, 46 chromosomes. Repository identification No. GM-589. Cytogenet. Cell Genet. 18:242.

Borisy, G. G. and E. W. Taylor. 1967. The mechanism of action of colchicine. Binding of colchicine-^{3}H to cellular protein. J. Cell Biol. 34:525–533.

Bosemark, N. O. 1957a. On accessory chromosomes in *Festuca pratensis*. V. Influence of accessory chromosomes on fertility and vegetative development. Hereditas 43:211–235.

Bosemark, N. O. 1957b. Further studies on accessory chromosomes in grasses. Hereditas 43:236–297.

Botstein, D., and N. Kleckner. 1977. Translocation and illegitimate recombination by the tetracycline resistance element Tn10. *In:* Bukhari et al., 1977. pp. 185–203.

Bourgin, J. P., and J. P. Nitsch. 1967. Obtention de *Nicotiana* haploïdes à partir d'étamines cultivées *in vitro*. Ann. Physiol. Veg. 9:377–382.

Boveri, T. 1887. Zellen-Studien. I. Die Bildung der Richtungskörper bei *Ascaris megalocephala* und *Ascaris lumbricoides*. Zeitschr. Naturwiss., Jena 21:423–515.

Boveri, T. 1888. Zellen-Studien. II. Die Befruchtung und Teilung des Eies von *Ascaris megalocephala*. Zeitchr. Mediz. Naturw. 22:685–882.

Boveri, T. 1892. Über die Entstehung des Gegensatzes zwischen den Geschlechtszellen und den somatichen Zellen bei *Ascaris megalocephala* nebst Bemerkungen zur Entstehungsgeschichte der Nematoden. Sitzungsber. Ges. Morph. Phys., München 8:114–125.

Boveri, T. 1895. Über das Verhalten der Centrosomen bei der Befruchtung des Seeigeleies nebst allgemeinen Bemerkungen über Centrosomen und Verwandtes. Verh. Phys. Med. Ges., Würzburg 29:1.

Boveri, T. 1902. Über mehrpolige Mitosen als Mittel zur Analyse des Zellkerns. Verh. Phys.-med. Ges., Würzburg, N.F. 35:67–90.

Boveri, T. 1904. Ergebnisse über die Konstitution der chromatischen Substanz des Zellkerns. G. Fischer, Jena.

Boveri, T. 1907. Zellen-Studien. VI. Die Entwicklung dispermer Seeigeleier. Ein Beitrag zur Befruchtungslehre und zur Theorie des Kerns. Zeitschr. Naturw., Jena 43:1-292.

Boyer, H. W. 1974. Restriction and modification of DNA: enzymes and substrates. Fed. Amer. Soc. Exp. Biol., Fed. Proc. 33:1125-1127.

Bozzini, A., and G. Martini. 1971. Analysis of desynaptic mutants induced in durum wheat (*Triticum durum* Desf.). Caryologia 24:307-316.

Brachet, J. 1969. Acides nucléiques et différenciation embryonnaire. Annales d'Embryologie et de Morphogénèse, Suppl. 1:21-37.

Bradbury, E. M., R. J. Inglis and H. R. Matthews. 1974a. Control of cell division by very lysine rich histone (F1) phosphorylation. Nature 247:257-261.

Bradbury, E. M., R. J. Inglis, H. R. Matthews and T. A. Langan. 1974b. Molecular basis of control of mitotic cell division in eukaryotes. Nature 249:553-556.

Bradley, D. E. 1967. Ultrastructure of bacteriophages and bacteriocins. Bacteriol. Rev. 31:230-314.

Bretschneider, L. H., and C. P. Raven. 1951. Structure and isochemical changes in the egg cells of *Limnaea stagnalis* L. during oogenesis. Arch. néerl. Zool. 10:1-31.

Breuer, M., and C. Pavan. 1955. Behavior of polytene chromosomes of *Rhynchosciara* at different stages of larval development. Chromosoma 7:371-386.

Brevet, J., D. J. Kopecko, P. Nisen, and S. N. Cohen. 1977. Promotion of insertions and deletions by translocating segments of DNA carrying antibiotic resistance genes. *In:* Bukhari et al., 1977. pp. 169-179.

Brewen, J. G., and R. D. Brock. 1968. The exchange hypothesis and chromosometype aberrations. Mutation Res. 6:245-255.

Bridges, C. B. 1917. Deficiency. Genetics 2:445-465.

Bridges, C. B. 1922. The origin of variations in sexual and sex-limited characters. Amer. Nat. 56:51-63.

Bridges, C. B. 1923. Aberrations in chromosome materials. Scient. Pap. 2nd Int. Congr. Eugenics 1:76.

Bridges, C. B. 1925. Haploidy in *Drosophila melanogaster*. Proc. Natl. Acad. Sci. U.S.A. 11:706-710.

Bridges, C. B. 1929. Variation in crossing over in relation to age of female in *Drosophila melanogaster*. Carnegie Inst. Washington Publ. 399:64-89.

Bridges, C. B. 1932. The genetics of sex in *Drosophila. In:* Sex and internal secretion. Baillière, London. pp. 53-93.

Bridges, C. B. 1935. Salivary chromosome maps with a key to the banding of the chromosomes of *Drosophila melanogaster*. J. Hered. 26:60-64.

Bridges, C. B. 1938. A revised map of the salivary gland X-chromosome. J. Hered. 29:11-13.

Bridges, C. B., and K. S. Brehme. 1944. The mutants of *Drosophila melanogaster*. Carnegie Inst. Wash. Publ. 552, Washington, D.C.

Briggs, F. N., and P. F. Knowles. 1967. Introduction to plant breeding. Reinhold Publ. Co., New York.

Briggs, R. 1952. An analysis of inactivation of the frog sperm nucleus by toluidine blue. J. Gen Physiol. 35:761-780.

Brink, R. A., and D. C. Cooper. 1931. The association of semisterile-1 in maize with two linkage groups. Genetics 16:595-628.

Brink, R. A. and R. A. Nilan. 1952. The relation between light variegated and medium variegated pericarp in maize. Genetics 37:519-544.

Brinkley, B. R., and W. N. Hittelman. 1975. Ultrastructure of mammalian chromosome aberrations. Intern. Rev. Cytol. 42:49-101.

Britten, R. J., and D. E. Kohne. 1968. Repeated sequences in DNA. Science 161:529-540.

Brown, D. D., and J. B. Gurdon. 1964. Absence of ribosomal RNA synthesis in the anucleolate mutant of *Xenopus laevis*. Proc. Natl. Acad. Sci., U.S.A. 51:139–146.

Brown, R. 1828. A brief account of microscopical observations on the particles contained in the pollen of plants, and on general resistance of active molecules in organic and inorganic bodies. Edinburgh New Phil. J. 5:358.

Brown, S. W. 1966. Heterochromatin. Science 151:417–425.

Brown, S. W., K. H. Walen, and G. E. Brosseau. 1962. Somatic crossing-over and elimination of ring X-chromosomes of *Drosophila melanogaster*. Genetics 47:1573–1579.

Brown, W. L. 1949. Numbers and distribution of chromosome knobs in United States maize. Genetics 34:524–536.

Brown, W. M., and J. Vinograd. 1974. Restriction endonuclease cleavage maps of animal mitochondrial DNA's. Proc. Natl. Acad. Sci. U.S.A. 71:4617–4621.

Brown, W. V. 1972. Textbook of Cytogenetics. C. V. Mosby Co., St. Louis.

Brown, W. V., and E. M. Bertke. 1969. Textbook of cytology. C. V. Mosby Co., St. Louis.

Brumfield, R. T. 1943. Cell-lineage studies in root meristems by means of chromosome rearrangements induced by x-rays. Amer. J. Bot. 30:101–110.

Bryan, J. H. D. 1951. DNA-protein relations during microsporogenesis of *Tradescantia*. Chromosoma 4:369–392.

Büchner, T., R. A. Pfeiffer, und E. Stupperich. 1965. Reduplikationsverhalten der Chromosomen der Gruppe D(13-15) und Identifikation des Extra-Chromosoms bei Trisomie D. Klin. Wochenschr. 43:1062–1063.

Budrin, K. Z. 1969. Obtaining potato haploids. Genetika 5:43–50. (Russian).

Bukhari, A., J. Shapiro, and S. Adhya. 1977. (eds.). DNA insertion elements, plasmids, and episomes. Cold Spring Harbor Laboratory. Cold Spring Harbor, N.Y.

Burgos, M. H., and D. W. Fawcett. 1955. Studies on the fine structure of the mammalian testis. I. Differentiation of the spermatids in the cat *(Felis domestica)*. J. Biophys. Biochem. Cytol. 1:287–300.

Burk, L. G. 1960. Male-sterile flow anomalies in interspecific tabacco hybrids. J. Hered. 51:27–31.

Burk, L. G. 1962. Haploids in genetically marked progenies of tabacco. J. Hered. 53:222–225.

Burnham, C. R. 1930. Genetical and cytological studies of semisterility and related phenomena in maize. Proc. Natl. Acad. Sci. U.S.A. 16:269–277.

Burnham, C. R. 1932. The association of non-homologous parts in the chromosomal interchange in maize. Proc. 6th Intl. Congr. Genet. 2:19–20 (Abstr.).

Burnham, C. R. 1946. A gene for "long" chromosomes in barley. Genetics 31:212–213 (Abstr.).

Burnham, C. R. 1954. Tester set of translocations. Maize Genetics Coop. Newsletter 28:59–80.

Burnham, C. R. 1956. Chromosomal interchanges in plants. Bot. Rev. 22:419–552.

Burnham, C. R. 1957. The use of chromosomal interchanges to test for the independence of linkage groups estabished genetically. Proc. Int. Genetics Symposia, 1956:453–456.

Burnham, C. R. 1962. Discussion in cytogenetics. Burgess Publ. Co., Minneapolis, Minn.

Burnham, C. R. 1966. Cytogenetics in plant improvement. In. Frey, K. J. (ed.). Plant breeding. A symposium held at Iowa State University. pp. 139–187. Iowa State University, Ames, Iowa.

Burnham, C. R. and J. L. Cartledge. 1939. Linkage relations between smut resistance and semi-sterility in maize. J. Amer. Soc. Agron. 31:924–933.

Burton, G. W. 1956. Utilization of heterosis in pasture plant breeding. Proc. 7th Internat. Grassl. Congr. 439–449.

Burton, G. W. 1973. Bermudagrass. *In:* Heath, M. E., D. S. Metcalfe, and R. E. Barnes. (eds.). Forages; the science of grassland agriculture. 3rd ed. Iowa State Univ. Press, Ames.

Buss, G. R., and R. W. Cleveland. 1968. Pachytene chromosomes of diploid *Medicago sativa* L. Crop Sci. 8:744–747.

Butler, L. J., C. Chantler, N. E. France, and C. G. Keith. 1969. A liveborn infant with complete triploidy (69,XXX). J. Med. Genet. 6:413–420.

Byers, B. and L. Goetsch. 1975. Electron microscopic observations on the meiotic karyotype of diploid and tetraploid *Saccharomyces cerevisiae*. Proc. Natl. Acad. Sci. U.S.A. 72:5056–5060.

Cairns, J. 1963. The chromosome of *Escherichia coli*. Cold Spring Harbor Symp. Biol. 28:43–46.

Caldecott, R. S., and L. Smith. 1952. A study of x-ray induced chromosomal aberrations in barley. Cytologia 17:224–242.

Callan, H. G. 1941. A trisomic grasshopper. J. Hered. 32:296–298.

Callan, H. G. 1942. Heterochromatin in *Triton*. Proc. Roy. Soc. London, B. 130:324–335.

Callan, H. G. 1966. Chromosomes and nucleoli of the axolotl, *Ambystoma mexicanum*. J. Cell Sci. 1:85–108.

Calos, M. P., L. Johnsrud, and J. H. Miller. 1978. DNA sequence of insertion sequence elements. *In:* Schlessinger, D. (ed.). Microbiology 1978. p. 22. Amer. Soc. Microbiol., Washington, D.C.

Camerarius, R. J. 1694. Letters on the sex of plants. Cited *in:* Oswald's Klassiker Exact. Wiss., Leipzig, 1899.

Campbell, A. 1971. Genetic structure. *In:* Hershey, A. D. (ed.). The bacteriophage lambda. pp. 13–44. Cold Spring Harbor Lab., New York.

Campbell, A., D. Berg, D. Botstein, E. Lederberg, R. Novick, P. Starlinger, and W. Szybalski. 1977. Nomenclature of transposable elements in prokaryotes. pp. 15–22. *In:* Bukhari et al., 1977.

Campos, F. F., and D. T. Morgan. 1960. Haploid pepper from a sperm. J. Hered. 49:134–137.

Carlson, J. G., and A. Hollaender. 1948. Mitotic effects of ultraviolet radiation of the 2250 region, with special reference to the spindle and cleavage. J. Cellular Comp. Physiol. 31:149–173.

Carlson, P. 1972. Effect of inversions and *c(3)G* on intragenic recombination. Genet. Res. 19:129–132.

Carniel, K. 1960. Beiträge zum Sterilitäts- und Befruchtungsproblem von *Rhoeo discolor*. Chromosoma 11:456–462.

Carothers, E. E. 1917. The segregation and recombination of homologous chromosomes as found in two genera of Acrididae (Orthoptera). J. Morph. 28:445–520.

Carothers, E. E. 1921. Genetical behavior of heteromorphic homologous chromosomes of *Circotettix* (Orthoptera). J. Morph. 35:457–483.

Carpenter, A. T. C. 1975. Electronmicroscopy of meiosis in *Drosophila melanogaster* females. II. The recombination nodule—a recombination associated structure at pachytene? Proc. Natl. Acad. Sci. U.S.A. 72:3186–3189.

Carpenter, A. T. C., and L. Sandler. 1974. On recombination defective meiotic mutants in *Drosophila melanogaster*. Genetics 76:453–475.

Carson, H. L. 1946. The selective elimination of inversion dicentric chromatids during meiosis in the eggs of *Sciara impatiens*. Genetics 31:95–113.

Carson, H. L. 1953. The effects of inversions on crossing over in *Drosophila robusta*. Genetics 38:168–186.

Carter, T. C., M. F. Lyon, and R. J. S. Phillips. 1955. Gene-tagged chromosome translocations in eleven stocks of mice. J. Genet. 53:154–156.

Casey, M. D., L. J. Segall, D. R. K. Street, and C. E. Blank. 1966. Sex chromosome abnormalities in two state hospitals for patients requiring special security. Nature 209:641–642.

Casey, M. D., D. R. K. Street, L. J. Segall, and C. E. Blank. 1968. Patients with sex chromatin abnormality in two state hospitals. Ann. Human Genet. 32:53–63.

Caspersson, T. 1936. Über den Aufbau der Strukturen des Zellkernes. Skand. Arch Physiol. 73(Suppl. 8):1–151.

Caspersson, T. 1947. The relations between nucleic acid and protein synthesis. S. E. B.-Symp. 1:127.

Caspersson, T., S. Farber, G. E. Foley, J. Kudynowsky, E. J. Modest, E. Simonsson, U. Wagh, and L. Zech. 1968. Chemical differentiation along metaphase chromosomes. Exp. Cell Res. 49:219–222.

Caspersson, T., L. Zech, E. J. Modest, G. E. Foley, U. Wagh, and E. Simonsson. 1969a. Chemical differentiation with fluorescent alkylating agents in *Vicia faba* metaphase chromosomes. Exp. Cell Res. 58:128–140.

Caspersson, T., L. Zech, E. J. Modest, G. E. Foley, U. Wagh, and E. Simonsson. 1969b. DNA-binding fluorochromes for the study of the organization of the metaphase nucleus. Exp. Cell Res. 58:141–152.

Caspersson, T., L. Zech, and C. Johannson. 1970a. Differential binding of alkylating fluorochromes in human chromosomes. Exp. Cell Res. 60:315–319.

Caspersson, T., L. Zech, and C. Johannson. 1970b. Analysis of the human metaphase chromosome set by aid of DNA-binding fluorescent agents. Exp. Cell. Res. 62:490–492.

Caspersson, T., L. Zech, C. Johannson, and E. J. Modest. 1970. Identification of human chromosomes by DNA-binding fluorescent agents. Chromosoma 30:215–227.

Caspersson, T., J. Linstein, L. Zech, K. E. Buckton, and W. H. Price. 1972. Four patients with trisomy 8 identified by the fluorescence and Giemsa banding techniques. J. Med. Genetics 9:1–7.

Castle, W. E. 1934. Possible cytoplasmic as well as chromosomal control of sex in haploid males. Proc. Natl. Acad. Sci. U.S.A. 20:101–102.

Catcheside, D. G. 1939. An asynaptic *Oenothera*. New Phytol. 38:323–334.

Catcheside, D. G. 1954. Genetical and cytological studies of semisterility and related phenomena in maize. Proc. Natl. Acad. Sci. U.S.A. 16:269–277.

Catcheside, D. G. 1977. The genetics of recombination. Edward Arnold, London.

Catcheside, D. G., D. E. Lea, and J. M. Thoday. 1946. The production of chromosome structural changes in *Tradescantia* microspores in relation to dosage intensity and temperature. J. Genet. 47:137–149.

Cattanach, B. M. 1964. Autosomal trisomy in the mouse. Cytogenetics 3:159–166.

Cavalcanti, A. L. G., D. N. Falcão, and L. E. Castro. 1958. The interaction of nuclear and cytoplasmic factors in the inheritance of the "sex-ratio" character in *Drosophila prosaltans*. Publ. Fac. Nac. Filosof. Rio de Janero, Sér. Cient. 1:1–54.

Chaganti, R. S. K. and J. German. 1974. Human male infertility, probably genetically determined due to defective meiotic pairing and chiasma formation. Amer. J. Human Gen. 26:19A.

Chaganti, R. S. K., S. Schonberg, and J. German. 1974. A manifold increase in sister chromatid exchange in Bloom's syndrome lymphocytes. Proc. Natl. Acad. Sci., U.S.A.71:4508–4512.

Chapelle, A. de la, P. Vuopio, A. Icén. 1976. Trisomy in the bone marrow associated with high red cell glutathione reductase activity. Blood 47:815–826.

Chaplin, J. F. 1964. TOB. Sci. 8:105–109.

Chase, S. S. 1947. Techniques for isolating monoploid maize plants. Amer. J. Bot. 34:582 (Abstr.).

Chase, S. S. 1949a. Monoploid frequencies in a commercial double cross hybrid maize and its component single cross hybrids and inbred lines. Genetics 34:328–332.

Chase, S. S. 1949b. The reproduction success of monoploid maize. Amer. J. Bot. 36:795–796.

Chase, S. S. 1949c. Spontaneous doubling of the chromosome complement in monoploid sporophytes of maize. Proc. Iowa Acad. Sci. 56:113–115.

Chase, S. S. 1952a. Monoploids in maize. *In:* J. W. Gowen. (ed.). Heterosis. Iowa State Coll. Press, Ames, Iowa. pp. 389–399.

Chase, S. S. 1952b. Production of homozygous diploids of maize from monoploids. Agron. J. 44:263–267.

Chase, S. S. 1969. Monoploids and monoploid-derivatives of maize (*Zea mays* L.). Bot. Rev. 35:117–167.

Chiarugi, A. 1926. Aposporia e apogamia in *Artemisia nitida* Bertol. Nuovo Giorn. Bot. Ital. 33:501–626.

Chicago Conference. 1966. Standardization in human cytogenetics. Birth defects: original article series, Vol. II, No. 2. New York: The National Foundation.

Chittenden, R. J. 1927. Cytoplasmic inheritance in flax. J. Hered. 18:337–343.

Christiansen, C., G. Christiansen, A. Leth Bak, and A. Stenderup. 1973. Extrachromosomal deoxyribonucleic acid in different Enterobacteria. J. Bacteriol. 114:367–377.

Chrustschoff, G. K., A. H. Andres, und W. Iljina-Kakujewa. 1931. Kulturen von Blutleukozyten als Methode zum Studium des menschlichen Karyotypus. Anat. Anz. 73:159–168.

Chrustschoff, G. K., and E. A. Berlin. 1935. Cytological investigations on the cultures of normal human blood. J. Genet. 31:243–261.

Chu, Y. 1967. Pachytene analysis and observations of chromosome associations in haploid rice. Cytologia 32:87–95.

Chu, Z. C., C. C. Wang, C. S. Sun, N. F. Chien, K. C. Yin, and C. Hsu. 1973. Investigations on the induction and morphogenesis of wheat (*Triticum aestivum*) pollen plants. Acta Bot. Sinica 15:1–11.

Cipar, M. S., and C. H. Lawrence. 1972. Scab resistance of haploids from *Solanum tuberosum* cultivars. Am. Potato J. 49:117–121.

Clapham, D. 1971. *In vitro* development of callus from the pollen of *Lolium* and *Hordeum*. Z. Pflanzenz. 65:285–292.

Clapham, D. 1973. Haploid *Hordeum* plants from anthers *in vitro*. Z. Pflanzenz. 69:142–145.

Clark, A. J., and E. A. Adelberg. 1962. Bacterial conjugation. Ann. Rev. Microbiol. 16:289–319.

Clark, F. J. 1940. Cytogenetic studies of divergent meiotic spindle formation in *Zea mays*. Amer. J. Bot. 27:547–559.

Clausen, R. E., and T. H. Goodspeed. 1924. Inheritance in *Nicotiana tabacum*. IV. The trisomic character "enlarged". Genetics 9:181–197.

Clausen, R. E., and T. H. Goodspeed. 1926. Inheritance in *Nicotiana tabacum*. VII. The monosomic character, "flutet". Univ. Calif. Pub. Bot. 11:61–82.

Clavier, Y., and A. Cauderon. 1951. Un cas d'haploïdie chez l'orge cultivée. Ann. Amélior. Plantes 1:332–335.

Clayberg, C. D. 1959. Cytogenetic studies of precocious meiotic centromere division in *Lycopersicon esculentum* Mill. Genetics 44:1335–1346.

Cleaver, J. E. 1974. Repair processes for photochemical damage in mammalian cells. Adv. Rad. Biol. 4:1.

Cleland, R. E. 1962. The cytogenetics of *Oenothera*. Advanc. Genet. 11:147–237.

Cleland, R. E. 1972. *Oenothera*—Cytogenetics and evolution. Acad. Press, London.

Cleveland, L. R. 1949. The whole life cycle of chromosomes and their coiling systems. Trans. Amer. Phil. Soc. 39:1–100.

Clever, U. 1964. Puffing in giant chromosomes of Diptera and the mechanisms of its control. In: J. Bonner and P. Tso (eds.). The nucleohistones. Holden-Day, San Francisco.

Close, H. G., A. S. R. Goonetilleke, P. A. Jacobe, and W. H. Price. 1968. The incidence of sex chromosomal abnormalities in mentally subnormal males. Cytogenetics 7:277–285.

Coe, E. H., Jr. 1959. A line of maize with high haploid frequency. Amer. Nat. 93:381–382.

Coe, G. E. 1953. Cytology of reproduction in *Cooperia pedunculata*. Amer. J. Bot. 40:335–343.

Cognetti, G. 1961a. Citogenetica della partenogenesi negli afidi. Arch. Zool. Ital. 46:89–122.

Cognetti, G. 1961b. Endomeiosis in parthenogenetic lines of aphids. Experientia 17:168–169.

Cognetti, G. 1961c. Endomeiosi e selezione in ceppi partenogenetici di afidi. Atti. Assoc. Genet. Ital. 6:449–454.

Cognetti, G. 1962. La partenogenesi negli afidi. Boll. Zool. 29:129–147.

Cohen, M. M., and M. W. Shaw. 1964. Effects of mitomycin C on human chromosomes. J. Cell Biol. 23:386–395.

Cohen, M. M., M. J. Marinello, and N. Back. 1967a. Chromosomal damage in human leucocytes induced by lysergic acid diethylamide. Science 155:1417–1419.

Cohen, M. M., K. Hirschhorn, and W. A. Frosch. 1967b. In vivo and in vitro chromosomal damage induced by LSD-25. New Engl. J. Med. 277:1043–1049.

Cohen, S. N., A. C. Y. Chang. H. W. Boyer, and R. B. Helling. 1973. Construction of biologically functional bacterial plasmids in vitro. Proc. Natl. Acad. Sci. U.S.A. 70:3240–3244.

Cohen, S. N. and D. J. Kopecko. 1976. Structural evolution of bacterial plasmids: role of translocating genetic elements and DNA sequence insertions. Feder. Proc. 35:2031.

Cohn, N. S. 1969. Elements of cytology. 2nd ed. Harcourt, Brace and World, Inc., New York.

Cold Spring Harbor Symposia on Quantitative Biology. 1974. Chromosome structure and function. Vol. 38. Cold Spring Harbor Laboratory, Cold Spring Harbor, New York.

Collins, G. B., and R. S. Sadasivaiah. 1972. Meiotic analysis of haploid and doubled haploid forms of Nicotiana otophora and N. tabacum. Chromosoma 38:387–404.

Collins, G. B., and N. Sunderland. 1974. Pollen-derived haploids of Nicotiana knightiana, N. raimondii, and N. attenuata. J. Exp. Bot. 25:1030–1036.

Colwin, A. L., and L. H. Colwin. 1961. Fine structure of the spermatozoon of Hydroides hexagonus (Annelida) with special reference to the acrosomal region. J. Biophys. Biochem. Cytol. 10:211–230.

Colwin, L. H., and A. L. Colwin. 1967. Membrane fusion in relation to sperm-egg association. In: Metz, C. B. and A. Monroy. (eds.). Fertilization. Comparative morphology, biochemistry and immunology. Acad. Press, New York.

Comings, D. E. 1968. The rationale for an ordered arrangement of chromatin in the interphase nucleus. Amer. J. Hum. Genet. 20:440–460.

Comings, D. E., and T. A. Okada. 1971. Triple chromosome pairing in triploid chickens. Nature 231:119–121.

Committee on Standardized Genetic Nomenclature for Mice. 1972. Standard karyotype of the mouse, Mus musculus. J. Hered. 63:69–72.

Cook, S. A. 1965. Reproduction, heredity and sexuality. Macmillan, London.

Cooke, P., and R. R. Gordon. 1965. Cytological studies on a human ring chromosome. Ann. Hum. Genet. 29:147–150.

Cooper, H. L., and K. Hirschhorn. 1962. Enlarged satellites as a familial chromosome marker. Amer. J. Human Genet. 14:107–124.

Cooper, K. W. 1939. The nuclear cytology of the grass mite, Pediculopsis graminum (Reut.) with special reference to karyomerokinesis. Chromosoma 1:51–103.

Cooper, K. W. 1941. Bivalent structure in the fly Melaphagus ovium L. Proc. Natl. Acad. Sci., U.S.A. 27:109–114.

Cooper, K. W. 1944. Analysis of meiotic pairing in Olfersia and consideration of the reciprocal chiasmata hypothesis of sex chromosome conjugation in male Drosophila. Genetics 29:537–568.

Coriell, L. L. 1973. Cell repository. Science 180:427.

Correns, C. 1900a. Gregor Mendels Versuche über "Pflanzenhybride" und die Bestätigung ihrer Ergebnisse durch die neuesten Untersuchungen. Bot. Zeit. 58.

Correns, C. 1900b. G. Mendel's Regel über das Verhalten der Nachkommenschaft der Rassenbastarde. Ber. Deutsch. Boton. Ges. 18:158–167.

Correns, C. 1902. Über Bastardierungsversuche mit Mirabilis-Sippen. Ber. Deutsch. Botan. Ges. 20:549–608.

Correns, C. 1904. Experimentelle Untersuchungen über die Gynodiocie. Ber. Deutsch. Bot. Ges. 22:506–517.

Correns, C. 1909. Vererbungsversuche mit blass (gelb) grünen und buntblättrigen Gruppen bei *Mirabilis, Urtica* und *Lunaria*. Z. Ind. Abst. Vererbungsl. 1:291.

Court Brown, W. M. 1967. Human population cytogenetics. North Holland Publ., Amsterdam.

Court Brown, W. M., D. J. Mantle, K. E. Buckton, and I. M. Tough. 1964. Fertility in an XY/XXY male married to a translocation heterozygote. J. Med. Genet. 1:35–38.

Court Brown, W. M., K. E. Buckton, P. A. Jacobs, I. M. Tough, E. V. Kuenssberg, and J. D. E. Knox. 1966. Chromosome studies on adults. Eugenics Lab., Mem. No. 42, Galton Lab., Univ. Coll. London, Cambridge Univ. Press, London.

Craig-Holmes, A. P. and M. W. Shaw. 1971. Polymorphism of human constitutive heterochromatin. Science 174:702–704.

Craig-Holmes, A. P., F. B. Moore, and M. W. Shaw. 1973. Polymorphism of human C-band heterochromatin. I. Frequency of variants. Amer. J. Human Genet. 25:181–192.

Cramer, P. J. S. 1954. Chimeras. Bibl. Genet. 16:193–381.

Creighton, H. B., and B. McClintock. 1931. A correlation of cytological and genetical crossing-over in *Zea mays*. Proc. Natl. Acad. Sci. U.S.A. 17:492–497.

Crick, F. H. C. 1966. Codon-anticodon pairing: the wobble hypothesis. J. Mol. Biol. 19:548–555.

Crick, F. H. C. 1971. General model for the chromosomes of higher organisms. Nature 234:25–27.

Crick, F. H. C., L. Barnett, S. Brenner, and R. J. Watts-Tobin. 1961. General nature of the genetic code of proteins. Nature 192:1227–1232.

Crouse, H. C. 1943. Translocations in *Sciara:* their bearing on chromosome behavior and sex determination. Univ. Missouri Agr. Exp. Sta. Res. Bull. 379:1–75.

Crouse, H. V., and H.-G. Keyl. 1968. Extra replications in the "DNA Puffs" of *Sciara coprophila*. Chromosoma 25:357–364.

Crumpacker, D. W., and V. M. Salceda. 1968. Uniform heterokaryotypic superiority for viability in a Colorado population of *Drosophila pseudoobscura*. Evolution 22:256–261.

Curran, J. P., F. L. Al-Salihi, and P. W. Allderdice. 1970. Partial deletion of the long arm of chromosome E-18. Pediatrics 46:721–729.

Curtis, G. J. 1967. Graft-transmission of male sterility in sugar beet (*Beta vulgaris* L.). Euphytica 16:419–424.

Dahlgren, K. V. O. 1915. Über die Überwinterungststadien der Pollensäcke und der Samenanlagen bei einigen Angiospermen. Sv. Bot. Tiedskr. 9:1–12.

Danna, K. J., G. H. Sack, and D. Nathans. 1973. Studies of Simian virus 40 DNA. VII. A cleavage map of the SV 40 genome. J. Molec. Biol. 78:363–376.

Darlington, C. D. 1929a. Chromosome behavior and structural hybridity in the Tradescantiae. J. Genet. 21:207–286.

Darlington, C. D. 1929b. Meiosis in polyploids. II. Aneuploid hyacinths. J. Genet. 21:17–36.

Darlington, C. D. 1929c. Ring-formation in *Oenothera* and other genera. J. Genet. 20:345–363.

Darlington, C. D. 1930. A cytological demonstration of genetic crossing over. Proc. Roy. Soc., London, B 107:50–59.

Darlington, C. D. 1931. Meiosis. Biol. Rev. 6:221.

Darlington, C. D. 1932. Recent advances in cytology. Churchill, London.

Darlington, C. D. 1933. The origin and behavior of chiasmata. VIII. *Secale cereale* (n,8). Cytologia 4:444–452.

Darlington, C. D. 1936. The external mechanics of chromosomes. Proc. Roy. Soc. (London), B 121:264.

Darlington, C. D. 1937. Recent advances in cytology. 2nd ed. Churchill, London.

Darlington, C. D. 1939a. Misdivision and the genetics of the centromere. J. Genet. 37:341–364.

Darlington, C. D. 1939b. The evolution of genetic systems. Cambridge Univ. Press, Cambridge.

Darlington, C. D. 1940. The origin of iso-chromsomes. J. Genet. 39:351–361.

Darlington, C. D. 1965. Cytology. Churchill, London.

Darlington, C. D., and S. O. S. Dark. 1932. The origin and behavior of chiasmata. II. *Stenobothrus parallelus*. Cytologia 3:169–185.

Darlington, C. D., J. B. Hair, and R. Hurcombe. 1951. The history of the garden hyacinths. Heredity 5:233–252.

Darlington, C. D., and E. K. Janaki-Ammal. 1945. Chromosome atlas of cultivated plants. Allen and Unwin, London.

Darlington, C. D., and L. LaCour. 1938. Differential reactivity of the chromosomes. Ann. Bot. 2:615–625.

Darlington, C. D., and L. LaCour. 1940. Nucleic acid starvation of chromosomes in *Trillium*. J. Genet. 40:185–213.

Darlington, C. D., and L. LaCour. 1950. Hybridity selection in *Campanula*. Heredity 4:217–248.

Darlington, C. D., and P. T. Thomas. 1941. Morbid mitosis and the activity of inert chromosomes in *Sorghum*. Proc. Roy. Soc. London, B 130:127–150.

Darlington, C. D., and M. B. Upcott. 1941. The activity of inert chromosomes in *Zea mays*. J. Genet. 41:275–296.

Darlington, C. D., and A. P. Wylie. 1955. Chromosome atlas of flowering plants. 2nd ed. George Allen and Unwin, Ltd., London.

Davidson, N., R. C. Deonier, S. Hu, and E. Ohtsubo. 1975. The DNA sequence organization of F and F-primes and the sequences involved in Hfr formation. *In:* Schlesinger, D. (ed). Microbiology-1974. p. 56. Amer. Soc. Microbiol., Washington, D.C.

Davidson, R. G., H. M. Nitowsky, and B. Childs. 1963. Demonstration of two populations of cells in the human female heterozygous for glucose-6 phosphate dehydrogenase variants. Proc. Natl. Acad. Sci. U.S.A. 50:481–485.

Davidson, W. M., and D. R. Smith. 1954. A morphological sex difference in the polymorphonuclear neutrophil leucocytes. Acta Cytol. 6:13–24.

Davidson, W. M., and D. R. Smith. 1963. The nuclear sex of leucocytes. *In:* Overzier. (ed). Intersexuality, pp. 72–85. Academic Press, New York.

Davis, R. W., and N. Davidson. 1968. Electron-microscopic visualization of deletion mutations. Proc. Natl. Acad. Sci. U.S.A. 60:243–250.

Davis, W. H., and L. M. Greenblatt. 1967. Cytoplasmic male sterility in alfalfa. J. Hered. 58:301–305.

Dawson, G. W. P. 1962. An introduction to the cytogenetics of polyploids. Blackwell Scientific Publications, Oxford.

Debergh, P., and C. Nitsch. 1973. Premiers résultats sur la culture *in vitro* de grains de pollen isolés chez la tomate. C. R. Acad. Sci. Paris 276:1281–1284.

De Graaf, R. 1672. De mulierum organis generationi inservientibus tractatus novus— Leiden.

De Grouchy, J. 1965. A complex familial chromosome translocation. Amer. J. Human Genet. 17:501–509.

De Grouchy, J., M. Aussannaire, H. E. Brissaud, and M. Lamy. 1966. Aneusomic de recombinaison: three further examples. Amer. J. Human Genet. 18:467–484.

De Grouchy, J., A. Herrault, and J. Cohen-Solal. 1968. Une observation de chromosome 18en anneau (18r). Ann. Genet., Semaine Hop. 11:33–38.

De Grouchy, J., P. Royer, C. Salmon, and M. Lamy. 1964. Délétion partielle des bras longs du chromosome 18. Path. Biol. Semaine Hop. [N.S.] 12:579–582.

De Grouchy, J., C. Turleau, and C. Léonard. 1971. Etude en fluorescence d'une trisomie C mosaïque, probablement 8: 46,XY/47,XY ? 8+. Ann. Génét. 14:69.

De Grouchy, J., F. Josso, S. Beguin, C. Turleau, P. Jalbert, and C. Laurent. 1974. Deficit en facteur VII de la coagulation chez trois sujects trisomiques 8. Ann. Genet. 17:105–108.

De Hann, I., and J. Doorenbos. 1951. Meded. Landb. Hoogesch., Wageningen 51:151.

De La Chapelle, A., J. Schröder et al. 1974. Pericentric inversions of human chromosomes 9 and 10. Am. J. Hum. Genet. 26:746–766.

Demerec, M. 1948. Induction of mutations in *Drosophila* by dibenzanthracene. Genetics 33:337–348.

Demerec, M., E. A. Adelberg, A. J. Clark, and P. E. Hartman. 1966. A proposal for uniform nomenclature in bacterial genetics. Genetics 54:61–76.

Demerec, M., and M. E. Hoover. 1936. Three related X-chromosome deficiencies in *Drosophila*. J. Hered. 27:207–212.

Dempster, L. T., and G. L. Stebbins. 1968. A cytotaxonomic revision of the fleshy-fruited *Galium* species of the Californias and Southern Oregon (Rubiaceae). Univ. Calif. Publ. Bot. 46:1–52.

Dermen, H. 1941. Simple and complex periclinal tetraploidy in peaches induced by colchicine. Proc. Amer. Soc. Hort. Sci. 38:141.

Dermen, H. 1945. The mechanism of colchicine-induced cytohistological changes in cranberry. Amer. J. Bot. 32:387–394.

Dermen, H. 1953. Periclinal cytochimeras and origin of tissues in stem and leaf of peach. Amer. J. Bot. 40:154–168.

Dermen, H. 1960. Nature of bud sports. Amer. Hort. Mag. 39:123–173.

DeRobertis, E. D. P., W. W. Nowinski, and F. A. Saez. 1965. Cell biology. 4th ed. W. B. Saunders, Philadelphia.

DeWeerd-Kastelein, E. A., W. Keijzer, G. Rainaldi, and D. Bootsma. 1977. Induction of sister chromatid exchanges in *Xeroderma pigmentosum* cells after exposure to ultraviolet light. Mutation Res. 45:253–261.

Dewey, D. R. 1965. Morphology, cytology and fertility of synthetic hybrids of *Agropyron spicatum* x *Agropyron dasystachyum-riparium*. Bot. Gaz. 126:269–275.

Dewey, D. R. 1970. Genome relations among diploid *Elymus junceus* and certain tetraploid and octoploid *Elymus* species. Amer. J. Bot. 57:633–639.

Dewey, D. R. 1975. The origin of *Agropyron smithii*. Amer. J. Bot. 62:524–530.

Dewey, W. C., A. Westra, H. H. Miller, and H. Nagasawa. 1971. Int. J. Radiat. Biol. 20:505.

Dhillon, T. S., and E. D. Garber. 1960. The genus *Collinsia*. X. Aneuploidy in *C. heterophylla*. Bot. Gaz. 121:125–133.

Diaz, M., and C. Pavan. 1965. Changes in chromosomes induced by microorganism infection. Proc. Natl. Acad. Sci. U.S.A. 54:1321–1328.

Diaz, M., C. Pavan, and R. Basile. 1969. Effects of a virus and a microspordian infection in chromosomes of various tissues of *Rhynchosciara angelae* (Nonato et Pavan, 1951). Rev. Brasil. Biol. 29:191–206.

Digby, L. 1910. The somatic, premeiotic and meiotic nuclear divisions of *Galtonia candicans*. Ann. Bot. 24:727–758.

Dippell, R. 1959. Distribution of DNA of *kappa* particles of *Paramecium* in relation to their bacterial affinities. Science 130:1415.

Dobzhansky, T. 1941. Genetics and the origin of species. 2nd ed. Columbia Univ. Press, New York.

Dobzhansky, T., and J. Schultz. 1934. The distribution of sex-factors in the X-chromosome of *Drosophila melanogaster*. J. Genet, 28:349–386.

Dodds, K. S., and N. W. Simmonds. 1938. Genetical and cytological studies of *Musa*. IX. The origin of an edible diploid and the significance of interspecific hybridization in the banana complex. J. Genet. 48:285–296.

Dougan, L., and H. J. Woodcliff. 1965. Presence of two Ph[1] chromosomes in cells from a patient with chronic granulocytic leukaemia. Nature 205:405–406.

Dover, G. A., and R. Riley. 1973. The effect of spindle inhibitors applied before meiosis on meiotic chromosome pairing. J. Cell Sci. 12:143–161.

Drets, M. E., and M. W. Shaw. 1971. Specific banding patterns of human chromosomes. Proc. Natl. Acad. Sci. U.S.A. 68:2073–2077.

Driscoll, C. J., and N. L. Darvey. 1970. Chromosome pairing: effect of colchicine on an isochromosome. Science 169:290–291.

Dronamraju, K. R. 1965. The function of the Y chromosomes in man, animals and plants. Adv. Genet. 13:227–310.

Dublin, P. 1973. Haploids of *Theobroma cacao*. Compt. R. Acad. Sci., Paris 276:757–759.

Dujon, B., and G. Michaelis. 1974. Extrakaryotic inheritance. Progr. Bot. 36:236–246.

Dunwell, J. M., and N. Sunderland. 1973. Anther culture of *Solanum tuberosom* L. Euphytica 22:317–323.

DuPraw, E. J. 1965a. The ultrastructure of human chromosomes. Amer. Zool. 5:648.

DuPraw, E. J. 1965b. Macromolecular organization of nuclei and chromosomes: a folded fibre model based on whole-mount electron microscopy. Nature 206:338–343.

DuPraw, E. J. 1965c. The organization of nuclei and chromosomes in honeybee embryonic cells. Proc. Natl. Acad. Sci. U.S.A. 53:161–168.

DuPraw, E. J. 1966. Evidence for a "folded fibre" organization in human chromosomes. Nature 209:577–581.

DuPraw, E. J. 1968. Cell and molecular biology. Academic Press, Inc., New York.

DuPraw, E. J., and P. M. M. Rae. 1966. Polytene chromosome structure in relation to the "folded fibre" concept. Nature 212:598–600.

DuPraw, E. J. 1970. DNA and chromosomes. Holt, Reinhart and Winston, Inc., New York.

Dussoix, D., and W. Arber. 1962. Host specificity on DNA produced by *Escherichia coli*. II. Control over acceptance of DNA from infecting phage λ. J. Molec. Biol. 5:37–49.

Dutrillaux, B. 1973. Nouveau systeme de marquage chromosomique: les bandes T. Chromosoma 41:395–402.

Dutrillaux, B., J. deGrouchy, C. Finaz and J. LeJeune. 1971. Evidence de la structure fine des chromosomes humains par digestion enzymatique (pronase) en particulier. C. R. Acad. Sci. 273:587–588.

Dutrillaux, B., and J. LeJeune. 1971. Sur une nouvelle technique d'analyse du caryotype humain. C. R. Acad. Sci. Paris 272:2638–2640.

Duvick, D. N. 1966. Influence of morphology and sterility on breeding methodology. *In:* Frey, K. J. (ed.). Plant Breeding. A symposium held at Iowa State University, pp. 85–138. Iowa State Univ. Press, Ames, Iowa.

Eberle, D. 1963. Meiotische Chromosomen des Mannes. Klin. Wochenschr. 41:848–856.

Ecochard, R., M. S. Ramanna, and D. de Nettancourt. 1969. Detection and cytological analysis of tomato haploids. Genetica 40:181–190.

Edwards, J. H., D. G. Harnden, A. H. Cameron, V. M. Crosse, and O. H. Wolff. 1960. A new trisomic syndrome. Lancet I:787–790.

Edwards, R. G. 1954. The experimental induction of pseudogamy in early mouse embryos. Experientia 10:499–500.

Edwards, R. G. 1957a. The experimental induction of gynogenesis in the mouse. I. Irradiation of the sperm by x-rays. Proc. Roy. Soc. (London), Ser. B. 146:469–487.

Edwards, R. G. 1957b. The experimental induction of gynogenesis in the mouse. II. Ultra-violet irradiation of the sperm. Proc. Roy. Soc. (London), Ser. B. 146:488–504.

Edwards, R. G. 1958. The experimental induction of gynogenesis in the mouse. III. Treatment of sperm with trypaflavine, toluidine blue, or nitrogen mustard. Proc. Roy. Soc. (London), Ser. B. 149:117–129.

Edwardson, J. R. 1970. Cytoplasmic male sterility. Bot. Rev. 36:341–420.

Edwardson, J. R., and M. K. Corbett. 1961. Asexual transmission of cytoplasmic male sterility. Proc . Natl. Acad. Sci. U.S.A. 47:390–396.

Edwardson, J. R., and H. E. Warmke. 1967. Fertility restoration in cytoplasmic male-sterile *Petunia*. J. Hered. 58:195–196.

Ehrendorfer, F. 1961. Zur Phylogenie der Gattung *Achillea*. V. Akzessorische Chromosomen bei *Achillea:* Struktur, cytologisches Verhalten, zahlenmässige Instabilität und Endstehung. Chromosoma 11:523–552.

Ehrensberger, R. 1948. Versuche zur Auslösung von Haploidie bei Blütenpflanzen. Biol. Zentralbl. 67:537–546.

Eigsti, O. J. 1937. Pollen tube behaviour in self-fertile and interspecific pollinated Rese-daceae. Amer. Nat. 71:520–521.

Eigsti, O. J. 1940. The effects of colchicine upon the division of the generative cell in *Polygonatum, Tradescantia,* and *Lilium.* Amer. J. Bot. 27:512–524.

Eigsti, O. J. 1942. A cytological investigation of *Polygonatum* using colchicine-pollen tube technique. Amer. J. Bot. 29:626–636.

Eigsti, O. J., and P. Dustin. 1955. Colchicine—in agriculture, medicine, biology, and chemistry. Iowa State College Press, Ames, Iowa.

Einset, J. 1943. Chromosome length in relation to transmission frequency of maize tri-somes. Genetics 28:349–364.

Einset, J. 1947. Chromosome studies in *Rubus.* Gentes Her. 7:181–192.

Einset, J., and B. Lamb. 1951. Chromosome numbers of apple varieties and sports. III. Proc. Amer. Soc. Hort. Sci. 58:103–108.

Eisen, G. 1900. The spermatogenesis of *Batrachoseps.* J. Morph. 17:1–117.

El Alfi, O. S., H. C. Powell, and J. J. Biesele. 1963. Possible trisomy in chromosome group 6-12 in a mentally retarded patient. Lancet I:700–701.

El'Darov, A. L. 1965. Obtaining haploid androgenesis in the axolotl *(Ambystoma mexi-canum)* by ultraviolet irradiation and a cytomorphological study of the mechanism by which it occurs. Tsitologiya 7:704–711. (Russian).

Elgin, S. C. R., and H. Weintraub. 1975. Chromosomal proteins and chromatin struc-ture. Ann. Rev. Biochem. 44:725–774.

Ellerström, S., and J. Sjödin. 1974. Studies on the use of induced autopolyploidy in the breeding of red clover. III. Frequency and behaviour of aneuploids in a tetraploid clo-ver *Ley.* Z. Pflanzenz. 71:253–263.

Elliott, F. C. 1958. Plant Breeding and cytogenetics. McGraw-Hill Book Co., Inc., New York.

Ellis, R. J. 1969. Chloroplast ribosomes: stereospecificity of inhibition by chlorampheni-col. Science 163:477–478.

Ellis, R. J. 1974. The biogenesis of chloroplasts: protein synthesis by isolated chloroplasts. Biochem. Soc. Transact., 544th Meetg., London 2:179–182.

Ellis, R. J., and L. S. Penrose. 1961. Enlarged satellites and multiple malformations in the same pedigree. Ann Human Genet. 25:159–162.

Emerson, R. A. 1914. The inheritance of a recurring somatic variation in variegated ears of maize. Amer. Nat. 48:87–115.

Emerson, R. A., and E. M. East. 1913. The inheritance of quantitative characters in maize. Nebraska Agr. Exp. Sta. Res. Bull. 2:1–120.

Emerson, R. A., G. W. Beadle, and A. C. Fraser. 1935. A summary of linkage studies in maize. Cornell Univ. Exp. Sta. Mem. 180:1–83.

Emery, W. H. P. 1957. A study of reproduction in *Setaria macrostachya* and its relatives in the southwestern United States and northern Mexico. Bull. Torrey Botan. Club 84:106–121.

Emery, W. H. P., and W. V. Brown. 1958. Apomixis in the Gramineae, tribe Andropo-goneae: *Heteropogon contortus.* Madrõno 14:238–246.

Emsweller, S. L. 1951. Recent developments in lily breeding techniques. Sci. Monthly 72:207–216.

Endrizzi, J. E. 1974. Alternate-1 and alternate-2 disjunctions in heterozygous reciprocal translocations. Genetics 77:55–60.

Endrizzi, J. E., and R. J. Kohel. 1966. Use of telosomes in mapping three chromosomes in cotton. Genetics 54:535–550.

Engvild, K. C. 1974. Plantlet ploidy and flower bud size in tobacco anther cultures. Her-editas 76:320–322.

Enriques, P. 1922. Hologynic heredity. Genetics 7:583.

Ephrussi, B. 1949. Action de l'acriflavine sur les levures. Unités biologiques dovées de continuité génétique. Centre Nat. Rech. Sci., Paris.

Ephrussi, B. 1953. Nucleo-cytoplasmic relations in micro-organisms. Clarendon Press, Oxford.

Ephrussi, B., H. Hottinguer, and A.-M Chimenes. 1949. Action de l'acriflavine sur les levures. I. La mutation < petite colonie >. Ann. Inst. Pasteur 76:351-367.

Ephrussi, B., and M. C. Weiss. 1969. Hybrid somatic cells. *In:* Srb, A. M., R. D. Owen, and R. S. Edgar. (eds.). Facets of genetics. Readings from Scientific American. W. H. Freeman and Co., San Francisco.

Erickson, J. 1965. Meiotic drive in *Drosophila* involving chromosome breakage. Genetics 51:555-571.

Erickson, J., and G. D. Hanks. 1961. Time of temperature sensitivity of meiotic drive in *Drosophila melanogaster*. Amer. Nat. 95:247-250.

Erickson, J. D. 1978. Downs syndrome, paternal age, maternal age and birth order. Ann. Hum. Genet. 41:289-298.

Ervin, C. D. 1941. A study of polysomaty in *Cucumis melo*. Amer. J. Bot. 28:113-124.

Evans, D. A., and E. F. Paddock. 1976. Comparisons of somatic crossing over frequency in *Nicotiana tabacum* and three other crop species. Can. J. Gen. Cyt. 18:57-63.

Evans, E. P., C. E. Ford, R. S. K. Chaganti, C. E. Blank, and H. Hunter. 1970. XY spermatocytes in an XYY male. Lancet I:19-20.

Evans, E. P., M. F. Lyon, and M. Daglish. 1967. A mouse translocation giving a meta-centric marker chromosome. Cytogenetics 6:105-119.

Evans, E. P., and J. S. Phillips. 1975. Inversion heterozygosity and the origin of XO daughters of *Bpa/+* female mice. Nature 256:40-41.

Evans, H. J. 1960. Supernumerary chromosomes in wild population of the snail *Helix pomatia* L. Heredity 15:129-138.

Evans, H. J. 1962. Chromosome aberrations induced by ionizing radiation. Intern. Rev. Cytol. 13:221-321.

Evans, H. J. 1967. Repair and recovery from chromosome damage induced by fraction-ated x-ray exposures. *In:* Silini, G. (ed.). Radiation research. Proc. 3rd Intern. Congr. Rad. Res. 1966. North-Holland Publ. Co., Amsterdam.

Evans, H. J. 1968. Repair and recovery of chromosome and cellular levels: similarities and differences. *In:* Recovery and regain mechanisms in radiobiology. Brookhaven Symp. Biol. 21:111-133.

Evans, H. J., and D. Scott. 1964. Influence of DNA synthesis on the production of chro-matid aberrations by x-rays and maleic hydrazide in *Vicia faba*. Genetics 49:17-38.

Evans, H. J., G.J. Neary, and F. S. Williamson. 1959. The relative biological efficiency of single doses of fast neutrons and gamma rays on *Vicia faba* roots and the effects of oxygen. II. Chromosome damage: the production of micronuclei. Ind. J. Radiat. Biol. 1:216-229.

Fahmy, O. G., and M. J. Fahmy. 1955. Cytogenetic analysis of the actions of carcinogens and tumour inhibitors in *Drosophila melanogaster*. III. Chromosome structural changes induced by 2:4:6-tri(ethyleneimino)-1:3:5-triazine. J. Genetics 53:181-199.

Falconer, D. S. 1960. Introduction to quantitative genetics. Oliver and Boyd, Edinburgh.

Falk, H., B. Liedvogel, and P. Sitte. 1974. Circular DNA in isolated chromoplasts. Z. Naturforsch. 29c:541-544.

Fankhauser, G. 1937. The sex of a haploid metamorphosed salamander (*Triton taeniatus* Laur.). Genetics 22:192-193.

Fankhauser, G. 1945. The effect of changes in chromosome number on amphibian devel-opment. Quart. Rev. Biol. 20:20-78.

Fankhauser, G., and R. B. Griffiths. 1939. Induction of triploidy and haploidy in the newt, *Triturus viridescens*, by cold treatment on unsegmented eggs. Proc. Natl. Acad. Sci. U.S.A. 25:233-238.

Faye, G., H. Fukuhara, C. Grandchamp, J. Lazoswka, F. Michel, J. Casey, G. S. Getz, J. Locker, M. Rabinowitz, M. Bolotin-Fukuhara, D. Coen, J. Deutsch, B. Dujon, P. Netter, and P. P. Slonimski. 1973. Mitochondrial nucleic acids in the petite colonie mutants: deletions and repetitions of genes. Biochimie 55:779-792.

Fedak, G. 1972. Production of haploids in barley. Barley Newsl. 16:36-37.

Fedorov, A. A. 1974. Chromosome numbers of flowering plants. Otto Koeltz Science Publishers, Koenigstein, Germany.

Feldman, M. 1966. The effect of chromosomes 5B, 5D, and 5A on chromosomal pairing in *Triticum aestivum*. Proc. Natl. Acad. Sci. U.S.A. 55:1447–1453.

Ferguson-Smith, M. A. 1964. The sites of nucleolus formation in human pachytene chromosomes. Cytogenetics 3:124–134.

Ferguson-Smith, M. A. 1967. Clinical cytogenetics. *In:* Crow, J. F., and J. V. Neel. (eds.). Proc. 3rd Intl. Congr. Human Genet., Chicago, 1966, pp. 69–77. The John Hopkins Press, Baltimore.

Ferguson-Smith, M. A., and S. D. Handmaker. 1961. Observations on the satellited human chromosomes. Lancet I:638–640.

Ferguson-Smith, M. A., B. F. Newman, P. M. Ellis, D. M. G. Thomson, and I. D. Riley. 1973. Assignment by deletion of human red cell acid phosphatase gene locus to the short arm of chromosome 2. Nature N.B. 243:271–274.

Fernandes, A. 1934. Bol. Soc. Brot. Ser. 2, 11:1.

Feulgen, R. 1914. Über die 'Kohlenhydratgruppe' in der echten Nucleinsäure. Vorläufige Mitteilungen. F. Hoppe Seyler's Zeitschr. Physiol. Chem. 92:154–158.

Feulgen, R., und H. Rossenbeck. 1924. Mikroskopisch-chemischer Nachweis einer Nukleinsäure vom Typus der Thymonukleinsäure und die darauf beruhende elektive Färbung von Zellkernen in mikroskopischen Präparaten. F. Hoppe-Seyler's Zeitschr. Physiol. Chem. 135:203–248.

Fiandt, M., W. Szybalski, and M. H. Malamy. 1972. Polar mutations in lac,gal and phage λ consist of a few DNA sequences inserted with either orientation. Molec. Gen. Genet. 119:223–231.

Finn, W. W. 1937. Entwicklungsgeschichte des männlichen Gametophyten der Angiospermen. Tra. Inst. Rech. Sci. Biol. Univ. Kiew, pp. 71–86.

Fisher, H. E. 1962. Über Vorkommen und Bedeutung verschiedener Genomstufen bei *Beta vulgaris* L. Züchter 32:40–48.

Flagg, R. O. 1958. A mutation and an inversion in *Rhoeo discolor*. J. Hered. 49:185–188.

Flamm, W. G., and M. L. Birnstiel. 1964. The nuclear synthesis of ribosomes in cell cultures. Biochem. Biophys. Acta 87:101–110.

Flavell, R. A., and P. O. Trampe. 1973. The absence of an integrated copy of mitochondrial DNA in the nuclear genome of *Tetrahymena pyriformis*. Biochim. Biophys. Acta 308:101–105.

Flemming, W. 1879. Contributions to the knowledge of the cell and its life phenomena. Arch. Mikr. Anat. 16:302–406.

Flemming, W. 1882. Zellsubstanz, Kern- und Zellteilung. Vogel, Leipzig.

Flovik, K. 1940. Chromosome numbers and polploidy within the flora of Spitzbergen. Hereditas 26:430–440.

Focke, W. O. 1881. Die Pflanzenmischlinge, ein Beitrag zur Biologie der Gewächse. Borntraeger, Berlin.

Ford, C. E., T. C. Carter, and J. L. Hamerton. 1956. Cytogenetics of a mouse translocation. Heredity 10:284.

Ford, C. E., K. W. Jones, O. J. Miller, U. Mittwoch, L. S. Penrose, M. A. C. Ridler, and A. Shapiro. 1959. The chromosomes in a patient showing both mongolism and the Kleinefelter syndrome. Lancet I:709–710.

Ford, L. 1970. Chromosome association in *Zea mays* monoploids. Nucleus 13:99–105.

Forer, A., and C. Koch. 1973. Influence of autosome movements and of sex-chromosome movements on sex-chromosome segregation in crane fly spermatocytes. Chromosoma 40:417–442.

Foster, T. J. 1976. R factor mediated tetracycline resistance in *Escherichia coli* K12 dominance of some tetracycline sensitive mutants and relief of dominance by deletion. Molec. Gen. Genetics 143:339–344.

Fraccaro, M., D. Ikkos, J. Lindstein, R. Luft, and K. Kaijser. 1960. A new type of chromosomal abnormality in gonadal dysgenesis. Lancet II:1144.

Frandsen, K. J. 1945. Tidsskr. Plateavl. 49:445.

Frandsen, N. O. 1967. Haploid production in potato breeding material with intensive backcrossing to wild species. Züchter 37:120–134.

Frankel, R. 1956. Graft-induced transmission to progeny of cytoplasmic male sterility in *Petunia*. Science 124:684–685.

Frankel, R. 1962. Further evidence of graft induced transmission to progeny of cytoplasmic male sterility in *Petunia*. Genetics 47:641–646.

Frankel, R. 1971. Genetical evidence on alternative maternal and Mendelian hereditary elements in *Petunia hybrida*. Heredity 26:107–119.

Franklin, J. 1973. The removal of mu particles from *Paramecium aurelia* (stock 540) by penicillin. J. Gen. Microbiol. 74:175–177.

Franklin, N. C., and S. E. Luria. 1961. Transduction by bacteriophage Pl and the properties of the 2ac genetic region in *E. coli* and *S. dysenteriae*. Virology 15:299–311.

Franklin, R. E., and R. G. Gosling. 1953. Evidence for 2-chain helix in crystalline structure of sodium deoxyribonucleate. Nature 172:156–157.

Fredericq, P. 1954. Transduction génétique des propriétés colicinogènes chez *Escherichia coli*. Compt. Rend. Soc. Biol. 148:399–402.

Fredericq, P., and M. Betz-Bareau. 1953. Transfert génétique de la propriété colicinogène en rapport avec la polarité F des parents. Compt. Rend. Soc. Biol. 147:2043–2045.

Freese, E. 1958. The arrangement of DNA in the chromosome. Cold Spring Harbor Symp. Quant. Biol. 23:13–18.

Frost, H. B. 1926. Polyembryony, heterozygosis and chimera in *Citrus*. Hilgardia 1:365–402.

Frost, H. B. 1938a. Nucellar embryony and juvenile characters in clonal varieties of *Citrus*. J. Hered. 29:423–432.

Frost, H. B. 1938b. The genetics of *Citrus*. Cur. Sci. (Special number on genetics):24–27.

Fröst, S. 1959. The cytological behaviour and mode of transmission of accessory chromosomes in *Plantago serraria*. Hereditas 45:191–210.

Fröst, S. 1969. The inheritance of accessory chromosomes in plants, especially in *Ranunculus acris* and *Phleum nodosum*. Hereditas 61:317–326.

Fujita, S., and K. Takamoto. 1963. Synthesis of messenger RNA on the polytene chromosomes of dipteran salivary gland. Nature 200:494–495.

Gagliardi, A. R. T., E. H. Tajara, M. Varella-Garcia and L. M. A. Moreira. 1978. Trisomy 8 syndrome. J. Med. Genet. 15:70–73.

Gall, J. G. 1961. Centriole replication. A study of spermatogenesis in the snail *Viviparus*. J. Biophys. Biochem. Cyt. 10:163–193.

Gall, J. G., and H. G. Callan. 1962. H^3-uridine incorporation in lampbrush chromosomes. Proc. Natl. Acad. Sci., U.S.A. 48:562–570.

Gall, J. G., and M. L. Pardue. 1969. Formation and detection of RNA-DNA hybrid molecules in cytological preparations. Proc. Natl. Acad. Sci., U.S.A. 63:378–383.

Galton, F. 1869. Hereditary genius. 3rd ed. (Introduction by C. D. Darlington, Fontana, 1962). Murray, London.

Galton, F. 1876. The history of twins as a criterion of the relative powers of nature and nurture. J. Anthrop. Inst. 5:391–406.

Garber, E. D. 1964. The genus *Collinsia*. XXII. Trisomy in *C. heterophylla*. Bot. Gaz. 125:46–50.

Gates, R. R. 1908. A study of reduction in *Oenothera rubinervis*. Bot. Gaz. 46:1–34.

Gates, R. R. 1911. Pollen formation in *Oenothera gigas*. Ann. Bot. 25:909–974.

Gauthier, F. M., and R. C. McGinnis. 1968. The meiotic behavior of a nulli-haploid plant in *Avena sativa* L. Can. J. Genet. Cytol. 10:186–189.

Gay, D., and C. Ozolins. 1968. Transmission héréditaire du virus *sigma* par des Drosophiles porteuses du géne "refractaire." I. Etude d'une souche virale p⁻g⁻. Ann. Inst. Pasteur 114:29–48.

Geitler, L. 1937. Die Analyse des Kernbaus und der Kernteilung der Wasserlaüfer *Gerris lateralis* und *Gerris lacustris* (Hemiptera, Heteroptera) und die Somadifferenzierung. Zellforsch. 26:641–672.

Geitler, L. 1939. Die Entstehung der polyploiden Somakerne der Heteropteren durch Chromosomenteilung ohne Kernteilung. Chromosoma 1:1–22.

Geitler, L. 1941. Das Wachstum des Zellkerns in tierischen und pflanzlichen Geweben. Ergebn. Biol. 18:1–54.

Gellert, M. 1967. Formation of covalent circles of lambda DNA by *E. coli* extracts. Proc. Natl. Acad. Sci. U.S.A. 57:148–155.

Gentcheff, G., and A. Gustafsson. 1939. The double chromosome reproduction in *Spinacia* and its causes. I. Normal behavior.Hereditas 25:349–358.

George, L., and S. Narayanaswamy. 1973. Haploid *Capsicum* through experimental androgenesis. Planta 78:467–470.

Geraci, G., A. Esen, and R. K. Soost. 1975. Triploid progenies from 2x x 2x crosses of *Citrus* cultivars. J. Hered. 66:177–178.

Gerassimova, H. 1936a. Experimentally produced haploid plants in *Crepis tectorum*. Biologitcheski J. 5:895–900.

Gerassimova, H. 1936b. Experimentell erhaltene haploide Pflanzen von *Crepis tectorum*. Planta 25:696–702.

Gerlach, W. L. 1977. N-banded karyotypes of wheat species. Chromosoma 62:49–56.

German, J. 1969. Bloom's syndrome. I. Genetical and clinical observations in the first twenty-seven patients. Am. J. Hum. Genet. 21:196–227.

German, J., R. Archibald, and D. Bloom. 1965. Chromosomal breakage in a rare and probably genetically determined syndrome of man. Science 148:506–507.

German, J., and L. P. Crippa. 1966. Chromosomal breakage in diploid cell lines from Bloom's syndrome and Fanconi's anaemia. Ann. Génét. 9:143–154.

Gey, W. 1967. Partial deletion of the long arm of chromosome 13-15:Dq-. Mammal. Chrom. Newsl. 8:267.

Giannelli, F. 1965a. Autoradiographic identification of the D(13-15) chromosome responsible for D_1 trisomic Patau's syndrome. Nature 208:669–672.

Giannelli, F. 1965b. Autoradiographic identification of the chromosomes of the D group (13-15) Denver: an autoradiographic and measurement study. Cytogenetics 6:420–435.

Giannelli, F., and R. M. Howlett. 1966. The identification of the chromosomes of the D group (13-15) Denver: an autoradiographic and measurement study. Cytogenetics 5:186–205.

Gibson, I. 1973. Transplantation of killer endosymbionts in *Paramecium*. Nature 241:127–129.

Gierer, A. 1966. Model for DNA and protein interactions and the function of the operator. Nature 212:1480–1481.

Gill, J. J. B. 1971. The cytology and transmission of accessory chromosomes in *Cochlearia pyrenaica* DC. (Cruciferae). Caryologia 24:173–181.

Gilles, C. B. 1972. Reconstruction of the *Neurospora crassa* pachytene karyotype from serial sections of synaptonemal complexes. Chromosoma 36:119–130.

Givens, J. F. 1974. Molecular hybridization and cytological characterization of plants partially hyperploid for different segments of the nucleolar organizer region of *Zea mays*. Ph.D. Thesis, Univ. Minnesota.

Givens, J. F., and R. L. Phillips. 1976. The nucleolus organizer of maize (*Zea mays* L.). Ribosomal RNA gene distribution and nucleolar interactions. Chromosoma 57:103–117.

Glass, H. B. 1933. A study of dominant mosaic eye-color mutants in *Drosophila melanogaster*. II. Tests involving crossing-over and non-disjunction. J. Genetics 28:69–112.

Gohil, R. N., and A. K. Koul. 1971. Desynapsis in some diploid and polyploid species of *Allium*. Can. J. Genet. Cytol. 13:723–728.

Goldschmidt, R. 1915. Vorläufige Mitteilung über weitere Versuche zur Vererbung und Bestimmung des Geschlechts. Biol. Zbl. 35:565.

Goldschmidt, R. 1916. Experimental intersexuality and the sex problem. Amer. Nat. 50:705–718.

Goldschmidt, R. 1920. Untersuchungen über Intersexualität. I. Z. Ind. Abst. Vererbungsl. 23:1–199.

Goldschmidt, R. 1922. Untersuchungen über Intersexualität. II. Z. Ind. Abst. Vererbungsl. 29:145–185.

Goldschmidt, R. 1923. Untersuchungen über Intersexualität. III. Z. Ind. Abst. Vererbungsl. 31:100–133.

Goldschmidt, R. 1929. Untersuchungen über Intersexualität. IV. Z. Ind. Abst. Vererbungsl. 49:168–242.

Goldschmidt, R. 1934. Untersuchungen über Intersexualität. VI. Z. Ind. Abst. Vererbungsl. 67:1–40.

Goldschmidt, R. B. 1934. *Lymantria*. Bibl. Genetica 11:1–186.

Goldschmidt, R. 1945. The structure of Podoptera, a homeotic mutant of *Drosophila melanogaster*. J. Morph. 77:71–301.

Golubovskaya, I. N. 1979. Meiotic mutation in maize. Maize Genet. Coop. Newslet. 53:66–70.

Goodenough, U. 1978. Genetics. 2nd ed. Holt, Rinehart and Winston, New York.

Goodspeed, T. H., and P. Avery. 1929. The occurrence of a *Nicotiana glutinosa* haplont. Proc. Natl. Acad. Sci. U.S.A. 15:502–504.

Goodspeed, T. H., and P. Avery. 1939. Trisomic and other types in *Nicotiana sylvestris*. J. Genet. 38:381–458.

Gottschalk, W. 1954. Die Chromosomenstruktur der Solanaceen unter Berücksichtigung phylogenetischer Fragestellung. Chromosoma 6:539–626.

Gottschalk, W. 1977. Mutation. Prog. Bot. 39:133–172.

Gottschalk, W. 1978. Problems in polyploidy research. Nucleus 21:99–112.

Gottschalk, W., and S. R. Baquar. 1971. Desynapsis in *Pisum sativum* induced through gene mutation. Can. J. Genet. Cytol. 13:138–143.

Gottschalk, W., und A. Jahn. 1964. Cytogenetische Untersuchungen an desynaptischen und männlich-sterilen Mutanten von *Pisum*. Z. Vererbungsl. 95:150–166.

Gottschalk, W., and M. L. H. Kaul. 1974. The genetic control of microsporogenesis in higher plants. Nucleus 17:133–166.

Gottschalk, W., and H. D. Klein. 1976. The influence of mutated genes on sporogenesis. A survey on the genetic control of meiosis in *Pisum sativum*. Theor. Appl. Genet. 48:23–34.

Gottschalk, W., und O. Konvička. 1972. Desynapsis bei *Brassica oleracea*. Biologia 27:179–186.

Gottschalk, W., and M. Milutinovic. 1973. Trisomics from desynaptic mutants. Nucleus 16:1–10.

Gottschalk, W., und N. Peters. 1955. Die Chromosomenstruktur diploider Wildkartoffel-Arten und ihr Vergleich mit der Kulturkartoffel. Z. Pflanzenz. 34:351–374.

Gottschalk, W., und N. Peters. 1956. Weitere Untersuchungen über die Morphologie der Pachytänchromosomen tubarer *Solanum*-Arten. Z. Pflanzenz. 36:421–433.

Gowen, J. W. 1933. Meiosis as a genetic character in *Drosophila melanogaster*. J. Exp. Morph. 65:83–106.

Graaf, R. de. 1672. De mulierum organis generationi inservientibus tractatus novus—Leiden.

Granick, S., and A. Gibor. 1967. The DNA of chloroplasts, mitochondria, and centrioles. Progr. Nucl. Acid Res. Mol. Biol. 6:143–186.

Grant, V. 1971. Plant speciation. Columbia University Press, New York.

Gray, J. E., D. E. Mutton, and D. W. Ashby. 1962. Pericentric inversion of chromosome 21: a possible further cytogenetic mechanism in mongolism. Lancet I:21–23.

Green, B. R., and H. Burton. 1970. *Acetabularia* chloroplast DNA: electron microscopic visualization. Science 168:981–982.

Greenberg, R. 1962. Dros. Inf. Serv. 36:70.

Greilhuber, J. 1977. Why plant chromosomes do not show G-bands. Theor. Appl. Genet. 50:121–124.

Grell, K. G., und A. Ruthmann. 1964. Über die Karyologie des Radiolars *Aulacantha scolymantha* und die Feinstruktur seiner Chromosomen. Chromosoma 15:185–211.

Grell, R. F. 1969. Meiotic and somatic pairing. *In:* E. W. Caspari and A. W. Raven. (eds.). Genetic organization. pp. 361–492. Academic Press, New York.

Grell, S. M. 1946a. Cytological studies in *Culex.* I. Somatic reduction divisions. Genetics 31:60–76.

Grell, S. M. 1946b. Cytological studies in *Culex.* II. Diploid and meiotic divisions. Genetics 31:77–94.

Gresshoff, P. M., and C. H. Doy. 1972. Development and differentiation of haploid *Lycopersicon esculentum* (tomato). Planta 107:161–170.

Grew, N. 1672. The anatomy of vegetables begun with a general account of vegetation founded thereon. London.

Grew, N. 1682. The anatomy of plants with an idea of a philosophical history of plants and several other lectures read before the Royal Society, London.

Griffen, A. B., and M. C. Bunker. 1964. Three cases of trisomy in the mouse. Proc. Natl. Acad. Sci. U.S.A. 52:1194–1198.

Griffen, A. B., and M. C. Bunker. 1967. Four further cases of autosomal primary trisomy in the mouse. Proc. Natl. Acad. Sci. U.S.A. 58:1146–1152.

Grindley, N. D. F. 1978. IS*1* insertion generates duplication of a nine base pair sequence at its target site. Cell 13:419.

Gripenberg, U. 1967. The cytological behavior of a human ring-chromosome. Chromosoma 20:284–289.

Grobstein, C. 1977. The recombinant-DNA debate. Sci. Amer. 237:22–33.

Gropp, A., and G. Flatz. 1967. Chromosome breakage and blastic transformation of lymphocytes in *Ataxia telangiectasia.* Humangen. 5:77–79.

Grunewaldt, J., and S. Malepszy. 1975. Observations on anther callus from *Hordeum vulgare* L. *In:* Gaul, H. (ed.). Barley genetics. III. Proc. Third Internatl. Barley Genet. Symp., Garching, pp. 367–373.

Guanti, G., and P. Petrinelli. 1974. Cell Differ. 2:319.

Guénin, H. A. 1949. L'évolution de la formule chromosomique dans le genre *Blaps* (Coleopt. Tenebr.). Rev. Suisse Zool. 56:336.

Guénin, H. A. 1953. Les chromosomes sexuels multiples de *Blaps polychresta* Forsk. (Col. Tenebr.). Rev. Suisse Zool. 60:462–466.

Guha, S., and S. C. Meheshwari. 1964. *In vitro* production of embryos from anthers of *Datura.* Nature 204:497.

Guha, S., and S. C. Maheshwari. 1966. Cell division and differentiation of embryos in the pollen grains of *Datura in vitro.* Nature 212:97–98.

Gunthardt, H., L. Smith, M. E. Haferkamp, and R. A. Nilan. 1953. Studies on aged seeds. II. Relation of age of seeds to cytogenetic effects. Agron. J. 45:438–441.

Gupta, S. B. 1969. Duration of mitotic cycle and regulation of DNA replication in *Nicotiana plumbaginifolia* and a hybrid derivative of *N. tabacum* showing chromosome instability. Can. J. Genet. Cytol. 11:133–142.

Gustafson, J. P., and C. O. Qualset. 1974. Genetics and breeding of 42-chromosome Triticale. I. Evidence for substitutional polyploidy in secondary Triticale populations. Crop Sci. 14:248–251.

Gustafson, J. P., and C. O. Qualset. 1975. Genetics and breeding of 42-chromosome Triticale. II. Relations between chromosomal and reproductive characters. Crop. Sci. 15:810–813.

Gustafsson, A. 1932. Zytologische und experimentelle Studien in der Gattung *Taraxacum.* Hereditas 16:41–62.

Gustafsson, A. 1935a. Studies on the mechanism of parthenogenesis. Hereditas 21:1–112.

Gustafsson, A. 1935b. The importance of apomicts for plant geography. Bot. Not. 1935:325–330.

Gustafsson, A. 1937. The occurrence of a sexual population within the apomictic *Taraxacum vulgare* group. Bot. Not. 1937:332–336.

Gustafsson, A. 1946. Apomixis in the higher plants. I. The mechanism of apomixis. Lunds Univ. Arsskr. N.F. Avd. 2, 42:1–66.

Gustafsson, A. 1947a. Apomixis in higher plants. II. The causal aspect of apomixis. Lunds Univ. Arsskr. N.F. Avd. 2, 43:71–178.

Gustafsson, A. 1947b. Apomixis in higher plants. III. Byotype and species formation. Lunds Univ. Arsskr., 44:183–370.

Gustavson, K.-H. 1964. Down's syndrome. A clincal and cytogenetical investigation. Almqvist and Wilksell, Uppsala.

Gustavsson, I. 1966. Chromosome abnormality in cattle. Nature 211:865–866.

Gustavsson, I., M. Fraccaro, L. Tiepolo, and J. Lindsten. 1968. Presumptive X-autosome translocation in the cow. Preferential inactivation of the normal X chromosome. Nature 218:183–184.

Gutherz, S. 1907. Zur Kenntis der Heterochromosomen. Arch Mikr. Anat. 69:491.

Gyulavari, O. 1970. Studies on the monoploid method and their results. *In:* Kovacs, I. (ed.). Some methodological achievements of the Hungarian hybrid maize breeding. Akademiai Kiado, Budapest.

Hadder, J. C., and G. B. Wilson. 1958. Cytological array of c-mitosis and prophase poison reactions. Chromosoma 9:99–104.

Haeckel, E. 1894. Systematische Phylogenie. Reimer, Berlin.

Haendle, J. 1971a. Röntgeninduzierte mitotische Rekombination bei *Drosophila melanogaster*. I. Ihre Abhängigkeit von der Dosis, der Dosisrate und vom Spektrum. Mol. Gen. Genet. 113:114–131.

Haendle, J. 1971b. Röntgeninduzierte mitotische Rekombination bei *Drosophila melanogaster*. II. Beweis der Existenz und Charakterisierung zweier von der Art des Spektrums abhängiger Reaktionen. Mol. Gen. Genet. 113:133–149.

Haendle, J. 1974. X-ray induced mitotic recombination in *Drosophila melanogaster*. III. Dose dependence of the "pairing" component. Mol. Gen. Genet. 128:233–239.

Haga, T. 1953. Meiosis in *Paris*. II. Spontaneous breakage and fusion of chromosomes. Cytologia 18:50–66.

Hagberg, A., and E. Åkerberg. 1962. Mutation and polyploidy in plant breeding. William Heinemann, Ltd., London.

Hageman, P. C., J. Links, and P. Bentvelzen. 1968. Biological properties of B particles from C3H and C3Hf mouse milk. J. Nat. Cancer Inst. 40:1319–1324.

Hair, J. B., and E. J. Benzenberg. 1961. High polyploidy in a New Zealand *Poa*. Nature 189:160.

Håkansson, A. 1948. Behavior of accessory rye chromosomes in the embryo sac. Hereditas 34:35–59.

Hall, B. D., and S. Spiegelman. 1961. Sequence complementarity of T2-DNA and T2-specific RNA. Proc. Natl. Acad. Sci. U.S.A. 47:137–146.

Hall, J. C. 1972. Chromosome segregation influenced by two alleles of the meiotic mutant c(3)G in *Drosophila melanogaster*. Genetics 71:367–400.

Hamerton, J. L. 1962. Cytogenetics of mongolism. *In:* Hamerton, J. L. (ed.). Chromosomes in medicine. pp. 140–183. Heinemann, London.

Hamerton, J. L. 1966. Chromosome segregation in three human interchanges. *In:* Darlington, C. D. and K. R. Lewis. (eds.). Chromosomes today. Vol. 1, pp. 237–252. Oliver and Boyd, Edinburgh.

Hamerton, J. L. 1971a. Human cytogenetics. Vol. 1. General cytogenetics. Academic Press, New York.

Hamerton, J. L. 1971b. Human cytogenetics. Vol. II. Clinical cytogenetics. Academic Press, New York.

Hamerton, J. L., V. A. Cowie, F. Giannelli, S. M. Briggs, and P. E. Polani. 1961. Differential transmission of Down's syndrome (mongolism) through male and female translocation carriers. Lancet II:956–958.

Hamilton, L. 1963. An experimental analysis of the development of the haploid syndrome in embryos of *Xenopus laevis*. J. Embryol. Exptl. Morphol. 11:267–278.

Hamilton, L. 1966. The role of the genome in the development of the haploid syndrome in *Anura*. J. Embryol. Exp. Morph. 16:559–568.

Hampel, K. E., and A. Levan. 1964. Breakage in human chromosomes induced by low temperature. Hereditas 51:315–343.

Handmaker, S. D. 1963. The satellited chromosomes of men with reference to the Marfan syndrome. Amer. J. Human Genet. 15:11–18.

Hanhart, E. 1960. 800 Fälle von Mongoloidismus in konstitutioneller Betrachtung. Arch. Julius Klaus-Stift, Vererbungsforsch. Sozialanthropol. Passenhyg. 35:1–312.

Harland, S. C. 1936. Haploids in polyembryonic seeds of Sea Island cotton. J. Hered. 27:229–231.

Harris, H., and J. F. Watkins. 1965. Hybrid cells derived from mouse and man: artificial heterokaryons of mammalian cells from different species. Nature 205:640–646.

Hartl, D. L. and S. W. Brown. 1970. The origin of male haploid genetic systems and their expected sex ratio. Theor. Pop. Biol. 1:165–190.

Hartl, D. L., Y. Hiraizumi and J. F. Crow. 1961. Evidence for sperm dysfunction as the mechanism of segregation distortion in *Drosophila melanogaster*. Proc. Natl. Acad. Sci. U.S.A. 58:2240.

Hartl, D. L. and Y. Hiraizumi. 1976. Genetics and biology of *Drosophila*. In: Ashburn, M. and E. Novitski. (eds.). Genetics and biology of *Drosophila*. Vol. 1b. pp. 615–666. Acad. Press, London.

Harvey, W. 1651. Exercitationes de generatione animalium. London.

Hasegawa, N. 1934. A cytological study on 8-chromosome rye. Cytologia 6:68–77.

Hasitschka, G. 1956. Bildung von Chromosomenbündeln nach Art der Speicheldrüsenchromosomen spiralisierter Ruhekernchromosomen und andere Struktureigentümlichkeiten in den endopolyploiden Riesenkernen der Antipoden von *Papaver rhoeas*. Chromosoma 8:87–113.

Hayes, W. 1964. The genetics of bacteria and their viruses. Wiley, New York.

Hayes, W. 1968. The genetics of bacteria and their viruses. 2nd ed. Wiley, New York.

Heber, U. 1962. Protein synthesis in chloroplasts during photosynthesis. Nature 195:91–92.

Hecht, F., R. D. Koler, D. A. Rigas, G. S. Dahnke, M. P. Case, V. Tisdale, and R. W. Miller. 1966. Leukaemia and lymphocytes in *Ataxia telangiectasia*. Lancet II:1193.

Hecht, F., J. S. Bryant, A. G. Motulsky, and E. R. Giblett. 1963. The no. 17-18(E) trisomy syndrome. J. Pediat. 63:605–621.

Hedges, R. W., and A. E. Jacob. 1974. Transposition of ampicillin resistance from RP4 to other replicons. Molec. Gen. Genet. 132:31–40.

Heemert, C. van. 1973. Androgenesis in the onion fly *Hylemya antiqua* (Meigen) demonstrated with a chromosomal marker. Nature, N.B. 246:21–22.

Heitz, E. 1928. Das Heterochromatin der Moose. I. Jahrb. Wiss. Botan. 69:762–818.

Heitz, E. 1929. Heterochromatin, Chromocentren, Chromomeren. Ber. Deutsch. Botan. Ges. 47:274–284.

Heitz, E. 1931. Die Ursache der gesetzmässigen Zahl, Lage, Form und Grösse pflanzlicher Nukleolen. Planta 12:775–844.

Heitz, E. 1931. Nukleolen und Chromosomen in der Gattung *Vicia*. Planta 15:495–505.

Heitz, E., und H. Bauer. 1933. Cytologische Untersuchungen an Dipteren. I. Beweise für die Chromosomennatur der Kernschleifen in den Knäuelkernen von *Bibio hortulanum* L. Zeitschr. Zellforsch. 17:67–82.

Helwig, E. R. 1941. Multiple chromosomes in *Philocleon anomalus* (Orthoptera: Acrididae). J. Morph. 69:317–327.

Helwig, E. R. 1942. Unusual integrations of the chromatin in *Machaerocera* and other genera of the Acrididae (Orthoptera). J. Morph. 71:1–33.

Henderson, A. S., D. Warburton, and K. C. Atwood. 1972. Location of ribosomal DNA in human chromosome complement. Proc. Natl. Acad. Sci. U.S.A. 69:3394–3398.

Henderson, R. G. 1931. Transmission of tobacco ring spot by seed of *Petunia*. Phytopath. 21:225–229.

Henderson, S. A. 1967a. A second example of normal coincident endopolyploidy and polyteny in salivary gland nuclei. Caryologia 20:181–186.

Henderson, S. A. 1967b. The salivary gland chromosomes of *Dasyneura crataegi* (Diptera: Cecidomyiidae). Chromosoma 23:38–58.

Heneen, W. K. 1965. On the meiosis of haploid rye. Hereditas 52:421–424.

Henking, H. 1891. Untersuchungen über die ersten Entwicklungsvorgänge in den Eiern der Insekten. II. Über Spermatogenese und deren Beziehung zur Entwicklung bei *Pyrrhocoris apterus* L. Zeitsch. Wiss. Zool. 51:685–736.

Herrmann, F. H., D. Martorin, K. Timofeev, T. Börner, A. B. Rubin, and R. Hagemann. 1974. Biochem. Physiol. Pflanz. 165:393–400.

Herschman, H. R., and D. R. Helinski. 1967. Purification and characterization of colicin E$_2$ and colicin E$_8$. J. Biol. Chem. 242:5360–5368.

Hershey, A. A., E. Burgi, and L. Ingraham. 1963. Cohesion of DNA molecules isolated from phage lambda. Proc. Natl. Acad. Sci. U.S.A. 49:748.

Herskowitz, I. H. 1965. Genetics. 2nd ed. Little, Brown and Co., Boston, Mass.

Herskowitz, I. H. 1967. Basic principles of molecular genetics. Little, Brown, Boston, Mass.

Hertwig, G. 1935. Die Vielwertigkeit der Speicheldrüsenkerne und Chromosomen bei *Drosophila melanogaster*. Zeitschr. Ind. Abstamm. Vererbungsl. 70:496–501.

Hertwig, O. 1877. Beiträge zur Kenntnis der Bildung, Befruchtung und Theilung des Thierischen Eies. I,II. Morph. Jahrb. 1:347–434, 3:1–86, 271–279.

Hertwig, O. 1890. Vergleich der Ei- und Samenbildung bei Nematoden. Eine Grundlage für celluläre Streitfragen. Arch Mikr. Anat. Entwicklungsmech. 36:1–138.

Hewitt, G. M., and B. John. 1965. The influence of numerical and structural chromosome mutations on chiasma conditions. Heredity 20:123–135.

Hinton, C. W. 1966. Enhancement of recombination associated with the c(3)G mutant of *Drosophila melanogaster*. Genetics 53:157–164.

Hirschhorn, K., H. L. Cooper, and I. L. Firschein. 1965. Deletion of short arms of chromosome 4-5 in a child with defects of midline fusion. Humangen. 11:479–482.

Hirota, Y., Y. Nishimura, F. Ørskov and I. Ørskov. 1964. Effect of drug resistance factor R on the F properties of *E. coli*. Bacteriol. 87:341–351.

Hoar, C. S. 1931. Meiosis in *Hypericum punctatum*. Bot. Gaz. 92:396–406.

Hoar, C. S., and E. J. Haertl. 1932. Meiosis in the genus *Hypericum*. Bot. Gaz. 93:197–204.

Hochman, B. 1974. Analysis of a whole chromosome in *Drosophila*. Cold Spring Harbor Symp. Quant. Biol. 38:581–589.

Hockett, E. A., and R. F. Eslick. 1971. Genetic male-sterile genes useful in hybrid barley production. Proc. Second Barley Genet. Symp., Washington State Univ. Press, Pullman, pp. 298–307.

Hodges, E. M., G. B. Klinger, J. E. McCaleb, O. C. Ruelke, R. J. Allen, Jr., S. C. Shank, and A. E. Kretschmer, Jr. 1967. Pangolagrass. Florida Agric. Exp. Sta. Bull. 718.

Hofmeister, W. 1851. Vergleichende Untersuchungen der Keimung, Entfaltung und Fruchtbildung höherer Kryptogamen (Moose, Farnen, Equisetaceen, Rhizocarpeen und Lycopodiaceen) und der Samenbildung der Coniferen. Hofmeister, Leipzig.

Hollenberg, C. P., P. Borst, R. A. Flavell, C. F. van Kreijl, E. F. J. van Bruggen, and A. C. Arnberg. 1972. The usual properties of mtDNA from a "low-density" petite mutant of yeast. Biochim. Biophys. Acta 277:44–58.

Holliday, R. 1968. Genetic recombination in fungi. Replication and recombination of genetic material. Australian Academy of Science, Canberra. pp. 157–174.

Holliday, R. 1977. Recombination and meiosis. Philos. Trans. R. Soc. London Ser. B. 277:359–370.

Holt, C. M. 1917. Multiple complexes in the alimentary tract of *Culex pipiens*. J. Morph. 29:607–618.

Hooke, R. 1665. Micrographia or some physiological descriptions on minute bodies by magnifying glasses. London. (Facsimilis edition published by R. T. Gunther *in:* Early science in Oxford. XIII. The life and works of R. Hooke. Oxford, 1938).

Hoskins, G. C. 1969. Electron micrographic observations of centromeres from unsectioned mammalian chromosomes isolated by micrurgy. Caryologia 22:229–247.

Hotta, Y., M. Ito, and H. Stern. 1966. Synthesis of DNA during meiosis. Proc. Natl. Acad. Sci. U.S.A. 56:1184–1191.

Hotta, Y., and H. Stern. 1971. Meiotic protein in spermatocytes of mammals. Nature N.B. 234:83–86.

Hougas, R. W., S. J. Peloquin, and R. W. Ross. 1958. Haploids of the common potato. J. Hered. 47:103–107.

Hougas, R. W., S. J. Peloquin, and A. C. Gabert. 1964. Effect of seed-parent and pollinator on frequency of haploids in *Solanun tuberosum*. Crop Sci. 4:593–595.

Howard, A., and S. R. Pelc. 1953. Synthesis of deoxyribonucleic acid in normal and irradiated cells and its relation to chromosome breakage. Heredity 6(Suppl.):261–273.

Hsu, T. C. 1969. Robertsonian fusion between homologous chromosomes in a natural population of the least cotton rat, *Sigmodon minimus* (Rodentia, Cricetidae). Experientia 25:205–206.

Hsu, T. C. 1973. Longitudinal differentiation of chromosomes. Ann. Rev. Genet. 7:153–176.

Hu, S., E. Ohtsubo, and N. Davidson. 1975a. Electron microscope heteroduplex studies of sequence relations among plasmids of *E. coli*. XI. The structure of F13 and related F-primes. J. Bacteriol. 122:749–763.

Hu, S., E. Ohtsubo, N. Davidson, and H. Saedler. 1975b. Electron microscope heteroduplex studies of sequence relations among bacterial plasmids. Identification and mapping of the insertion sequences IS*1* and IS*2* in *F* and *R* plasmids. J. Bacteriol. 122:764–775.

Hu, S., K. Ptashne, S. N. Cohen, and N. Davidson, 1975c. The $\alpha\beta$ sequence of *F* is IS*3*. J. Bacteriol. 123:687–692.

Huang, R. C., and J. Bonner. 1962. Histone, a suppressor of chromosomal RNA synthesis. Proc. Natl. Acad. Sci. U.S.A. 48:1216–1222.

Huberman, J. A., and A. D. Riggs. 1968. On the mechanism of DNA replication in mammalian chromosomes. J. Mol. Biol. 32:327–341.

Hughes, A. F. W. 1952. The mitotic cycle. Academic Press, New York.

Hughes, A. F. W., and M. M. E. Preston. 1949. J. Roy. Microsc. Soc. 69:121–131. Cited *in:* Mazia, D. 1961.

Hughes, E. C., and T. V. Csermely. 1966. Chromosomal constitution of human endometrium. Nature 209:326.

Hughes-Schrader, S. 1925. Cytology of hermaphroditism in *Icerya purchasi* (Coccoidae). Zeitschr. Zellforsch. 2:264–292.

Hughes-Schrader, S. 1927. Origin and differentiation of the male and female germ cells in the hermaphrodite of *Icerya purchasi* (Coccoidae). Zeitschr. Zellforsch. 6:509–540.

Hughes-Schrader, S. 1943a. Meiosis without chiasmata in diploid and tetraploid spermatocytes of the mantid *Callimantis antillarum* Saussure. J. Morph. 73:111–141.

Hughes-Schrader, S., 1943b. Polarization, kinetochore movements, and bivalent structure in the meiosis of male mantids. Biol. Bull. 85:265–300.

Hughes-Schrader, S. 1948. Cytology of Coccids (Coccoidae-Homoptera). Adv. Genet. 2:127–203.

Hughes-Schrader, S. 1950. The chromosomes of mantids (Orthoptera, Manteidae) in relation to taxonomy. Chromosoma 4:1–55.

Hughes-Schrader, S., and H. Ris. 1941. The diffuse spindle attachment of coccids, verified by the mitotic behavior of induced chromosome fragments. J. Zool. 87:429–451.

Hultén, M. 1970. Meiosis in XYY men. Lancet I:717–718.

Hultén, M., and J. Lindsten. 1970. The behaviour of structural aberrations at male meiosis; information from man. *In:* Jacobs, P. A., W. H. Price, and P. Law. (eds.). Human population cytogenetics. pp. 24–61. Edinburgh Univ. Press, Edinburgh.

Hultén, M., and J. Lindsten. 1973. Cytogenetic aspects of human male meiosis. *In:* Harris, H. and K. Hirschhorn. (eds.). Advances in human genetics. pp. 327–387. Plenum Press, New York.

Hungerford, D. A., and A. M. Hungerford. 1978. Chromosome structure and function in man. VI. Pachytene chromosome maps of 16, 17 and 18. Pachytene as a reference standard for metaphase banding. Cytogenet. Cell Genet. 21:212–230.

Hurd, E. A. 1976. Plant breeding for resistance. *In:* Kozlowski, T. T. (ed.). Water deficits and plant growth. Vol. 4. Soil water measurement, plant responses, and breeding for drought resistance. pp. 317–353. Academic Press, New York.

Huskins, C. L. 1932. A cytological study of Vilmorin's unfixable dwarf wheat. J. Genet. 25:113–124.

Huskins, C. L. 1948. Segregation and reduction in somatic tissues. J. Hered. 39:311–325.

Huskins, C. L., and L. M. Steinitz. 1948a. The nucleus in differentiation and development. I. J. Hered. 39:34–43.

Huskins, C. L., and L. M. Steinitz. 1948b. The nucleus in differentiation and development. II. J. Hered. 39:66–77.

Hyde, B. J. 1951. Forsythia polyploids. J. Arnold Arboretum (Harvard Univ.) 32:155–156.

Ikeda, H. 1965. Interspecific transfer in the "sex ratio" agent of *Drosophila willistoni,* in *D. bifasciata* and *D. melanogaster.* Science 147:1147–1148.

Illies, Z. M. 1974. Induction of haploid parthenogenesis in *Populus tremula* by male gametes inactivated with toluidine blue. *In:* Kasha, 1974. pg. 136.

Imai, Y. 1936. Geno-and plasmotypes of variegated pelargoniums. J. Genet. 33:169–195.

Inoué, S., and H. Sato. 1966. Deoxyribonucleic acid arrangement in living sperm. *In:* Hayashi, T. and A. G. Szent-Györgyi. (eds.). Molecular architecture in cell physiology. p. 209. Prentice-Hall, Englewood Cliffs, New Jersey.

International System, 1978. An international system for human cytogenetic nomenclature (1978). Cytogen. Cell Genet. 21:309–404.

Irikura, Y., and S. Sakaguchi. 1972. Induction of 12-chromosome plants from anther culture in a tuberous *Solanum.* Potato Res. 15:170–173.

Iseki, S., and T. Sakai. 1953. Artificial transformation of O antigens in *Salmonella* E group. I. Transformation by antiserum and bacterial autolysate. Proc. Jap. Acad. 29:121.

Ising, U., and A. Levan. 1957. The chromosomes of two highly malignant human tumours. Acta Pathol. Microbiol. Scand. 40:13–24.

Ivanov, M. A. 1938. Experimental production of haploids in *Nicotiana rustica* L. Genetica 20:295–397.

Ivanovskaja, E. V. 1939. A haploid plant of *Solanum tuberosum* L. Comp. Rend. (Doklady) Acad. Sci. U.R.S.S. 24:517.

Jacob, F., and S. Brenner. 1963. Sur la régulation de la synthèse du DNA chez les bactéries: L'hypothèse du replicon. C. R. Acad. Sci., Paris 256:298.

Jacob, F., A. Lwoff, L. Siminovitch, and E. L. Wollman. 1953. Définition de quelques termes relatifs à la lysogénie. Ann. Inst. Pasteur 84:222–224.

Jacob, F., and J. Monod. 1961a. On the regulation of gene activity. Cold Spring Harbor Symp. Quant. Biol. 3:193–211.

Jacob, F., and J. Monod. 1961b. Genetic regulatory mechanisms in the synthesis of proteins. J. Mol. Biol. 3:318–356.

Jacob, F., and E. L. Wollman. 1957. Genetic aspects of lysogeny. *In:* McElroy, W. D. and B. Glass. (eds.). The chemical basis of heredity. pp. 468–500. Johns Hopkins Press, Baltimore, Maryland.

Jacob, F., and E. L. Wollman. 1958. Les épisomes, éléments génétiques ajoutés. C. R. Acad. Sci., Paris 247:154.

Jacobj, W. 1925. Über das rythmische Wachstum der Zellen durch Verdopplung ihres Volumens. Roux Arch. 106:124–192.

Jacobs, P. A., R. R. Angell, I. M. Buchanan, T. J. Hassold, A. M. Mattsuyama, and B. Manuel. 1978. The origin of human triploids. Ann. Hum. Genet. 42:49–57.

Jacobs, P. A., G. Cruickshank, M. J. W. Faed, A. Frackiewicz, E. B. Robson, H. Harris and T. Sutherland. 1968. Pericentric inversion of a group C autosome. A study of three families. Ann. Human Genet. 31:219–230.

Jacobs, P. A., D. G. Harnden, W. M. Court Brown, J. Goldstein, H. G. Close, T. N. MacGregor, N. Maclean, and J. A. Strong, 1960. Abnormalities involving the X chromosome in women. Lancet I:1213–1216.

Jacobs, P. A., M. Melville, S. Ratcliffe, A. J. Keay, and J. Syme. 1974. A cytogenetic survey of 11,680 newborn infants. Ann Hum. Genet. 31:359–376.

Jacobs, P. A., and A. Ross. 1966. Structural abnormalities of the Y chromosome in men. Nature 210:352–354.

Jacobs, P. A., and J. A. Strong. 1959. A case of human intersexuality having a possible XXY sex-determining mechanism. Nature 183:302–303.

Jacobsen, P., M. Mikkelsen and F. Rosleff. 1974. The trisomy 8 syndrome: report of two further cases. Ann. Génét. 17:87–94.

James, S. H. 1965. Complex hybridity in *Isotoma petraea*, I. The occurrence of interchange heterozygosity, autogamy and a balanced lethal system. Heredity 20:341–353.

James, S. H. 1970. Complex hybridity in *Isotoma petraea*. II. Components and operation of a possible evolutionary mechanism. Heredity 25:53–77.

Janick, J., D. L. Mahoney, and D. L. Pfahler. 1959. The trisomics of *Spinacea oleracea*. J. Hered. 50:47–50.

Janssen, Z., and H. Janssen. 1591. Cited in: Njordenskiöld, 1935.

Janssens, F. A. 1905. Spermatogénèse dans les batraciens. III. Évolution des auxocytes mâles du *Batracoseps attenuatus*. La Cellule 22:378–428.

Janssens, F. A. 1909. Spermatogénèse dans les batraciens. V. La théorie de la chiasmatypie. Nouvelle interprétation des cinèses de maturation. La Cellule 25:389–411.

Jenkins, B. C. 1969. History of the development of some presently promising hexaploid Triticales. Wheat Inf. Serv. 28:18–20.

Jensen, C. J. 1974. Chromosome doubling techniques in haploids. Cited in: Kasha. 1974. pp. 153–190.

Jewell, D. C. 1979. Chromosome banding in *Triticum aestivum* cv. Chinese Spring and *Aegilops variabilis*. Chromosoma 71:129–134.

Johannsen, W. L. 1896. Om arvelighed og variabilitet. Copenhagen.

Johannsen, W. L. 1903. Über Erblichkeit in Populationen und reinen Linien. Fischer, Jena.

Johannsen, W. L. 1905. Arvelighedslaerens elementer. Copenhagen.

Johannsen, W. L. 1909. Elemente der exakten Erblichkeitslehre. Fischer, Jena.

John, B., and G. M. Hewitt. 1966. Karyotype stability and DNA variability in the Acrididae. Chromosoma 20:155–172.

John, B., and G. M. Hewitt. 1968. Patterns and pathways of chromosome evolution within the Orthoptera. Chromosoma 25:40–74.

John, B., and K. R. Lewis. 1958. Studies on *Periplaneta americana*. III. Selection for heterozygosity. Heredity 12:185–197.

John, B., and K. R. Lewis. 1968. The chromosome complement. Protoplasmatologia VIa. Springer-Verlag, Vienna.

John, B., and K. R. Lewis. 1969. The chromosome cycle. Protoplasmatologia 6(B):1–125.

Johnson, H. 1944. Meiotic aberrations and sterility in *Alopecurus myosuroides* Huds. Hereditas 30:469–566.

Joly, B. and R. Cluzel. 1975. The role of heavy metals and their derivatives in the selection of antibiotic resistant gram-negative rods. Ann. Microbiol. 126:51.

Jones, D. F. 1937. Somatic segregation and its relation to atypical growth. Genetics 22:484–522.

Jones, H. A. 1946. Cytology, genetics and breeding. Problems and progress in onion breeding. Herbertia 11:275–294.

Jones, H. A., and A. E. Clarke. 1943. Inheritance of male sterility in the onion and the production of hybrid seed. Proc. Amer. Soc. Hort. Sci. 43:189–194.

Jones, H. A., and G.N. Davis. 1944. Inbreeding and heterosis and their relation to the development of new varieties of onions. USDA Tech. Bull. 874.

Jones, H. A., and S. L. Emsweller. 1937. A male-sterile onion. Proc. Amer. Soc. Hort. Sci. 34:582–585.

Jones, H. A., D. R. Porter, and L. D. Leach. 1939. Breeding for resistance to onion downy mildew caused by *Peronospora destructor*. Hilgardia 12:531–550.

Jones, R. N. 1975. B-chromosome systems in flowering plants and animal species. Intern. Rev. Cytol. 40:1–100.

Jordan, E., H. Saedler, and P. Starlinger. 1968. 0° and strong polar mutations in the *gal* operon are insertions. Molec. Gen. Genet. 102:353–363.

Jørgensen, C. A. 1928. The experimental formation of heteroploid plants in the genus *Solanum*. J. Genet. 19:133–211.

Judd, B. H., M. W. Shen, and T. C. Kaufman. 1972. The anatomy and function of a segment of the X chromosome of *Drosophila melanogaster*. Genetics 71:139–156.

Jurand, A., B. M. Rudman, and J. R. Preer. 1971. Prelethal effects of killing action by stock 7 of *Paramecium aurelia*. Exptl. Cell Res. 177:365–388.

Käfer, E. 1977. Meiotic and mitotic recombination in *Aspergillus* and its chromosomal aberrations. Adv. Genet. 19:33–131.

Kahn, R. P. 1956. Seed transmission of the tomato-ringspot virus in the Lincoln variety of soybeans. Phytopathology 46:295.

Kakati, S., M. Nihill, and A. K. Sinha. 1973. An attempt to establish trisomy 8 syndrome. Humangen. 19:293–300.

Kalmus, H., J. Kerridge, and F. Tattesfield. 1954. Occurrence of susceptibility to carbon dioxide in *Drosophila melanogaster* from different countries. Nature 173:1101–1102.

Kamanoi, M., and B. C. Jenkins. 1962. Trisomics in common rye, *Secale cereale* L. Seiken Zihô 13:118–123.

Karakashian, M. W., and S. J. Karakashian. 1973. Intracellular digestion and symbiosis in *Paramecium bursaria*. Exp. Cell Res. 81:111–119.

Karpechenko, G. D. 1928. Polyploid hybrids of *Raphanus sativa* L. x *Brassica oleracea* L. Z. Ind. Abst. Vererbungsl. 39:1–7.

Kasamatsu, H., D. L. Robberson, and J. Vinograd. 1971. A novel closed-circular mitochondrial DNA with properties of a replicating intermediate. Proc. Natl. Acad. Sci. U.S.A. 68:2252–2257.

Kasha, K. J. (ed.). 1974. Haploids in higher plants. Advances and potential. Proc. First Internatl. Symp., Guelph, Ontario, Canada.

Kasha, K. J. 1974. Haploids from somatic cells. *In:* Kasha, 1974. pp. 67–87.

Kasha, K. J., and K. N. Kao. 1970. High frequency haploid production in barley, *Hordeum vulgare* L. Nature 225:874–876.

Katayama, T. 1961. Cytogenetical studies on asynaptic rice plant *(Oryza sativa)* induced by x-rays. La Kromosomo 48:1591–1601.

Katayama, Y. 1934. Haploid formation by x-rays in *Triticum monococcum*. Cytologia 5:234–237.

Kato, H., and H. F. Stich. 1976. Sister chromatid exchanges in aging and repair-deficient human fibroblasts. Nature 260:447–448.

Kato, K. 1930. Cytological studies of pollen mother cells of *Rhoeo discolor* Hance with special reference to the question of the mode of syndesis. Mem. Coll. Sci. Kyoto Imp. Univ., Ser. B. 5:139–161.

Kaufmann, B. P. 1943. A complex induced rearrangement of *Drosophila* chromosomes and its bearing on the problem of chromosome recombinations. Proc. Natl. Acad. Sci. U.S.A. 29:8–12.

Kaufmann, B. P. 1948. Chromosome structure in relation to the chromosome cycle. II. Bot. Rev. 14:57–126.

Kaul, C. L. 1975. Cytology of a spontaneously recurring desynaptic *Allium cepa*. Cytologia 40:243–248.

Kayano, H. 1957. Cytogenetic studies in *Lilium callosum*. III. Preferential segregation of a supernumerary chromosome in EMC's. Proc. Jap. Acad. 33:553–558.

Kelly, T. J., and H. O. Smith. 1970. A restriction nuclease from *Haemophilus influenzae*. II. Base sequence of the recognition site. J. Molec. Biol. 51:393–409.

Khush, G. S. 1973. Cytogenetics of aneuploids. Academic Press, New York.

Khush, G. S., and C. M. Rick. 1966. The origin, identification and cytogenetic behavior of tomato monosomics. Chromosoma 18:407–420.

Khush, G. S., and C. M. Rick. 1967a. Haplo-triplo-disomics of the tomato: origin, cytogenetics and utilization as a source of secondary trisomics. Biol. Zentralbl. 86:257–265.

Khush, G. S., and C. M. Rick. 1967b. Tomato tertiary trisomics: origin, identification, morphology, and use in determining position of centromeres and arm location of markers. Can. J. Genet. Cytol. 9:610–631.

Khush, G. S., and C. M. Rick. 1968a. Cytogenetic analysis of the tomato genome by means of induced deficiencies. Chromosoma 23:452–484.

Khush, G. S., and C. M. Rick. 1968b. The use of secondary trisomics in the cytogenetic analysis of the tomato genome. Proc. 12th Intern. Congr. Genet., Tokyo 1:117.

Khush, G. S., and C. M. Rick. 1968c. Tomato telotrisomics: origin, identification, and use in linkage mapping. Cytologia 33:137–148.

Khush, G. S., and C. M. Rick. 1969. Tomato secondary trisomics: origin, identification and use in cytogenetic analysis of the genome. Heredity 24:129–146.

Kihara, H. 1930. Genomanalyse bei *Triticum* und *Aegilops*. Cytologia 1:263–270.

Kihara, H. 1951a. Substitution of nucleus and its effects on genome manifestations. Cytologia 16:177–193.

Kihara, H. 1951b. Triploid watermelons. Proc. Amer. Soc. Sci. 58:217–230.

Kihara, H., und T. Ono. 1926. Chromosomenzahlen und systematische Gruppierung der *Rumex*-Arten. Z. Zellforsch. 4:475.

Kihara, H., and K. Tsunewaki. 1962. Use of alien cytoplasm as a new method of producing haploids. Jap. J. Genet. 37:310–313.

Kihlman, B. A. 1970 *In:* Hollaender, A. (ed.). Chemical Mutagens: principles and methods for their deletion. pp. 489–514. Plenum Press, New York.

Kihlman, B. A. 1975. Sister chromatid exchanges in *Vicia faba*. II. Effects of thiotepa, caffeine, and 8-ethoxy caffeine on the frequency of SCE's. Chromosoma 51:11–18.

Kilgour, R., and A. N. Bruere. 1970. Behaviour patterns in chromatin-positive Klinefelter's syndrome of sheep. Nature 225:71–72.

Kimber, G., and R. Riley. 1963. Haploid angiosperms. Bot. Rev. 29:480–531.

Kimber, G., and E. R. Sears. 1968. Nomenclature for the description of aneuploids in the Triticinae. Third Intern. Wheat Genet. Symp., Canberra, 1968. pp. 468–473. Plenum Press, New York.

King, R. C. 1965. Genetics. 2nd ed. Oxford Univ. Press, New York.

King, R. C. 1970. The meiotic behavior of the *Drosophila* oocyte. Intern. Rev. Cytol. 28:125–168.

Kirillova, G. A. 1965. The production of independent somatic mutations in haploid tomatoes and the transition of the latter into a diploid state. Genetika, Moskow 1965: No. 3:65–69. (Russian).

Kirk, J. T. O. 1966. Nature and function of chloroplast DNA. *In:* Goodwin, T. W. (ed.). Biochemistry of chloroplasts. Vol. I. pp. 319–340. Academic Press, New York.

Kirschner, R. H., D. R. Wolstenholme, and N. J. Gross. 1968. Replicating molecules of circular mitochondrial DNA. Proc. Natl. Acad. Sci. U.S.A. 60:1466–1472.

Kiss, A. 1966. Neue Richtung in der *Triticale*-Züchtung. Z. Pflanzenz. 55:309–329.

Kitzmiller, J. B. 1976. Genetics, cytogenetics, and evolution of mosquitoes. Adv. Genet. 18:315–433.

Kjéssler, B. 1964. Meiosis in a man with a D/D translocation and clinical sterility. Lancet I:1421–1423.

Kjéssler, B. 1966. Karyotype, meiosis and spermatogenesis in a sample of men attending an infertility clinic. Monograph in Human Genetics, Vol. 2. Karger, Basel.

Kjéssler, B. 1970. Meiotic studies on the human male. *In:* Emerery, A. E. H. (ed.). Modern trends in human genetics. Chap. 7. pp. 214–240. Butterworths, London.

Kleckner, N., R. K. Chan, B. K. Tye, and D. Botstein. 1975. Mutagenesis by insertion of a drug-resistance element carrying an inverted repetition. J. Mol. Biol. 97:561–575.

Klein, H. D. 1969. Desynapsis und Chromosomenbrüche in einer Mutante von *Pisum sativum*. Mutat. Res. 8:277–284.

Klein, H. D. 1971. Eine *Pisum*-Mutante mit zahlreichen meiotischen Störungen. Cytologia 36:15–25.

Klein, H. D., and S. R. Baquar. 1972. Genetically controlled chromosome breakage and reunions in the meiosis. Chromosoma 37:223–231.

Knoll, M., und E. Ruska, 1932. Das Elektronenmikroskop. Zeitschr. Physik 78:318–339.

Kodani, M. 1957a. Three diploid chromosome numbers of man. Proc. Natl. Acad. Sci. U.S.A. 43:285–292.

Kodani, M. 1957b. The karyotype of man with the diploid chromosome number of 48. Cytologia, Suppl. Vol. 103–107.

Kodani, M. 1958a. Three chromosome numbers in Whites and Japanese. Science 127:1339–1340.

Kodani, M. 1958b. The supernumerary chromosome of man. Amer. J. Human Genet. 10:125–140.

Koduru, P. R. K., and M. K. Rao. 1978. Chromosome pairing and desynapsis in spontaneous autopolyploids of *Pennisetum typhoides*. Cytologia 43:445–452.

Koller, P. C. 1936. Structural hybridity in *Drosophila pseudoobscura*. J. Genet. 32:79–102.

Koller, P. C. 1938. Asynapsis in *Pisum sativum*. J. Genet. 36:275–306.

Kolodner, R., and K. K. Tewari. 1973. Replication of circular chloroplast DNA. J. Cell. Biol. 59:174a.

Kölreuter, J. G. 1761–1766. Vorläufige Nachricht von einigen das Geschlecht der Pflanzen betreffenden Versuchen, und Beobachtungen, nebst Fortsetzungen 1, 2 und 3. Engelmann, Leipzig.

Koltzoff, N. 1934. The structure of the chromosomes in the salivary glands of *Drosophila*. Science 80:312–313.

Komai, T., and T. Takahu. 1942. On the effect of the X-chromosome on crossing over in *Drosophila virilis*. Cytologia 23:109–111.

Konvička, O., und W. Gottschalk. 1971. Cytologische Untersuchungen an strahleninduzierten Mutanten von *Brassica oleracea* var. *capitata*. Biol. Plant 13:325–332.

Kornberg, R. 1974. Chromatin structure: a repeating unit of histones and DNA. Science 184:868–871.

Kostoff, D. 1930. Discoid structure of the spireme. J. Hered. 21:323–324.

Kostoff, D. 1938. Heterochromatin, somatic "crossing over," and the interchange hypothesis between nonhomologous chromosomes. Proc. Indian Acad. Sci., Sec. B., 8:11–44.

Koswig, C., and A. Shengun. 1974. Intraindividual variability of chromosome IV of *Chironomus*. J. Hered. 38:235–259.

Koul, A. K. 1962. Desynapsis and spontaneous chromosome breakage in *Allium cepa*. Phyton 19:115–120.

Koul, A. K. 1970. Supernumerary cell divisions following meiosis in the spider plant. Genetica 41:305–310.

Kozinski, A. W., P. B. Kozinski, and R. James. 1967. Molecular recombination in T4 bacteriophage deoxyribonucleic acid. J. Virol. 1:758–770.

Kramer, H. H., R. Veyl, and W. D. Hanson. 1954. The association of two genetic linkage groups in barley with one chromosome. Genetics 39:159–168.

Kroon, A. M., P. Borst, E. F. J. van Bruggen, and G. J. C. M. Ruttenberg. 1966. Mitochondrial DNA from sheep heart. Proc. Natl. Acad. Sci. U.S.A. 56:1836–1843.

Kuliev, A. M., V. I. Kukharenko, K. N. Grinberg, A. T. Mikhailov and A. D. Tamarkina. 1975. Human triploid cell strain, phenotype on cellular level. Humangen. 30:127–134.

Kung, C. 1971. Aerobic respiration of *kappa* particles from *Paramecium aurelia*. J. Protozool. 18:328–332.

Kuo, J. S., Y. Y. Wang, N. F. Chien, S. J. Ku, M. L. Kung, and H. C. Hsu. 1973. Investigations on the anther culture *in vitro* of *Nicotiana tabacum* L. and *Capsicum annuum* L. Acta Bot. Sinica 15:37–50.

Kusanagi, A., and N. Tanaka. 1960. Japan. J. Genet. 35:67–70.

Kuwada, Y. 1925. On the number of chromosomes in maize. Bot. Mag. (Tokyo) 39:227–234.

Lacadena, J.-R. 1974. Spontaneous and induced parthenogenesis and androgenesis. *In:* Kasha, 1974. pp. 13–32.

Lacadena, J.-R. and A. Ramos. 1968. Meiotic behaviour in a haploid plant of *Triticum durum* Desf. Genét. Ibérica 20:1–26.

Laird, C. D. 1971. Chromatid structure: relationship between DNA content and nucleotide sequence diversity. Chromosoma 32:378–406.

Lamarck, J. B. de. 1809. Philosophie zoologique. 2 vols., Paris.

Lamm, R. 1936. Cytological studies on inbred rye. Hereditas 22:217–240.

Lamm, R. 1944. A case of abnormal meiosis in *Lycopersicon esculentum*. Hereditas 30:253.

Lamm, R. 1945. Cytogenetic studies in *Solanum*, Sect. Tuberarium. Hereditas 31:1–128.

Lammerts, W. E. 1934. On the nature of chromosome association in *Nicotiana tabacum* haploids. Cytologia 6:38–50.

Lange, W. 1971. Crosses between *Hordeum vulgare* L. and *H. bulbosum* L. II. Elimination of chromosomes in hybrid tissues. Euphytica 20:181–194.

Langlet, O. 1927. Zur Kenntnis der polysomatischen Zellkerne in Wurzelmeristem. Svensk. Bot. Tids. 21:397–422.

Larter, E. N., T. Tsuchiya, and L. E. Evans. 1968. Breeding and cytology of *Triticale*. Third Int. Wheat Genet. Symp., Proc., pp. 213–221.

Latt, S. A. 1973. Microfluorometric detection of deoxyribonucleic acid replication in human metaphase chromosomes. Proc. Natl. Acad. Sci. U.S.A. 70:3395–3399.

Laurent, C., J. B. Cotton, A. Nivelon, and M.-T. Freycon. 1967. Délétion partielle du bras long d'un chromosome du group D(13-15):Dq-. Ann Génét. (Paris) 10:25–31.

Laurent, C., J. M. Robert, J. Grambert, and B. Dutrillaux. 1971. Observations clinques et cytogénétiques de deux adultes trisomiques en mosaïque. Individualisation du chromosome surnuméraire par la technique moderne de dénaturation: 47, XY,?8+. Lyon Méd. 226:827–833.

Lawrence, W. J. C. 1931a. The genetics and cytology of *Dahlia variabilis*. J. Genet. 24:257–306.

Lawrence, W. J. C. 1931b. The secondary association of chromosomes. Cytologia 2:352–384.

Leder, A., H. I. Miller, D. H. Hammer, J. G. Seidman, B. Norman, M. Sullivan, and P. Leder. 1978. Comparison of cloned mouse alpha- and beta-globin genes.—Conservation of intervening sequence locations and extragenic homology. Proc. Natl. Acad. Sci. U.S.A. 75:6187–6191.

Leder, P., S. M. Tilghman, D. C. Tiemeier, F. I. Polsky, J. G. Seidman, M. H. Edgell, L. W. Enquist, A. Leder, and B. Norman. 1977. The cloning of mouse globin and surrounding gene sequences in bacteriophage λ. Cold Spring Harbor Symp. Quant. Biol. 42:915–920.

Lederberg, J. 1952. Cell genetics and hereditary symbiosis. Physiol. Rev. 32:403–430.

Lederberg, J. 1955. Recombination mechanisms in bacteria. J. Cell. Comp. Physiology 45(Suppl. II):75–107.

Lee, C. S. N., P. Bowen, H. Rosenblum, and L. Linsao. 1964. Familial chromosome-2, 3 translocation ascertained through an infant with multiple malformations. New Engl. J. Med. 271:12–16.

Leeuwenhoek, A. van. 1674. Phil. Trans. Roy. Soc. London 9:23.

Lefevre, G., Jr. 1974. The one band-one gene hypothesis: evidence from a cytogenetic analysis of mutant-nonmutant rearrangement breakpoints in *Drosophila melanogaster.* Cold Spring Harbor Symp. Quant. Biol. 38:591–599.

Lejeune, J., R. Berger, J. Lafourcade, and M. Réthoré. 1966. La délétion partielle du bras long du chromosome 18. Individualisation d'un nouvel état morbide. Ann. Génét. 9:32.

Lejeune, J., J. Lafourcade, R. Berger, J. Vialette, M. Boeswillwald, P. Seringe, and R. Turpin. 1963. Trois cas de délétion partielle du bras court d'un chromosome 5. C. R. Acad. Sci. 277:3098–3102.

Lejeune, J., J. Lafourcade, R. Berger, J. Cruveiller, M.-O. Réthoré, B. Dutrillaux, D. Abonyi, and H. Jerome. 1968a. Le phénotype (Dr). Etude de trois cas de chromosomes D en anneau. Ann. Génét. Semaine Hop. 11:79–87.

Lejeune, J., J. Lafourcade, M.-O. Réthoré, R. Berger, O. Abonyi, B. Dutrillaux, and P. Cayroche. 1968b. Translocation t(lp+;2p−) identique chez une femme et son fils arriérés mentaux. Ann. Génét. Semaine Hop. 11:177–180.

Lejeune, J., M. O. Réthoré, B. Dutrillaux, and G. Martin. 1972. Translocation 8-22 sans changement de longeur et trisomie partielle 8q. Détection par dénaturation ménagée. Exp. Cell Res. 74:294–295.

Léonard, A., and G. H. Deknudt. 1967. A new marker for chromosome studies in the mouse. Nature 214:504–505.

Lesley, J. W. 1932. Trisomic types of the tomato and their relations to the genes. Genetics 17:545–559.

Lesley, M. M., and H. B. Frost. 1927. Mendelian inheritance of chromosome shape in *Matthiola.* Genetics 12:449–460.

Levan, A. 1938. The effect of colchicine on root mitosis in *Allium.* Hereditas 24:471–486.

Levan, A. 1942a. Plant breeding by induction of polyploidy and some results in clover. Hereditas 28:245–246.

Levan, A. 1942b. Studies on the meiotic mechanism of haploid rye. Hereditas 28:177–211.

Levan, A. 1956. Chromosome studies in some human tumors and tissues of normal origin, grown *in vivo* and *in vitro* at the Sloan-Kettering Institute. Cancer 9:648–663.

Levan, A. 1966. Non-random representation of chromosome types in human tumor stemlines. Hereditas 55:28–38.

Levan, A., and T. S. Hauschka. 1953. Endomitotic reduplication mechanisms in ascites tumors of the mouse. J. Nat. Cancer Inst. 14:1–43.

Leventhal, E. 1968. The sex ratio condition in *Drosophila bifasciata;* its experimental transmission to several species in *Drosophila.* J. Invertebr. Pathol. 11:170–183.

Levine, R. P., and E. E. Levine. 1955. Variable crossing over arising in different strains of *Drosophila pseudoobscura.* Genetics 40:399–405.

Lewin, B. 1977. Gene expression. Vol 3. Plasmids and phages. John Wiley and Sons, New York.

Lewis, E. B. 1945. The relation of repeats to position effect in *Drosophila melanogaster.* Genetics 30:137–166.

Lewis, E. B. 1950. The phenomenon of position effect. Adv. Genetics 3:73–116.

Lewis, K. R., and B. John. 1957. The organization and evolution of the sex multiple in *Blaps mucronata.* Chromosoma 9:69–80.

Lewis, K. R., and B. John. 1959. Breakdown and restoration of chromosome stability following inbreeding in a locust. Chromosoma 10:589–618.

Lewis, K. R., and B. John. 1963. Chromosome marker. Churchill, London.

Lewontin, R. C., and M. J. D. White. 1960. Interaction between inversion polymorphism of two chromosome pairs in the grasshopper *Moroba scurra.* Evolution 14:116–129.

L'Héritier, P. 1962. Les relations du virus héréditaire de la drosophile avec son hôte. Ann. Inst. Pasteur 102:511–526.

L'Héritier, P., and G. Teissier. 1937. Une anomalie physiologique héréditaire chez la drosophile. C. R. Acad. Sci. Paris 205:1099–1101.

Li, H., W. Pao, and C. H. Li. 1945. Desynapsis in the common wheat. Amer. J. Bot. 32:92–101.

Liebeskind, D., R. Bases, F. Mendez, F. Elequin, and M. Koenigsberg. 1979. Sister chromatid exchanges in human lymphocytes after exposure to diagnostic ultrasound. Science 205:1273–1275.

Lima-de-Faria, A. 1949. The structure of the centromere of the chromosomes of rye. Hereditas 35:77–85.

Lima-de-Faria, A. 1952. The chromosome size gradient of the chromosomes of rye. Hereditas 38:246–248.

Lima-de-Faria, A. 1953. Pairing and transmission of a small accessory iso-chromosome in rye. Chromosoma 6:142–148.

Lima-de-Faria, A. 1954. Chromosome gradient and chromosome field in *Agapanthus*. Chromosoma 6:330–370.

Lima-de-Faria, A., and P. Sarvella. 1958. The organization of telomeres in species of *Solanum, Salvia, Scilla, Secale, Agapanthus*, and *Ornithogalum*. Hereditas 44:337–346.

Lin, B.-Y. 1974. TB-10 breakpoints and marker genes on the long arm of chromosome 10. A seed-size effect associated with certain B-10 translocations. Maize Genetics Coop. News Letter 48:182–186.

Lin, Y. J. 1979. Chromosome distribution and catenation in *Rhoeo spathacea* var. *concolor*. Chromosoma 71:109–127.

Linde-Laursen, I. 1978. Giemsa C-banding of barley chromosomes. I. Banding pattern polymorphism. Hereditas 88:55–64.

Lindqvist, K. 1960. Inheritance studies in lettuce. Hereditas 46:387–470.

Lindsley, D. L., and E. H. Grell. 1968. Genetic variations of *Drosophila melanogaster*. Carnegie Inst. Wash. Publ. No. 677.

Lindsley, D. L., and W. J. Peacock. 1976. Unpublished results. Cited *in:* Baker et al., 1976.

Lindsley, D. L., L. Sandler, B. Nicoletti, and G. Trippa. 1968. Genetic control of recombination in *Drosophila. In:* Peacock, W. J. and R. D. Brock. (eds.). Replication and recombination of genetic material. Australian Acad. Sci., Canberra.

Lindsley, D. L., L. Sandler et al. 1972. Segmental aneuploidy and the genetic gross structure of the *Drosophila* genome. Genetics 71:157–184.

Linnert, G. 1961a. Cytologische Untersuchungen an Arten und Artbastarden von *Aquilegia*. I. Struktur und Polymorphismus der Nucleolenchromosomen, Quadrivalente und B-Chromosomen. Chromosoma 12:449–459.

Linnert, G. 1961b. Cytologische Untersuchungen an Arten und Artbastarden von *Aquilegia*. II. Die Variabilität der Pachytänchromosomen. Chromosoma 12:585–606.

Linnert, G. 1962. Untersuchungen an hemiploiden Nachkommen Autotetraploider. I. Der Verteilungsmodus einer reziproken Translokation und die daraus folgende Erhöhung der Hertozygoten-Frequenz in der Nachkommenschaft von Duplex-Heterozygoten. Z. Vererbungl. 93:389–398.

Litardière, R. de. 1925. Sur l'existence de figures diploïdes dans le méristème radiculaire du *Cannabis sativa* L. La Cellule 35:21–25.

Logachev, E. D. 1956. On the mutual relations between the nucleus and the cytoplasm in growing egg-cells of *Platyhlminthes*. C. R. Acad. Sci. URSS 111:507–509.

Longley, A. E. 1927. Supernumerary chromosomes in *Zea mays*. J. Agric. Res. 35:769–784.

Longley, A. E. 1937. Morphological characters of teosinte chromosomes. J. Agric. Res. 54:835–862.

Longley, A. E. 1938. Chromosomes of maize from North American Indians. J. Agr. Res. 56:177–196.

Longley, A. E. 1939. Knob positions on corn chromosomes. J. Agric. Res. 59:475–490.

Longley, A. E. 1945. Abnormal segregation during megasporogenesis in maize. Genetics 30:100–113.

Longo, F. J., and E. Anderson. 1968. The fine structure of pronuclear development and fusion in the sea urchin *Arbacia punctulata*. J. Cell Biol. 39:339–368.

Lorz, A. 1937. Cytological investigations on five chenopodiaceous genera with special emphasis on chromosome morphology and somatic doubling in *Spinacia*. Cytologia 8:241–276.

Löve, A., and D. Löve. 1948. Chromosome numbers of northern plant species. Univ. Inst. Appl. Sci. Dpt. Agric. Reports, Series B, No. 3, Reykjavik, 9–131.

Lubs, H. A., and F. H. Ruddle. 1970. Chromosomal abnormalities in the human population: estimation of rates based on New Haven newborn study. Science 169:495–497:

Luria, S. E. 1963. Molecular and genetic criteria in bacterial classification. Recent Progr. Microbiol. 8:604.

Lwoff, A., and A. Gutmann. 1950. Recherches sur un *Bacillus megatherium* lysogène. Ann. Inst. Pasteur 78:711–739.

Lyon, M. F. 1961. Gene action in the X-chromosome of the mouse (*Mus musculus* L.). Nature 190:372–373.

Lyon, M. F. 1962a. Sex chromatin and gene action in the mammalian X-chromosome. Am. J. Human Genet. 14:135–148.

Lyon, M. F. 1962b. Attempts to test the inactive-X theory of dosage compensation in mammals. Ann. Human Genet. 25:423.

Lyon, M. F. 1963. Attempts to test the inactive-X theory of dosage compensation in mammals. Genet. Res. 4:93–103.

Lyon, M. F. 1966. X-chromosome inactivation in mammals. *In:* Woollam, D. H. (ed.). Advances in teratology. Vol. 1, pp. 25–54. Academic Press, New York.

Lyon, M. F. 1970. Genetic activity of sex chromosomes in somatic cells of mammals. Phil Trans. Roy. Soc. London B. 259:41–52.

Lyon, M. F. 1971. Possible mechanisms of X-chromosome inactivation. Nature N.B. 232:229–232.

Lyon, M. F. 1972. X-chromosome inactivation and developmental patterns in mammals. Biol. Rev. 47:1–35.

McClintock, B. 1929a. Chromosome morphology of *Zea mays*. Science 69:629–630.

McClintock, B. 1929b. A cytological study of triploid maize. Genetics 14:180–222.

McClintock, B. 1929c. A 2n-1 chromosomal chimera in maize. J. Hered. 20:218.

McClintock, B. 1930. A cytological demonstration of the location of an interchange between two non-homologous chromosomes of *Zea mays*. Proc. Natl. Acad. Sci. U.S.A. 16:791–796.

McClintock, B. 1932. A correlation of ring-shaped chromosomes with resegregation in *Zea mays*. Proc. Natl. Acad. Sci. U.S.A. 18:677–681.

McClintock, B. 1933. The association of non-homologous parts of chromosomes in the midprophase of *Zea mays*. Z. Zellf. Mikr. Anat. 19:191–237.

McClintock, B. 1934. The relation of a particular chromosomal element to the development of the nucleoli in *Zea mays*. Z. Zellf. Mikr. Anat. 21:294–328.

McClintock, B. 1938a. The fusion of broken ends of sister half-chromatids following chromatid breakage at meiotic anaphase. Missouri Agri. Exp. Sta. Res. Bull. 290:1–48.

McClintock, B. 1938b. The production of homozygous deficient tissues with mutant characteristics by means of the aberrant behavior of ring-shaped chromosomes. Genetics 23:315–376.

McClintock, B. 1941a. Spontaneous alterations in chromosome size and form in *Zea mays*. Cold Spring Harbor Symp. Quant. Biol. 9:72–81.

McClintock, B. 1941b. The association of mutants with homozygous deficiencies in *Zea mays*. Genetics 26:542–571.

McClintock, B. 1941c. The stability of broken ends of chromosomes in *Zea mays*. Genetics 26:234–282.

McClintock, B. 1942. The fusion of broken ends of chromosomes following nuclear fusion. Proc. Natl. Acad. Sci. U.S.A. 28:458–463.

McClintock, B. 1944. The relation of homozygous deficiencies to mutations and allelic series in maize. Genetics 29:478–502.

McClintock, B. 1950a. The origin and behavior of mutable loci in maize. Proc. Natl. Acad. Sci. U.S.A. 36:344–355.

McClintock, B. 1950b. Mutable loci in maize. Carnegie Inst. Wash. Yearb. 49:157–167.

McClintock, B. 1951. Chromosome organization and genic expression. Cold Spring Harbor Symp. Quant. Biol. 16:13–47.

McClintock, B. 1953. Induction of instability at selected loci in maize. Genetics 38:579–599.

McClintock, B. 1961. Some parallels between gene control systems in maize and in bacteria. Amer. Nat. 95:265–277.

McClintock, B. 1965. The control of gene action in maize. Brookhaven Symp. Biol. 18:162–184.

McClintock, B., and H. E. Hill. 1931. The cytological identification of the chromosome associated with the R-G linkage group in *Zea mays*. Genetics 16:175–190.

McClung, C. E. 1905. The chromosome complex of orthopteran spermatocytes. Biol. Bull., Wood's Hole 9:304.

McClure, H. M., K. H. Belden, and W. A. Pieper. 1969. Autosomal trisomy in a chimpanzee: resemblance to Down's syndrome. Science 165:1010–1012.

MacDonald, M. D. 1961. Barley pachytene chromosomes and the localization of the erectoides 7 translocation point on chromosome 5. Canad. J. Genet. Cytol. 3:13–17.

Mace, S. E., M. N. Macintyre, K. B. Turk, and W. E. Johnson. 1978. The trisomy 9 syndrome: multiple congenital anomalies and unusual pathological findings. J. Pediat. 92:446–448.

McFee, A. F., M. W. Banner, and J. M. Rary. 1966. Variation in chromosome number among European wild pigs. Cytogenetics 5:77–81.

McGinnis, R. C., G. Y. Andrews, and R. I. H. McKenzie. 1963. Determination of chromosome arm carrying a gene for chlorophyll production in *Avena sativa*. Can. J. Genet. Cytol. 5:57–59.

McGregor, J. F. 1970. The chromosomes of the maskinonge *(Esox mosquinongy)*. Canad. J. Genet. Cytol. 12:224–229.

McIlree, M., W. S. Tulloch, and J. E. Newsam. 1966. Studies on human meiotic chromosomes from testicular tissue. Lancet I:679–682.

McKay, R. D. G. 1973. The mechanism of G and C banding in mammalian metaphase chromosomes. Chromosoma 44:1–14.

Mackensen, O. 1935. Locating genes on salivary chromosomes. J. Hered. 26:163–174.

MacKnight, R. H. 1937. Crossing-over in the sex chromosome of racial hybrids of *Drosophila pseudoobscura*. Genetics 22:249–256.

MacKnight, R. H., and K. W. Cooper. 1944. The synapsis of the sex chromosomes of *Drosophila miranda* in relation to their directed segregation. Proc. Natl. Acad. Sci. U.S.A. 30:384–387.

McKusick, V. A. 1964. Human genetics. Prentice-Hall, Inc., Englewood Cliffs, New Jersey.

McKusick, V. A., and F. H. Ruddle. 1977. The status of the gene map of the human chromosomes. Science 196:390–405.

McLennan, H. A. 1947. Cytogenetic studies of a strain of barley with long chromosomes. M.S. Thesis. Univ. Minnesota, Minneapolis.

McLennan, H. A., and C. R. Burnham. 1948. Cytogenetic studies of "long chromosomes" in barley. Amer. Soc. Agron. 1948 Ann. Mtgs., Fort Collins, Colorado.

Madjolelo, S. D. P., C. O. Grogan, and P. A. Sarvella. 1966. Morphological expression of genetic male sterility in maize (*Zea mays* L.). Crop Sci. 6:379–380.

Magenis, R. E., R. D. Koler, E. Lovrien, R. H. Bigley, M. C. DuVal, and K. M. Overton. 1975. Birth defects, Orig. Artic. Ser. 12:326.

Magni, G. E. 1954. Thermic cure of cytoplasmic sex-ratio in *Drosophila bifasciata.* Carologia 6(Suppl.):1213–1216.

Magoon, M. L., and K. R. Khanna. 1963. Haploids. Caryologia 16:191–235.

Magoon, M. L., M. S. Ramanna, and Y. Shambulingappa. 1961. Desynapsis and spontaneous chromosome breakage in *Sorghum purpureoseviceum.* Indian J. Genet. Plant Breed. 21:87–97.

Magoon, M. L., and K. G. Shambulingappa. 1960. Karyomorphological studies in *Sorghum ankolib* var. *annalib red,* A. Eu-Sorghum. Indian J. Genet. Plant Breed. 20:166–177.

Magoon, M. L., and K. G. Shambulingappa. 1961. Karyomorphology of *Sorghum propinquum* and its bearing on the origin of 40-chromosome *Sorghum.* Chromosoma 12:460–465.

Maguire, M. P. 1974. A new model for homologous chromosome pairing. Caryologia 27:349–357.

Maguire, M. P. 1977. Homologous chromosome pairing. Phil. Trans. R. Soc. Lond. B. 277:245–258.

Maheshwari, P. 1949. The male gametophyte of angiosperms. Bot. Rev. 15:1–75.

Makino, S. 1951. An atlas of the chromosome numbers in animals. Iowa State Coll. Press, Ames.

Makino, S., and H. Nakahara. 1953. Z. Krebsforsch. 59:298. Cited in: Mazia, D. 1961.

Makino, S., and M. Sasaki. 1961. A study of somatic chromosomes in a Japanese population. Am. J. Human Genet. 13:47–63.

Makino, S., K. Yamada, and T. Kajii. 1965. Chromosome aberrations in leucocytes of patients with aseptic meningitis. Chromosoma 16:372–380.

Malawista, S. E., H. Sato, and K. G. Bensch. 1968. Vinblastine and griseofulvin reversibly disrupt the living mitotic spindle. Science 160:770–772.

Malik, C. P., and R. C. Tripathi. 1970. Mode of chromosome pairing in the polyhaploid tall fescue (*Festuca arundinacea* Shreb. 2n = 42). Z. Biol. 116:332–339.

Malpighi, M. 1661. De pulmonibus observationes anatomicae. Bolognia.

Malpuech, G., B. Dutrillaux, Y. Fonck, J. Gaulme, and B. Bovche. 1972. Trisomie 8 en mosaïque. Arch. Franç. Pédiat. 29:853–859.

Malogolowkin, C. 1958. Maternally inherited "sex ratio" conditions in *Drosophila willistoni* and *D. paulistorum.* Genetics 43:274–286.

Malogolowkin, C. 1959. Temperature effects on maternally inherited "sex-ratio" conditions in *Drosophila willistoni* and *D. equinoxialis.* Amer Nat. 93:365–368.

Malogolowkin, C., G. G. Carvalho, and M. Da Paz. 1960. Interspecific transfer of "sex-ratio" condition in *Drosophila.* Genetics 45:1553–1557.

Manga, V. 1976. Chiasma frequencies in primary trisomics of pearl millet. Can. J. Genet. Cytol. 18:11–15.

Manga, V., and J. V. Pantalu. 1971. The meiotic behavior of a haploid pearl millet. Genetica 42:319–328.

Mange, E. J. 1961. Meiotic drive in natural populations of *Drosophila melanogaster.* VI. A preliminary report on the presence of segregation-distortion in a Baja, California population. Amer. Nat. 95:87–96.

Manning, J. E., D. R. Wolsteinholme, R. S. Ryan, J. A. Hunter, and O. C. Richards. 1971. Circular chloroplast DNA from *Euglena gracilis.* Proc. Natl. Acad. Sci. U.S.A. 68:1169–1173.

Marberger, E., R. A. Boccabella, and W. O. Nelson. 1955. Oral smear as a method of chromosomal sex determination. Proc. Soc. Exptl. Biol. Med. 89:488–489.

Marden, P. M., D. W. Smith, and M. J. McDonald. 1964. Congenital anomalies in the newborn infant, including minor variations. J. Pediat. 64:357–371.

Marks, G. E. 1973. An introduction to the embryology of angiosperms. McGraw-Hill, New York.

Martini, G., and A. Bozzini. 1965. Analisi di un mutante "sticky" in dotto da radiazioni in frumento duro. Genet. Agrar. 19:184–194.

Masters, M., and P. Broda. 1971. Evidence for the bidirectional replication of the *Escherichia coli* chromosome. Nature N.B. 232:137–140.

Mather, K. 1933. The relations between chiasmata and crossing-over in diploid and triploid *Drosophila melanogaster*. J. Genet. 27:243–259.

Mather, K. 1937. The experimental determination of the time of chromosome doubling. Proc. Roy. Soc. B. 124:97–106.

Mather, K. 1938. Crossing over. Biol. Rev. 13:252–292.

Matsuda, T. 1970. On the accessory chromosomes of *Aster*. I. The accessory chromosomes of *Aster ageratoides* group. J. Sci. Hiroshima Univ. Ser. B, Div. 2 (Bot.) 13:1.

Matsui, S., and M. Sasaki. 1973. Differential staining of nucleolus organisers in mammalian chromosomes. Nature 246:148–150.

Matthey, R. 1965. Un type nouveau de chromosomes sexuels multiples chez une souris Africaine du groupe *Mus* (Leggada) *minutoides* (Mammalia-Rodentia). Male:X_1X_2Y. Female:X_1X_2/X_1X_2. Chromosoma 16:351–364.

Matuszewski, B. 1964. Polyploidy and polyteny induced by a hymenopteran parasite in *Dasyneura urticae* (Diptera, Cecidomyiidae). Chromosoma 15:31–35.

Matuszewski, B. 1965. Transition from polyteny to polyploidy in salivary glands of Cecidomyiidae. Chromosoma 16:22–24.

Maude, P. F. 1939. The Merton Catalogue. A list of the chromosome numerals of species of British flowering plants. New Phytol. 38:1–31.

Maupertuis, P. L. M. de. 1745. Vénus physique, contenant deux dissertations, l'une sur l'origine des hommes et des animaux, et l'autre sur l'origine des noirs. La Haye. Also in: Oeuvres, (Lyons) 2:1–134.

Maupertuis, P. L. M. de. 1751. Système de la nature. Oeuvres (Lyons) 2:135–184.

Maupertuis, P. L. M. de. 1752. Lettres. Oeuvres (Lyons) 2:185–340.

Mazia, D. 1961. Mitosis and the physiology of cell division. *In:* Brachet, T. and A. E. Mirsky. (eds.). The cell. Vol. 3. Academic Press, Inc., New York.

Mehlquist, G. A. L. 1947. Polyploidy in the genus *Paphiopedilum* Pfitz (*Cypripedium* Hort.) and its practical implications. Missouri Botan. Garden Bull. 35:211.

Mehra, R. C., and K. S. Rai. 1970. Cytogenetic studies of meiotic abnormalities in *Collinsia tinctoria*. I. Chromosome stickiness. Can. J. Genet. Cytol. 12:560–569.

Mehra, R. C., and K. S. Rai. 1972. Cytogenetic studies of meiotic abnormalities in *Collinsia tinctoria*. II. Desynapsis. Can. J. Genet. Cytol. 14:637–644.

Melander, Y. 1950a. Studies on the chromosomes of *Ulophysema oresundense*. Hereditas 36:233–255.

Melander, Y. 1950b. Accessory chromosomes in animals, especially in *Polycelis tenuis*. Hereditas 36:261–296.

Melander, Y. 1959. The mitotic chromosomes of some cavicorn mammals (*Bos taurus* L., *Bison bonasus* L., and *Ovis aries* L.). Hereditas 45:649–664.

Melander, Y., and O. Knutsen. 1953. The spermiogenesis of the bull from a karyological point of view. Hereditas 39:505–517.

Mendel, G. 1866. Versuche über Pflanzenhybriden. Verh. Naturfosch. Ver., Brünn., Bd. 4:3–47.

Mendelson, P., and D. Zohary. 1972. Behaviour and transmission of supernumerary chromosomes in *Aegilops speltoides*. Heredity 29:329–339.

Menzel, M. Y., and J. M. Price. 1966. Fine structure of synapsed chromosomes in F_1 *Lycopersicon esculentum-Solanum lycopersicoides* and its parents. Amer. J. Bot. 53:1079–1086.

Menzel, M. Y., and F. D. Wilson. 1963. An allododecaploid hybrid of *Hibiscus diversifolius*. J. Hered. 54:55–60.

Metz, C. W. 1931. Chromosomal differences between germ cells and soma in *Sciara*. Biol. Zentralbl. 51:119–124.

Metz, C. W. 1933. Monocentric mitosis with segregation of chromosomes in *Sciara* and its bearing on the mechanism of mitosis. Biol. Bull. Woods Hole 54:333–347.

Metz, C. W. 1934. Evidence indicating that in *Sciara* the sperm regularly transmits two sister sex chromosomes. Proc. Natl. Acad. Sci. U.S.A. 20:31–36.

Metz, C. W. 1936. Factors influencing chromosome movements in mitosis. Cytologia 7:219–231.

Metz, C. W. 1938a. Chromosome behavior, inheritance and sex determination in *Sciara*. Amer. Nat. 72:485–520.

Metz, C. W. 1938b. Observations on evolutionary changes in the chromosomes of *Sciara* (Diptera). Carnegie Inst. Publ. 501:275–293.

Metz, C. W., M. Moses, and E. Hoppe. 1926. Chromosome behavior and genetic behavior in *Sciara*. I. Chromosome behavior in the spermatocyte divisions. Z. Indukt. Abst. Vererbl. 42:237–270.

Metz, C. W., and M. S. Moses. 1928. Observations on sex-ratio determination in *Sciara* (Diptera). Proc. Natl. Acad. Sci. U.S.A. 14:930–932.

Metz, C. W., and M. L. Schmuck. 1929. Unisexual progenies and the sex chromosome mechanism in *Sciara*. Proc. Natl. Acad. Sci. U.S.A. 15:863–866.

Metz, C. W., and M. L. Schmuck. 1931. Differences between chromosome groups of soma and germ-line in *Sciara*. Proc. Natl. Acad. Sci. U.S.A. 17:272–275.

Meurman, O. 1933. Chromosome morphology, somatic doubling and secondary association in *Acer platanoides* L. Hereditas 18:145–173.

Meyer, G. 1964. A possible correlation between submicroscopic structure of meiotic chromosomes and crossing over. *In:* Proc. 3rd Eur. Conf. Electron Microscopy. Publ. House Czech Acad. Sci., Prague, pp. 461–462.

Meyer, V. G. 1971. Cytoplasmic effects on anther numbers in interspecific hybrids of cotton. II. *Gossypium herbaceum* and *G. harknessii*. J. Hered. 62:77–78.

Meyers, L. L., F. S. Newman, G. R. Warren, J. E. Catlin and C. K. Anderson. 1975. The calf ligated intestinal segment test to detect enterotoxigenic *Escherichia coli*. Infect. Immun. 11:588–591.

Michie, D. 1953. Affinity: a new genetic phenomenon in the house mouse. Evidence from distant crosses. Nature 171:26–27.

Michie, D. 1955. Genetical studies with "vestigal tail" mice. I. The sex differences in crossing-over between vestigal and rex. J. Genet. 53:270–279.

Miescher, F. 1871. Über die chemische Zusammensetzung der Eiterzellen. F. Hoppe-Seyler's Med. Chem. Untersuch. 4:441–460.

Mikelsaar, A. V. N. 1967. The mosaicity with aspect of the deletion of a part of the long arm of one chromosome of group D (in man). Genetika 4:142–145.

Miller, D. A., R. Tantravahi, U. G. Dev, and O. J. Miller. 1977. Frequency of satellite association of human chromosomes is correlated with amount of Ag-staining of the nucleolus organizer region. Amer. J. Hum. Genet. 29:490–502.

Miller, O. J., W. R. Berg, B. Mukherjee, A. Van, N. Gamble, and A. C. Cristakos. 1963. Non-random distribution of chromosomes in metaphase figures from cultured human leucocytes. II. The peripheral location of chromosomes 13, 17–18 and 21. Cytogenetics 2:152–168.

Miller, O. L. 1963. Cytological studies in asynaptic maize. Genetics 48:1445–1466.

Miller, O. L. 1964. Extrachromosomal nucleolar DNA in amphibian oocytes. J. Cell Biol. 23:60A.

Miller, O. L. 1966. Structure and composition of peripheral nucleoli of salamander oocytes. Natl. Cancer Inst. Monogr. 23:53–66.

Miller, O. L., B. A. Hamkalo, and C. A. Thomas. 1970. Visualization of bacterial genes in action. Science 169:392–395.

Misra, R. N., and S. V. S. Shastry. 1969. Desynapsis and intragenomic differentiation in cultivated species of *Oryza*. Cytologia 34:1–5.

Mitchell, M. B., and H. K. Mitchell. 1952. A case of "maternal" inheritance in *Neurospora crassa*. Proc. Natl. Acad. Sci. U.S.A. 38:442–449.

Mittwoch, U. 1973. Genetics of sex differentiation. Academic Press, New York.

Moens, P. 1965. The transmission of a heterochromatic isochromosome in *Lycopersicon esculentum*. Canad. J. Genet. Cytol. 7:296–303.

Moens, P. B. 1968. Synaptinemal complexes of *Lilium tigrinum* (triploid) sporocytes. Can. J. Genet. Cytol. 10:799–807.

Moens, P. B. 1969a. Genetic and cytological effects of three desynaptic genes in the tomato. Can. J. Genet. Cytol. 11:857–869.

Moens, P. B. 1969b. The fine structure of meiotic chromosome polarization and pairing in *Locusta migratoria* spermatocytes. Chromosoma 23:1–25.

Moens, P. B. 1973. Quantitative electron microscopy of chromosome organization at meiotic prophase. Cold Spring Harb. Symp. Quant. Biol. 38:99–107.

Mogensen, H. L., 1977. Ultrastructural analysis of female pachynema and the relationship between synaptonemal complex length and crossing-over in *Zea mays*. Carlsberg Res. Commun. 42:475–497.

Moh, C. C., and R. A. Nilan. 1954. "Short" chromosome—A mutant in barley induced by atomic bomb irradiation. Cytologia 19:48–53.

Mohl, H. von. 1835. Über die Vermehrung der Pflanzenzelle durch Theilung. Dissertation, Tübingen.

Mohr, O. L. 1924. Z. Ind. Abst. Vererbl. 32:118. Cited *in:* Lindsley and Grell, 1967.

Montgomery, T. H. 1904. Some observations and considerations upon the maturation phenomena of the germ cells. Biol. Bull. 6:137–158.

Montgomery, T. H. 1906. Chromosomes in the spermatogenesis of the Hemiptera-Heteroptera. Trans. Am. Phil. Soc. (N.S.) 31:97–173.

Mookerjea, A. 1956. A cytological study of several members of the Liliaceae and their interrelationships. Sudmal. El. Kasv. Sevra. Vanamo. Kasv. Julk. 29:1–44.

Moore, D. H. 1967. Correlation of Bittner virus activity with milk fractions separated by density gradient ultracentrifugation. Proc. Amer. Assoc. Cancer Res. 8:48.

Moore, J. E. S. 1895. On the structure changes in the reproductive cells during spermatogenesis of Elasmobranchs. Quart. J. Micro. Sci. 38.

Moore, K. L., and M. L. Barr. 1955. Smears from the oral mucosa in the detection of chromosomal sex. Lancet II:57–58.

Moorhead, P. S., and E. Saksela. 1963. Non-random chromosomal aberrations in SV_{40}-transformed human cells. J. Cell. Comp. Physiol. 62:57–72.

Morgan, D. T., and R. Rappleye. 1950. Twin and triplet pepper seedlings. A study of polyembryony in *Capsicum frutescens*. J. Hered. 41:91–95.

Morgan, D. T., and R. Rappleye. 1954. A cytogenetic study of the origin of multiple seedlings of *Capsicum frutescens*. Amer. J. Bot. 41:576–586.

Morgan, L. V. 1922. Non-criss-cross inheritance in *Drosophila melanogaster*. Biol. Bull. 42:267–274.

Morgan, L. V. 1933. A closed X-chromosome in *Drosophila melanogaster*. Genetics 18:250–283.

Morgan, T. H. 1910a. Chromosomes and heredity. Amer. Nat. 44:449–496.

Morgan, T. H. 1910b. Sex limited inheritance in *Drosophila*. Science 32:120–122.

Morgan, T. H. 1911. Chromosomes and associative inheritance. Science 34:636–638.

Morgan, T.H., C. B. Bridges, and J. Schultz. 1932. Constitution of the germinal material in relation to heredity. Carnegie Inst. Wash. Yearbook 31:303–307.

Morgan, T. H., C. B. Bridges, and J. Schultz. 1933. Constitution of the germinal material in relation to heredity. Carnegie Inst. Yearbook 32:298–302.

Morgan, T. H., C. B. Bridges, and J. Schultz. 1935. Constitution of germinal material in relation to heredity. Carnegie Inst. Wash. Yearbook 34:284–291.

Morinaga, T. 1934. Interspecific hybridization in *Brassica*. VI. The cytology of F_1 hybrids of *B. juncea* and *B. nigra*. Cytologia 6:62–67.

Morris, R., and E. R. Sears. 1967. The cytogenetics of wheat and its relatives. *In:* Quisenberry, K. S. and L. P. Reitz. (eds.). Wheat and wheat improvement. Amer. Soc. Agron., Washington, D.C.

Morrison, G. 1932. The occurrence and use of haploid plants in tomato with especial reference to the variety Marglobe. Proc. VI Int. Congr. Genet. 2:137.

Morse, M., E. Lederberg, and J. Lederberg. 1956a. Transduction in *Escherichia coli* K-12. Genetics 41:121–156.

Morse, M., E. Lederberg, and J. Lederberg. 1956b. Transductional heterogenotes in *Escherichia coli*. Genetics 41:758–779.

Moseman, H. G., and L. Smith. 1954. Gene location by three point test and telocentric half chromosome fragment in *Triticum monococcum*. Agron. J. 46:120–124.

Moses, M. J. 1956. Chromosomal structure in crayfish spermatocytes. J. Biophys. Biochem. Cytol. 2:215–218.

Moses, M. J. 1968. Synaptinemal complex. Ann. Rev. Genet. 2:363–412.

Mosharrafa, E., W. Pilacinski, J. Zissler, M. Fiandt, and W. Szybalski. 1976. Insertion sequence IS2 near the gene for prophage λ excision. Molec. Gen. Genetics 147:103–110.

Mosier, H. D., L. W. Scott, and L. H. Cotter. 1960. The frequency of the positive sex-chromatin pattern in males with mental deficiency. Pediatrics 25:291–297.

Moutschen, J. 1965. Analyse des effects du L-threitol-l,4-bismethansulfonate sur les chromosomes de *Vicia faba* L. La Cellule 55:161–189.

Muller, A. J. 1965. Über den Zeitpunkt der Mutationsauslösung nach Einwirkung von N-Nitroso-N-methylharnstoff auf quellende Samen von *Arabidopsis*. Mutation Res. 2:426–437.

Müller, D. 1975. Tertiäre Trisomie für Chromosom 3^5 bei *Pisum sativum*. Caryologia 28:111–125.

Müller, U., H. Zentgraf, I. Eicken, and W. Keller. 1978. Higher order structure of Simian virus 40 chromatin. Science 201:406–415.

Muller, H. J. 1914a. A factor for the fourth chromosome of *Drosophila*. Science 39:906.

Muller, H. J. 1914b. A gene for the fourth chromosome in *Drosophila*. J. Exptl. Zool. 17:325–336.

Muller, H. J. 1914c. A new mode of segregation in Gregory's tetraploid primulas. Amer. Nat. 48:508–512.

Muller, H. J. 1915. The mechanics of crossing over. Ph.D. Thesis.

Muller, H. J. 1916. The mechanism of crossing-over. I and II. Amer. Nat. 50:193–221, 284–305, 350–366, 421–434.

Muller, H. J., 1917. An *Oenothera*-like case in *Drosophila*. Proc. Natl. Acad. Sci., U.S.A. 3:619.

Muller, H. J. 1925. Why polyploidy is rarer in animals than in plants. Amer. Nat. 59:346–353.

Muller, H. J. 1927. Artificial transmutation of the gene. Science 66:84–87.

Muller, H. J. 1928. The production of mutations by X-rays. Proc. Natl. Acad. Sci. U.S.A. 14:714–726.

Muller, H. J. 1932a. Further studies on the nature and causes of gene mutations. Proc. VI Int. Genet. Congr., Ithaca 1:213–255.

Muller, H. J. 1932b. Some genetic aspects of sex. Amer. Nat. 66:118–138.

Muller, H. J. 1938. The re-making of chromosomes. Collecting Net, Woods Hole 13:181–195 and 198.

Muller, H. J. 1940a. Bearings of the *Drosophila* work on systematics. *In:* Huxley, J. (ed.). The new systematics. pp. 185–268. Clarendon Press, Oxford.

Muller, H. J. 1940b. Analysis of the process of structural change in chromosomes of *Drosophila*. J. Genet. 40:1–66.

Muller, H. J. 1947–1948. Evidence of the precision of genetic adaptation. Harvey Lectures Ser. 43:165–229.

Muller, H. J., and I. H. Herskowitz. 1954. Concerning the healing of chromosome ends produced by breakage in *Drosophila melanogaster*. Amer. Nat. 88:177–208.

Muller, H. J., B. B. League, and C. A. Offerman. 1931. Effects of dosage changes of sex-linked genes, and compensatory effects of other gene differences between male and female. Anat. Rec. 51(Suppl.):110.

Muller, H. J., and T. S. Painter. 1929. The cytological expression of changes in gene alignment produced by x-rays in *Drosophila*. Amer. Nat. 63:193–200.

Müntzing, A. 1936. The evolutionary significance of autopolyploidy. Hereditas 21:263–378.

Müntzing, A. 1937. Polyploidy from twin seedlings. Cytologia Fujii Jubilaei Volumen 1937:211–227.

Müntzing, A. 1946. Cytological studies of extra fragment chromosomes in rye. III. The mechanism of nondisjunction at the pollen mitosis. Hereditas 32:97–119.

Müntzing, A. 1948a. Cytological studies of extra fragment chromosomes in rye. IV. The positions of various fragment types in somatic plates. Hereditas 34:161–180.

Müntzing, A. 1948b. Accessory chromosomes in *Poa alpina*. Heredity 2:49–61.

Müntzing, A. 1951. Cyto-genetic properties and practical value of tetraploid rye. Hereditas 37:17–84.

Müntzing, A. 1954. An analysis of hybrid vigour in tetraploid rye. Hereditas 40:265–277.

Müntzing, A. 1957. Polyploidiezuchtung. *In:* Roemer, W. und W. Rudorf. (eds.). Handbuch der Pflanzenzüchtung. 2nd ed., Vol. 1, pp. 700–731.

Müntzing, A. 1961. Genetic research. A survey of methods and main results. Lts Förlag, Stockholm.

Müntzing, A. 1963. Effects of accessory chromosomes in diploid and tetraploid rye. Hereditas 49:371–426.

Müntzing, A., and S. Adkik. 1948. Cytological disturbances in the first inbred generation of rye. Hereditas 34:485–509.

Müntzing, A., and A. Lima-de-Faria. 1949. Pachytene analysis of standard and large isofragments in rye. Hereditas 35:253–268.

Müntzing, A., and A. Lima-de-Faria. 1952. Pachytene analysis of a deficient accessory chromosome in rye. Hereditas 38:1–10.

Müntzing, A., and A. Lima-de-Faria. 1953. Pairing and transmission of a small accessory isochromosome in rye. Chromosoma 6:145–148.

Murakami, M., N. Takahashi, and K. Harada. 1972. Induction of haploid plants by anther culture in maize. I. On the callus formation and root differentiation. Sci. Rep. Kyoto Prefect Univ. Agric. 24:1–8.

Myers, W. M. 1939. Colchicine induced tetraploidy in perennial ryegrass, *Lolium perenne* L. J. Hered. 30:499–504.

Nabholtz, M., V. Miggiano and W. Bodmer. 1969. Genetic analysis with human-mouse somatic cell hybrids. Nature 223:358–363.

Nabors, M. W. 1976. Using spontaneously occurring and induced mutations to obtain agriculturally useful plants. Bioscience 26:761–768.

Nagao, S., and K. Saki. 1939. Association of chromosomes in *Chelidonium majus* L. Jap. J. Genet. 15:23–28.

Nägeli, C. von. 1842. Zur Entwicklungsgeschichte des Pollens. Zürich.

Nägeli, C. von 1884. Mechanisch-physiologische Theorie der Abstammungslehre. München.

Nagl, W. 1962. 4096-Ploidy und "Riesenchromosomen" in Suspensor von *Phaseolus coccineus*. Naturwiss. 49:261–262.

Nagl, W. 1965. Die SAT-Riesenchromosomen der Kerne des Suspensors von *Phaseolus coccineus* und ihr Verhalten während der Endmitose. Chromosoma 16:511–520.

Nagl, W. 1967. Die Riesenchromosomen von *Phaseolus coccineus* L.: Baueigentümlichkeiten, Strukturmodifikationen, zusätzliche Nukleolen und Vergleich mit den mitotischen Chromosomen. Oesterr. Bot. Z. 114:171–182.

Nagl, W. 1969a. Banded polytene chromosomes in the legume *Phaseolus vulgaris*. Nature 221:70–71.

Nagl, W. 1969b. Puffing of polytene chromosomes in a plant *(Phaseolus vulgaris)*. Naturwiss. 56:221–222.

Nagl, W. 1970. Differentielle RNS an pflanzlichen Riesenchromosomen. Ber. Deut. Bot. Ges. 83:301–309.

Nagley, P., and A. W. Linnane. 1972. Biogenesis of mitochrondria. XXI. Studies on the nature of the mitochrondrial genome in yeast: the degenerative effects of ethidium bromide on mitochondrial genetic information in a respiratory competent strain. J. Mol. Biol. 66: 181–193.

Nagley, P., E. B. Gingold, H. B. Lukins, and A. W. Linnane. 1973. Biogenesis of mitochondria. XXV. Studies on the mitochondrial genomes of petite mutants of yeast using ethidium bromide as a probe. J. Mol. Biol. 78:335-350.

Nakata, K., and M. Tanaka. 1968. Differentiation of embryoids from developing germ cells in anther culture of tabacco. Jap. J. Genet. 43:65-71.

Nance, W. E., and E. Engel. 1967. Autosomal deletion mapping in man. Science 155:692-694.

Narayanaswamy, S., and L. P. Chandy. 1971. In vitro induction of haploid, diploid and triploid androgenetic embryoids and plantlets in Datura metel L. Ann. Bot. 35:535-542.

Nass, M. M. K. 1966. The circularity of mitochondrial DNA. Proc. Natl. Acad. Sci. U.S.A. 56:1215-1222.

Nass, M. M. K., and S. Nass. 1962. Fibrous structures within the matrix of developing chick embryo mitochondria. Exptl. Cell Res. 26:424-437.

Nass, M. M. K., and S. Nass. 1963. Intramitochondrial fibers with DNA characteristics. J. Cell Biol. 19:593-596.

Navashin, M. S., Z. V. Bolkhovskikh, and L. M. Makushenko. 1959. A morphophysiological investigation of pollen tubes. Abstr. 1st Conf. Cytol., Leningrad, pp. 101-103 (in Russian).

Navashin, M., and H. Gerassimova. 1935. Nature and cause of mutation. Biol. Zhurn. 4:627-633.

Neary, G. J., and H. J. Evans. 1958. Chromatid breakage by irradiation and the oxygen effect. Nature 182:890-891.

Nebel, B. R. 1937. Mechanisms of polyploids through colchicine. Nature 140:1101.

Nebel, B. R. 1941. Structure of Tradescantia and Trillium chromosomes with particular emphasis on number of chromonemata. Cold Spring Harbor Symp. Quant. Biol. 9:7-12.

Neilson-Jones, W. 1969. Plant chimeras. 2nd ed. Methuen and Co., Ltd., London.

Nel, P. M. 1973. Reduced recombination associated with the asynaptic mutant of maize. Genetics 74:s193-194.

Nemec, B. 1910. Das Problem der Befruchtungsvorgänge und andere zytologische Fragen. Borntraeger, Berlin.

Nestel, B. L., and M. J. Creek. 1962. Pangolagrass. Herb. Abstr. 32:265-271.

Neuffer, M. G., L. Jones, and M. S. Zuber. 1968. The mutations of maize. Crop Science Soc. Amer., Madison, Wisconsin.

Newman, L. J. 1967. Meiotic chromosomal aberrations in wild populations of Podophyllum peltatum. Chromosoma 22:258-273.

Newmeyer, D., and C. W. Taylor. 1967. A pericentric inversion in Neurospora with unstable duplication progeny. Genetics 56:771-791.

Nichols, W. W., A. Levan, B. Hall, and G. Östergren. 1962. Measles-associated chromosome breakage. Preliminary communication. Hereditas 48:367-370.

Nichols, W. W. 1970. Virus-induced chromosome abnormalities. Ann. Rev. Microbiol. 24:479-500.

Nicoletti, B., G. Trippa, and A. DeMarco. 1967. Reduced fertility in SD males and its bearing on segregation distortion in Drosophila melanogaster. Atti Acad. Naz. Lincet 43:383-392.

Niebuhr, E. 1974. Triploidy in man. Cytogenetical and clincal aspects. Humangen. 21:103-125.

Nielsen, J., A. Homma, F. Christiansen, K. Rasmussen, and P. Saldaña-Garcia. 1977. Deletion in long arm 13. Hum. Genet. 13:339-345.

Niizeki, H., and K. Oono. 1968. Induction of haploid rice plant from anther culture. Proc. Jap. Acad. 44:554-557.

Niizeki, H., and K. Oono. 1971. Rice plants obtained by anther culture. In: Les Cultures de Tissus de Plantes Colloques Internationaux du C.N.R.S., No. 193. Paris.

Nikajima, G., and A. Zenyozi. 1966. Cytogenetics of wheat and rye hybrids. Seiken Zihô 18:39-48.

Nilan, R. A. 1956. Factors governing plant radiosensitivity. Proc. Conf. Radioact. Isotopes Agric., East Lansing, Michigan, 1956.

Ninan, C. A. 1958. Studies on the cytology and phylogeny of the Pteridophytes. VI. Observations on the Ophioglossaceae. Cytologia 23:291–316.

Ninan, C. A., and T. G. Raveendrananatth. 1965. A natural occurring haploid embryo in the coconut palm (*Cocos nucifera* L.). Caryologia 18:619–623.

Nishiyama, I., and M. Tabata. 1964. Cytogenetic studies in *Avena*. XII. Meiotic chromosome behaviour in a haploid cultivated oat. Jap. J. Genet. 38:311–316.

Nitsch, C. 1974. Pollen culture—a new technique for mass production of haploid and homozygous lines. *In:* Kasha. 1974. pp. 123–135.

Nitsch, J. P. 1972. Haploid plants from pollen. Z. Pflanzenz. 67:3–18.

Nitsch, J. P., C. Nitsch, and S. Hamon. 1968. Réalisation expérimentale de l'androgénèse chez divers *Nicotiana*. C. R. Séanc. Soc. Biol. 162:369–372.

Nitsch, J. P., and C. Nitsch. 1969. Haploid plants from pollen grains. Science 163:85–87.

Njordenskiöld, E. 1935. The history of biology. A survey. Tudor Publishing Co., New York.

Nonidez, J. F. 1915. Estudios sobre las celulas sexuales. I. Los chromosomas, goniales y las mitosis de maduracion en *Blaps lusitanica* y *B. waltli*. Mem. R. Soc. Esp. Hist. Nat. 10:149–190.

Nöth, M. H., and W. O. Abel. 1971. Zur Entwicklung haploider Pflanzen aus unreifen Microsporen verschiedener *Nicotiana*-Arten. Z. Pflanzenz. 65:277–284.

Novick, R. P. 1969. Extrachromosomal inheritance in bacteria. Bacteriol. Rev. 33:210–263.

Novikoff, A. B., and E. Holtzman. 1976. Cells and organelles. 2nd ed. Holt, Rhinehart and Winston, New York.

Novitski, E. 1977. Human genetics. Macmillan Publ. Co., Inc., New York.

Novitski, E., and G. D. Hanks. 1961. Analysis of irradiated *Drosophila* populations for meiotic drive. Nature 190:989–990.

Novitski, E., W. J. Peacock, and J. Engel. 1965. Cytological basis of sex ratio in *Drosophila pseudoobscura*. Science 148:516–517.

Nowell, P. C., and D. A. Hungerford, 1960. Chromosome studies on normal and leukemic human leukocytes. J. Natl. Cancer Inst. 25:85–108.

Ogur, M., R. O. Erickson, G. V. Rosen, K. B. Sax, and C. Holden. 1951. Nucleic acids in relation to cell division in *Lilium longiflorum*. Expt. Cell Res. 2:73–89.

Ohno, S. 1967. Sex chromosomes and sex-linked genes. *In:* Labhart, A., T. Mann, L. T. Samuels, and J. Zander. (eds.). Monographs on endocrinology. Springer-Verlag, Berlin.

Ohno, S., and T. S. Hauschka. 1960. Allocycly of the X-chromosome in tumors and normal tissues. Cancer Res. 20:541–545.

Ohno, S., J. M. Trujillo, W. D. Kaplan, and R. Kinosita. 1961. Nucleolus-organizers in the causation of chromosomal anomalies in men. Lancet II:123–126.

Ohtsubo, E., R. C. Deonier, H. J. Lee and N. Davidson. 1974. Electron microscope heteroduplex studies of sequence relations among plasmids of *Escherichia coli*. J. Mol. Biol. 89:565.

Oishi, K., and D. F. Poulsen. 1970. A virus associated with SR-spirochaetes of *Drosophila nebulosa*. Proc. Natl. Acad. Sci. U.S.A. 67:1565–1572.

Okazaki, R., T. Okazaki, K. Sakabe, K. Sugimoto, and A. Sugino. 1968. Mechanism of DNA chain growth. I. Possible discontinuity and unusual secondary structure of newly synthesized chains. Proc. Natl. Acad. Sci. U.S.A. 59:598–605.

Okuno, S. 1952. Jap. J. Genet. 27:107.

Olins, A. L., and D. E. Olins. 1974. Spheroid chromatin units (ν bodies). Science 183:330–332.

Olivera, B. M., and I. R. Lehman. 1967. Linkage of polynucleotides through phospodiester bonds by an enzyme from *Escherichia coli*. Proc. Natl. Acad. Sci. U.S.A. 57:1426–1433.

Olmo, H. P. 1952. Breeding tetraploid grapes. Amer. Soc. Hort. Sci. 59:285–290.

O'Mara, J. G. 1939. Observations on the immediate effects of colchicine. J. Hered. 30:35–37.

O'Mara, M. K., and M. D. Hayward. 1978. Asynapsis in *Lolium perenne*. Chromosoma 67:87–96.

Øster, J. 1953. Mongolism: a clinical and genealogical investigation comprising 526 mongols living on Seeland and neighboring islands in Denmark. Danish Sci. Press, Copenhagen.

Östergren, G. 1943. Elastic chromosome repulsion. Hereditas 29:444–450.

Östergren, G. 1947. Heterochromatic B-chromosomes in *Anthoxanthum*. Hereditas 33:261–296.

Oudet, P., M. Gross-Bellard, and P. Chambon. 1975. Electron microscopic and biochemical evidence that chromatin structure is a repeating unit. Cell 4:281–300.

Ouyang, T. W., H. Hu, C. C. Chuang, and C. C. Tseng. 1973. Induction of pollen plants from anthers of *Triticum aestivum* L. cultivated *in vitro*. Scientia Sinica 16:79–90.

Pagliai, A. 1961. L'endomeiosi in *Toxoptera aurantiae* (Boyer de Foscolombe) (Homoptera, Aphididae). Rend. Acad. Naz. Lincei Ser. VIII, 31:455–457.

Pagliai, A. 1962. La maturazione dell'uovo partenogenetico e dell'uovo anfigonico in *Brevicoryne brassicae*. Carologia 15:537–544.

Painter, T. S. 1933. A new method for the study of chromosome rearrangements and the plotting of chromosome maps. Science 78:585–586.

Painter, T. S. 1934a. A new method for the study of chromosome aberrations and the plotting of chromosome maps in *Drosophila melanogaster*. Genetics 19:175–188.

Painter, T. S. 1934b. Salivary chromosomes and the attack on the gene. J. Hered. 25:465–476.

Painter, T. S., and H. J. Muller, 1929. Parallel cytology and genetics of induced translocations and deletions in *Drosophila*. J. Hered 20:287–298.

Palmer, R. G. 1974. A desynaptic mutant in the soybean. J. Hered. 65:280–286.

Palmer, R. G. 1976. Chromosome transmission and morphology of three primary trisomics in soybeans *(Glycine max)*. Can. J. Genet. Cytol. 18:131–140.

Pardue, M. L., and J. G. Gall. 1970. Chromosome localization of mouse satellite DNA. Science 168:1356–1358.

Paris Conference (1971). 1972. Standardization in human cytogenetics. Cytogenetics 11:317–362.

Park, S. J., E. Reinbergs, E. J. Walsh and K. J. Kasha. 1974. Comparison of the haploid recharge with pedigree and single seed descent methods in barley breeding. *In:* Kasha, K. J. 1974. pp. 278–279.

Parker, J. S. 1975. Chromosome-specific control of chiasma formation. Chromosoma 49:391–406.

Parry, D. M. 1973. A meiotic mutant affecting recombination in female *Drosophila melanogaster*. Genetics 73:465–486.

Parsons, P. A. 1959. Possible affinity between linkage groups V and XIII of the house mouse. Genetica 29:304–311.

Patau, K. A. 1961. Chromosome identification and the Denver Report. Lancet I:933–934.

Patau, K. A., D. W. Smith, E. M. Therman, S. L. Inhorn, and H. P. Wagner. 1960. Multiple congenital anomaly caused by an extra autosome. Lancet I:790–793.

Patil, S. R., S. Merrick, and H. A. Lubs. 1971. Identification of each human chromosome with a modified Giemsa stain. Science 173:821–822.

Paton, G. R., J. P. Jacobs, and F. T. Perkins. 1965. Chromosome changes in human diploid-cell cultures infected with *Micoplasma*. Nature 207:43–45.

Patterson, J. T., and H. J. Muller. 1930. Are "progressive" mutations produced by x-rays? Genetics 15:495–577.

Patterson, J. T., and W. S. Stone. 1952. Evolution in the genus *Drosophila*. Macmillan, New York.

Pavan, C. 1967. Chromosomal changes induced by infective agents. Triangle 8:42–48.

Pavan, C., and R. Basile. 1966. Chromosome changes induced by infections in tissues of *Rhynchosciara angelae*. Science 151:1556–1558.

Pavan, C., and M. E. Breuer. 1952. Polytene chromosomes in different tissues of *Rhynchosciara*. J. Hered. 43:151–157.

Pavan, C., and A. B. DaCunha. 1968. Chromosome activities in normal and in infected cells of Sciaridae. The Nucleus: Seminar on Chromosomes, pp. 183–196.

Pavan, C., and A. B. DaCunha. 1969. Chromosomal activities in *Rhynchosciara* and other Sciaridae. Ann. Rev. Genet. 3:425–450.

Pavlovsky, O., and T. Dobzhansky. 1966. Genetics of natural populations. XXXVII. The coadapted system of chromosomal variants in a population of *Drosophila pseudoobscura*. Genetics 53:843–854.

Peacock, W. J. 1965. Chromosome replication. J. Nat. Cancer Inst. Monog. 18:101.

Peacock, W. J., and J. Erickson. 1965. Segregation-distortion and regularly nonfunctional products of spermatogenesis in *Drosophila melanogaster*. Genetics 51:313–328.

Pelletier, G., C. Raquin, and G. Simon. 1972. La culture *in vitro* d'anthères d'asperge *(Asparagus officinalis)*. C. R. Acad. Sci. Paris 274:848–851.

Pelling, C. 1964. Ribonukleinsäuresynthese der Riesenchromosomen, autoradiographische Untersuchungen an *Chironomus tentans*. Chromosoma 15:71–122.

Pelling, C. 1966. A replicative and synthetic chromosome unit—the modern concept of the chromomere. Proc. Roy. Soc. Lond. B. 164:279–289.

Penrose, L. S. 1961. Mongolism. Brit. Med. Bull. 17:184–189.

Penrose, L. S., and G. F. Smith. 1966. Down's anomaly. Churchill, London.

Perkins, D. D. 1972. Special-purpose *Neurospora* stocks. Neurospora. Newsl. 19:30–32.

Perkins, D. D., and E. G. Barry. 1977. The cytogenetics of *Neurospora*. Adv. Genet. 19:134–285.

Perkins, D. D., D. Newmeyer, C. W. Taylor, and D. C. Bennett. 1969. New markers and map sequences in *Neurospora crassa*, with a description of mapping by duplication coverage, and of multiple translocation stocks for testing linkage. Genetics 40:247–278.

Perry, P. and H. J. Evans. 1975. Cytological detection of mutagen-carcinogen exposure by sister chromatid exchange. Nature 258:121–125.

Person, C. 1956. Some aspects of monosomic wheat breeding. Canad. J. Bot. 34:60–70.

Peters, N. 1954. Zytologische Untersuchungen an *Solanun tuberosum* und polyploiden Wildkartoffelarten. Z. Vererbungsl. 86:373–398.

Peterson, P. A. 1970. Controlling elements and mutable loci in maize: their relationship to bacterial episomes. Genetica 41:33–56.

Peto, F. H., and J. W. Boyes. 1940. Comparison of diploid and triploid sugar beets. Canad. J. Res., Sect. C, 18:273–282.

Petrov, D. F., and B. E. Yudin. 1973. Haploid apomixis and its importance for selection of hybrid corn. Izo. Sib. Ord. Akad. Nauk. SSSR Ser. Biol. Med. Nauk. 5:52–62 (Russian).

Pfeifer, D., D. Kubai-Maroni, and P. Habermann. 1977. Specific sites for integration of IS-elements within the transferase gene of the galactose operon. *In:* Bukhari et al., 1977.

Philip, J., and C. L. Huskins. 1931. The cytology of *Matthiola incana* R. Br., especially in relation to the inheritance of double flowers. J. Genet. 24:359–404.

Phillips, R. L., E. B. Patterson, and P. J. Buescher. 1977. Cytogenetic mapping of genes in the nucleolus organizer-satellite region of chromosome 6. Maize Genet. Coop. News Letter 51:49–52.

Picard, E., and J. de Buyser. 1973. Obtention de plantules haploïdes de *Triticum aestivum* L. à partir culture d'anthères *in vitro*. C. R. Acad. Sci. Paris 277:1463–1466.

Pincus, G. 1939a. The breeding of some rabbits produced by recipients of artificially activated ova. Proc. Natl. Acad. Sci. U.S.A. 25:557–559.

Pincus, G. 1939b. The comparative behavior of mammalian eggs *in vivo* and *in vitro*. IV. The development of fertilized and artificially activated rabbit eggs. J. Exptl. Zool. 82:85–129.

Pipkin, S. B. 1940. Multiple sex genes in the X-chromosome of *Drosophila melanogaster*. Univ. Texas Publ. 4032:126–156.

Pipkin, S. B. 1942. Intersex modifying genes in wild strains of *Drosophila melanogaster*. Genetics 27:286–298.

Pipkin, S. B. 1947. A search for sex genes in the second chromosome of *Drosphila melanogaster*, using the triploid method. Genetics 32:592–607.

Pipkin, S. B. 1960. Sex balance in *Drosophila melanogaster:* aneuploidy of long regions of chromosome 3, using the triploid method. Genetics 45:1205–1216.

Pissarev, V. 1963. Different approaches in *Triticale* breeding. 2nd Int. Wheat Genet. Symp., Proc., Hereditas Suppl. 2:279–290.

Piza, S. de T. 1939. Comportamento dos cromossômios na primeira divisão do espermatocito do *Tityus bahiensis*. Sci. Genet. 1:255–261.

Piza, S. de T. 1950. Observações cromossômicas en escorpiões brasileiros. Ciênc. Cultura 2:202–206.

Plessers, A. G. 1963. Haploids as a tool in flax breeding. Cereal News 8:3–6.

Plewa, M. J., and D. F. Weber. 1975. Monosomic analysis of fatty acid composition in embryo lipids of *Zea mays*. Genetics 81:277–286.

Plus, N. 1962. Comportement du virus *Sigma* de la Drosophile en centrifugation à gradient de densité de saccharose. C. R. Acad Sci. (Paris) 255:3498–3500.

Plus, N. 1963. Action de la 5-fluorodésozyuridine sur la multiplication du virus *Sigma* de la Drosophile. Biochim. Biophys. Acta 72:92–105.

Polani, P. E., J. L. Hamerton, F. Gianelli, and C. O. Carter. 1965. Cytogenetics of Down's syndrome (mongolism). III. Frequency of interchange trisomics and mutation rate of chromosome interchanges. Cytogenetics 4:193–206.

Polansky, D., and J. Ellison. 1970. 'Sex ratio' in *Drosophila pseudoobscura:* spermiogenic failure. Science 169:886–887.

Pontecorvo, G. 1958. Trends in genetic analysis. Columbia University Press, New York.

Porter, K. R. 1961. The ground substance; observations from electron microscopy. *In:* Brachet, J. and A. E. Mirsky. (eds.). The cell. Vol. 2, pp. 621–675. Academic Press, New York.

Poulson, D. F. 1963. Cytoplasmic inheritance and hereditary infections in *Drosophila. In:* Burdette, W. J. (ed.). Methodology in basic genetics. pp. 404–424. Holden-Day, San Francisco.

Poulson, D. F. 1968. Nature, stability and expression of hereditary SR infections in *Drosophila*. Proc. 12th Int. Congr. Genet. 2:91–92.

Poulson, D. F., and B. Sakaguchi. 1961. Nature of sex-ratio agent in *Drosophila*. Science 133:1489–1490.

Powell, J. B. 1969. Haploids in pearl millet, *Pennisetum typhoides*. Agron. Abst. 1969:15.

Prakken, R. 1943. Studies of asynapsis in rye. Hereditas 29:475–495.

Prasad, G. 1976. Translocation trisomics in barley. Nucleus 19:49–51.

Prasad, G., and K. Das. 1975. Interchange trisomics in a 6-rowed barley. Cytologia 40:627–632.

Preer, J. R., Jr. 1950. Microscopically visible bodies in the cytoplasm of the "killer" strains of *Paramecium aurelia*. Genetics 35:344–362.

Preer, J. R., Jr. 1971. Extrachromosomal inheritance: hereditary symbionts, mitochondria, chloroplasts. Ann. Rev. Genet. 5:361–406.

Preer, J. R., Jr., L. A. Hufnagel, and L. B. Preer. 1966. Structure and behavior of 'R' bodies from killer Paramecia. J. Ultrastruct. Res. 15:131–143.

Preer, L. B., A. Jurand, J. R. Preer, Jr., and B. M. Rudman. 1972. The classes of kappa in *Paramecium aurelia*. J. Cell Sci. 11:581–600.

Prescott, D. M. 1970. Cited *in:* Crick, 1971.

Pring, D. R., C. S. Levings, III, W. W. L. Hu and D. H. Timothy. 1977. Unique DNA associated with mitochondria in the "S"-type cytoplasm of male-sterile maize. Proc. Natl. Acad. Sci. U. S. A. 74:2904–2908.

Ptashne, K., and S. N. Cohen. 1975. Occurrence of insertion sequence (IS) regions on plasmid deoxyribonucleic acid as direct and inverted nucleotide sequence duplications. J. Bacteriol. 122:776–781.

Purkinje, J. E. 1825. Subjectae sunt symbolae ad ovi avium historiam ante incubationem. Ed. 1. Vratislaviae.

Puteyevsky, E., and D. Zohary. 1970. Behaviour and transmission of supernumerary chromosomes in diploid *Dactylis glomerata*. Chromosoma 32:135–141.

Raina, S. K., and R. D. Iyer. 1973. Differentiation of diploid plants from pollen callus in anther cultures of *Solanun melongena* L. Planta 70:275–280.

Rajhathy, T. 1975. Trisomics of *Avena strigosa*. Can. J. Genet. Cytol. 17:151–166.

Rajhathy, T., and G. Fedak. 1970. A secondary trisomic in *Avena strigosa*. Can. J. Genet Cytol. 12:358–360.

Rakha, F. A., and D. S. Robertson. 1970. A new technique for the production of A-B translocations and their use in genetic analysis. Genetics 65:223–240.

Ramage, R. T. 1960. Trisomics from interchange heterozygotes in barley. Agron. J. 52:156–159.

Ramage, R. T. 1964. Chromosome aberrations and their use in genetics and breeding— translocations. Proc. First Intl. Barley Genet. Symp., Wageningen, 1963, pp. 99–115.

Ramage, R. T. 1965. Balanced tertiary trisomics for use in hybrid seed production. Crop Sci. 5:177–178.

Ramage, R. T. 1975. Hybrid barley. Barley Genetics III. Proc. Third Intern. Barley Genet. Symp., Garching, pp. 761–770.

Ramage, R. T., and C. A. Suneson. 1958. A gene marker for the g chromosome in barley. Agron. J. 50:114.

Ramanujam, S., and D. Srinivasachar. 1943. Cytogenetic investigations in the genus *Brassica* and the artificial synthesis of *B. juncea*. Indian J. Genet. Plant Breed. 3:73–88.

Ramarkrishnan, T., and E. A. Adelberg. 1965. Regulatory mechanisms in the biosynthesis of isoleucine and valine. III. Map order of the structural genes and operator genes. J. Bacteriol. 89:661–664.

Randolph, L. F. 1928. Types of supernumerary chromosomes in maize. Anat. Rec. 41:102–105.

Randolph, L. F. 1932. Some effects of high temperature on polyploidy and other variations in maize. Proc. Natl. Acad. Sci. U.S.A. 18:222–229.

Randolph, L. F. 1936. Developmental morphology of the caryopsis in maize. J. Agric. Res. 53:881–916.

Randolph, L. F. 1941a. An evaluation of induced polyploidy as a method of breeding crop plants. Amer. Nat. 75:347–365.

Randolph, L. F. 1941b. Genetic characteristics of the B-chromosomes in maize. Genetics 26:608–631.

Randolph, L. F., and H. E. Fischer. 1939. The occurrence of parthenogenetic diploids in tetraploid maize. Proc. Natl. Acad. Sci. U.S.A. 25:161–164.

Randolph, L. F., and D. B. Hand. 1940. Relation between carotenoid content and number of genes per cell in diploid and tetraploid corn. J. Agric. Res. 60:51–64.

Rao, D. R., and S. V. S. Shastry. 1961. Current Sci. 30:232–233.

Rao, M. K., and P. R. K. Koduru. 1978. Asynapsis and spontaneous centromeric breakage in an inbred line of *Pennisetum americanum* (L.) Leeke. Proc. Inc. Acad. Sci. 87(B):29–35.

Rattner, J. B., and M. W. Berns. 1976. Centriole behavior in early mitosis of kangaroo rat cells (PTK_2). Chromosoma 54:387–395.

Reddi, V. R. 1968. Chromosome pairing in haploid *Sorghum*. Cytologia 33:471–476.

Redfield, H. 1955. Recombination increase due to heterologous inversions and the relation to cytological length. Proc. Natl. Acad. Sci. U.S.A. 41:1084–1091.

Redfield, H. 1957. Egg mortality and interchromosomal effects on recombinations. Genetics 42:712–728.

Rees, H. 1952. Asynapsis and spontaneous chromosome breakage in *Scilla*. Heredity 6:89–97.

Rees, H. 1958. Differential behavior of chromosomes in *Scilla*. Chromosoma 9:185–192.

Rees, H., and A. Jamieson. 1954. A supernumerary chromosome in *Locusta*. Nature 173:43–44.

Rees, H., and J. B. Thompson. 1955. Localization of chromosome breakage at meiosis. Heredity 9:399–407.

Renner, O. 1914. Befruchtung und Embryobildung bei *Oenothera lamarckiana* und einigen verwandten Arten. Flora 107:115–151.

Renner, O. 1916. Zur Terminologie des pflanzlichen Generationswechsels. Biol. Zentralbl. 36:337.

Renner, O. 1917. Versuch über die genetische Konstitution der Oenotheren. Z. Ind. Abst. Vererbl. 18:121.

Renner, O. 1921. Heterogamie im weiblichen Geschlecht und Embryosackentwicklung bei den Oenotheren. Zeitschr. Bot. 13:609–621.

Revell, S. H. 1955. A new hypothesis for 'chromatid' changes. *In:* Bacq, Z. M. and P. Alexander. (eds.). Proceedings of the radiobiology symposium, Liege 1954, pp. 243–253. Butterworth, London.

Revell, S. H. 1959. The accurate estimation of chromatid breakage, and its relevance to a new interpretation of chromatid aberrations induced by ionizing radiations. Proc. Roy. Soc. Lond. B, 150:563–589.

Revell, S. H. 1960. Some implications of a new interpretation of chromatid aberrations induced by ionizing radiations and chemical agents. *In:* Erwin Baur Memorial Lectures. Abhandl. Deutsch Akad. Wiss. Berlin, Kl. Med., pp. 45–46.

Revell, S. H. 1963. Chromatid aberrations—the generalized theory. *In:* Wolff, S. (ed.). Radiation induced chromosome aberrations. pp. 41–72. Columbia Univ. Press, New York.

Revell, S. H. 1966. Evidence for a dose-squared term in the dose-response curve for real chromatid discontinuities induced by x-rays and some theoretical consequences thereof. Mutation Res. 3:34–53.

Rhoades, M. M. 1933. A secondary trisomic in maize. Proc. Natl. Acad. Sci. U.S.A. 19:1031–1038.

Rhoades, M. M. 1936. A cytogenetical study of a chromosome fragment in maize. Genetics 21:491–505.

Rhoades, M. M. 1938. On the origin of a secondary trisome through the doubling of a half-chromosome fragment. Genetics 23:163–164.

Rhoades, M. M. 1940. Studies of a telocentric chromosome in maize with reference to the stability of its centromere. Genetics 25:483–520.

Rhoades, M. M. 1942. Preferential segregation in maize. Genetics 27:395–407.

Rhoades, M.M. 1947. Crossover chromosomes in unreduced gametes of asynaptic maize. Genetics 32:101 (Abstract).

Rhoades, M. M. 1952. Preferential segregation in maize. *In:* Gowen, J. W. (ed.). Heterosis. pp. 66–80. Iowa State College Press, Ames.

Rhoades, M. M. 1955. The cytogenetics of maize. *In:* Sprague, G. F. (ed.). Corn and corn improvement. pp. 123–219. Academic Press, New York.

Rhoades, M. M., and W. E. Kerr. 1949. A note on centromere-organization. Proc. Natl. Acad. Sci. U.S.A. 35:129–132.

Rhoades, M. M., and B. McClintock. 1935. The cytogenetics of maize. Bot. Rev. 1:292–325.

Ricciuti, R., and F. H. Ruddle. 1973. Assignment of three gene loci (PGK, HGPRT, G6PD) to the long arm of the human X chromosome by somatic cell genetics. Genetics 74:661–678.

Richardson, M. M. 1935. Meiosis in *Crepis*. II. Failure of pairing in *Crepis capillaris*. J. Genet. 31:119–143.

Richmond, M. H. 1970. Plasmids and chromosomes in prokaryotic cells. *In:* Charles, H. P. and C. J. G. Knight. (eds.). Organzation and control in prokaryotic and eucaryotic cells. pp. 249–278. University Press, Cambridge.

Richmond, T. R., and R. J. Kohel. 1961. Analysis of a completely male-sterile character in American upland cotton. Crop Sci. 1:397–401.

Rick, C. M. 1948. Genetics and development of nine male-sterile tomato mutants. Hilgardia 18:599–633.

Rick, C. M., W. H. Dempsey, and G. S. Khush. 1964. Further studies on the primary trisomics of the tomato. Canad. J. Genet. Cytol. 6:93–108.

Rieger, R. 1957. Inhomologenpaarung und Meioseablauf bei haploiden Formen von *Antirrhinum majus* L. Chromosoma 9:1–38.

Rieger, R. 1966. Alternative Integrationsversuche zur Entstehung chromosomaler Strukturumbauten: Bruch-Reunion- und Austausch-Hypothese. Biol. Zbl. 85:29–46.

Rieger, R., und A. Michaelis. 1958. Genetisches und cytogenetisches Wörterbuch. Zweite. Aufl., Springer-Verlag, Berlin.

Rieger, R., A. Michaelis, and M. M. Green. 1968. A glossary of genetics and cytogenetics. Classical and molecular. 3rd ed. Springer-Verlag, New York.

Rieger, R., A. Michaelis, and M. M. Green. 1976. Glossary of genetics and cytogenetics. Classical and molecular. 4th ed. Springer-Verlag, Berlin.

Riley, R. 1960. The diploidization of polyploid wheat. Heredity 15:407–429.

Riley, R. 1974. The status of haploid research. *In:* Kasha, 1974.

Riley, R., and V. Chapman. 1958. Genetic control of the cytologically diploid behavior of hexaploid wheat. Nature 182:1244–1246.

Riley, R., and C. N. Law. 1965. Genetic variation in chromosome pairing. Adv. Genet. 13:57–114.

Ris, H. 1942. A cytological and experimental analysis of the meiotic behaviour of the univalent X-chromosome in the bearberry aphid *Tamalia* (= *Phyllapis*) *coweni* (Ckll.) J. Exp. Zool. 90:267–326.

Ris, H. 1945. The structure of meiotic chromosomes in the grasshopper and its bearing on the nature of "chromomeres" and "lamp-brush chromosomes." Biol. Bull. 89:242–257.

Ris, H. 1961. Ultrastructure and molecular organization of genetic systems. Canad. J. Genet. Cytol. 3:95–120.

Ris, H., and B. L. Chandler. 1963. The ultrastructure of genetic systems in prokaryotes and eukaryotes. Cold Spring Harbor Symp. Quant. Biol. 28:1–8.

Ris, H. and W. Plaut. 1962. Ultrastructure of DNA-containing areas in the chloroplast of *Chlamydomonas*. J. Cell Biol. 13:383–391.

Ritossa, F. M., and S. Spiegelman. 1965. Localization of DNA complementary to RNA in the nucleolus organizer region of *D. melanogaster*. Proc. Natl. Acad. Sci. U.S.A. 53:737–745.

Röbbelen, G. 1960. Beiträge zur Analyse des *Brassica*-Genoms. Chromosoma 11:205–228.

Robberson, D. L., D. A. Clayton, and J. F. Morrow. 1974. Cleavage of replacing forms of mitochondrial DNA by Eco RI endonuclease. Proc. Natl. Acad. Sci. U.S.A. 71:4447–4451.

Roberts, P. A., R. F. Kimball, and C. Pavan. 1967. Response of *Rhynochosciara* chromosomes to a microsporidian infection: increased polyteny and generalized puffing. Exp. Cell Res. 47:408–422.

Robertson, D. W., G. A. Wiebe and F. R. Immer. 1941. A summary of linkage studies in barley. J. Amer. Soc. Agron. 33:47–64.

Robertson, W. R. B. 1916. Chromosome studies. I. Taxonomic relationships shown in the chromosomes of Tettigradae and Acrididae V-shaped chromosomes and their significance in Acrididae, Locustidae and Gryllida: chromosomes and variation. J. Morph. 27:179–331.

Rocchi, A. 1967. Sulla presenza di cromosomi soprannumerari in una popolazione di *Asellus waxalis*. Caryologia 20:107–113.

Rogers, J. S., and J. R. Edwardson. 1952. The utilization of cytoplasmic male-sterile inbreds in the production of corn hybrids. Agron. J. 44:8–13.

Rogers, O., and J. Ellis. 1966. Pollen nuclear division prevented with toluidine blue in *Vinca rosea* L. Hort. Sci. 1:62–63.

Rohloff, H. 1970. Die Spermatocytenteilungen der Tipuliden. IV. Mitteilung. Analyse der Orientierung röntgenstrahleninduzierter Quadrivalente bei *Pales ferruginea*. Dissertation, Eberhard-Karls-Universität, Tübingen.

Roman, H. 1947. Mitotic non-disjunction in the case of interchanges involving the B-type chromosome in maize. Genetics 32:391–409.

Roman, H. 1948. Directed fertilization in maize. Proc. Natl. Acad. Sci. U.S.A. 34:36–42.

Roman, H., and A. J. Ullstrup. 1951. The use of A-B translocations to locate genes in maize. Agron. J. 43:450–454.

Rosenberg, O. 1904. Über die Individualität der Chromosomen im Pflanzenreich. Flora 93:251.

Rosenberg, O. 1906. Über die Embryobildung in der Gattung *Hieracium*. Ber. Deutsch. Botan. Ges. 24:157–161.

Rosenberg, O. 1907. Cytological studies on the apogamy of *Hieracium*. Svedish Botan. Tidskr. 28:143–170.

Rosenberg, O. 1917. Die Reductionsteilung und ihre Degeneration in *Hieracium*. Sved. Bot. Tidskr. 11:145–206.

Rosenberg, O. 1930. Apogamie und Parthenogenesis bei Pflanzen. Handb. Vererbl. 2.

Ross, A. 1968. The substructure of centriole subfibers. J. Ultrastruct. Res. 23:537–539.

Ross, A., and I. P. Gormley. 1973. Examination of surface topography of Giemsa-banded human chromosomes by light and electron microscopic techniques. Exp. Cell. Res. 81:79–86.

Ross, W. M. 1971. News release. Kansas and Nebraska Agricultural Experiment Stations and Plant Science Research Divisions, A.R.S., U.S.D.A., Beltsville, Maryland.

Roux, W. 1883. Über die Bedeutung der Kernteilungsfiguren. Leipzig.

Rowe, P. R. 1974. Methods of producing haploids: parthenogenesis following interspecific hybridization. *In:* Kasha, 1974.

Rückert, T. 1892. Zur Entwicklungsgeschichte des Ovarialeies bei Seelachseiern. Anat. Anz. 7:107

Rudkin, G. T., and S. L. Corlette. 1957. Disproportionate synthesis of DNA in polytene chromosome region. Proc. Natl. Acad. Sci. U.S.A. 43:964–968.

Rudorf-Lauritzen, M. 1958. The trisomics of *Antirrhinum majus*. Proc. Tenth Inter. Congr. Genet. 2:243–244.

Runquist, E. W. 1968. Meiotic investigations in *Pinus sylvestris* L. Hereditas 60:77–128.

Rupp, W. P., and P. Howard-Flanders. 1968. Discontinuities in the DNA synthesized in an excision-defective strain of *E. coli* following ultraviolet radiation. J. Mol. Biol. 31:291–304.

Ruska, E. 1934. Über Fortschritte im Bau und in der Leistung des magnetischen Elektronenmikroskops. Z. Physik 87:580–602.

Russel, L. B. 1961. Genetics of mammalian sex chromosomes. Science 133:1795–1803.

Rutishauser, A. 1956. Genetics of fragment chromosomes in *Trillium grandiflorum*. Heredity 10:195–204.

Ruzicka, F., and H. G. Schwarzacher. 1974. The ultrastructure of human mitotic chromosomes and interphase nuclei treated by Giemsa banding techniques. Chromosoma 46:443–454.

Sadasivaiah, R. S., and K. J. Kasha. 1971. Meiosis in haploid barley—an interpretation of non-homologous chromosome associations. Chromosoma 35:247–263.

Sadasivaiah, R. S., and K. J. Kasha. 1973. Non-homologous associations of haploid barley chromosomes in the cytoplasm of *Hordeum bulbosum* L. Can. J. Genet. Cytol. 15:45–52.

Sadasivaiah, R. S., and M. L. Magoon. 1965. Studies of desynapsis in *Sorghum*. Genetica 36:307–314.

Saedler, H., and B. Heiss. 1973. Multiple copies of the insertion DNA sequences IS*1* and IS*2* in the chromosome of *E. coli* K-12. Molec. Gen. Genet. 122:267–277.

Saedler, H., D. Kubai, M. Nomura, and S. R. Jaskunas. 1975. IS*1* and IS*2* mutations in the ribosomal protein genes of *E. coli* K-12. Mol. Gen. Genet. 141:85–89.

Sakaguchi, B., and K. Oishi. 1965. Nature of the substance produced by nebulosa-SR agent that kills willistoni-SR agent in *Drosophila*. Ann. Rep. Nat. Inst. Gen. Jap. 15:38–39.

Sakaguchi, B., K. Oishi and S. Kobayashi. 1965. Interference between "sex-ratio" agents of *Drosophila willistoni* and *Drosophila nebulosa*. Science 147:160–162.

Sakharov, V. V., and V. V. Kuvarin. 1970. Aneuploidy in tetraploid rye. Genetika USSR 6:17–22.

Sanders, J. P. M., C. Heyting, and P. Borst. 1975a. The organization of genes in yeast mitochondrial DNA. I. The genes for large and small ribosomal RNA are far apart. Biochem. Biophys. Res. Commun. 65:699–707.

Sanders, J. P. M., P. Borst, and P. J. Weijers. 1975b. The organization of genes in yeast mitochondrial DNA. II. The physical map of Eco RI and Hind II + III fragments. Molec. Gen. Genet. 143:53–64.

Sandler, L., Y. Hiraizumi, and I. Sandler. 1959. Meiotic drive in natural populations of *Drosophila melanogaster*. I. The cytogenetic basis of segregation-distortion. Genetics 44:233–250.

Sandler, L., and Y. Hiraizumi. 1960. Meiotic drive in natural populations of *Drosophila melanogaster*. IV. Instability at the segregation-distorter locus. Genetics 45:1269–1287.

Sandler, L., and Y. Hiraizumi. 1961. Meiotic drive in natural population of *Drosophila melanogaster*. VIII. A heritable aging effect on the phenomenon of segregation-distortion. Canad. J. Genet. Cytol. 3:34–46.

Sandler, L., and E. Novitski. 1957. Meiotic drive as an evolutionary force. Amer. Nat. 91:105–110.

Sarvella, P., J. B. Holmgren, and R. A. Nilan, 1958. Analysis of barley chromosomes. Nucleus 1:183–204.

Sasaki, M. 1964. Notes on polyploidy in human male germ cells. Proc. Jap. Acad. 40:553.

Sasaki, M., and S. Makino. 1965. The meiotic chromosomes of man. Chromosoma 16:637–651.

Satina, S., and A. F. Blakeslee. 1935. Cytological effects of a gene in *Datura* which causes dyad formation in sporogenesis. Bot. Gaz. 96:521–532.

Satina, S., and A. F. Blakeslee. 1941. Periclinal Chimeras in *Datura stramonium* in relation to development of leaf and flower. Amer. J. Bot. 28:862–871.

Satina, S., A. F. Blakeslee, and A. G. Avery. 1940. Demonstration of the three germ layers in the shoot apex of *Datura* by means of induced polyploidy in periclinal chimeras. Amer. J. Bot. 27:895–905.

Savosjkin, I. P., and V. S. Cheredeyeva. 1969. Certain characteristic cytogenetical features of radish tetraploids. Genetika USSR 5:22–31.

Sax, K. 1931. Chromosome ring formation in *Rhoeo discolor*. Cytologia 3:36–53.

Sax, K. 1938. Chromosome aberrations induced by x-rays. Genetics 23:494–516.

Sax, K. 1941. Types and frequencies of chromosomal aberrations induced by x-rays. Cold Spring Harbor Symp. Quant. Biol. 9:93–101.

Sax, K., and E. Anderson. 1933. Segmental interchange in chromosomes of *Tradescantia*. Genetics 18:53–67.

Sax, K., and H. W. Edmonds. 1933. Development of the male gametophyte in *Tradescantia*. Bot. Gaz. 95:156–163.

Scaife, J. 1963. Conjugation and the sex factor in *Escherichia coli*. Sci. Progr. 51:566.

Schertz, K. F. 1963. Chromosomal, morphological and fertility characteristics of haploids and their derivatives in *Sorghum vulgare* Pers. Crop. Sci. 3:445–447.

Scheuring, J. F., D. R. Clark, and R. T. Ramage. 1976. Coordinators Report: desynaptic genes. Barley Genet. Newsl. 6:108–109.

Schleiden, M. J. 1838. Beiträge zur Phytogenesis. Joh. Müllers Arch. Anat. Physiol. Wiss. Med. 137–176.

Schlösser, L. A. 1936. Frosthärte und Polyploidie. Züchter 8:75–80.

Schmid, W. 1967. Pericentric inversions. Report on two malformation cases suggestive of parental inversion heterozygosity. J. Genet. Humaine 16:89–96.

Schmidt, J. 1920. Racial investigations. IV. The genetic behavior of a secondary sexual character. Compt. Rend. Trav. Lab. Carlsberg Ser. Physiol. 14:1–12.

Schmuck, M. L., and C. W. Metz. 1932. The maturation divisions and fertilization in eggs of *Sciara coprophila* Lint. Proc. Natl. Acad. Sci. U.S.A. 18:349–352.

Schnarf, K. 1929. Embryologie der Angiospermen. Handb. Pflanzenanat. 10(2), pgs. 1–5.

Schnedl, W. 1971. Analysis of the human karyotype using a reassociation technique. Chromosoma 34:448–454.

Schrader, F. 1940a. The formation of tetrads and the meiotic mitosis in the male of *Rhytidolomia senilis* Say (Hemiptera-Heteroptera). J. Morph. 67:123–141.

Schrader, F. 1940b. Touch-and-go pairing in chromosomes. Proc. Natl. Acad. Sci. U.S.A. 26:634–636.

Schrader, F. 1941. The sex chromosomes: heteropycnosis and its bearing on some general questions of chromosome behavior. *In:* Cytology, genetics and evolution. Univ. Penn. Press, Philadelphia.

Schrader, F. 1944. Mitosis. Columbia University Press, New York.

Schrader, F. 1953. Mitosis. The movement of chromosomes in cell division. Columbia Univ. Press, New York.

Schroeder, T. M., F. Auschwitz, and A. Knapp. 1964. Spontane Chromosomenaberrationen bei familiarer Panmylopathie. Humangen. 1:194–196.

Schultz, J. 1936. Variegation in *Drosophila* and the "inert" chromosome regions. Proc. Natl. Acad. Sci. U.S.A. 22:27–33.

Schultz, J., and D. G. Catcheside. 1937. The nature of closed X-chromosomes in *Drosophila melanogaster*. J. Genet. 35:315–320.

Schultz, J., and H. Redfield. 1951. Interchromosomal effects on crossing-over in *Drosophila*. Cold Spring Harbor Symp. Quant. Biol. 16:175–197.

Schultz, J., and P. St. Lawrence. 1949. A cytological basis for a map of the nuclear chromosomes in man. J. Hered. 40:31–38.

Schulz-Schaeffer, J. 1956. Cytologische Untersuchungen in der Gattung *Bromus* L. Z. Pflanzenz. 35:297–320.

Schulz-Schaeffer, J. 1960. Cytological investigations in the genus *Bromus*. J. Hered. 51:269–277.

Schulz-Schaeffer, J. 1970. A possible source of cytoplasmic male sterility in intermediate wheatgrass, *Agropyron intermedium* (Host) Beauv. Crop Sci. 10:204–205.

Schulz-Schaeffer, J. 1978. Registration of Montana-1 male sterile intermediate wheatgrass germplasm (Reg. No. GP 1). Crop Sci. 18:920.

Schulz-Schaeffer, J., P. W. Allderdice, and G. C. Creel. 1963. Segmental allopolyploidy in tetraploid and hexaploid *Agropyron* species of the crested wheatgrass complex (section Agropyron). Crop Sci. 3:525–530.

Schulz-Schaeffer, J., R. I. Baeva, and J. H. Kim. 1971. Genetic control of chromosome pairing in *Triticum* x *Agropyron* derivatives. Z. Pflanzenz. 65:53–67.

Schulz-Schaeffer, J., and C. R. Haun. 1961. The chromosomes of hexaploid common wheat, *Triticum aestivum* L. Z. Pflanzenz. 46:112–124.

Schulz-Schaeffer, J., and P. Jurasits. 1962. Biosystematic investigations in the genus *Agropyron*. I. Cytological studies of species karyotypes. Amer. J. Bot. 49:940–953.

Schulz-Schaeffer, J., J. H. Kim, and S. R. Chapman. 1973. Meiotic studies of the second substitution backcross to the amphidiploid hybrid, *Triticum durum* Desf. x *Agropyron intermedium* (Host) Beauv. Wheat Inf. Serv. 37:21–24.

Schulz-Schaeffer, J., and F. H. McNeal. 1977. Alien chromosome addition in wheat. Crop Sci. 17:891–896.

Schwann, T. 1839. Mikroskopische Untersuchungen über die Übereinstimmung in Struktur und dem Wachstum der Tiere und Pflanzen. Ostwalds Klassiker Nr. 176, Engelmann, Leipzig.

Schwartz, D. 1958. A new temperature-sensitive allele at the sticky locus in maize. J. Hered. 49:149–152.

Schwarz, D. 1960. Deoxyribonucleic acid and chromosome structure. In: Mitchell, J. S. (ed.). The cell nucleus. p. 227. Academic Press, New York.

Schwartz, H. 1932. Der Chromosomenzyclus von Tetraneura ulmi de Geer. Z. Zellforsch. 15:645–686.

Seabright, M. 1972. The use of proteolytic enzymes for the mapping of structural rearrangements in the chromosomes of man. Chromosoma 36:204–210.

Sears, E. R. 1944. Cytogenetic studies with polyploid species of wheat. II. Additional chromosomal aberrations in Triticum vulgare. Genetics 29:232–246.

Sears, E. R. 1952a. The behavior of isochromosomes and telocentrics in wheat. Chromosoma 4:551–562.

Sears, E. R. 1952b. Misdivision of univalents in common wheat. Chromosoma 4:535–550.

Sears, E. R. 1952c. Homoeologous chromosomes in Triticum aestivum. Genetics 37:624.

Sears, E. R. 1953. Nullisomic analysis in common wheat. Amer. Nat. 87:245–252.

Sears, E. R. 1954. The aneuploids of common wheat. Univ. Missouri Agr. Exp. Sta. Res. Bull. No. 572, Columbus, Missouri.

Sears, E. R. 1958. The aneuploids of common wheat. Proc. First Internat. Wheat Genet. Symp., Winnipeg, Manitoba, Canada, pp. 221–229.

Sears, E. R. 1962. The use of telocentrics in linkage mapping. Genetics 47:983.

Sears, E. R. 1966. Chromosome mapping with the aid of telocentrics. Proc. 2nd Wheat Genet. Symp., Lund 1963, Hereditas, Suppl. Vol. 2:370–381.

Sears, E. R. 1969. Wheat cytogenetics. Ann. Rev. Genet. 3:451–468.

Sears, E. R., and A. Camara. 1952. A transmissible dicentric chromosome. Genetics 37:125–135.

Seecof, R. L. 1962. CO_2 sensitivity in Drosophila as a latent virus infection. Cold Spring Harbor Symp. Quant. Biol. 16:99–111.

Seecof, R. L. 1968. The Sigma virus infection of Drosophila melanogaster. In: Maramorosch, K. (ed.). Insect virsues. pp. 59–93. Springer-Verlag, New York.

Sehested, J. 1974. A simple method for R banding of human chromosomes, showing a pH-dependent connection between R and G bands. Humangen. 21:55–58.

Selman, G. G. 1958. An ultra-violet light method for producing haploid amphibian embryos. J. Embryol. Exptl. Morphol. 6:634–637.

Sen, N. K. 1952. Isochromosome in tomato. Genetics 37:227–241.

Sen, S. K. 1970. Synaptonemal complexes in haploid Petunia and Antirrhinum spp. Naturwissensch. 57:550.

Sen, S. 1973. Polysomaty and its significance in Liliales. Cytologia 38:737–751.

Sendino, A. M., and J.-R. Lacadena. 1974. Cytogenetic studies of aneuploidy in the varieties Senatore Capelli and Bidi 17 of Triticum durum Desf. In: Genetics and breeding of durum wheat. Proc. Eucarpia Meetings, Bari, Italy, 1973.

Serebrovsky, A. S. 1929. A general scheme for the origin of mutations. Amer. Nat. 63:374–378.

Shapiro, J. A., and S. L. Adhya. 1969. The galactose operon of E. coli K-12. II. A deletion analysis of operon structure and polarity. Genetics 62:249–264.

Shapiro, L., L. I. Grossman, J. Marmur, and A. K. Kleinschmidt. 1968. Physical studies on the structure of yeast mitochondrial DNA. J. Mol. Biol. 33:907–922.

Sharma, G. P., M. L. Gupta, and G. S. Randhawa. 1967. Polysomy in Chrotogonus trachypterus (Blanchard) (Orthoptera: Acridoidea:Pyrgomorphidae) from three more populations and its possible role in animal speciation. Res. Bull. (N.S.) Panjab Univ. 18:157–163.

Sharp, W. R., D. K. Dougall, and E. F. Paddock, 1971. Haploid plantlets and callus from immature pollen grains of *Nicotiana* and *Lycopersicon*. Bull. Torrey Bot. Club 98:219–222.

Sharp, W. R., R. S. Raskin, and H. E. Sommer. 1972. The use of tissue culture in the development of haploid clones in tomato. Planta 104:357–361.

Shastry, S. V. S., W. K. Smith, and D. C. Cooper. 1960a. Chromosome differentiation in several specis of *Melilotus*. Amer. J. Bot. 47:613–621.

Shastry, S. V. S., D. R. R. Rao, and R. N. Misra. 1960b. Pachytene analysis in *Oryza*. I. Chromosome morphology in *Oryza sativa*. Indian J. Genet. 20:15–21.

Shaw, M. W. 1970. Human chromosome damage by chemical agents. Ann. Rev. Med. 21:409–432.

Shaw, M. W., and R. S. Krooth. 1966. Chromosomal instability in cultured fibroblasts derived from an infant with ring chromosome. Proc. 3rd Intern. Cong. Human. Genet., Chicago, 1966.

Shelton, E.R., P. M. Wasserman, and M. L. DePamphilis. 1978. Structure of Simian virus 40 chromosomes in nuclei from infected monkey cells. J. Mol. Biol. 125:491–514.

Sheridan, W. F., and H. Stern. 1967. Histones of meiosis. Exp. Cell Res. 45:323–330.

Shimada, K., R. A. Weisberg, and M. E. Gottesman. 1973. Prophage λ at unusual chromosomal locations. II. Mutations induced by bacteriophage λ in *Escherichia coli* K-12. J. Mol. Biol. 80:297–314.

Shinji, O. 1931. The evolutional significance of the chromosomes of the Aphididae. J. Morph. 51:373–433.

Shiraishi, Y., A. J. Freeman, and A. A. Sandberg. 1976. Increased sister chromatid exchange in bone marrow and blood cells from Bloom's syndrome. Cytogenet. Cell Genet. 17:162–173.

Siegel, R. W. 1953. A genetic analysis of the mate-killing trait in *Paramecium aurelia*, variety 8. Genetics 38:550–560.

Siemens, H. W. 1921. Konstitutions- und Vererbungspathologie. Springer, Berlin.

Simmonds, N. W. 1945. Meiosis in tropical *Rhoeo discolor*. Nature 155:731.

Simpson, C. E., and E. C. Bashaw. 1969. Cytology and reproductive characteristics in *Pennisetum setaceum*. Amer. J. Bot. 56:31–37.

Sinclair, J. H., and B. J. Stevens. 1966. Circular DNA filaments from mouse mitochondria. Proc. Natl. Acad. Sci. U.S.A. 56:508–514.

Sinet, P. M., D. Allard, J. Lejeune, and H. Jerome. 1975. Gene dosage effect in trisomy 21. Lancet I:276.

Singh, R. B., B. D. Singh, V. Laxmi, and R. M. Singh. 1977. Meiotic behaviour of spontaneous and mutagen induced partial desynaptic plants in pearl millet. Cytologia 42:41–47.

Singh, R. J. and T. Tsuchiya. 1977. Morphology, fertility, and transmission in seven monotelotrisomics of barley. Z. Pflanzenz. 78:327–340.

Sinnott, E. W., L. C. Dunn, and E. W. Dobszhansky. 1958. Principles of Genetics. 5th ed. McGraw-Hill Book Co., Inc., New York.

Sitte, P. 1965. Bau und Feinbau der Pflanzenzelle. Fischer, Jena.

Sjödin, J. 1970. Induced asynaptic mutants in *Vicia faba* L. Hereditas 66:215–232.

Skalinska, M. 1950. Studies on chromosome numbers of Polish angiosperms. Acta Soc. Bot. Polon. 20:45–68.

Skovsted, A. 1934. Cytological studies in cotton. I. The mitosis and the meiosis in diploid and triploid Asiatic cotton. Ann. Bot. 47:227–251.

Slizinsky, B. M. 1955. Chiasmata in the male mouse. J. Genet. 53:597–605.

Slizinsky, B. M. 1957. Cytological analysis of translocations in the mouse. J. Genet. 55:122–130.

Slizynska, H. 1938. Salivary gland analysis of the white-facet region of *Drosophila melanogaster*. Genetics 23:291–299.

Slizynska, H. 1963. Mutagenic effects of x-rays and formaldehyde food in spermatogenesis of *Drosophila melanogaster*. Genet. Res. 4:248–257.

406   References

Slonimski, P. 1949. Action de l'acriflavine sur les levures. IV. Mode d'utilisation du glucose par les mutants < petite colonie >. Ann. Inst. Pasteur 76:510–530.

Smith, D. W., J. M. Doctor, R. E. Ferrier, J. L. Prias, and A. Spock. 1968. Possible localization of the gene for cystic fibrosis of the pancreas to the short arm of chromosome 5. Lancet II:309–312.

Smith, F. J., J. H. de Jong, and J. L. Oud. 1975. The use of primary trisomics for the localization of genes on the seven different chromosomes of *Petunia hybrida*. I. Triplo V. Genetica 45:361–370.

Smith, L. 1939. Mutants and linkage studies in *Triticum monococcum* and *T. aegilopoides*. Univ. Mo. Agric. Exp. Sta. Res. Bull. 298:1–26.

Smith, L. 1942. Cytogenetics of a factor for multiploid sporocytes in barley. Amer. J. Bot. 29:451–456.

Smith, P. A., and R. C. King. 1968. Genetic control of synaptonemal complexes in *Drosophila melanogaster*. Genetics 60:335–351.

Smith-Sonneborn, J. E., and W. J. Van Wagtendonk. 1964. Purification and chemical characterization of *kappa* of stock 51, *Paramecium aurelia*. Exp. Cell Res. 33:50–59.

Smith-White, S. G., W. J. Peacock, B. Turner, and G. M. Den Dulk. 1963. A ring chromosome in man. Nature 197:102–103.

Snow, R. 1969. Permanent translocation heterozygosity associated with an inversion system in *Paeonia boronii*. J. Hered. 60:103–106.

Soldo, A. T., and G. A. Godoy. 1973. Molecular complexity of *Paramecium* symbiont *lambda* deoxyribonucleic acid: evidence for the presence of a multicopy genome. J. Mol. Bio. 73:93–108.

Soller, M., M. Wysoki, and B. Padeh. 1966. A chromosomal abnormality in phenotypically normal Saanen goats. Cytogenetics 5:88–93.

Somaroo, B. H., and W. F. Grant. 1971. Meiotic chromosome behavior in induced autotetraploids and amphidiploids in the *Lotus corniculatus* group. Can. J. Genet. Cytol. 13:663–671.

Sonneborn, T. M. 1938. Mating types in *P. aurelia:* diverse conditions for mating in different stocks; occurrence, number, and interrelations of the types. Proc. Amer. Phil. Soc. 79:411.

Sonneborn, T. M. 1943. Gene and cytoplasm. I. The determination and inheritance of the killer character in variety 4 of *Paramecium aurelia*. Proc. Natl. Acad. Sci. U.S.A. 29:329–343.

Sonneborn, T. M. 1959. Kappa and related particles in *Paramecium*. Adv. Virus Res. 6:229–356.

Soost, R. K. 1951. Comparitive cytology and genetics of asynaptic mutants in *Lycopersicon esculentum* Mill. Genetics 36:410–434.

Southern, D. T. 1969. Stable telocentric chromosomes produced by centric misdivision in *Myrmeleotettix maculatus* (Thunb.). Chromosoma 26:140–147.

Sparrow, A. H., C. L. Huskins, and G. B. Wilson. 1941. Studies on the chromosome spiralization cycle in *Trillium*. Can. Jour. Res. 19:323–350.

Sparvoli, E., H. Gay, and B. P. Kaufmann. 1965. Number and pattern of association of chromonemata in the chromosomes of *Tradescantia*. Chromosoma 16:415–435.

Spencer, W. P. 1940. Subspecies hybrids and speciation in *Drosophila hydei* and *Drosophila virilis*. Amer. Nat. 74:157–179.

Spencer, W. P., and C. Stern. 1948. Experiments to test the validity of linear r-dose mutation frequency relation in *Drosophila* at a low dosage. Genetics 33:43–74.

Sperber, M. A. 1975. Schizophrenia and organic brain syndrome with trisomy 8 (Group C trisomy 8 [47, XX, 8+]). Biol. Psychiat. 10:27–43.

Sprey, B., und N. Gietz. 1973. Isolierung von Etioplasten und elektronenmikroskopische Abbildung membranassoziierter Etioplasten-DNA. Z. Pflanzenphysiol. 68:397–414.

Srb, A. M., R. D. Owen, and R. S. Edgar. 1965. General genetics. 2nd ed. W. H. Freeman and Co., San Francisco.

Sreenath, P. R., and S. K. Sinha. 1968. An analysis of chromosomal behaviour during meiosis in asynaptic maize. I. Distribution of bivalents. Cytologia 33:69–72.

Stadler, L. J. 1928. Mutations in barley induced by x-rays and radium. Science 68:186–187.

Stadler, L. J. 1931. The experimental modification of heredity in crop plants. I. Induced chromosome irregularities. Sci. Agric. 11:557–572.

Stadler, L. J. 1932. On the genetic nature of induced mutations in plants. Proc. VI. Int. Congr. Genet. 1:274.

Stadler, L. J. 1941. The comparison of ultraviolet and x-ray effects on mutations. Cold Spring Harbor Symp. Quant. Biol. 9:169–178.

Stadler, L. J. 1942. Chromosome length in relation to transmission frequency in maize trisomes. Genetics 28:349–364.

Stadler, L. J., and H. Roman. 1948. The effect of x-rays upon mutation of the gene A in maize. Genetics 33:273–303.

Stahl, F. W. 1969. The mechanics of inheritance. 2nd ed. Prentice-Hall, Englewood Cliffs, New Jersey.

Stanford, E. H., and W. M. Clement. 1955. A haploid plant of Medicago sativa. Abstr. Ann. Meet. Amer. Soc. Agron.

Stanford, E. H., W. M. Clement, Jr., and E. T. Bingham. 1972. Cytology and evolution of Medicago sativa-falcata complex. In: Hanson, C. H. (ed.). Alfalfa science and technology. pp. 87–101. Amer. Soc. Agron., Inc., Madison.

Starlinger, P. 1977. DNA rearrangements in procaryotes. Ann. Rev. Genet. 11:103–126.

Starlinger, P., and H. Saedler. 1976. IS-elements in microorganisms. Curr. Top. Microbiol. Immunol. 75:111–152.

Stearn, W. T. 1957. The boat-lily (Rhoeo spathacea). Baileya 5:195–198.

Stebbins, G. L., Jr. 1941. Apomixis in the angiosperms. Bot. Rev. 7:507–542.

Stebbins, G. L., Jr. 1947a. The origin of the complex of Bromus carinatus and its phytogeographic implications. Contr. Gray Herb. 165:42–55.

Stebbins, G. L., Jr. 1947b. Types of polyploids: their classification and significance. Adv. Genet. 1:403–429.

Stebbins, G. L., Jr. 1950. Variation and evolution in plants. Columbia Univ. Press, New York.

Stebbins, G. L., Jr. 1957. Artificial polyploidy as a tool in plant breeding. Brookhaven Symp. Biol. 9:37–52.

Stebbins, G. L., Jr., and S. Ellerton. 1939. Structural hybridity in Paeonia californica and P. brownii. J. Genet. 38:1–36.

Stebbins, G. L., Jr., and J. A. Jenkins. 1939. Apsoporic development in North American species of Crepis. Genetics 21:1–34.

Stebbins, G. L., Jr., and H. A. Tobgy. 1944. The cytogenetics of hybrids in Bromus. I. Hybrids within the section Ceratochloa. Amer. J. Bot. 31:1–11.

Steffen, K. 1963. Male gametophyte. In: Maheshwari, P. (ed.). Recent advances in the embryology of angiosperms. pp. 15–40. Intl. Soc. Plant Morph., Univ. Delhi.

Steinberg, A. G., and F. C. Fraser. 1944. Studies on the effect of the X-chromosome inversions on crossing over in the third chromosome of Drosophila melanogaster. Genetics 29:83–103.

Stephens, J. S., and K. F. Schertz. 1965. Asynapsis and its inheritance in Sorghum vulgare Pers. Crop Sci. 5:337–339.

Stephens, S. G. 1942. Colchicine-produced polyploids in Gossypium. I. An auto-tetraploid Asiatic cotton and certain of its hybrids with wild diploid species. J. Genet. 44:272–295.

Stern, C. 1931. Zytologisch-genetische Untersuchungen als Beweise für die Morgansche Theorie des Faktorenaustausches. Biol. Zbl. 51:547–587.

Stern, C. 1936. Somatic crossing over and segregation in Drosophila melanogaster. Genetics 21:625–730.

Stern, C. 1973. Principles of human genetics. 3rd ed. Freeman and Co., San Francisco.

Stettler, R. 1968. Irradiated mentor pollen: its use in remote hybridization of black cottonwood. Nature 219:746–747.

Stettler, R., and K. Bawa. 1971. Experimental induction of haploid parthenogenesis in black cottonwood (*Populus trichocarpa* T. & G. ex Hook.). Silv. Genet. 20:15–25.

Stevens, L. 1974. Cited *in:* Allderdice et al., 1975.

Stevens, N. 1905. J. Exp. Zool. 2:313.

Stevens, N. M. 1908. A study of the germ cells of certain Diptera with reference to the heterochromosomes and the problem of synapsis. J. Exp. Zool. 5:359–374.

Stevenson, I. 1965. The biochemical status of MV particles in *Paramecium aurelia*. J. Gen. Microbiol. 57:61–75.

Stomps, T. J. 1910. Kernteilung und Synapsis bei *Spinacia oleracea* L. Biol. Zbl. 31:267–309.

Storms, R. and R. J. Hastings. 1977. A fine structure analysis of meiotic pairing in *Chlamydomonas reinhardii*. Exp. Cell Res. 104:39–46.

Strasburger, E. 1884. Neue Untersuchungen über den Befruchtungsvorgang bei den Phanerogamen als Grundlage für eine Theorie der Zeugung. Jena.

Strasburger, E. 1904. Über Reductionsteilung. Sitz. König. Preuss. Akad. Wiss. 18:1–28.

Strasburger, E. 1905. Typische und allotypische Kernteilung. Jahresber. Wiss. Bot. 42:1–71.

Streicher, S., and R. C. Valentine. 1977. The genetic basis of dinitrogen fixation in *Klebsiella pneumoniae*. *In:* Hardy, R. W. F. and W. S. Silver. (eds.). A treatise on dinitrogen fixation. Section III. Biology. pp. 623–655. John Wiley and Sons, New York.

Streisinger, G., Y. Okada, J. Emrich, J. Newton, A. Tsugita, E. Terzaghi, and M. Inouye. 1966. Frame shift mutations and the genetic code. Cold Spring Harbor Symp. Quant. Biol. 31:77–84.

Strickberger, M. W., and C. J. Wills. 1966. Monthly frequency changes of *Drosophila pseudoobscura* third chromosome gene arrangements in a California locality. Evolution 20:592–602.

Stringham, G. R. 1970. A cytochemical analysis of three asynaptic mutants in *Brassica campestris*. Can. J. Genet. Cytol. 12:743–749.

Stringham, G. R., and R. K. Downey. 1973. Haploid frequencies in *Brassica napus*. Can. J. Plant Sci. 53:229–232.

Sturtevant, A. H. 1915. The behaviour of chromosomes as studied through linkage. Zeitschr. Ind. Abstam. Vererbungsl. 13:234–287.

Sturtevant, A. H. 1919. Contributions to the genetics of *Drosophila melanogaster*. III. Inherited linkage variations in the second chromosome. Carnegie Inst. Wash. Publ. 278:305–341.

Sturtevant, A. H. 1925. The effects of unequal crossing-over at the Bar-locus in *Drosophila*. Genetics 10:117–147.

Sturtevant, A. H. 1936. Preferential segregation in Triplo-IV females of *Drosophila melanogaster*. Genetics 21:444–466.

Sturtevant, A. H., and T. Dobzhansky. 1936. Geographical distribution and cytology of "sex ratio" in *Drosophila pseudoobscura* and related species. Genetics 21:473–490.

Sturtevant, A. H., and F. Randolph. 1945. Iris genetics. Amer. Iris Soc. Bull. 99:52–66.

Subrahmanyam, N. C., and K. J. Kasha. 1973. Selective chromosomal elimination during haploid formation in barley following interspecific hybridization. Chromosoma 42:111–125.

Šubrt, T., and B. Blehová. 1974. Robertsonian translocation between the chromosomes Y and 15. Humangen. 23:305–309.

Sullivan, J. T., and W. M. Myers. 1939. Chemical composition of diploid and tetraploid *Lolium perenne* L. J. Amer. Soc. Agron. 31:869–871.

Summers, A. O., and S. Silver. 1972. Mercury resistance in a plasmid bearing strain of *Escherichia coli*. J. Bacteriol. 112:1228.

Summitt, R. L. 1966. Familial 2/3 translocation. Am. J. Human Genet. 18:172–186.

Sumner, A. T., H. J. Evans, and R. A. Buckland. 1971. A new technique for distinguishing between human chromosomes. Nature 232:31–32.

Sun, C. S., C. C. Wang, and C. C. Chu. 1974. Cell division and differentiation of pollen grains in *Triticale* anthers cultured *in vitro*. Scientia Sinica 17:47–51.

Sunderland, N. 1970. Pollen plants and their significance. New Scient. 47:142–144.

Sunderland, N. 1974. Anther culture as a means of haploid induction (Tenth Anniversary Lecture). *In:* Kasha, 1974. pp. 91–122.

Sunderland, N., G. B. Collins, and J. M. Dunwell. 1974. Nuclear fusion in pollen embryogenesis of *Datura innoxa* Mill. Planta 117:227–241.

Sunderland, N., and F. M. Wicks. 1971. Embryoid formation in pollen grains of *Nicotiana tabacum*. J. Exp. Bot. 22:213–226.

Suneson, C. A. 1940. A male sterile character for barley. A new tool for the plant breeder. J. Hered. 31:213–214.

Sutton, W. S. 1902. On the morphology of the chromosome group in *Brachystola magna*. Biol. Bull. IV:24–39.

Sutton, W. S. 1903. The chromosomes in heredity. Biol. Bull. 4:231–251.

Suzuki, H. 1959. Karyomorphological studies in barley. Bull. Brew. Sci. 5:43–57 (Japanese).

Swammerdam, J. 1752. Bibel der Natur. Leipzig.

Swanson, C. P. 1942. Some considerations on the phenomenon of chiasma terminalization. Amer. Nat. 76:593–610.

Swanson, C. P. 1957. Cytology and cytogenetics. Prentice-Hall, Inc., Englewood Cliffs, New Jersey.

Swanson, C. P., T. Merz, and W. J. Young. 1967. Cytogenetics. Prentice-Hall, Inc., Englewood Cliffs, New Jersey.

Swift, H. H. 1950. The constancy of desoxyribose nucleic acid in plant nuclei. Proc. Natl. Acad. Sci. U.S.A. 36:643–654.

Swift, H. 1962. Nucleic acids and cell morphology in Dipteran salivary glands. *In:* Allen, J. M. (ed.). The molecular control of cellular activity. pp. 73–125. McGraw-Hill, New York.

Swomley, B. A. 1957. Genetic linkage analysis involving chromosomal interchanges in *Hordeum vulgare*. M.S. Thesis, Purdue University.

Sybenga, J. 1966. The quantitative analysis of chromosome pairing and chiasma formation based on the relative frequencies of MI configurations. V. Interchange trisome. Genetica 37:481–510.

Sybenga, J. 1972. General cytogenetics. American Elsevier Publ. Co., New York.

Sybenga, J. 1975. Meiotic configurations. Springer-Verlag, New York.

Sykes, M. G. 1908. Nuclear division in *Funkia*. Arch. Zellforsch. 1:380–398.

Tabak, H. F., P. Borst, and A. J. H. Tabak. 1973. Search for mitochondrial DNA sequences in chick nuclear DNA. Biochim. Biophys. Acta 294:184–191.

Tai, W. 1970. Multipolar meiosis in diploid crested wheatgrass, *Agropyron cristatum*. Am. J. Bot. 57:1160–1169.

Tan, Y. H., E. L. Schneider, J. Tischfield, C. J. Epstein, and F. H. Ruddle. 1974. Human chromosome 12 dosage: effect on the expression of the interferon induced antiviral state. Science 186:61–63.

Tanaka, M., and K. Nakata. 1969. Tobacco plants obtained by anther culture and experiments to get diploid seeds from haploids. Jap. J. Genet. 44:47–54.

Tannreuther, G. 1907. History of the germ cells and early embryology of certain aphids. Zool. Jahrb. 24:609–642.

Taylor, A. I. 1968. Autosomal trisomy syndromes: a detailed study of 27 cases of Edwards' Syndrome and 27 cases of Patau's Syndrome. J. Med. Genet. 5:227–252.

Taylor, A. I., and E. C. Moores. 1967. A sex chromatin survey of newborn children in two London hospitals. J. Med. Genet. 4:258–259.

Taylor, A. M. R., J. A. Metcalf, J. M. Oxford, and D. G. Harnden. 1976. Is chromatid-type damage in *Ataxia telangiectasia* after irradiation at G_0 a consequence of a defective repair? Nature 260:441–443.

Taylor, A. L., and C. D. Trotter. 1972. Linkage map of *Escherichia coli* strain K-12. Bacteriol. Rev. 36:504–524.

Taylor, J. H. 1958. The duplication of chromosomes. Sci. Amer. 198:36–42.

Taylor, J. H. 1963. The replication and organization of DNA in chromosomes. *In:* Taylor, J. H. (ed.). Molecular genetics. Part 1. pp. 65–111. Academic Press, New York.

Taylor, J. H., P. S. Woods, and W. L. Hughes. 1957. The organization and duplication of chromosomes as revealed by autoradiographic studies using tritium-labeled thymidine. Proc. Natl. Acad. Sci. U.S.A. 43:122–128.

Taylor, W. R. 1926. Chromosome morphology in *Fritillaria, Alstroemeria, Silphium* and other genera. Amer. J. Bot. 13:179–193.

Telfer, M. A., D. Baker, G. R. Clark, and C. E. Richardson. 1968. Incidence of gross chromosomal errors among tall criminal American males. Science 159:1249–1250.

Tewari, K. K., and S. G. Wildman. 1970. Information content in the chloroplast DNA. *In:* Miller, P. L. (ed.). Control of organelle development. pp. 147–179. Soc. Exp. Biol. Symp. 24. Academic Press, New York.

Therman, E., K. Patau, R. I. De Mars, D. W. Smith, and S. L. Inhorn. 1963. Iso-telo-D mosaicism in a child with an incomplete D_1 trisomy syndrome. Port. Act. Biol. A7:211–224.

Theurer, J. C., and G. K. Ryser. 1969. Double-cross sugarbeet hybrids utilizing pollen restorers. Crop. Sci. 9:610–612.

Thoday, J. M. 1951. The effect of ionizing radiation on the broad bean root. Part IX. Chromosome breakage and the lethality of ionizing radiations to the root meristem. Brit. J. Radiol. N.S. 24:572–576, 622–628.

Thomas, H., and T. Rajhathy. 1966. A gene for desynapsis and aneuploidy in tetraploid *Avena*. Can. J. Genet. Cytol. 8:506–515.

Thomas, P. T. 1936. Genotypic control of chromosome size. Nature 138:402.

Thompson, D. H. 1931. The side chain theory of the structure of the gene. Genetics 16:267–290.

Thompson, D. L. 1954. Combining ability of homozygous diploids of corn relative to lines derived by inbreeding. Agron. J. 46:133–136.

Thompson, K. F. 1969. Frequencies of haploids in spring oil seed rape *(Brassica napus)*. Heredity 24:318–319.

Ting, Y. C. 1966. Duplications and meiotic behavior of the chromosomes in haploid maize *(Zea mays* L.). Cytologia 31:324–329.

Ting, Y. C. 1969. Fine structure of the meiotic first prophase chromosomes in haploid and diploid maize. Genetics Suppl. 61:58.

Ting, Y. C. 1971. Fate of the synaptonemal complex during microsporocyte divisions of haploid maize. Amer. J. Bot. 58:461. (Abstract).

Tjio, J. H., and A. Levan. 1950. The use of oxiquinoline in chromosome analysis. Ann. Est. Exp. Aula Dei 2:21–64.

Tjio, J. H., and A. Levan. 1954. Chromosome analysis of three hyperploid ascite tumors of the mouse. Lunds Univ. Arsskr. N.F. 2:50, No. 15.

Tjio, J. H., T. T. Puck, and A. Robinson. 1960. The human chromosomal satellites in normal persons and in two patients with Marfan's syndrome. Proc. Natl. Acad. Sci. U.S.A. 46:532–539.

Tough, I. M., W. M. Court Brown, A. G. Baikie, K. E. Buckton, D. G. Harnden, P. A. Jacobs, M. J. King, and J. A. McBride. 1961. Cytogenetic studies in chromic myeloid leukaemia and acute leukaemia associated with mongolism. Lancet I:411–417.

Tschermak, E. 1900a. Über künstliche Kreuzung bei *Pisum sativum*. Ber. Deutsch. Bot. Gesellsch. 18:232–239.

Tschermak, E. 1900b. Über kuństliche Kreuzung bei *Pisum sativum.* Z. Landw. Versuchsw. Österr. 3:465–555.

Tschermak-Seysenegg, E. von. 1951. The rediscovery of Gregor Mendel's work. J. Heredity 42:163–171.

Tschermak-Woess, E. 1947. Cytologische und embryologische Untersuchungen an *Rhoeo discolor.* Osterr. Bot. 94:128–135.

Tschermak-Woess, E. 1956. Chromosoma 8:114–134.

Tschermak-Woess, E. 1963. Strukturtypen der Ruhekerne von Pflanzen und Tieren. Protoplasmotologia, 1. Springer-Verlag, Vienna.

Tschermak-Woess, E., and G. Hasitschka. 1954. Oesterr. Botan. Z. 101:79.

Tsuchiya, T. 1958a. Karyotype studies in barley—a collective review. La Kromosomo 37–38:1294–1312 (Japanese).

Tsuchiya, T. 1958b. Studies on the trisomics in barley. I. Origin and the characteristics of primary simple trisomics in *Hordeum spontaneum* C. Koch. Seiken Ziho 9:69–86.

Tsuchiya, T. 1959a. Studies on trisomics in barley. Ph.D. Thesis, Kihara Inst. Biol. Res., Yokahama.

Tsuchiya, T. 1959b. Genetic studies in trisomic barley. I. Relationship between trisomics and genetic linkage groups in barley. Jap. J. Bot. 17:14–28.

Tsuchiya, T. 1959c. Preliminary note on cytological abnormalities in barley. Seiken Ziho 10:49–56.

Tsuchiya, T. 1960. Cytogenetic studies of trisomics in barley. Jap. J. Bot. 17:177–213.

Tsuchiya, T. 1961. Studies on the trisomics in barley. II. Cytological identification of the extra chromosomes in crosses with Burnhams's translocation testers. Japan. J. Genet. 36:444–451.

Tsuchiya, T. 1962. Haploid plants in barley. Chrom. Inf. Serv. 3:14–15.

Tsuchiya, T., J. Hayashi, and R. Takahashi. 1959. Genetic studies in trisomic barley. I. Relationship between trisomics and genetic linkage groups of barley. Japan. J. Bot. 17:19–28.

Tsuchiya, T., J. Hayashi, and R. Takahashi. 1960. Genetic studies in trisomic barley. II. Further studies on the relationship between trisomics and the genetic linkage groups. Japan. J. Genet. 35:153–160.

Tuleen, N. A. 1972. Rings of six chromosomes in barley. Barley Newsletter 15:66–67.

Turcotte, E. L., and C. V. Feaster. 1969. Semigametic production of haploids in Pima cotton. Crop Sci. 9:653–655.

Turner, B. C., C. W. Taylor, D. D. Perkins, and D. Newmeyer. 1969. New duplication-generating inversions in *Neurospora.* Can. J. Genet. Cytol. 11:622–638.

Turner, F. R. 1968. An ultrastructural study of plant spermatogenesis. J. Cell Biol. 37:370–393.

Tyler, A. 1948. Fertilization and immunity. Physiol. Rev. 28:180–219.

Tyler, A. 1967. Masked messenger RNA and cytoplasmic DNA in relation to protein synthesis and processes of fertilization and determination in embryonic development. Devl. Biol. Suppl. 1:170–226.

U, N. 1935. Genome analysis in *Brassica* with special reference to the experimental formation of *B. napus* and peculiar mode of fertilization. Japan. J. Genet. 7:389–452.

Uchida, A., and K. Suda. 1973. Ethidium bromide-induced loss and retention of cytoplasmic drug resistance factors in yeast. Mutat. Res. 19:57–63.

Upcott, M. 1937. Timing unbalance at meiosis in the pollen-sterile *Lathyrus odoratus.* Cytologia, Fujii Jub. Vol.:299–310.

Upcott, M. 1938. The genetic structure of *Tulipa.* III. Structural hybridity. J. Genet. 37:303–339.

Upcott, M., and L. La Cour. 1936. The genetic structure of *Tulipa.* I. A chromosome survey. J. Genet. 33:237–254.

Ursprung, H. K., D. Smith, W. H. Sofer, and D. T. Sullivan. 1968. Assay systems for the study of gene function. Science 160:1075–1081.

Vaarama, A. 1950. Cases of asyndesis in *Matricaria indora* and *Hyoscyamus niger*. Hereditas 36:342–362.

Valentin, J. 1973. Characterization of a meiotic control gene affecting recombination in *Drosphila melanogaster*. Hereditas 75:5–22.

Van Kreijl, C. F., D. Borst, R. A. Flavell, and C. P. Hollenberg. 1972. Pyrimidin tract analysis of mtDNA from a "low-density" petite mutant of yeast. Biochim. Biophys. Acta 277:61–70.

Van't Hof, J. 1974. The duration of chromosomal DNA synthesis of the mitotic cycle, and of meiosis of higher plants. *In:* King, R. C. (ed.). Handbook of genetics. Vol. 2. Plants, plant viruses, and protists. pp. 363–377. Plenum Press, New York.

Vasek, F. C. 1956. Induced aneuploidy in *Clarkia unguiculata* (Onagraceae). Amer. J. Bot. 43:366–371.

Vasek, F. C. 1962. "Multiple spindle"—a meiotic irregularity in *Clarkia exillis*. Am. J. Bot. 49:536–539.

Vasek, F. C. 1963. Phenotypic variation in trisomics of *Clarkia unguiculata*. Am. J. Bot. 50:308–314.

Vengerov, Y. Y., V. I. Popenko, H. Lang, and A. S. Tikhonenko. 1978. Organization of nucleosomes in chromatin fibres. Biol. Zbl. 97:29–38.

Vig, B. K. 1973a. Somatic crossing over in *Glycine max* (L.) Merrill: effect of inhibitors of DNA synthesis on the induction of somatic crossing over and point mutations. Genetics 73:583–596.

Vig, B. K. 1973b. Somatic crossing over in *Glycine max* (L.) Merrill: mutagenicity of sodiumazide and lack of synergistric effect with caffeine and mitomycin C. Genetics 75:265–277.

Villa-Komarov, L., A. Efstratiadis, S. Broome, P. Lomedico, R. Tizard, S. P. Naber, W. L. Chick, and W. Gilbert. 1978. A bacterial clone synthesizing proinsulin. Proc. Natl. Acad. Sci. U.S.A. 75:3727–3731.

Virchow, R. 1858. Die Cellularphathologie in ihrer Begründung auf physiologische und pathologische Gewebelehre. Berlin.

Visheveshwara, S. 1960. Occurrence of a haploid in *Coffea arabica* L., cultivar "Kents." Indian Coffee 24:123–124.

Vogel, F. 1964. A preliminary estimate of the number of human genes. Nature 201:847.

Vosa, C. G. 1962. Chromos. Inform. Serv. 3:26.

Vosa, C. G., and P. Marchi. 1972. Quinacrine fluorescence and Giemsa staining in plants. Nature 237:191–192.

Vries, H. de. 1900a. Sur la loi de disjonction des hybrides. C. R. Acad. Sci. Paris. 130:845–847.

Vries, H. de. 1900b. Sur les unités des caractères spécifiques et leur application á l'étude des hybrides. Rev. Génér. Botan. 12:257–271.

Vries, H. de. 1900c. Das Spaltungsgesetz der Bastarde. Ber. Deutsch. Bot. Gesellsch. 18:83–90.

Vries, H. de. 1901. Die Mutationstheorie. I. Veit und Co., Leipzig.

Wachtel, S. S., G. C. Koo, E. E. Zuckerman, V. Hammerling, M. P. Scheid, and E. A. Boyse. 1974. Serological crossreactivity between H-Y (male) antigens of mouse and man. Proc. Natl. Acad. Sci. U.S.A. 71:1215–1218.

Wachtel, S. S., G. C. Koo et al. 1976. Serological detection of a Y-linked gene in XX males and XX true hermaphrodites. New Engl. J. Med. 295:750–754.

Wahrman, J., and A. Zahavi. 1955. Cytological contributions to the phylogeny and classification of the rodent genus *Gerbillus*. Nature 175:600–602.

Walker, P. M. B., and A. McClaren. 1965. Fraction of mouse deoxyribonucleic acid on hydroxyapatite. Nature 208:1175–1179.

Wallace, B. Cited *in:* Lindsley and Grell, 1968.

Wallace, M. E. 1953. Affinity: a new genetic phenomenon in the house mouse. Evidence from within laboratory stocks. Nature 171:27–28.

Wallace, M. E. 1957. The use of affinity in chromosome mapping. Biometrics 13:98–110.

Wallace, M. E. 1958. Experimental evidence for a new genetic phenomenon. Proc. Royal Soc. Phil. Trans. London, Ser. B, 241:211–254.

Wallace, M. E. 1959. An experimental test of the hypothesis of affinity. Genetica 29:243–255.

Wallace, M. E. 1961. Affinity: evidence from crossing inbred lines of mice. Heredity 16:1–23.

Walters, J. L. 1942. Distribution of structural hybrids in *Paeonia californica.* Genetica 29:243–255.

Walters, M. S., and D. U. Gerstel. 1948. A cytological investigation of a tetraploid *Rhoeo discolor.* Amer. J. Bot. 35:141–150.

Walzer, S., B. Favara, P.-M. L. Ming, and P. S. Gerald. 1966. A new chromosomal syndrome (3/B). New Engl. J. Med. 275:290–298.

Wang, C. C., C. C. Chu, C. S. Sun, S. H. Wu, K. Cyin, and C. Hsu. 1973. The androgenesis in wheat *(Triticum aestivum)* anthers cultured *in vitro.* Scientia Sinica 16:218–222.

Wang, Y. Y., C. S. Sun, C. C. Wang, and N. F. Chieng. 1973. The induction of the pollen plantlets of *Triticale* and *Capsicum annuum* from anther culture. Scientia Sinica 16:147–151.

Waris, H. 1950. Physiol. Plant 3:1. Cited *in:* Mazia, D., 1961.

Warmke, A. E. 1954. Apomixis in *Panicum maximum.* Amer. J. Bot. 41:5–11.

Warmke, H. E. 1946. Sex determination and sex balance in *Melandrium.* Amer. J. Bot. 33:648–660.

Wassermann, F. 1926. Zur Analyse der mitotischen Kern- und Zellteilung. Z. Ges. Anat. I. Z. Anat. Entw. Gesch. 80:344.

Watanabe, T. 1963. Infective heredity of multiple drug resistance in bacteria. Bacteriol. Rev. 27:87–115.

Watanabe, T., and T. Fukasawa. 1961. Episome mediated transfer of drug resistance in Enterobacteriaceae. I. Transfer of resistance factors by conjugation. J. Bacteriol. 81:669–678.

Watson, J. D. 1970. Molecular biology of the gene. 2nd ed. W. A. Benjamin, Inc., New York.

Watson, J. D. 1976. Molecular biology of the gene. 3rd ed. W. A. Benjamin, Inc., Menlo Park, California.

Watson, J. D., and F. H. C. Crick. 1953a. Genetical implications of the structure of desoxyribonucleic acid. Nature 171:964–967.

Watson, J. D., and F. H. C. Crick. 1953b. Molecular structure of nucleic acids. A structure for deoxyribose nucleic acid. Nature 171:737–738.

Watts, V. M. 1962. A marked male-sterile mutant in watermelon. Proc. Amer. Soc. Hort. Sci. 81:498–505.

Weaver, J. B. 1971. An asynaptic character in cotton inherited as a double recessive. Crop Sci. 11:927–928.

Webber, J. M. 1933. A haploid plant of durum wheat, *Triticum durum.* Comp. Rend. Acad. Sci. URSS 10:243–244.

Weber, D. F. 1974. A monosomic mapping method. Maize Genet. Coop. News Letter 48:49–52.

Weber, D. F., and D. E. Alexander. 1972. Redundant segments in *Zea mays* detected by translocation of monoploid origin. Chromosoma 39:27–42.

Weigle, J., M. Meselson, and K. Paigen. 1959. Density alterations associated with transducing ability in the bacteriophage lambda. J. Mol. Biol. 1:379–386.

Weismann, A. 1883. Über die Vererbung. Jena.

Weismann, A. 1885. Die Kontinuität des Keimplasmas als Grundlage einer Theorie der Vererbung. *In:* Aufsätze über Vererbung und verwandte biologische Fragen. Fischer, Jena.

Weiss, M. C., and H. Green. 1967. Human-mouse hybrid cell lines containing partial complements of human chromosomes and functioning human genes. Proc. Natl. Acad. Sci. U.S.A. 58:1104–1111.

Welch, J. P. 1974. Cited in: Allderdice et al., 1975.

Welshons, W. J. 1965. Analysis of a gene in Drosophila. Science 150:1122–1129.

Westergaard, M. 1948. The relation between chromosomal constitution and sex in the offspring of triploid Melandrium. Hereditas 34:257–279.

Westergaard, M. 1958. The mechanism of sex determination in dioecious flowering plants. Adv. Genet. 9:217–281.

Westergaard, M., and D. von Wettstein. 1970. Studies on the mechanism of crossing over. IV. The molecular organization of the synaptinemal complex in Neottiella (Cooke) Saccardo (Ascomycetes). C. R. Lab. Carlsberg 37:239–268.

Westmoreland, B. C., W. Szybalski, and H. Ris. 1969. Mapping of deletions and substitutions in heteroduplex DNA molecules of bacteriophage lambda by electron microscopy. Science 163:1343–1348.

Wettstein, F. von. 1924. Cited in: Edwardson, J. R. Cytoplasmic male-sterility. Bot. Rev. 22:696–738.

Wettstein, F. von. 1928. Über plasmatische Vererbung und über das Zusammenwirken von Genen und Plasma. Ber. Deutsch. Botan. Ges. 46:32–49.

Wettstein, R. and J. R. Sotelo. 1967. Electron microscope serial reconstruction of the spermatocyte I nuclei at pachytene. J. Microscopie 6:557–576.

Whaley, W. G., H. H. Mollenhauer, and J. H. Leech. 1960. Some observation on the nuclear envelope. J. Biophys. Biochem. Cytol. 8:233.

White, B. J., and J. H. Tjio. 1968. A mouse translocation with 38 and 39 chromosomes but normal N.F. Hereditas 58:284–296.

White, M. J. D. 1935. Eine neue Form von Tetraploidie nach Röntgenbestrahlung. Naturwiss. 23:390–391.

White, M. J. D. 1938. A new and anomalous type of meiosis in a mantid, Callimantis antillarum. Proc. Roy. Soc. London B, 125:516–523.

White, M. J. D. 1945. Animal cytology and evolution. Univ. Press, Cambridge.

White, M. J. D. 1946. The cytology of the Cecidomyiidae (Diptera). I. Polyploidy and polyteny in salivary gland cells of Lestodiplosis spp. J. Morph. 78:201–219.

White, M. J. D. 1948. The cytology of the Cecidomyiidae (Diptera). IV. The salivary gland chromosomes of several species. J. Morph. 82:53–80.

White, M. J. D. 1954. Animal cytology and evolution. 2nd ed. Univ. Press, Cambridge.

White, M. J. D. 1956. Adaptive chromosomal polymorphism in an Australian grasshopper. Evolution 10:298–313.

White, M. J. D. 1957. Some general problems of chromosomal evolution and speciation in animals. Surv. Biol. Progr. 3:109–147.

White, M. J. D. 1958. Restrictions on recombination in grasshopper populations and species. Cold Spring Harbor Symp. Quant. Biol. 23:307–317.

White, M. J. D. 1973. Animal cytology and evolution. 3rd ed. University Press, Cambridge.

White, M. J. D., R. E. Blackith, R. M. Blackith, and J. Cheney. 1967. Cytogenetics of the viatica group of morabine grasshoppers. I. The coastal species. Aust. J. Zool. 15:263–302.

White, M. J. D., and F. H. W. Morley. 1955. Effects of pericentric rearrangements on recombination in grasshopper chromosomes. Genetics 40:604–619.

White, M. J. D., R. C. Lewontin, and L. E. Andrew. 1963. Cytogenetics of the grasshopper Moraba scurra. VII. Geographic variation of adaptive properties of inversions. Evolution 17:147–162.

White, T. G., and J. E. Endrizzi. 1965. Test for the association of marker loci with chromosomes in Gossypium hirsutum by the use of aneuploides. Genetics 51:605–612.

Whitehouse, H. L. K. 1963. A theory of crossing-over by means of hybrid deoxyribonucleic acid. Nature 199:1034–1040.

Whitehouse, H. L. K. 1965. Crossing over. Sci. Progr. 53:285–296.

Whitehouse, H. L. K., and P. J. Hastings. 1965. The analysis of genetic recombination on the polaron hybrid DNA model. Genet. Res. 6:27–92.

Whittinghill, M. 1937. Induced crossing over in *Drosophila* males and its probable nature. Genetics 22:114–129.

Whittinghill, M. 1947. Spermatogonial crossing over between the third chromosomes in the presence of the curly inversions in *Drosophila melanogaster*. Genetics 32:608–614.

Wiens, D. 1975. Chromosome numbers in African and Madagascan Loranthaceae and Viscaceae. Bot. J. Linn. 71:295–310.

Wiens, D., and B. A. Barlow. 1973. Unusual translocation heterozygosity in an East African mistletoe *(Viscum fisheri)*. Nature N.B. 243:93–94.

Wiens, D., and B. A. Barlow. 1975. Permanent translocation heterozygosity and sex determination in East African mistletoe. Science 187:1208–1209.

Wilkins, M. H. F., and J. T. Randal. 1963. Crystallinity in sperm heads: molecular structure of nucleoprotein *in vivo*. Biochem. Biophys. Acta 10:192–194.

Wilkins, M. H. F., W. E. Seeds, A. R. Stokes, and H. R. Wilson. 1953. Helical structure of crystalline deoxypentose nucleic acid. Nature 172:759–762.

Williamson, D. L., K. Oishi, and D. F. Poulson. 1977. Viruses of *Drosophila* sex ratio spiroplasm. *In:* Maramorosch, C. (ed.). An atlas on the ultrastructure of insect and plant viruses. pp. 465–472. Academic Press, New York.

Williamson, M., W. Jacobson, and C. C. Stock. 1952. Testing of chemicals for inhibition of the killing action of *Paramecium aurelia*. J. Biol. Chem. 197:763–770.

Wilson, D. A., and C. A. Thomas, Jr. 1974. Palindromes in chromosomes. J. Mol. Biol. 84:115–144.

Wilson, E. B. 1896. The cell in development and inheritance. Macmillan Co., New York.

Wilson, E. B. 1911. The sex chromosome. Arch. Mikr. Anat. 77:249.

Wilson, E. B. 1925. The cell in development and heredity. 3rd ed. Macmillan Co., New York.

Wilson, G. B., and E. R. Boothroyd. 1941. Studies in differential reactivity. I. The rate and degree of differentiation in the somatic chromosomes of *Trillium erectum* L. Canad. J. Res. C. 19:400–412.

Wilson, G. B., and E. R. Boothroyd. 1944. Temperature-induced differential contraction in the somatic chromosomes of *Trillium erectum* L. Canad. Jour. Res. C. 22:105–119.

Wilson, J. A., and W. M. Ross. 1961. Haploidy in twin seedlings of winter wheat, *Triticum aestivum*. Crop. Sci. 1:82.

Wilson, J. A., and W. M. Ross. 1962. Cross breeding in wheat, *Triticum aestivum* L. II. Hybrid seed set on a cytoplasmic male-sterile winter wheat composite subjected to cross-pollination. Crop Sci. 2:415–417.

Wimber, D. E. 1968. The nuclear cytology of bivalent and ring-forming rhoeos and their hybrids. Amer. J. Bot. 55:572–574.

Winge, O., and O. Lautsen. 1940. On a cytoplasmic effect of inbreeding in homozygous yeast. C. R. Lab. Carlsberg 23:17–40.

Winkler, H. 1916. Über die experimentelle Erzeugung von Pflanzen mit abweichenden Chromosomenzahlen. Z. Bot. 8:417–531.

Winton, L. L., and D. Einspahr. 1968. The use of heat-treated pollen for aspen haploid induction. Forest Sci. 14:406–407.

Winton, L. L., and R. F. Stettler. 1974. Utilization of haploidy in tree breeding. pp. 259–273. *In:* Kasha, 1974.

Wipf, L., and D. C. Cooper. 1938. Chromosome numbers in nodules and roots of red clover, common vetch and garden pea. Proc. Natl. Acad. Sci. U.S.A. 24:87–91.

Witkin, H. A., S. A. Mednick et al. 1976. Criminality in *XYY* and *XXY* Men. Science 193:547–555.

Wolf, U., R. Porsch, H. Baitsch, and H. Reinwein. 1965a. Deletion on short arms of a B-chromosome without "cri du chat" syndrome. Lancet I:769.

Wolf, U., H. Reinwein, R. Porsch, R. Schröter und H. Baitsch. 1965b. Defizienz an den kurzen Armen eines Chromosoms Nr. 4. Humangen. 1:397–413.

Wolf, U., und H. Reinwein. 1967. Klinische und cytogenetische differentiale Diagnose der Defizienzen an den kurzen Armen der B-Chromosomen. Z. Kinderhk. 98:235–246.

Wolfe, E. 1941. Die Chromosomen in der Spermatogenese einiger Nematoceren. Chromosoma 2:192–246.

Wolfe, S. L. 1972. Biology of the cell. Wadsworth Publ. Co., Inc., Belmont, California.

Wolff, S. 1961. Radiation genetics. In: Errera, M. and A. Fossberg. (eds.). Mechanisms in radiobiology. Vol. 1:419–476. Academic Press, New York.

Wolff, S. 1969. Strandedness of chromosomes. Int. Rev. Cytol. 25:279–296.

Wolff, S., and B. Rodin. 1978. Saccharin-induced sister chromatid exchanges in Chinese hamster and human cells. Science 200:543–545.

Wollman, E. L. 1953. Sur le déterminisme génétique de la lysogénie. Ann. Inst. Pasteur 84:281–284.

Wollman, E. L., F. Jacob, and W. Hayes. 1956. Conjugation and genetic recombination in Escherichia coli K12. Cold Spring Harbor Symp. Quant. Biol. 21:141.

Woolam, D. H. M., J. W. Millen, and E. H. R. Ford. 1967. Points of attachment of pachytene chromosomes to the nuclear membrane in mouse spermatocytes. Nature 213:298–299.

Woolley, G. W., L. W. Law, and C. C. Little. 1943. Increase in mammary carcinoma incidence following inoculations of whole blood. Proc. Natl. Acad. Sci. U.S.A. 29:22–24.

Worton, R. G., M. M. Aronson, A. E. Greene, and L. L. Coriell. 1977. A(2;8) translocation, 46 chromosomes. Repository identification No. GM-327. Cytogen. Cell Genet. 18:243.

Yamashita, K. 1947. X-ray induced reciprocal translocations in Triticum aegilopoides and T. monococcum. Jap. Jour. Genetics 22:34–37.

Yamashita, K. 1950. A synthetic complex heterozygote in Einkorn wheats: Aegilopoides monococcum. Proc. Jap. Acad. 26:66–71.

Yamashita, K. 1951. Studies on x-ray induced reciprocal translocations in Einkorn wheats. III. A newly synthesized ring of 14 chromosomes in a complex heterozygote, Aegilopoides monococcum. Cytologia 16:164–176.

Yamasaki, N. 1956. Differentielle Färbung der somatischen Metaphase-Chromosomen von Cypripedium debile. I. Mitt. Chromosoma 7:620–626.

Yefeiken, A. K., and B. I. Vasilev. 1936. Artificial induction of haploid durum wheats with x-rayed pollen. Bull. Appl. Bot., Genet. Plant Breed. II. 9:39–45.

Yeh, B. P., S. J. Peloquin, and R. W. Hougas. 1964. Meiosis in Solanum tuberosum haploids and haploid-haploid F_1 hybrids. Canad. Jour. Genet. Cytol. 6:393–402.

Yerganian, G. 1957. Cytologic maps of some isolated human pachytene chromosomes. Am. J. Human Genet. 9:42–54.

Yosida, T. H., and K. Amano. 1965. Autosomal polymorphism in laboratory bred and wild Norway rats, Rattus norvegicus, found in Misima. Chromosoma 16:658–667.

Yosida, T. H., A. Nakamura, and T. Fukaya. 1965. Autosomal polymorphism in Rattus rattus (L.) collected in Kusudomari and Misima. Chromosoma 16:70–78.

Yotsuyangi, Y. 1962. Etudes sur le chondriome de la levure. J. Ultrastruct. Res. 7:121–140.

Young, B. A., R. T. Sherwood, and E. C. Bashaw. 1978. Cytological observations on embryo sacs of buffelgrass. Agronomy Abstr. 1978:68.

Yudin, B. F., and M. N. Khvatova. 1966. The recognition and utilization of haploidy and polyploidy in maize breeding. Genetika 3:60–67 (Russian).

Yunis, J. J., and E. B. Hook. 1966. Deoxyribonucleic acid replication and mapping of the D_1 chromosome. A study of two patients with trisomy D_1. Am. J. Diseases in Children 111:83–89.

Yunis, J. J., and O. Sanchez. 1973. G-banding and chromosome structure. Chromosoma 44:15–23.

Yunis, J. J., and W. G. Yasmineh. 1970. Satellite DNA in constitutive heterochromatin of the guinea pig. Science 168:263–265.

Yunis, J. J., and W. G. Yasmineh. 1971. Heterochromatin satellite DNA, and cell function. Science 174:1200–1205.

Zepp, H. D., J. H. Conover, K. Hirschhorn, and H. L. Hodes. 1971. Human-mosquito somatic cell hybrids induced by ultraviolet-inactivated Sendai virus. Nature N.B. 22k:119–121.

Zickler, D. 1977. Development of the synaptonemal complex and the "recombination nodules" during meiotic prophase in the seven bivalents of the fungus *Sordaria macrospora* Auersw. Chromosoma 61:289–316.

Zillinsky, F. J., and N. E. Borlaug. 1971. Progress in developing *Triticale* as an economic crop. CIMMYT Res. Bul. 17.

Zimmering, S., L. Sandler, and B. Nicoletti. 1970. Mechanisms of meiotic drive. Ann. Rev. Genet. 4:409–436.

Zimmermann, F. K., R. Schwaier, and U. V. Laer. 1967. Mitotic recombination induced in *Saccharomyces cerevisiae* with nitrous acid, diethyl sulfate, and carcinogen alkylating nitrosamides. Z. Vererbungsl. 98:230–246.

Zinder, N. D., and J. Lederberg. 1952. Genetic exchange in *Salmonella*. J. Bacteriol. 64:679–699.

Zubay, G., and P. Doty. 1959. The isolation and properties of deoxyribonucleoprotein particles containing single nucleic acid particles. J. Mol. Biol. 1:1–20.

Index

Abbé 11
Abel, W. O. 251
Abo El-Nil, M. M. 248
Absate, M. 232
acentric fragment 163, 183, 208
Acetabularia 331
achiasmatic mechanism 216
acquired characteristics 5, 9
acriflavin dye 331
acrocentrics 166
acrosomal granule 136
acrosome 136, 137, 146
Activator-Dissociation system 22, 195, 199, 222
Adelberg, E. A. 339
adenine 52
Adhya, S. L. 344
Adik, S. 304
adventitious embryony 320
agamospermy 319, 320, 323
Ag-As ammoniacal silver staining method 42
Ahloawalia, B. S. 302
Åkerberg, E. 257
akinetic site 97
Al Aish, M. S. 278
Albers, F. 174
Alexander, D. E. 249
Alexander, G. 130, 131
Alexander, M. L. 208
Alfert, M. 299
Alfi, O. S. 192
algae 128
Allard, R. W. 307
Allderdice, P. W. 75, 191, 192, 201, 204, 205, 248
allele 14
Allen, N. S. 97
allocycly 84
alloploids (*See* polyploidy) 249, 252, 257
alternating generations 128

Al-Yasari, S. 245
Amano, K. 208
Ames, B. N. 188
amino acids 21
amphidiploid (*See* diploid)
amphimixis 318, 323
amphiploid 269
anaphase 90, 99, 100, 127
  bridge 101, 208
  I 107, 122
  meiotic 122
  mitotic 90
Anders, G. 82
Anders, J. M. 286
Anderson, E. 145, 222, 237, 238, 306
Anderson, E. G. 239
anemia,
  Fanconi 180
aneuploid 17, 259
aneuploidy 272
  multiform 273
animalculists 4
anther 131
  culture 251
antibiotics 339, 340, 345
  chloramphenol 340
  kanamycin 340
  neomycin 340
  penicillin 340
  streptomycin 340
  sulphonamide 340
  tetracycline 340
antimutator polymerase 181
antipodals 134
aphid 324, 326
apogamety 322, 323
  aposporic 323
  diplosporic 323
apomixis 318, 323
  aposporic 320, 321
  diplosporic 320, 321

apomixis [cont.]
  facultative   318
  gametophytic   320
  obligate   318
  somatic apospory   321
Arber, W.   27, 28, 341
Archebald, E. E. A.   321
archespore cell   321
arm ratio   35
Asay, K. H.   269
Ascaris   9, 11, 146
asexual reproduction   318, 320
Asimov, I.   348
Asker, S.   281
assortment
  independent, *See* Independent
    Assortment (Law of)
  random   120, 263
asters   94, 143
Atanasoff, D.   325
Atlee, W.   258, 261
Aula, P.   165, 180
Austin, C. R.   146
autoradiography   52
autosome   62
autosynapsis   167
autotetraploid, (*See* tetraploid, *See Also*
    polyploidy)
Avanzi, S.   152
Avery, A. G.   17, 256, 283, 287, 288,
    291
Avery, O. T.   22
Avery, P.   245, 291
axial filament   137
Azael, A.   281

Babcock, E. B.   321
bacteriophage
  lambda   340
  lytic   337, 340
  temperate   337
Badenhuizen, N. P.   245
Baer, K. E.   5
Baglioni, C.   196
Baikie, A. G.   193
Bajer, A.   97
Baker, B. S.   301, 303
Baker, R. J.   314
Baker, R. L.   302
Baker, W. K.   200
Balbiani, E. G.   148
Balbiani rings   152
Baldwin, J. T.   254, 272
Balinsky, B. I.   117, 134, 136, 138, 139,
    141, 143, 144

Baltzer, F.   348
Bamford, R.   257
bands   17, 44, 74, 150, 154
  C-   33, 44, 51
  chromosome   27, 32, 44, 150, 151,
    153, 154
  Drosophila   17
  G-   27, 33, 45, 49, 51
  H-   50
  miscellaneous   50
  N-   50
  Q-   27, 33, 46, 47, 50
  R-   33, 48
  T-   50
band heteromorphy   45
banding
  patterns   32
  techniques   43
Baquar, S. R.   302, 306
Barclay   248
barley   289, 304, 305
Barlow, B. A.   238
Barnum, C. P.   335
Barr, M. L.   22, 84, 85, 86
Barr body   22, 25, 44, 84, 86, 168
Barrow, J. R.   251
Barry, E. G.   241
basal membrane   135
base
  nitrogenous-purine   53
  nitrogenous-pyrimidine   53
  pairs   52
Bashaw, E. C.   321, 322
basic chromosome number (x)   244
basic genome   123, 128
Basile, R.   154
Bateson, W.   14
Battaglia, E. R.   33, 172, 173, 322
Bauer, H.   20, 118, 148, 150, 228
Beadle, G. W.   21, 174, 221, 300, 302,
    303, 304, 306
bean
  faba   183, 297
Beasley, J. O.   106
Beatty, R. A.   326, 327
Beck, H.   165
Beckett, J. B.   241
Beckman, L.   284
bee chromosomes   27
Beermann, W.   25, 153, 154, 157
Belcheva, R. G.   212
Belhova, B.   231
Belling, J.   17, 18, 108, 234
Belling's hypothesis   62, 64
Benda, C. E.   284

Beneden, E., van 9, 93, 102, 146
Bennett, M. D. 107, 113, 248
Bentvelzen, P. 335
Benzenberg, E. J. 268
Benzer, S. 188
Berg, D. E. 345
Berger, C. A. 97, 153, 299, 300
Berger, R. 45
Bergner, A. D. 302
Bergsma, D. 192
Berkaloff, A. 334
Berlin, E. A. 19
Bernard, U. 332
Bernhard, W. 93, 95, 335
Berns, M. W. 95, 183
Bertke, E. M. 173
Bertram, E. G. 22, 84
Beutler, E. 84
Bhaduri, P. N. 238
Bhavanandan, K. V. 301
Bianchi, K. 335
Bijlsma, J. B. 284
Bingham, E. T. 247, 250
bipartitioning 124
Birnstiel, M. L. 141
Bittner, J. J. 335
bivalents 10, 103
  cross 118
  interlocking 222
  open ring 118, 119
  rod 118
Blackwood, M. 171, 174
Blakeslee, A. F. 17, 256, 262, 280, 281,
    282, 287, 290, 291, 297, 298, 299,
    305
Blanco, J. L. 304
blastomere 12
Blickenstaff, J. 304
Blochman, F. 325
Bloom, S. E. 244
Böcher, T. W. 323
Bogart, J. P. 270
Bogdanov, Y. F. 111
Bollum, F. J. 57
Bonner, J. 25, 105
Böök, J. A. 260, 273
Boothroyd, E. R. 50
Bopp-Hassenkamp, G. 132
Borgaonkar, D. S. 163, 232, 233
Borisy, G. G. 95
Borlaug, N. E. 269
Bosemark, N. O. 174
Botstein, D. 345
bouquet stage 108
Bourgin, J. P. 247

Boveri, T. 11, 93, 102, 260
Boyer, H. W. 29
Boyes, J. W. 256
Bozzini, A. 302, 306
Brachet, J. 141
Bradbury, E. M. 93
Bradley, D. E. 339
breakage-fusion-bridge cycle 36, 176,
    186, 194
breakage-reunion hypothesis 65, 176
Brehme, K. S. 200
Brenner, S. 92
Bretschneider, L. H. 141
Breuer, M. 153
Brevet, J. 177
Brewen, J. G. 177
bridge 210, 211
  elimination mechanism 212, 214, 216
  formation 35
Bridges, C. B. 17, 71, 79, 81, 152, 153,
    183, 200, 244
Bridges' balance theory 79, 270
Briggs, F. N. 256
Briggs, R. 244
Brink, R. A. 239, 240, 295
Brinkley, B. R. 179
Britten, R. J. 86
Brock, R. D. 177
Broda, P. 162
Brown, D. D. 41
Brown, R. 5, 6
Brown, S. W. 44, 85, 165, 324
Brown, W. L. 42
Brown, W. M. 331, 332
Brown, W. V. 99, 166, 168, 173, 219,
    318, 321, 348
Brownian movement 6
Bruere, A. N. 286
Brumfield, R. T. 297
Bryan, J. H. D. 132
Bryophyta 130
Büchner, T. 284
Budrin, K. Z. 245
bulbils 320
bulbs 320
Bunker, M. C. 286
Burgos, M. H. 136, 137
Burk, L. G. 246, 336
Burnham, C. R. 43, 189, 220, 222,
    223, 225, 239, 241, 287, 288, 290,
    304
Burton, G. W. 320
Burton, H. 331
Buss, G. R. 41
Bütschli, O. 11

Buyser, J. de   248
Byers, B.   111

C-value   104
caffeine   180
  8-ethoxycaffeine   180
Cairns, J.   159, 162
Caldecott, R. S.   169
Callan, H. G.   50, 139, 157, 287
Calos, M. P.   345
Camara, A.   36
Camerarius, R. J.   4
Campbell, A.   342, 345
Campos, F. F.   246
carcinogens   180
  sodium saccharin   180
Carlson, J. G.   97
Carlson, P.   303
Carniel, K.   237
Carothers, E. E.   307
Carpenter, A. T. C.   303
Carson, H. L.   211, 218
Carter, T. C.   227
Cartledge, J. L.   239
Casey, M. D.   286
Caspersson, T. O.   21, 27, 46, 47, 286
Castle, W. E.   244
Catcheside, D. G.   165, 183, 288, 302,
    303, 304
Cattanach, B. M.   286
Cauderon, A.   272
Cavalcanti, A. L. G.   334
cell
  cycle   90
  division   9
  donor   337
  duet   127
  egg   134
  F⁺   337, 338
  F⁻   337, 338
  F   337
  F′   337, 339
  fusion (See hybridization, somatic cell)
  generative   132
  Hfr   337, 338
  lineage   6
  living   8
  plate   101
  quartet (See quartet cell)
  recipient   337
  theory   6, 7
  tube   132
  vegetative   132
central element   110

centric
  fusion   229, 230
  segment   183
centriole   93, 136, 145
  distal   137
  duplexes   93
  proximal   136, 137
  ring   137
  tubulin   95
centromere   60, 165
  holocentromere   34, 35
  localized   34
  neo-   34, 41, 307
  non-localized   34, 35
  poly-   34
  ultrastructure   60
centromere orientation   97, 228, 235
  adjacent   225, 227
  alternate   227, 235
  co-orientation   228, 261
  non-coorientation   228, 261, 276
  quadruple   223, 226, 227
  re-orientation   228
  zigzag   238
centromeric index   35
centrosome   11, 93
cerebellum   299
Chaganti, R. S. K.   180, 303
Chandler, G. L.   58, 340
Chandy, L. P.   251
Chapelle, A. de la   75, 206
Chaplin, J. F.   336
Chapman, V.   250
Chase, S. S.   245, 246
Cheredeyeva, V. S.   262
Chiarugi, A.   321
chiasma   15, 62, 65, 114, 303
  chiasmatype theory   15, 62, 63
  end-   63, 116, 120
  interference, (See chromosome
      interference)
  interstitial   116
  localized   115
  movement   116
chiasma terminalization   20, 115
  coiling hypothesis   116
  electrostatic hypothesis   116
  repulsion hypothesis   116
chimera   296, 297
  gene differential   297
  mericlinal   297
  periclinal   297
  sectorial   297
Chittenden, R. J.   336

Chlamydomonas 331
Chlorella 331
chloroplasts 331
Christiansen, C. 344
chromatid 20, 95
  aberration 179
  exchange (sister) 179, 180, 181, 186
  non-sister 15
  sister 56, 103
  sister chromatid reunion 194
  unit 27, 56, 103
chromatin 8, 12, 17, 21
  euchromatin 21, 32, 37, 38, 39, 85,
    86, 90, 184
  heterochromatin 21, 32, 37, 38, 39,
    40, 42, 45, 85, 184, 307, 317
    constitutive 44, 85
    facultative 44, 85, 86
    redundant 86
    repetitive 86
    sex 84, 86
  loops 156
  sex-chromatin, (*See Also* Barr
    body) 22
chromocenters 38, 90, 101
chromomere 32, 39, 108, 151, 165
  macrochromomere 39, 40
  microchromomere 39
  size gradient 39
chromonemata 60, 150, 155
chromosome 8, 9, 12, 134, 167
  A- 170, 173
  accessory 170, 171
  acrocentric 36, 166, 167
  akinetic site 98
  attached-X 167
  B- 159, 170, 171, 173
  breakage 176, 306
  breakage and reunion, 176 (*See also*
    breakage-reunion hypothesis)
  cancer 23
  chimeras 296, 299
  coiling 98
  compensation 291
  congression 97
  contraction 95
  co-orientation 226
  C-pairs 99
  despiralization 117
  dicentric 35, 186, 210
  dimunition, (*See* chromosome
    elimination)
  distribution 98
  dyad 103, 122, 125, 126

early replicating 90
elimination 268, 316
extra 17, 170
extranuclear 26
fibers 100
four-strand stage 294
fragments, (*See* chromomal fragments)
giant salivary gland 15, 17, 20, 32,
  70, 71, 150, 152
homologous 65, 67, 103, 143
homomorphic 78
honey bee 27
human 9, 19, 23, 75, 76
intercalary segment 185
interdependence 310
interference 68
isochromatid break 180
isochromosomes 159, 165, 167, 202
isochromosome Turner syndrome 168
kinetic sites 97
laggard 273
lampbrush 117, 139, 148, 156, 158
loops 158
late replicating 56, 91
limited 314
metacentric 35, 167
mitochondrial 136, 137
mitotic metaphase 32, 33
molecular structure 23
monad 56, 103, 127
monocentric 35
mosaics 271, 299
neocentric 36
neo-Y 313
nucleolus 139
Philadelphia 192
plant 51
polymorphism 208, 214
polytene 148, 150, 151
primary structural changes 195
prochromosomes 38
repulsion 117
reunion 176
ribosomal 141
ring 159, 162, 163, 165, 185, 186
satellite 36
secondary 274, 280
sex 80, 311
sister-strand reunion 176
size 33
spermatogonial 134
stability 62
stickiness 176, 221, 340
structure, changes in 13, 56, 195

chromosome [cont.]
  submetacentric 35
  substitution 274
  substitution lines 276
  sub-telocentric 35
  super 154
  supernumerary 170, 171, 173, 174
  tandem satellites 37
  telocentric 35, 159, 165, 166
  telocentric, natural 166
  tertiary 274, 280
  tetrad 103
  Theory of Inheritance 11, 12, 14
  triradial 180
  ultrastructure 55
  vertebrate 51
chromosome aberrations 92, 179, 182, 185
  primary 187
  secondary 187
chromosome analysis
  metaphase 32
  pachytene 39
chromosome mapping 39, 239
  assignment test 71
  B-A translocation 241
  cytogenetic 71, 188
  cytological 17
  deletion 71, 75, 176, 188
  duplication 75
  genetic linkage 17, 69, 72, 75, 76, 281
  regional 75
  somatic cell hybridization 26
  synteny test 71
  transcript 27
  translocation 69, 239–241
chromosome models
  folded fiber 27, 56
  general 59, 154
  molecular 57, 58
  multistranded 57
  polyneme 58
  Watson-Crick 24, 56
chromosome mosaicism, (See mosaicism)
chromosome number 244
  basic 34, 244
  gametic 122, 244
  genome 34, 123
  n 32, 122
  somatic 122
chromosome stability 62
chromosome substitution, alien 276
chromosome ultrastructure 55, 63
Chrustschoff, G. K. 19
Chu, Y. 249

Chu, Z. 248
cine records 100
Cipar, M. S. 245
cisternae 136
Citrus 320
Clapham, D. 248
Clark, A. J. 339
Clark, F. J. 304
Clarke, A. E. 336
Clarkia 305
Clausen, R. E. 64, 277, 281
Clavier, Y. 272
Clayberg, C. D. 306
ClB method 216
cleavage 101
Cleaver, J. E. 182
Cleland, R. E. 236
Clement, W. M. 250
Cleveland, L. R. 166
Cleveland, R. W. 42
Clever, U. 154
Close, H. G. 286
Cluzel, R. 339
codon 24
Coe, E. H., Jr. 245
Coe, G. E. 318
Cognetti, G. 324
Cohen, M. M. 179, 180
Cohen, S. N. 28, 344, 345
cohesive ends 340
Cohn, N. S. 94, 123, 299, 348
coiling 108
  cycle 84, 95, 116
  hypothesis 116
  level 98
  paranemic 95
  plectonemic 95
  relational 62, 95, 98
coils
  major 108
  minor 95
  somatic 95
colchicine 17, 27, 97, 99, 132, 169, 251,
  256, 281, 298
  binding protein 113
Cold Spring Harbor Symposia on
  Quantitative Biology 55
colicinogenic factors 339
  nontransmissible 339
  sex factor 339
colicins 339
Collins, G. B. 247, 249, 251
collochores 118
Colwin, A. L. 145

Colwin, L. H.   145
Comings, D. E.   113, 256
Committee on Standardized Genetic
  Nomenclature for Mice   47
conjugation
  bacterial   338
constrictions
  primary   36
  secondary   36
  tertiary   37
contact points   109
conversion   64
Cook, S. A.   128, 129, 130
Cooke, P.   165
Cooper, D. C.   150, 239, 240
Cooper, H. L.   202
Cooper, K. W.   118, 313
co-orientation   97, 120, 228
copy-choice hypothesis   64, 65
Corbett, M. K.   335
Coriell, L. L.   75
Corlette, S. L.   153
Correns, C. F. J.   12, 330, 336
Court Brown, W. M.   206, 229, 230, 231
c-pairs   99
Craig-Holmes, A. P.   45
Cramer, P. S.   296
Creek, M. J.   320
Creighton, H. B.   19, 67
Crepis   297
Crick, F. H. C.   23, 52, 59, 108, 153
Crippa, L. P.   180
cross
  reciprocal   10
  three-point   69, 239, 240
crossing over   15, 18, 62, 63, 64, 66, 101,
    113, 120, 178, 303
  fourstrand   63
  four-strand double   64, 207, 210
  mechanism of   18, 62, 178
  parental classes   69
  recombinant classes   69
  sister strand   64
  somatic   294, 300, 303
  three-strand double   64, 207
  two-strand double   206, 210
  two-strand single   209, 210
  unequal   195, 197
crossing over mechanisms   178
  Belling's hypothesis   63
  copy-choice hypothesis   64
  partial chiasmatype theory   63
    (*See Also* chiasmatype theory)
  polaron hybrid DNA model   64

crossover
  frequency   216
  interference   68, 218
  position interference, model of   111
  suppressor   216, 217, 218
Crouse, H. C.   315
Crouse, H. V.   153
Crumpacker, D. W.   216
Csermely, T. W.   299
Curran, J. P.   192
Curtis, G. J.   335
cystic fibrosis   189, 190
cytogenetics   14
cytokinesis   90, 101, 124, 132
cytological coloring   210, 211
cytology   11
  experimental   8
  history of   1–29
  plants   6
cytomixis   305
cytoplasmic
  factors   80
  inheritance   26
  male sterility (CMS)   335, 336
cytosine   52
cytotaxonomic studies   98

D-loops   331
DaCunha, A. B.   153, 154
Dahlgren, K. V. O.   132
Dampier, W. C.   348
Danna, K. J.   27
Dark, S. O. S.   116
Darlington, C. D.   19, 50, 63, 97, 98,
    115, 116, 120, 123, 167, 172, 174,
    176, 203, 222, 237, 252, 253, 259,
    323, 330, 348
Darvey, N. L.   169
Darwin, C.   13
Das, K.   288
Datura   17, 287, 297, 298, 302, 305
Davidson, N. R.   344
Davidson, R. G.   84
Davidson, W. M.   86
Davis, G. N.   336
Davis, R. J.   141
Davis, R. W.   344
Davis, W. H.   336
Dawes, B.   348
Dawson, G. W. P.   261, 262, 263,
    271
DeBeer, G.   348
Debergh, P.   248
decapitation   255

deficiency 17, 183, 185, 200
  terminal 183
DeGrouchy, J. 163, 192, 206, 207, 231, 284, 286
DeHaan, I. 257
DeHarven, E. 93
Deknudt, G. H. 231
deletion 71, 74, 176, 183, 185, 187
  B-group 191
  C-group 192
  E-group 192
  G-group 192
  homozygous 42
  interstitial 185
  X 193
Demerec, M. 185, 217
Dempster, L. T. 268
Deonier, C. 344
deoxyribonuclease 57
deoxyribonucleic acid, (See DNA)
deoxyribonucleotides 52
Derman, H. 298
DeRobertis, E. D. P. 96
development
  parthenogenetic 245
DeWeerd-Kastelein, A. E. 180
Dewey, D. R. 269
Dewey, W. C. 180
Dhillon, T. S. 281
diakinesis 63, 106, 107, 117, 118
Diaz, M. 154
dicentric 101, 186, 210
  bridges 212
dictyotene 117
Digby, L. 156
dihybrid ratio 265
diplochromosomes 148
diplohaplont 128, 134, 322
diploid 128, 251
  amphi- 251, 252
  functional 128, 251
  good 252
diploidization 251
diploidy 128
diplont 128, 134
diplontic 128
diplophase 128, 131, 134
diplotene 63, 66, 106, 107, 114, 117, 139, 156
Dippell, R. 333
disomic 83, 274
disphermy 260
displacement loops 331
ditelosomic 166

division
  equational 102, 103
  extra 174
  meiotic, (See meiosis)
  mitotic, (See mitosis)
  polymitotic 174, 300
  precocious chromosome 306
  reductional 102, 103, 300
DNA 8, 17, 22, 23, 52, 56, 57, 59
  chloroplast 330, 331
  cloning 28
  complementary strand 92
  double helix 53, 56, 57
  excision repair 182
  fragments 92
  heteroduplex 344
  highly repetitious 45
  ligase 92
  mitochondrial 26, 332
  molecular structure 24, 52
  nucleolar 139
  plastid 330, 331
  polymerase 56
  postreplication repair 182
  puffs, (See puffing)
  repetitious 28, 45
  satellite 86
  supercoiled configuration 56
  synthesis 91, 114, 141
  turn-around region 29
  unique 45
  Watson-Crick Model 22, 24
DNA replication 64, 92, 162
  bidirectional model 162
  late 56
  point of initiation 113
  unidirectional model 159
DNA-RNA hybridization 344
Dobzhansky, T. 81, 203, 216, 311, 312
Dodds, K. S. 257
dominance 263
  complete 263
  incomplete 263
  pseudo- 71
Doorenbos, J. 257
dosage compensation 83, 84
  genes 83
dosage factors, cytoplasmic 80
dosage sensitive loci 184
Doty, P. 54
doublets 150
Dougan, L. 193
Dover, G. A. 113

Down's syndrome   230, 283, 286, 304
Downey, R. K.   244
Doy, C. H.   248
Dp-Df gametes   206, 227
Drets, M. E.   45
Driscoll, C. J.   169
Drosophila   15, 17, 18, 67, 69, 71, 74,
    79, 81, 82, 83, 150, 165, 167, 182,
    184, 185, 195, 196, 197, 200, 211,
    214, 216, 218, 219, 221, 244, 278,
    286, 294, 296, 301, 302, 303, 309,
    311, 313, 314, 333, 334, 335
  chromosome map   70, 74
  non-stabilized lines   334
  stabilized lines   334
drumsticks   86, 87
Dublin, P.   244
Dujon, B.   332
Dunn, L. C.   348
Dunwell, J. M.   248
duplex   262, 263, 266
duplicate gene expression   265, 266
duplications   17, 176, 186, 194, 200, 201
  displaced   194
  intrachromosomal   194
  reverse tandem   194, 203
  tandem   194, 195
duplication-deficiency (Dp-Df)   206
DuPraw, E. J.   26, 56, 138, 141, 151,
    152, 158, 331
Dussoix, D.   27
Dustin, P.   133
Dutrillaux, B.   45, 48, 50
Duvick, D. N.   336
dyad, (*See* chromosomes, dyad)
dysfunctional sperm hypothesis   309

East, E. M.   16
Eberle, D.   38
Ecochard, R.   247, 249
edema, ventral   327
Edmonds, H. W.   132
Edwards, J. H.   284
Edwards, R. G.   244, 327
Edwardson, J. R.   335, 336
egg apparatus   134
Ehrendorfer, F.   174
Ehrensberger, R.   245
Eigsti, O. J.   97, 132, 134
Einset, J.   278, 318
Einspahr, D.   245
Eisen, G.   108
El Alfi, O. S.   284
El Darov, A. L.   244

elastic chromosome repulsion
    hypothesis   117
elastic connecter model   113
electron microscopy   20, 52
  whole mount   26
electrostatic hypothesis   116, 117
Elgin, S. C. R.   55
Ellerström, S.   262
Ellerton, S.   238
Elliott, F. C.   167, 253, 255
Ellis, J.   245
Ellis, R. J.   202, 331
Ellison, J.   311
embryo sac   134
  mother cell   132
Emerson, R. A.   16, 307
Emery, W. H. P.   321
emigrantes   324
Emsweller, S. L.   336
endometrium   299
endomitosis, (*See* mitosis)
endoplasmic reticulum   95
endopolyploidy   148, 149, 150
endosperm nucleus   134
Endrizzi, J. E.   166, 226, 279
Engel, E.   188
Engvild, K. C.   251
Enriques, P.   82
enzymatic proofreading mechanism   182
enzymes
  DNA polymerase   92
  F1 histone phosphokinase   93
  HKC   93
  proteolytic   57
  sperm lysine   145
Ephrussi, B.   268, 331
episomes   337
  conjugons   337
  F   337, 338, 339, 340
  Hfr   337, 338
epistasis   14
epithelium
  corneal   9
equatorial plate   120
Erickson, J.   309, 311, 312
Erickson, J. D.   284
Erigeron   322
Ervin, C. D.   299
Escherichia   339
Eslick, R. F.   307
euchromatin, (*See* chromatin)
eugenics   7
eukaryotes   159
euploidy   272

eutelomere 41
Evans, D. A. 296
Evans, E. P. 217, 231, 304
Evans, H. J. 173, 176, 177, 178, 179, 180
exchange
  hypothesis 177, 178
  initiation 178
  mechanical exchange process 177, 178
  point of partner 259
  points 114
extrachromosomal genetic factors 138
exules 324

F-agent, (*See* episome, F)
F-genote 339
Fahmy, O. G. 179
Fahmy, M. 179
Falconer, D. S. 326
Falk, H. 331
Fankhauser, G. 244, 271
Fastnaught-McGriff, C. E. 109, 115, 120, 225, 234
Fawcett, D. W. 136, 137
Faye, G. 331, 332
Feaster, C. V. 247
Fedak, G. 272, 287
Fedorov, A. A. 123
Feldman, M. 288
female
  promoting regions 78
  suppressor regions 78
Ferguson-Smith, M. A. 38, 75, 192, 193, 206
Fernandes, A. 257
fertility inhibition 340
fertilization 8, 9, 10, 102, 143, 145
  cross 14, 270
  double 11, 134
  preferential 171, 174
  self- 14, 236
  substages 145
Feulgen, R. J. 17, 333
  reaction 17, 333
  negative regions 178
Fiandt, M. 344
fibroblasts 100
Finch 248
Finn, W. W. 132
Fisher, H. E. 246
flagellum 137
Flagg, R. O. 237
Flamm, W. G. 142
Flatz, G. 180

Flavell, R. A. 332
Flemming, W. 8, 38
Flovik, K. 323
Focke, W. O. 322
Fogel, S. 348
follicle cells 141
Ford, C. E. 82, 227
Ford, L. 249
Forer, A. 310
Foster, T. J. 345
Fraccaro, M. 193
fragments
  chromosomal 208, 209, 211
Frandsen, K. J. 257
Frandsen, N. O. 245
Frankel, R. 335
Franklin, J. 333
Franklin, N. C. 344
Franklin, R. E. 52
Fraser, F. C. 218
Fredericq, P. 339
Freese, E. 57
Frost, H. B. 304, 320
Fröst, S. 117, 174
fuchsin 17
Fujita, S. 153
Fukasawa, T. 340
functional pole hypothesis 309
fundatrices 324
fusion 230
  ascomycetous 111
  fungi 97, 128
  heterochromatic 195
  nucleus 134

G_1 period 90, 91
  biochemical events 91
  physiological condition 91
G_2 period 90, 91
Gabriel, M. L. 348
Gagliardi, A. R. T. 286
Gall, J. G. 44, 93, 157
Galton, F. 7
gamete
  female 134
  male 131
gametic chromosome number 122
gametocyte
  female 134
  male 132
gametogenesis 104, 131, 134, 139
gametophyte 129, 131, 320
  female 134
  male 132, 237
gamma radiation 180, 245

gaps   178, 179
Garber, E. D.   281
Gardner, E. J.   348
Gates, R. R.   234, 305
Gauthier, F. M.   251
Gay, D.   334
Geitler, L.   148, 150, 299
Geitz, N.   331
Gellert, M.   181
gene   14, 62
    amplification   153
    expression   28
    holandric   82
    insulin   28
    nitrogen-fixing   28
    operator   24, 200
    regulator   24, 200
    size   55
    structural   24, 200
gene position
    all-arms marker method   239
    linked marker method   239
gene transmission
    father-son   88
genetic affinity   310
genetic factor complexes   234
genetic lettering   210, 211
genetic maps, (See chromosome mapping)
genetic recombination   128
genetics   14
genome
    analysis   18
    balance   272, 286
    basic   34
    formulas   252
genophores   159, 331
    heteroduplex   189
    mitochondrial   331
genotype   14, 330
Gentcheff, G.   299
George, L.   248
Geraci, G.   259
Gerassimova, H.   245
Gerlach, W. L.   51
German, J.   180, 303
germinal epithelium   141
germinal vesicles   139
germination   134
germ line   104, 134
germplasm
    theory of   9
Gerstel, D. U.   237, 238
Geschwindt, I. I.   299
Gey, W.   191
Giannelli, F.   169, 284

Gibor, A.   331
Gibson, I.   333
Giemsa banding, (See banding techniques)
Giemsa C-banding   45
Gierer, A.   60
Gietz, N.   331
Gill, J. J. B.   174
Gilles, C. B.   111, 250
Givens, J. F.   41
Glass, H. B.   200, 218, 348
Godoy, G.   333
Goetsch, L.   111
Gohil, R. N.   302
Goldschmidt, R.   80, 330, 348
Goldschmidt's theory   80
Golgi apparatus   136
Golubovskaya, I. N.   307
Goodenough, U.   55, 182
Goodspeed, T. H.   69, 245, 277, 281, 291
Gordon, R. R.   165
Gormley, I. P.   45
Gosling, R. G.   52
Gottschalk, W.   38, 39, 40, 256, 267, 281, 301, 302, 306
Gowen, J. W.   302
Graaf, R. de   3
Graafian follicles   3, 5
Granick, S.   331
Grant, V.   323
Grant, W. F.   262
Gray, J. E.   206
Green, B. R.   331
Green, H.   248, 268
Greenberg, R.   309
Greenblatt, L. M.   336
Greilhuber, J.   51
Grell, E. H.   312
Grell, K. G.   368
Grell, R. F.   113
Grell, S. M.   300
Gresshoff, P. M.   248
Grew, N.   3
Griffen, A. B.   286
Griffiths, R. B.   244
Grindley, N. D. F.   345
Gripenberg, U.   165
Gropp, A.   180
Grunewaldt, J.   247
guanine   52
Guanti, G.   75
Guénin, H. A.   314
Guha, S.   247
Gunthardt, H.   222
Gupta, S. B.   248

Gurdon, J. B. 41
Gustafson, J. P. 269
Gustafsson, A. 299, 320, 322, 323
Gustavson, K.-H. 284
Gustavsson, I. 231
Gutherz, S. 38
Gutman, A. 337
Guyer, 344
gynandromorphs 80
gynogenesis 326
Gyulavari, O. 251

Hadder, J. C. 97
Haeckel, E. 305
Haendle, J. 296
Haertl, E. J. 238
Haga, T. 208
Hagberg, A. 257
Hageman, P. C. 335
Hair, J. B. 268
Håkansson, A. 173
Hall, B. D. 344
Hall, J. C. 303
Hamerton, J. L. 65, 88, 163, 165, 168,
    169, 193, 228, 229, 230, 262, 270,
    278, 284, 286, 348
Hamilton, L. 244, 326
Hampel, K. E. 180
Hand, D. B. 256
Handmaker, S. D. 193, 202
Hanhart, E. 284
Hanks, G. D. 311
haploid 128, 244–251
    allopoly- 250
    autopoly- 250
    male 324
    mono- 244, 249
    poly- 244, 249, 276
haploidy syndrome, (See syndromes,
    haploidy)
haplont 128
haplophase 128, 134
Harland, S. C. 244
Harris, H. 26, 71, 189, 268
Hartl, D. L. 309, 324
Hartman, P. E. 188
Harvey, W. 2
Hasegawa, N. 173
Hasitschka, G. 152
Hastings, P. J. 64, 111
Haun, C. R. 38, 277
Hauschka, T. S. 83, 150
Hayes, W. 337
Hayward, M. D. 302
Hb-Lepore hemoglobin 196

heat treatment 251
Heber, U. 331
Hecht, F. 180, 284
Hedges, R. W. 345
Heemert, C. van 244
Heiss, B. 344
Heitz, E. 20, 38, 148
Helinski, D. R. 339
Helwig, E. R. 313
hemizygous 83, 189, 274
Henderson, A. S. 75
Henderson, R. G. 335
Henderson, S. A. 149
Heneen, W. K. 249
Henking, H. 12, 78, 312
Herrmann, F. H. 331
Herschman, H. R. 339
Hershey, A. A. 340
Herskowitz, I. H. 53, 176
Hertwig, G. 8, 299
Hertwig, R. 8
Hertwig, W. A. O. 10, 11, 14, 102
heterochromatin, (See chromatin)
heteroduplex technique 344
heterogametic sex 78
heterogenote 339, 341
heteromorphic 45, 78
heteroploid 255
    tissue 297
heteropycnosis 22, 38, 84, 87
    negative 36, 84
    positive 84, 314
heteropycnotic knob 41
heterotypic 102
heterozygosity 14, 203, 218
    complex 233, 234, 237, 239
Hewitt, G. M. 166, 287
hexaple 233
Hildebrandt, A. C. 248
Hill, H. E. 281, 283
Hinton, C. W. 303
Hiraizumi, Y. 307, 309
Hirota, Y. 340
Hirschhorn, K. 191, 202
histone 26, 105
    HKG 93
    meiotic 105
Hittelman, W. N. 179
HKG, F1 histone phosphokinase 93
Hoar, C. S. 238
Hochman, B. 153
Hockett, E. A. 307
Hodges, E. M. 320
Hofmeister, W. 129
Hollaender, A. 97

Hollenberg, C. P.   332
Holliday, R.   111, 113
Holt, C. M.   299
Holtzman, E.   335
homeostatic effect   218
homoeologous groups   250, 291
homoeology   250, 252, 266
homogametic sex   78
homologues, (*See* chromosomes,
    homologous)
homology
    partial   252
    residual   252
homotypic   102
homozygosity   13, 14
    complete   251
Hook, E. B.   284
Hooke, R.   3
Hoover, M. E.   185
Hoppe-Seyler   8
Hoskins, G. C.   60
Hotta, Y.   114
Hougas, R. W.   245, 246
Howard, A.   90
Howard-Flanders, P.   182
Howlett, R. M.   284
Hsu, T. C.   45, 47, 50, 169, 314
Hu, S.   344
Huang, R. C.   25, 105
Hubermann, J. A.   92
Hughes, A. F. W.   100
Hughes, E. C.   299, 348
Hughes-Schrader, S.   35, 118, 300, 313,
    314
Hultén, M.   229, 304
human chromosome instability
    syndromes   180
human chromosome mapping
    assignment test   71
    in situ hybridization   75
    regional mapping   75
    synteny test   71
Human Chromosome Study Group   45
    Chicago Conference   35
    International System   32
    Paris Conference   32, 45
human gene loci
    ACP-1   75
    G6PD   75
    HGPRT   75
    If-1   75
    PGK   75
    SOD-1   75
Human Genetic Mutant Cell
    Repository   232

human mutant cell bank   75
Hungerford, A. M.   38, 192
Hungerford, D. A.   38, 192
Hurd, E. A.   277
Huskins, C. L.   250, 252, 300, 301
hybrid duplexes   65, 71
hybrid molecules   25
hybridization   13, 245
    DNA   189
    in situ   75
    intergeneric   245
    interspecific   245, 246
    intraspecific   245
    nucleic acid   75
    plant   7
    somatic cell   26, 268
    wide   268
hybrids   5
Hyde, B. J.   257
hyperploid   273, 280
hypoploid   273

idiogram   33
idioplasm   10
idiotype   330
Ikeda, H.   334
Illies, Z. M.   246
Imai, Y.   330
independent assortment, Law of   7, 16,
    122
indolacetic acid   255
inheritance
    crisscross pattern of   88
    cytoplasmic   26
    mitochondrial   331
    noncrisscross   168
    quantitative characters   16
    tetrasomic   262, 266
Inoué, S.   139, 140
insertion sequence, (IS)   342, 345
insulin   28
integument   132
interbands   108, 150, 154
interchange hypothesis   234
interchromosomal transposition   195
interkinesis   123
International Atomic Energy
    Agency   183
interphase   90, 123
    meiotic   123
    mitotic   90, 92
    premeiotic   105
intersex   17, 80, 81
    time law   81
    turning point   81

interstitial chromosome segments   169
intervening sequences   28
inversion   174, 203
  complex   213
  direct tandem   214
  heterozygote   203, 206, 208, 209, 210,
    216, 218
  included   214, 217.
  independent   214
  loop   204, 205, 206, 208, 210, 216
  multiple, (*See* inversions,
    complex)
  overlapping   214
  paracentric   208
  pericentric   201, 203, 210
  polymorphism   212
  reversed tandem   194, 203, 214
Irikura, Y.   248
Iseki, S.   339
Ising, U.   165
isolabeling   58
Isotoma   238
Ivanov, M. A.   245
Ivanovskaja, E. V.   250
Iyer, R. D.   248

Jacob, A. E.   345
Jacob, F.   24, 92, 200, 337, 340, 344
Jacobj, W.   150
Jacobs, P. A.   82, 169, 193, 206, 229,
  260
Jacobsen, P.   286
Jaffe, B.   348
Jahn, A.   302
James, S. H.   238
Jamieson, A.   173
Janaki-Ammal, E. K.   123, 252, 253
Janick, J.   283
Janssen, J. S.   2
Janssen, Z.   2
Janssens, F. A.   14, 62
Jenkins, B. C.   268, 269, 281, 291
Jenkins, J. A.   318
Jensen, C. J.   248
Jewell, D. C.   51
Johannsen, W. L.   14
John, B.   91, 157, 166, 238, 287, 314
Johnson, H.   301, 306
Johnson, D.   121
Joly, B.   339
Jones, D. F.   296
Jones, H. A.   336
Jones, R. N.   172, 174
Jordan, E. H.   342
Jorgensen, C. A.   245

Judd, B. H.   108
Jurand, A.   333
Jurasits, P.   33

Käfer, E.   241
Kahn, R. P.   335
Kakati, S.   284, 286
Kalmus, H.   334
Kamanoi, M.   281, 291
Kao, K. N.   247, 248
kappa particles   332, 333
Karakashian, M. W.   333
Karakashian, S. J.   333
Karpechenko, G. D.   267
karyogamy   145
karyokinesis   90, 132
karyotype   33, 46
Kasamatsu, H.   331
Kasha, K. J.   247, 248, 249, 272
Katayama, T.   302
Katayama, Y.   245
Kato, H.   180
Kato, K.   237
Kaufmann, B. P.   38
Kaul, C. L.   302
Kaul, M. L. H.   306
Kayano, H.   174
Kelly, T. J.   27
Kerr, W. E.   41
Keyl, H. G.   153
Kezer, J.   66
Khanna, K. R.   245
Khush, G. S.   166, 274, 287, 288, 291
Khvatova, M. N.   251
Kihara, H.   18, 246, 269, 336
Kihlman, B. A.   178, 180
Kilgour, R.   286
Kimber, G.   245, 250, 279
kinetic sites   97, 101
kinetochore   35
King, R. C.   70, 111, 112, 114, 199, 302,
  348
Kirillova, G. A.   251
Kirk, J. T. O.   331
Kirschner, R. H.   322
Kiss, A.   269
Kitzmiller, J. B.   212
Kjéssler, B.   135, 229
Kleckner, N.   345
Klein, H. D.   301, 304, 306
knobs, heterochromatic   42, 44, 308
Knoll, M.   20
Knowles, P. F.   256
Knutsen, O.   166
Koch, C.   310

Kodani, M. 173
Kodburu, P. R. K. 303
Kohel, R. J. 166, 279, 307
Kohne, D. E. 86
Koller, P. C. 216, 302, 304
Kolodner, R. 331
Kölreuter, J. G. 5, 10, 330
Koltzoff, N. 148
Komai, T. 218
Konvička, O. 302
Kopecko, D. J. 345
Kornberg, R. 54
Kostoff, D. 222
Koswig, C. 153
Koul, A. K. 301, 302, 306
Kozinski, A. W. 181
Kramer, H. H. 240
Kroon, A. M. 332
Krooth, R. S. 164
Kuliev, A. M. 260
Kung, C. 333
Kuo, J. S. 248
Kusanagi, A. 38
Kuvarin, V. V. 262
Kuwada, Y. 170

labeling 57
    iso- 58
    radioactive 58
    thymidine 57
Lacadena, J. R. 244, 246
LaCour, L. 50
laggards 99
Laird, C. D. 60
Lamarck, J. B. de 5, 9, 13
Lamb, B. 257
lambda particles 333
Lamm, R. 304, 306
Lammerts, W. E. 251
Lange, W. 248
Langlet, O. 299
Larter, E. N. 269
laser microbeam 183
lateral elements 110
Latt, S. A. 180
Laurent, C. 191, 284
Lautsen, O. 331
Law, C. N. 109, 195, 250
Lawrence, C. H. 245
Lawrence, W. J. C. 97
Leder, A. 28
Leder, P. 28
Lederberg, J. 21, 64, 336, 341
Lee, C. S. N. 231
Leeuwenhoek, A. van 4

Lefevre, G. 153
Lehman, I. R. 181
Lejeune, J. 48, 163, 192, 231
Leonard, A. 231
leptotene 106, 107, 108
lesions
    primary 177, 178, 179
Lesley, J. W. 231
Lesley, M. M. 117, 304
lethality
    balanced 236, 237, 238
    zygotic 236
Levan, A. 23, 99, 150, 165, 166, 180, 305
Leventhal, E. 334
Levine, E. E. 218
Levine, R. P. 218
Lewin, B. 341, 343
Lewis, E. B. 150, 197
Lewis, K. R. 91, 157, 166, 238, 287, 314
Lewontin, R. C. 208
L'Heritier, P. 333, 334
Li, H. 302, 304
Liebeskind, D. 180
ligases
    DNA 181
    polynucleotide 180
Lima-de-Faria, A. 39, 41, 173
Lin, B.-Y. 241
Lin, Y. J. 238
Linde-Laursen, I. 45
Lindqist, K. 307
Lindsley, D. L. 303, 312
Lindsten, J. 229, 304
linear quartet cell 134
linkage 14, 62, 122
    alternative linkage theory 16
    group 62, 122
    maps 62
    parental classes 69
    points 65
    recombinant classes 69
Linnane, A. W. 331
Linnert, G. 38, 245
Litardiere, R. de 299
Logachev, E. D. 141
Longley, A. E. 38, 42, 170, 307, 308
Longo, F. J. 145
loops, lampbrush chromosome, (*See* chromosomes, lampbrush)
lordosis 327
Lorz, A. 299
Löve, A. 323
Löve, D. 323
LSD 180

Lubs, H. A.   286
Luria, S. E.   337, 344
Lwoff, A.   24, 337
Lymantria   80
Lyon, M. F.   25, 83, 84
Lyon hypothesis   25, 83
lysine   21

M-5 method   217
McClaren, A.   86
McClintock, B.   16, 18, 22, 36, 38, 41,
    42, 43, 44, 67, 162, 166, 177, 185,
    186, 187, 194, 196, 199, 200, 221,
    229, 278, 281, 283, 301
McClung, C. E.   108
McClure, H. M.   286
MacDonald, M. D.   38
Mace, S. E.   286
McFee, A. F.   231
McGinnis, R. C.   251, 279
McGregor, J. F.   166
McIlree, M.   135
McKay, R. D. G.   45
Mackensen, O.   71, 188
MacKnight, R. H.   218, 313
McKusick, V. A.   71, 74, 75, 82
McLennan, H. A.   304
McNeal, F. H.   118, 269
Madjolelo, S. D. P.   306
Magenis, R. E.   75
Magni, G. E.   334
Magoon, M. L.   38, 245, 302, 306
Maguire, M. P.   113
Maheshwari, P.   247
maize   16, 22, 41, 42, 69, 72, 111, 162,
    165, 170, 171, 177, 200, 211, 223,
    240, 241, 245, 246, 249, 277, 283,
    291, 300, 303, 307, 336
Makino, S.   123, 134, 135, 173, 180
Malawista, S. E.   97
male
    gametes   132
    promoting regions   78
    suppressor regions   78
Malepszy, S.   247
Malik, C. P.   251
Malogolowkin, C.   334
Malpighi, M.   3
Malpuech, G.   284
Manga, V.   249, 281
Mange, E. J.   309
Manning, J. E.   331
Marberger, E.   86
Marchi, P.   45

Marden, P. M.   284
Mariano, G.   304
marijuana   180
Marks, G. E.   244
Martini, G.   302, 306
Masters, M.   162
maternal effects   138
Mather, K.   179
Matsuda, T.   171
Matsui, S.   50
Matthey, R.   313
Matuszewsky, B.   149
Maude, P. F.   323
Maupertuis, P. L. M. de   5
Mazia, D.   92, 97, 100, 101
megasporangium   131, 132
megaspore   127, 134
megasporocyte   132, 236
megasporogenesis   127, 132
Mehlquist, G. A. L.   257
Mehra, R. C.   306
meiocyte   131
meiosis   10, 102, 120, 128, 235, 301
    I   103
    II   103
    apo-   321, 322
    duration   107
    interkinesis   123
    rudimentary   325
    stages of   104
    substages   104
    variations of   301–306
meiospores   131
meiotic
    a-   320
    division   102
    drive   306, 312
    segregation   120, 307
Melander, Y.   166, 172, 173
Melandrium   78, 82
Mendel, G.   7, 10, 12, 13
    laws   7, 13, 14, 122
Mendelism   13
Mendelson, P.   173
Menzel, M. Y.   249
meristem
    apical   298
merogenote   339
merozygote   338
metakinesis   90, 95
metaphase   90, 98, 100, 127
    I   107, 120, 127
    meiotic   104, 120
    mitotic   32, 90, 98

Metz, C. W.   314, 316, 317
Meurman, O.   299
Meyer, V. G.   302, 336
Meyers, L. L.   340
Michaelis, A.   98, 118, 217, 218, 265, 268, 332
Michie, D.   310
microcephaly   327
micropyle   134
microscopes   3, 4
  electron   20
  light   11
microsporangium   131
microspores   131
microsporocytes   131, 236
microsporogenesis   127, 131
microtubules   93
middle piece, spermatozoon   136, 137
Miescher, F.   7
Mihailova, P. V.   212
Mikelsaar, A. V. N.   191
milk factor   335
Miller, D. A.   42
Miller, O. J., Jr.   98
Miller, O. L.   139, 142, 159, 302
Milutinovic, M.   281
Mirabilis   330
Misra, R. N.   302
Mitchell, H. K.   332
Mitchell, M. B.   332
mitochondria   136, 331
mitochondrial spiral   137
mitosis   8, 11, 90, 300
  duration   100
  endo-   148, 149, 299
  monocentric   314
  pollen   132
  poly-   37, 300
  reductional   300
  variation   300, 301
mitotic elimination   171
Mittwoch, U.   141
mixoploidy   164, 165, 169, 271
Moens, P. B.   108, 112, 256, 287, 304, 306
Mogensen, H. L.   111
Moh, C. C.   304
Mohl, H. von   6
Mohr, O. L.   74
Molè-Bajer, G.   97
monaster   314
mongolism   283
Monod, J.   24, 200, 344
monoisodisomics   287

monoploid method   251
monosomic   69, 83, 273, 274
  double   274
  primary   274
  secondary   274
  series   69, 276
  tertiary   274
monotelodisomic   279
monotelosomic   279
Montgomery, T. H.   38, 84, 156
Mookerjea, A.   257
Moore, D. H.   335
Moore, J. E. S.   109
Moore, K. L.   86
Moores, E. C.   284
Moorhead, P. S.   180
Morgan, D. T.   246, 302
Morgan, L. V.   167, 168
Morgan, T. H.   15, 62, 69, 87, 88, 165, 182, 198, 200, 218
Morinaga, T.   254
Morley, F. H. W.   218
Morris, R.   166
Morrison, G.   244, 251
Morse, M.   339, 341
Morton, J. B.   209
mosaic   299
mosaicism   165, 199
  somatic   199
Mosby, C. V.   322
Moseman, H. G.   291
Moses, M. J.   110, 111
Moses, M. S.   317
Mosharrafa, E.   344
Mosier, H. D.   286
Moutschen, J.   179
mu particles   333
Müller, A. J.   179
Müller, D.   288
Muller, H. J.   18, 82, 83, 176, 182, 200, 201, 216, 219, 270
Müller, U.   55
multiples   234, 238
Müntzing, A.   150, 172, 173, 254, 255, 258, 268, 304
Murakami, M.   247
mutable genes   17
mutation   13, 185
  back-   185
  point   13, 185
  point polarity   344
  polarity   344
  rate   18
  spontaneous   18

mycoplasma 335
Myers, W. M. 256

n-number 32, 122
Nabholtz, M. 71
Nabors, N. W. 183
Nagao, S. 238
Nägeli, C. von 10
Nagl, W. 152
Nagley, P. 331
Nakahara, H. 100
Nakata, K. 248
Nance, W. E. 188
Narayanaswamy, S. 251
Nass, M. M. K. 26, 332
Nass, S. 26, 332
Nathans, D. 27, 28
Navashin, M. S. 132, 222
Neary, G. J. 179
Nebel, B. R. 58
Neilson-Jones, W. 297, 298
Nel, P. M. 302
Nemec, B. 103, 156, 165, 271
Nestel, B. L. 320
Neuffer, M. G. 72
neuroblasts 100
Neurospora 21, 332
Newman, L. J. 306
Newmeyer, D. 208
Nichols, W. 179, 180
Nicoletti, B. 309
Niebuhr, E. 260
Nielsen, J. 191, 192
Niizeki, A. 251
Nikajima, G. 269
Nilan, R. A. 295, 304
Nilsson, H. 222
Ninan, C. A. 244
Nishiyama, I. 251
Nitsch, C. 247
Nitsch, J. P. 247, 251
noncoorientation 228
nondisjunction 171, 173, 174, 276, 315
    Lillium type 174
    Secale type 173
    Sorghum type 174
    Zea type 174
Nonidez, J. F. 314
Nordenskiold, E. 348
Nöth, M. H. 251
Novick, R. P. 337
Novikoff, A. B. 335
Novitski, E. 180, 308, 309, 311
Nowell, P. C. 192
nu-bodies 54

nucellus 132
nuclear envelope 97, 107, 141
nuclear envelope homologue attachment
    site model 113
nuclear phases 128
nuclear sap 139
nucleic acid hybridization 25
nucleic acids 21
nucleoid 159
nucleoli 6, 41, 90, 95, 118, 120, 139
    ring 139
nucleolus chromosomes (See
    chromosomes, nucleolus)
nucleolus organizer bodies 41, 42
nucleolus organizer chromosomes 21,
    32, 36, 101
nucleolus organizer region 25, 36, 41, 139
    silver staining 42
nucleoprotein 22, 54
nucleosome 54
    internucleosomal DNA linkers 55
    nucleosomal fiber 55
    package 55
nucleotide 52
    sequencing 27
nucleus 7, 10
    daughter 101
    fusion 134
    generative 132, 174
    polar 133, 134
    restitution 322
    sperm 145
    supernumerary generative 174
    tube 132
    vegetative 132, 174
nulliplex 263
nullisomic analysis 274
nullisomic series 274
nullisomic-tetrasomics 291
nullisomics 273, 274, 291
nullisomy 273
nurse cells 141

Oenothera 234, 237, 238
    balanced lethals 236, 237, 238
    curvans 236, 237
    gaudens 235, 236
    lamarckiana 13, 236
    muricata 236
    Renner complex 234, 236
    Renner effect 236, 237
    rigens 236, 237
    rubrinervis 234
    velans 235, 236
    zygotic lethality 236

Ogur, M.   132
Ohno, S.   42, 83
Ohtsubo, E.   344
Oishi, K.   334
Okada, T. A.   256
Okazaki, R.   92, 181
Okazaki pieces   92
Okuno, S.   307
Olins, A. L.   54
Olins, D. E.   54
Olivera, B. M.   181
Olmo, H. P.   257
O'Mara, J. G.   97
O'Mara, M. K.   302
one band-one gene concept   153
one gene-one enzyme hypothesis   21
onion   336
Ono, T.   18
oocyte   117, 143
    diplotene   117
    primary   117, 139, 156
    secondary   143
oogenesis   127, 139
oogonia
    primary   139
    secondary   139
Oono, K.   251
ootid   143
Opuntia   321
orientation, (*See* centromere orientation)
Øster, J.   284
Östergren, G.   116, 117, 174
Oudet, P.   54
Ouyang, T. W.   248
ovarian disgenesis   193
ovary   131, 134, 139
oviparous   141
ovists   3
ovotid   143
ovule   132
oxyquinoline squash technique   23
Ozolins, C.   334

P-particles   332
pachytene   38, 41, 63, 106, 107, 113, 120
    chromosomes   40, 44
pachytene analysis   38, 43, 113, 304
Paddock, E. F.   296
Pagliai, A.   324
Painter, T. S.   17, 148, 182, 185, 200,
    219
pairing of chromosomes
    asynaptic   223
    fraternal   167
    incomplete   216

initiation   260
intergenomic   223, 249, 250
internal   167
intragenomic   250
meiotic   18
nonspecific   195
normal   139, 167
somatic   150, 156
suppressor   288
touch-and-go   313
palindromes   28
Pallister, P.   285
Palmer, R. G.   281, 302
Pantalu, J. V.   249
Paramecium   332, 333
    killer races   333
    sensitives   333
paramecin   333
Pardue, M. L.   44, 139
Park, S. J.   251
Parker, J. S.   304
Parry, D. M.   303
Parsons, P. A.   310
parthenogenesis   245, 246, 251, 322
    adult   327
    aposporic   323
    cyclical   324
    diploid   323, 324, 326
    diplosporic   323
    facultative   324
    haploid   323
    obligatory   324
    viviparous   324
partial chiasmatype theory   63
Patau, K. A.   230, 284
Patil, S. R.   45
Paton, G. R.   180
Patterson, J. T.   200, 201, 208
Pavan, C.   153, 154, 155
Pavlovsky, O.   216
pea   7, 306
    endosperm   100
Peacock, W. J.   58, 303, 309
Pelc, S. R.   90
Pelletier, G.   248
Pelling, C.   152, 153
Penrose, L. S.   202, 284
periblem   299
Perkins, D. D.   241
Perry, P.   180
Person, C.   272
Peters, N.   38, 40
Peterson, P. A.   200
Peto, F. H.   256
Petrinelli, P.   75

Petrov, D. F.   251
Pfeifer, D.   344
phages
   transducing   341
phenotype   14
Philip, J.   302
Phillips, J. S.   217
Phillips, R. L.   37, 41
phosphorus   8
photomicrography
   ultraviolet   21
Picard, E.   248
Pincus, G.   327
Pipkin, S. B.   81
Pipkin's theory   81
Pissarev, V.   269
pistil   134
pistillody   306
Piza, S. de T.   118, 238
plasmagene   320
plasmids   336, 339
   recombinant   28
plasmogamy   146
plasmon   329, 330
plasmotype   330
plastids   330, 331, 344
Plaut, W.   26, 331
plectonemically   52, 95
Plessers, A. G.   244
Plewa, M. J.   278
Plus, N.   334
points of initiation   113
Polani, P. E.   169
Polansky, D.   311
polar bodies   127, 143
   first   143
   second   143
   third   143
polaron   64, 65, 66
pollen   132
   mitosis   132
   mother cells   131
   tube   132, 134
polocyte   143
polydactyly   5
polymerase
   antimutator   181
   DNA   92
   repair   181
polymorphism   45, 171
   chromosome   208, 214
polynemy   57, 60
polynucleotide   52, 57
polyploidization   267

polyploidy, (ploidy)   33, 134, 250, 252,
   254, 270, 323, 324
   allo-   18, 249, 250, 253, 268
   allohexa-   252, 253
   allopenta-   252
   allotetra-   253, 267
   allotriploidy   254
   auto-   18, 249, 250, 252, 254, 255,
      266, 267, 268
   autotetra-   245, 252, 256, 260, 262,
      263
   autotri-   252, 256
   dibasic   254
   diploid like   252
   genome allo-   253, 266, 267
   monobasic   254
   polybasic   254
   segmental allo-   250, 252, 254, 266
   tribasic   254
polysomaty   299
polyteny   32, 148, 154
   hypothesis   150
Pontecorvo, G.   296
Porter, K. R.   97
position effect   197
   S-type   197
   V-type   197, 199
Poulson, D. F.   334, 335
poultry   14
Powell, J. B.   244
Prakken, R.   302, 303
Prasad, G.   288
precocity theory   19, 249
Preer, J. R., Jr.   332, 333, 334
premeiotic mitotic pairing   113
Prescott, D. M.   60
Price, J. M.   249
primary trisomic   281, 282, 283, 284,
   285
   multiple   256
Pring, D. R.   336
proacrosomal granules   136
procentriole   93
prokaryotes   159
prometaphase   97
pronuclei   146
propagules   320
prophage   337, 340
prophage sites   337
prophase   32, 39, 90, 95, 100
   meiotic   139
   mitotic   90, 95
prophase I   106, 120
protamine   8

protein 52
  linkers 57
  synthesis 91, 331
protelomere 41
protoplasm 6
protoplast 41
protozoa 97
pseudodominance 71
pseudogamy 322
  matroclinal 245
pseudoisochromosomes 169
Ptashne, K. 344, 345
Pteridophyta 129, 130
puffing 152, 158
  DNA 152
  RNA 152
Punnett, R. C. 14
Punnett square 264
pure line 14
Purkinje, J. E. 139
Purkinje cells 299
Puteyevsky, E. 174
Pyrrhocoris 78, 84, 311

quadripartitioning 124
quadrivalents 223, 260
quadruples 223, 226, 227
quadruplex 262, 263
quadruplochromosomes 149
Qualset, C. O. 269
quartet
  linear quartet cell 134
  radial quartet cell 127, 131
quinacrine staining 27, 47

R-bodies 333
R-factors 339
radial quartet cells 127, 131
radiaton
  alpha ray 180
  beta ray 180
  gamma ray 180, 245
  ionizing 281
  neutron 180
  proton 180
  ultraviolet 331, 341
  x 18, 21, 180, 182, 216, 217, 245, 296
Rae, P. M. M. 152
Rai, K. S. 306
Raina, S. K. 248
Rajhathy, T. 281, 287
Rakha, F. A. 241
Ramage, R. T. 240, 288, 289, 290
Ramanujam, S. 251

Ramarkrishnan, T. 339
Ramos, A. 244
Randall, T. J. 52
Randolph, L. F. 134, 170, 171, 246, 255, 256
Rao, D. R. 38
Rao, M. K. 303
Rappleye, R. 246
Rattner, J. B. 95
Raveendrananatth, T. G. 244
Raven, C. P. 141
recombination 16, 18, 120, 143
  genetic 126
  high frequency 338
  nodules 111
Reddi, V. R. 249
Redfield, H. 218
reduction 264
  double 264
  somatic 300
reduction division, (*See* meiosis)
reductional mitosis 300
Rees, H. 173, 208, 306
regression, concept of 7
Reinwein, H. 191
Renner, O. 128, 234, 237, 321
Renner complexes 234
Renner effect 237
repairase 181
repair polymerases 181
repair replication 182
replication
  early 90
  late 91
replicons 92
repulsion 114
  generalized 116
  localized 116
resistance
  antibiotics 345
  multiple drug 340
resistance factors, (R-factors) 340
  nontransmissible 340
  transmissible 340
resistance transfer factor, (RTF) 340
resolution 20
resolving power 11
restitution 176
restriction
  effect 27
  enzymes 27
  maps 27
Revell, S. H. 177, 178, 179
rhizomes 320

Rhoades, M. M.   35, 41, 166, 170, 172, 281, 287, 291, 303, 307
Rhoeo   237, 238
ribonuclease   57
ribonucleic acid, (*See* RNA)
ribosomes   25
Ricciuti, R.   75
Richardson, M. M.   302, 304
Richmond, M. H.   337
Richmond, T. R.   307
Rick, C. M.   166, 283, 287, 288, 291, 307
Rieger, R.   39, 41, 66, 98, 116, 117, 156, 170, 177, 188, 195, 217, 218, 228, 249, 260, 265, 268, 274, 276, 295, 300, 318, 337, 348
Riggs, A. D.   92
Riley, R.   109, 113, 195, 245, 247, 250, 274
Rimpau, W.   268
Ris, H.   26, 35, 58, 108, 159, 326, 331, 340
Ritossa, F. M.   25
RNA   18, 24, 53, 91
    messenger   25, 54
    puffs, (*See* puffing)
    ribosomal   25, 54
    structure   53
    synthesis   139, 152
    transfer   54
Röbbelen, G.   38
Robberson, D. L.   332
Roberts, P. A.   155
Robertson, D. S.   241
Robertson, D. W.   240
Robertson, W. R. B.   229, 240
Rocchi, A.   173
Rodin, B.   180
Rogers, J. S.   336
Rogers, O.   245
Rohloff, H.   228
Roman, H.   174, 241
Rosenberg, O.   38, 318, 321, 322
Ross, A.   45
Ross, W. M.   246, 336
Rossenbeck, H.   17
Roux, W.   9
Rowe, P. R.   245
Rückert, J.   157
Rückregulierung   268
Ruddle, F. H.   71, 75, 82, 286
Rudkin, G. T.   153
Rudorf-Lauritzen, M.   283
runners   320
Runquist, E. W.   304
Rupp, W. D.   182
Ruska, E. A. F.   20

Russel, L. B.   25
Ruthmann, A.   368
Rutishauser, A.   174
Ruzicka, F.   45
rye   171, 174, 304
Ryser, G. K.   336

S period   90, 91
    DNA synthesis   91
Saccharomyces   331
Sadasivaiah, R. S.   249, 251, 302
Saedler, H.   344
Sakaguchi, B.   334
Sakai, T.   339
Sakharov, V. V.   262
Saki, K.   238
Saksela, E.   180
salamander   8
Salceda, V. M.   216
salmon sperm   8
Sanchez, O.   45
Sanders, J. P. M.   332
Sandler, L.   303, 307, 308, 309
Santesson, B.   260
Sarvella, P.   38, 41
Sasaki, M.   50, 134, 135, 173
Satina, S.   298, 305
Sato, H.   139, 140
Savosjkin, I. P.   262
Sax, K.   132, 176, 222, 237, 238
Scaife, J.   338
SCE method   180
Schertz, K. F.   249, 302
Scheuring, J. F.   303
Schimke, R. N.   192
Schleiden, M. J.   6
Schlosser, L. A.   257
Schmid, W.   206
Schmidt, J.   82
Schmuck, M. L.   314
Schnarf, K.   321, 322
Schnedl, W.   45
Schott, O.   11
Schrader, F.   98, 100, 118, 313, 348
Schroeder, T. M.   180
Schultz, J.   38, 81, 165, 199, 218
Schultz-Redfield effect   218
Schulz-Schaeffer, J.   33, 36, 37, 38, 42, 99, 109, 113, 114, 115, 118, 119, 123, 124, 125, 126, 127, 266, 267, 269, 272, 277, 320, 336
Schwann, T.   6
Schwartz, D.   300
Schwartz, H.   324
Schwarz, D.   57

Schwarzacher, H. G.   45
Sciara   314, 317
Scott, D.   179
sea urchin   8, 11
Seabright, M.   45
Sears, E. R.   166, 273, 274, 275, 276,
    279, 280, 281, 287
sea urchin   8, 11
secondary trisomics   287, 288
Seecof, R. L.   334
seeds, polyembryonic   246
segregation   227, 309
    directed   227
    distortion   307
    meiotic   307
    preferential   227, 307, 308, 309
segregation, Law of   7
Sehested, J.   48
Selman, G. G.   326
seminiferous tubules   135
semisterility   227, 229, 240
Sen, N. K.   167, 287
Sen, S. K.   299
Sendino, A. M.   244
sensitives   333
separation
    postequational   102
    postreductional   102
    preequational   102
    prereductional   102
    reductional   102
sequences   27
    intervening   28
Serebrovsky, A. S.   185
sex
    chromatin   22, 86, 87
    diagnosis   76
    factors   337
    heterogametic   78
    heteromorphic   78
    homogametic   78
    homomorphic   78
    inheritance   88
    linkage   15, 88
    ratio   311, 334
    ratio gene   311
    time law   81
    turning point   81
sex chromosomes   78–88
    differential region   78
    homologous region   78
    multiple sex chromosome systems   313
sexuales   324
sexuparae   324
Shambulingappa, K. G.   38

Shapiro, J. A.   344
Shapiro, L.   332
Sharma, G. P.   287
Sharp, W. R.   248
Shastry, S. V. S.   38, 302
Shaw, M. W.   45, 164, 179, 180
Shelton, E. R.   55
Shengun, A.   153
Sheridan, W. F.   105
Shigella   340
Shimada, K.   344
Shinji, O.   314
Shiraishi, Y.   180
shoot
    apex   298
Siegel, R. W.   333
Siemens, H. W.   330
Silver, S.   339
Simmonds, N. W.   237
simplex   263
Simpson, C. E.   321, 322
Sinclair, J. H.   332
Sinet, P. M.   75
Singh, R. B.   291, 303
Single active X hypothesis   83
Sinha, S. K.   302
sister chromatid exchange   180, 186
sister-strand reunion   176, 194
Sitte, P.   93
Sjodin, J.   262, 302, 304
Skalinska, M.   323
Skovsted, A.   34
Slizinsky, B. M.   227, 304
Slizynska, H.   7, 74, 179, 188
Slonimski, P.   331
Smith, D. R.   86
Smith, D. W.   189
Smith, F. J.   283
Smith, G. F.   284
Smith, H. O.   27, 28
Smith, J. B.   107
Smith, L.   291, 302, 305
Smith, P. A.   302
Smith-Sonneborn, J. E.   333
Smith-White, S. G.   165
Snow, R.   208
Soldo, A. T.   333
Soller, M.   231
Somaroo, B. H.   262
somatic chromosome number   122
somatic crossing over   300, 303
    double spot   294
    single spot   294
    twin patch   294
    twin spot   294

Sonneborn, T. M.   332, 333
Soost, R. K.   302, 304
Sotelo, J. R.   112
Southern, D. I.   184
Sparrow, A. H.   34, 95
Sparvoli, E.   58
specific restriction endonuclease   27
Spencer, W. P.   217, 314
Sperber, M. A.   286
sperm   136
  lysine   145
  nucleus   145
  salmon   8
spermatids   127, 136
spermatocytes
  primary   117, 136
  secondary   136
spermatogenesis   127, 131, 134, 136, 324
spermatogonia   134, 135
  primary   134
  secondary   134
Spermatophyta   129, 130, 131
spermatozoa   136
  head   137
  middle piece   137, 145
  tail   137, 145
spermiogenesis   136
Spiegelman, S.   25, 36, 75, 344
spindle   94, 304
  fiber apparatus   304
  microtubulin   95
  spherule   60
spindle fibers   97, 100, 101, 120, 304, 307
  chromosomal   100
  continuous   100
spirochaetes   335
sporocyte
  mega-   132
  micro-   131
  multiple   305
sporogenesis   131
sporophyte   129, 131, 134
Sprey, B.   331
sprigging   320
Srb, A. M.   204
Sreenath, P. R.   302
Srinivasachar, D.   251
St. Lawrence, P.   38
Stadler, L. J.   18, 176, 246
Stahl, F. W.   342
staining methods
  C   44, 45, 46
  Giemsa   44, 45

Stanford, E. H.   250
Starlinger, P.   345
Stearn, W. T.   237
Stebbins, G. L.   34, 238, 250, 252, 318, 319, 320, 321, 323
Steffen, K.   132
Steinberg, A. G.   218
Steinitz, L. N.   300
Stephens, J. S.   302
Stephens, S. G.   252
sterility
  cytoplasmic male   335, 336
  genetic   307
  male   306, 307
  semi-   227, 229, 240
sterility genes   306
Stern, C.   19, 67, 68, 190, 217, 296, 301, 348
Stern, H.   67, 105
Stettler, R. F.   245
Stevens, B. J.   332
Stevens, L.   201
Stevens, N. M.   156, 325
Stevenson, I.   333
Stich, H. F.   180
sticky ends   340
stigma   134
stolons   320
Stomps, T. J.   299
Stone, W. S.   208
Storms, R.   111
Strasburger, E.   10, 14, 102, 156
Streicher, S.   28
Streisinger, G.   182
Strickberger, M. W.   216
Stringham, G. R.   244, 302, 306
Strong, J. A.   82
Stubbe, H.   348
Sturtevant, A. H.   69, 195, 201, 218, 286, 311, 312
style   134
Subrahmanyam, N. C.   247, 248
Šubrt, T.   231
Suda, K.   332
Sullivan, J. T.   256
Summers, A. O.   339
Summitt, R. L.   231
Sumner, A. T.   45
Sun, C. S.   248
Sunderland, N.   247
Suneson, C. A.   240, 307
Sutton, W. S.   14
Suzuki, H.   272
Swammerdam, J.   4

Swanson, C. P.   63, 64, 67, 69, 92, 105, 115, 116, 117, 123, 124, 165, 179, 194, 236, 252, 294, 296, 300, 301, 306, 348
Swift, H. H.   132, 153
Swomley, B. A.   240
Sybenga, J.   227, 238, 288, 308
Sykes, M. G.   156
symbionts   332, 333
synapsis   10, 11, 103, 109, 110, 156, 256
  a-   223, 274, 302, 322
  de-   302
  meiotic   156
  somatic   156
synaptomeres   110, 111
synaptomere-zygosome hypothesis   111
synaptonemal complex   110, 114, 249, 302
syncytes   305
syncytium   305
syndromes   193
  Bloom's   180
  chromosome instability   153, 180
  cri du chat   190
  C-trisomy   284
  double Y   286
  Down's   23, 230, 283, 286, 304
  duplication-deletion   201
  Edward's   284
  haploidy   326
  isochromosome Turner   168
  Klinefelter   23, 286
  Louis-Bar   153, 180
  ovarian disgenesis   193
  Patau's   284
  translocation   228, 229, 230, 231, 232
  Turner   23, 168, 193, 278
synergids   134
syngamy   128, 131, 132, 134, 143
synizesis   108
synizetic knot   108
synteny test   71

Tabak, H. F.   332
Tabata, M.   251
Tai, W.   305
Takahashi, R.   283
Takahu, T.   218
Takamoto, K.   153
Tan, Y. H.   75
Tanaka, M.   38, 248
Tandy, M.   270
Tannreuther, G.   325
Tatum, E. L.   21
Taylor, A. I.   284

Taylor, A. L.   160
Taylor, A. M. R.   180
Taylor, C. W.   208
Taylor, E. W.   95
Taylor, G. R.   348
Taylor, J. H.   57, 58
Taylor, W. R.   36
Tedesco, T. A.   75
Teissier, G.   333
Telfer, M. A.   286
telocentrics   165, 166
telomere   41, 111, 163
  attachment sites   112
telophase   90, 100, 101, 127, 163
  meiotic   107, 122
  mitotic   90, 122
telophase I   107, 123
telophase II   107, 127
telosomes   279, 291
telosomic   279
  ditelosomic   280
  monotelosomic   279
telotrisomic   166
teosinte   320
tertiary trisomics   288, 289, 290
tetrad chromosomes   103
Tetraneura   324
tetraploidy   245
tetrasomic   291
tetrasomy   291
tetraster   11
tetrazolium salts   331
Tewari, K. K.   331
Therman, E.   169
Theurer, J. C.   336
thiotepa   180
Thoday, J. M.   179
Thomas, C. A.   28, 341
Thomas, H.   302
Thomas, P. T.   304
Thompson, D. H.   337
Thompson, D. L.   251
Thompson, J. B.   208
Thompson, K. F.   244
three-point cross   239, 240
thymidine   92
thymine   52
Ting, Y. C.   249
Tjio, J. H.   23, 37, 166, 202, 231
tobacco   277, 336
Tobgy, H. A.   34
toluidine blue   245
Tough, I. M.   193
Tradescantia   5, 183
Trampe, P. O.   332

transcript maps 27
transduction 341
transformation 22
translocation 17, 44, 176, 219, 230, 239,
    240, 241, 266
  centric fusion type 229, 230
  complex 220, 221, 234
  D/D 230
  D/G 230
  G/G 230
  hypothesis (interchange) 234
  interchromosomal shift 221
  intrachromosomal shifts 220
  reciprocal 218, 220, 228, 231, 233,
    234, 237, 266
  Robertsonian 229, 230, 231
  shift type 220
  simple 218
translocation complex 234
translocation heterozygote 223, 227,
    238, 239
transposition 195, 199
transposons 337, 345
transverse elements 110
Trillium 33, 50
Tripathi, R. C. 251
triplex 262
triploidy 79, 156
trisomy 17, 69, 280
  balanced tertiary 289
  C-group 284
  compensating 280, 290
  D-group 284
  double 280
  E-group 284
  G-group 283
  method 283
  monotelo- 291
  primary 280, 281
  secondary 280, 287
  telosomic 280, 291
  tertiary 280, 288
Triticale 268
  substitutional hexaploid 269
trivalents 256
Trotter, C. D. 160
Tschermak, E. 13
Tschermak-Seysenegg, E. 14
Tschermak-Woess, E. 150, 152, 237
Tsuchiya, T. 244, 272, 281, 283, 291
Tsunewaki, K. 246
Tuleen, N. A. 234
tunica 298
Turcotte, E. L. 247

Turner, B. C. 208
Turner, F. R. 94
twin seedlings 255
twin spot 294
Tyler, A. 141

U,N. 254
Uchida, A. 332
Ullstrup, A. J. 241
ultrasound
  diagnostic 180
uninemy 57
unisexual progenies 317
univalent 103
  ring 167
  false 261
Upcott, M. 261, 304
uracil 53
Ursprung, H. 151

Vaarama, A. 306
vacuole 136
vacuolizer 333
Valentin, J. 303
Valentine, R. C. 28
Van Kreijl, C. F. 332
Van't Hof, J. 92, 105
Van Wagtendonk, W. J. 333
variegation, (*See* mosaicism, somatic)
Vasek, F. C. 281, 305
Vasilev, B. I. 245
vegetative reproduction 319
Vengerov, Y. Y. 55
ventral edema 327
Vig, B. K. 296
Vigfusson, N. V. 163, 164, 231, 232
Villa-Komarov, L. 28
Vinograd, J. 332
Virchow, R. L. C. 6
virion 335
virus 334
  bacterial, (*See* bacteriophage)
  sigma 333, 334
  tobacco ringspot 335
  tomato ringspot 335
Visca 238
Visheveshwara, S. 244
vivipary 320, 323
Vogel, F. 60
Vosa, C. G. 45, 174
Vries, H. de 13

Wachtel, S. S. 82
Wahrman, J. 314

Walker, P. M. B.   86
Wallace, B.   312
Wallace, M. E.   310
Walters, J. L.   238
Walters, M. S.   237, 238
Walzer, S.   231
Wang, C. C.   248
Wang, Y. Y.   251
Waris, H.   100
Warmke, A. E.   321
Warmke, H. E.   82
Wasserman, F.   97
Watanabe, T.   340
Watkins, J. F.   71, 189, 268
Watson, J. D.   23, 52, 54, 55, 57
Watson-Crick model   22, 24, 52, 56
Watts, V. M.   307
Weaver, J. B.   302
Webber, J. M.   245
Weber, D. F.   249, 278
Weigle, J.   341
Weintraub, H.   55
Weismann, A.   9, 10, 134
Weiss, M. C.   248, 268
Welch, J. P.   201
Welshons, W. J.   153
Westergaard, M.   79, 82, 110
Westmoreland, B. C.   189, 344
Wettstein, D. von   110
Wettstein, F. von   329
Wettstein, R.   112, 336
Whaley, W. G.   97
wheat   35, 269, 279, 291, 336
wheatgrass   266, 320
White, B. J.   231
White, M. J. D.   118, 148, 149, 158,
     165, 166, 173, 184, 208, 218, 270,
     271, 312, 313, 324, 325
White, T. G.   279
Whitehouse, H. L. K.   64, 66, 218
Whittinghill, M.   303
Wicks, F. M.   247
Wiens, D.   238
Wildman, S. G.   331
Wilkins, M. H. F.   24
Williamson, D. L.   335
Williamson, M.   333
Wills, C. J.   216
Wilson, D. A.   28
Wilson, E. B.   12, 98, 313, 348
Wilson, F. D.   268
Wilson, G. B.   50, 97
Wilson, J. A.   246, 336
Wimber, D. E.   237

Winge, O.   331
Winkler, H.   255
Winton, L. L.   245
Wipf, L.   150
Witkin, H. A.   286
Witkus, E. R.   97
wobble hypothesis   24
Wolf, U.   191
Wolfe, E.   313
Wolfe, S. L.   114, 313
Wolff, S.   58, 176, 180
Wollman, E. L.   337, 340
Woodcliff, H. J.   193
Woolam, D. H. M.   112
Woolley, G. W.   335
Worton, R. G.   233
Wylie, A. P.   123, 323

X-chromosome   12, 18, 74, 184, 216
   attached   167, 168
   heterochromatic   22
   late replicating   83
   loci
      G6PD   75
      HGPRT   75
      PGK   75
   trisomic   286
X-linked   88
x-number   122
X-O system   12, 78, 82, 278
x-ray diffraction   52
X-Y system   78

Y-chromosome   18, 45
   double Y syndrome   286
   H-Y   82
   TDF   82
Yamasaki, N.   50
Yamashita, K.   239
Yasmineh, W. G.   85
Yefeiken, A. K.   245
Yeh, B. P.   250
Yerganian, G.   38
yolk platelets   141
Yosida, T. H.   208
Yotsuyangi, Y.   331
Young, B. A.   318, 321
Yu, C. W.   46, 49, 50, 181
Yudin, B. F.   251
Yunis, J. J.   45, 85, 284

Zahavi, A.   314
Zenyozi, A.   269
Zepp, H. D.   268

Zickler, D.   111
Zillinsky, F. J.   269
Zimmering, S.   309
Zimmerman, F. K.   296
Zinder, N. D.   341

Zohary, D.   173, 174
Zubay, G.   54
zygosomes   111
zygote   10, 134, 143
zygotene   106, 107, 108, 120